Approximation Algorithms for Traveling Salesman Problems

The traveling salesman problem (TSP) is a central topic in discrete mathematics and theoretical computer science. It has been one of the driving forces in combinatorial optimization. The design and analysis of better and better approximation algorithms for the TSP has proved challenging but very fruitful. This is the first book on approximation algorithms for the TSP, featuring a comprehensive collection of all major results and an overview of the most intriguing open problems. Many of the presented results have been discovered only recently, and some are published here for the first time, including better approximation algorithms for the asymmetric TSP and its path version. This book constitutes and advances the state of the art and makes it accessible to a wider audience. Featuring detailed proofs, over 170 exercises, and 100 color figures, this book is an excellent resource for teaching, self-study, and further research.

VERA TRAUB has been Professor at the University of Bonn since 2023. Her research has received multiple awards, particularly her work on approximation algorithms for network design and the traveling salesman problem, including in 2023 the Maryam Mirzakhani New Frontiers Prize and the Heinz Maier-Leibnitz Prize. She is a member of the Hausdorff Center for Mathematics.

JENS VYGEN has been Professor at the University of Bonn since 2003. His work comprises many aspects of combinatorial optimization and its applications, notably to chip design and vehicle routing. He has co-authored two textbooks, organized several workshops and conferences, and has been co-editor of several scientific journals and books. He is a member of the Hausdorff Center for Mathematics.

'This is a fantastic book! Extensive coverage unifies the wide range of attacks on TSP complexity made over the past decade. A great read for experienced researchers and for those looking to join the field.'

William Cook, *University of Waterloo*

'Thoroughly Simplified Presentation of the latest approximation results on key variants of the TSP. A gem and a must-read for both novice and experts in the area! Like the 4 C's of a diamond: Clear, Comprehensive, Careful and Captivating.'

Michel Goemans, *Massachusetts Institute of Technology*

'The wonderful new textbook by Traub and Vygen is a pleasure to read. There have been very successful textbooks on the TSP, and on approximation algorithms, but this is the first to focus on approximation algorithms for the TSP. At first, this might seem like a too repetitive diet, but the richness of the developments of the past decade or so, all elegantly presented here to the last detail, demonstrates the wealth of variety of algorithmic thinking that has produced these advances.'

David B. Shmoys, *Cornell University*

'This is an amazing book by world-leading experts Vera Traub and Jens Vygen. It comprehensively covers and simplifies all recent developments on approximation algorithms for the traveling salesman problem. The clarity and extensive treatment of these sophisticated methods make this book a must-read for anyone interested in advanced algorithmic techniques and approximation algorithms.'

Ola Svensson, *École Polytechnique Fédérale de Lausanne*

'This book is a very welcome addition to the literature on the fascinating and addicting traveling salesman problem. It gives a consistent and unified treatment of approximation algorithms, starting with a very thorough treatment of the basics and extending through the most recent developments, including work by these two authors. This volume will be valued by researchers and graduate instructors alike.'

David P. Williamson, *Cornell University*

Approximation Algorithms for Traveling Salesman Problems

VERA TRAUB
University of Bonn

JENS VYGEN
University of Bonn

CAMBRIDGE
UNIVERSITY PRESS

CAMBRIDGE
UNIVERSITY PRESS

Shaftesbury Road, Cambridge CB2 8EA, United Kingdom

One Liberty Plaza, 20th Floor, New York, NY 10006, USA

477 Williamstown Road, Port Melbourne, VIC 3207, Australia

314–321, 3rd Floor, Plot 3, Splendor Forum, Jasola District Centre, New Delhi – 110025, India

103 Penang Road, #05–06/07, Visioncrest Commercial, Singapore 238467

Cambridge University Press is part of Cambridge University Press & Assessment, a department of the University of Cambridge.

We share the University's mission to contribute to society through the pursuit of education, learning and research at the highest international levels of excellence.

www.cambridge.org
Information on this title: www.cambridge.org/9781009445412

DOI: 10.1017/9781009445436

When citing this work, please include a reference to the DOI 10.1017/9781009445436

First published 2025

A catalogue record for this publication is available from the British Library.

Library of Congress Cataloging-in-Publication Data
Names: Traub, Vera, author. | Vygen, Jens, author.
Title: Approximation algorithms for traveling salesman problems / Vera Traub, Jens Vygen.
Description: Cambridge ; New York, NY : Cambridge University Press, 2024. |
Includes bibliographical references and index.
Identifiers: LCCN 2024011584 (print) | LCCN 2024011585 (ebook) | ISBN 9781009445412 (hardback) | ISBN 9781009445436 (epub)
Subjects: LCSH: Traveling salesman problem. | Approximation algorithms. Classification: LCC QA164 .T727 2024 (print) | LCC QA164 (ebook) | DDC 511/.5–dc23/eng/20240620
LC record available at https://lccn.loc.gov/2024011584
LC ebook record available at https://lccn.loc.gov/2024011585

ISBN 978-1-009-44541-2 Hardback

Preface

The traveling salesman problem (TSP) can be stated very easily: Given a finite set of cities and the distance of every pair, visit all of them (and return to the starting point) while minimizing the total distance. How can we find an optimum tour?

The TSP is probably the best-known combinatorial optimization problem. There are many reasons why:

- It can be described very easily to anybody, but it is not easy to solve. There are many exciting open questions that are easy to understand. Although studied intensively for more than 60 years, the TSP continues to pose grand challenges.
- While trying to make progress on the TSP, many related problems have been studied and important general techniques have been developed. A substantial part of the combinatorial optimization theory has been developed with the TSP in mind.
- The TSP is also relevant in many applications, most notably of course in vehicle routing. The techniques that have been developed originally for the TSP are being used in many diverse applications every day.

The first thing to note is that the TSP is *NP*-hard (Karp [1972]). This makes us conjecture that there is no polynomial-time algorithm that always finds an optimum tour. In fact, such an algorithm exists if and only if $P = NP$. Most researchers believe (and work with the assumption) that $P \neq NP$, and so do we, but this remains one of the most famous open problems in mathematics and computer science.

Assuming that there is no polynomial-time algorithm that always finds an optimum tour (i.e., $P \neq NP$), one needs to make compromises. There are three natural ways:

- One can relax the running time requirement – that is, accept that the algorithm is not guaranteed to terminate in polynomial time. Applegate et al. [2006] solved instances with tens of thousands of cities optimally, but there is no guarantee that their (or any known) algorithm will solve every instance with, say, 100 cities in reasonable time.
- One can relax the quality requirement – that is, accept that the algorithm could perhaps produce a bad tour. There are excellent heuristics (based on Lin and Kernighan [1973]) that run very fast and usually compute a solution that is very close to optimal (less than 1% longer). However, there is no guarantee, and on some instances, the computed tour could be 100 times longer than optimal or even worse.
- One can ask for guarantees both in the running time and in the quality of the tour but not require optimality. An *α-approximation algorithm* (for some constant $\alpha > 1$) is an algorithm that is guaranteed to terminate in polynomial time and is guaranteed to compute a tour at most α times longer than optimal.

In practice, one of the first two alternatives is usually chosen. The third one, however, is the one that has proved most fruitful from a theoretical point of view, and this is the focus of this book. So, in one sentence, our question is: How good tours can we guarantee to find in polynomial time?

There are several interesting variants of the TSP that we will study. The first question is whether or not distances are symmetric – that is, whether the distance from a to b equals the distance from b to a for every pair $\{a, b\}$ of cities. Christofides [1976], and independently Serdyukov [1978], devised a $\frac{3}{2}$-approximation algorithm for the SYMMETRIC TSP: It always finds a tour that is at most 50% longer than optimum. Recently, a slightly better approximation algorithm was found by Karlin, Klein, and Oveis Gharan [2021,2023], but we are still far from the ratio $\frac{4}{3}$, which is widely believed to be achievable. Interestingly, the new algorithm is based on sampling a random spanning tree, a technique that has first been used for the ASYMMETRIC TSP (Asadpour et al. [2017]).

For the ASYMMETRIC TSP, no constant-factor approximation algorithm was known until 2017, when Svensson, Tarnawski, and Végh [2020] found one. We could improve this to a $(22 + \varepsilon)$-approximation algorithm for any $\varepsilon > 0$ (Traub and Vygen [2022]), and in this book, we improve the ratio further to $17 + \varepsilon$, but this is still far from what we expect to be possible for the ASYMMETRIC TSP.

Many approximation algorithms start by solving a relaxation of the problem. Often, this is a linear program (LP) in which every tour corresponds to a feasible solution but infinitely more "fractional solutions" are allowed. Such *LP relaxations* have been studied since the 1950s (Dantzig, Fulkerson, and Johnson [1954]) and can be solved in polynomial time. In the end, the cost of a tour is compared to the cost of an optimum LP solution, which is always a lower bound on the cost of an optimum tour. Therefore, an important question is how strong these LP relaxations are: What is the worst possible ratio of the cost of an optimum tour to the cost of an optimum LP solution? In most cases, this so-called *integrality ratio* is unknown. The worst examples (that yield the best-known lower bounds on the integrality ratio) are quite special: Here, the distance of any pair of cities is simply the distance in an unweighted graph. Therefore, such instances have received special attention, and better approximation algorithms have been found, including a $\frac{7}{5}$-approximation algorithm for Graph TSP (Sebő and Vygen [2014]).

A generalization of the TSP with many interesting properties is the Path TSP: Here, the start and end of the tour are given and not necessarily identical. An, Kleinberg, and Shmoys [2015] were the first to beat Christofides' algorithm (for Path TSP with symmetric distances) and renewed interest in this problem. This led to the discovery of interesting new techniques and results, culminating in a general reduction to Symmetric TSP that loses only an arbitrarily small constant in the approximation guarantee (Traub, Vygen, and Zenklusen [2022]). For Asymmetric Path TSP, the known reductions lose more, but we can still achieve the ratio $17 + \varepsilon$ (another new result in this book). We will also mention some further generalizations, including the classical vehicle routing problem.

As mentioned, approximation algorithms for the TSP are not used often in practice. There are many reasons for this: Many instances occurring in practice are small enough and can be solved optimally. For others, the best heuristics usually compute better tours than approximation algorithms. There are exceptions, but they rarely occur in practice. Finally, the running time of some approximation algorithms, though bounded by a polynomial in the number of cities, can be too large for practical purposes.

However, interestingly, many techniques that have been developed during attempts to find better algorithms for the TSP or better upper bounds on the integrality ratios are being used in practice in a different context. We do not address practical aspects in this book, but this does not mean that the content of this book is irrelevant for practical purposes.

We try to present the state of the art as comprehensively as possible and include complete proofs, at least of the most important results, with very few exceptions. Most of these results have been obtained since 2010, some only while we were writing this book. This makes us hope that we will learn more in the near future. We will of course mention the most intriguing open questions. Hopefully, this book can serve as a basis for future research.

We also hope that this book might prove useful for teaching and learning. The open questions on the TSP have been a major driving force in combinatorial optimization, and much of the combinatorial optimization theory will be needed in this book. Although we are not attempting to write another book on combinatorial optimization, we will cover quite a few topics (including flows, connectivity, matching, matroids, and of course linear programming). For many of the classical results, we will not give proofs but rather refer to textbooks such as Schrijver [2003] or Korte and Vygen [2018]. Most of our notation will be consistent with these books.

This is not the first book on the TSP, but it is the first one that focuses on approximation algorithms. Lawler et al. [1985] wrote the first TSP book, and this already contains a chapter on approximation algorithms. Although of course outdated, it is still worth reading. Most of the chapters in the book edited by Gutin and Punnen [2002] focus on exact algorithms and heuristics, but there is also one by Arora on his approximation scheme for geometric instances of the TSP (Arora [1998]), which we do not cover here again. Applegate et al. [2006] focused on computational aspects and described the (still) leading algorithm to compute optimum tours. Cook's [2012] entertaining book gives an excellent introduction to the TSP and its history. There are many other books and book chapters, of course; in particular, almost every book on combinatorial optimization contains a chapter on the TSP.

Parts of this book are based on an earlier and much shorter survey of the second author (Vygen [2012]) and the PhD thesis of the first author (Traub [2020a]), as well as on several of our papers. We would like to thank Susanne Armbruster, Jannis Blauth, Martin Drees, Antonia Ellerbrock, Michel Goemans, Corinna Gottschalk, Swati Gupta, Dorothee Henke, Satoru Iwata, Volker Kaibel, Marcin Mucha, Martin Nägele, Meike Neuwohner, Luise Puhlmann, Stefan Rabenstein, R. Ravi, Niklas Schlomberg, Malte Schürks, András Sebő, David Shmoys, Ola Svensson, and several students in our courses for carefully reading parts of this book and providing useful remarks that helped improve the presentation. We also profited from many discussions with Sylvia Boyd, Bill Cook, Michel Goemans, Chien-Chung Huang, Nathan Klein, András Sebő, David Shmoys, Mohit Singh, Ola Svensson, László Végh, David Williamson, Rico Zenklusen, and Anke van Zuylen.

Most of this work was done at the University of Bonn, but part of it was also done at ETH Zurich. We thank FIM at ETH Zurich and HIM of the Hausdorff Center for Mathematics for their support. We are also grateful to our colleagues and our industrial cooperation partner Greenplan. Last but not least, we thank Cambridge University Press, in particular Arman Chowdhury and Katie Leach, for the excellent cooperation.

Vera Traub and Jens Vygen

Structure of This Book and Possible Courses

Chapters 1–4 of this book contain mostly classical results from combinatorial optimization, of course with a focus on the traveling salesman problem. We start with the basics and the classical algorithms of Christofides and Serdyukov and Frieze, Galbiati, and Maffioli. Then we discuss the most important linear programming relaxations in detail. These initial chapters also contain a few less-well-known results, such as the lower bound 2 on the integrality ratio of the classical LP relaxation of the ASYMMETRIC (GRAPH) TSP in Section 3.6, a different LP relaxation in Section 3.3, and Karzanov's uncrossing algorithm in Section 4.2. Nevertheless, readers who are familiar with the classical theory can skip these chapters initially.

Chapter 5 contains the $O(\frac{\log n}{\log \log n})$-approximation algorithm for the ASYMMETRIC TSP by Asadpour, Goemans, Mądry, Oveis Gharan, and Saberi. Although it is not the state of the art anymore, it introduced interesting new techniques and has inspired other researchers. The algorithm is based on sampling a spanning tree from a maximum entropy distribution, and this will be used again for the SYMMETRIC TSP in Chapter 10.

Chapters 6–8 give a complete description of a constant-factor approximation algorithm for the ASYMMETRIC TSP. Such an algorithm was first obtained by Svensson, Tarnawski, and Végh; however, we present our improved and simplified version. We start with the special case of the GRAPH ASYMMETRIC TSP in Chapter 6, because this relatively short proof (mostly due to Svensson) already contains many of the key ideas needed in the general case. The entire proof of the constant-factor approximation for the general ASYMMETRIC TSP ends with Section 8.5; the final Section 8.6 is needed only to improve the constant from 21 to 17.

Chapter 9 discusses the asymmetric path version. After the Feige–Singh black-box reduction in Section 9.2, which loses a factor 2, we present reductions that yield a constant upper bound on the integrality gap also for this more general problem, as well as the same approximation ratio that we obtained for the ASYMMETRIC TSP.

Then we turn to the SYMMETRIC TSP. Chapters 10 and 11 describe the main ideas of Karlin, Klein, and Oveis Gharan, who recently obtained the first approximation algorithm for the SYMMETRIC TSP that has an approximation ratio better than $\frac{3}{2}$. In Chapter 10, we describe the reduction to their main payment theorem in full detail. We can only sketch the proof of that theorem in Chapter 11, but we hope to convey the main ideas.

Chapters 12 and 13 deal with the GRAPH TSP. Chapter 12 is devoted to the removable pairing technique, introduced by Mömke and Svensson, while

Chapter 13 discusses the ear-decomposition approach by Sebő and Vygen. This algorithm has the currently best-known approximation ratio for Graph TSP: $\frac{7}{5}$.

Chapters 14–16 address the symmetric path version, which we simply call Path TSP. The basic theory presented in Sections 14.1–14.3 is used in the following two chapters. Hoogeveen observed that the natural generalization of Christofides' algorithm has only approximation ratio $\frac{5}{3}$ for Path TSP. Chapter 15 presents the result by An, Kleinberg, and Shmoys, who obtained the first improvement on this, as well as various subsequent improvements and generalizations (to the so-called T-tour problem) by Sebő, Gottschalk, van Zuylen, and the authors. This also leads to the best-known upper bounds on the integrality ratios for these problems. Chapter 16 presents a general black-box reduction to the Symmetric TSP that loses only an arbitrarily small constant; this is a variant of the reduction that the authors obtained with Zenklusen.

Chapter 17 discusses a selection of related results and applications to problems that are closely related to, or generalizations of, the traveling salesman problem. This chapter contains pointers to further reading rather than a comprehensive account; new results are being discovered while we write these lines. Finally, Chapter 18 summarizes the state of the art of the main variants of the traveling salesman problem that we studied in this book, as well as a list of open problems.

We taught several graduate courses on various parts of this book at the University of Bonn. Our detailed proofs with many explanations and figures should provide a solid basis for teaching; moreover, we included many exercises in this book. Here are some ideas how courses could be designed.

An introductory (graduate or advanced undergraduate) course on approximation algorithms for the TSP could cover Chapters 1 and 2, Sections 4.1–4.3, Chapter 12, and then either Sections 3.1 and 3.4, Chapter 6, and possibly Sections 9.1 and 9.2 on the Asymmetric TSP, or Sections 14.1–14.3 and 15.1–15.2 on Path TSP. If there is time, Sections 17.4–17.6 can conclude the course.

A more advanced course focusing on the Symmetric TSP could first review Chapter 2 and Sections 4.1, 4.3, and 4.4, then proceed to Chapters 12 and 13 on the Graph TSP, continue with Chapter 14 and (at least) one of Chapters 15 and 16 on the Path TSP, and save the most difficult part, Chapters 10 and 11, for the end. If there is time, Sections 5.3 and 5.4 can be included to show how the maximum entropy sampling can be implemented.

An advanced course on the Asymmetric TSP would start with Section 1.5, then cover Chapter 3 and Sections 4.1 and 4.2. Chapter 5 could be next (but can also be omitted). Then we would suggest the entire material in Chapters 6–9, possibly omitting Section 8.6. Sections 17.4 and 17.6 can be discussed at the end, if there is time.

Contents

1

Introduction

In this introductory chapter, we will formally introduce the main variants of the traveling salesman problem – symmetric and asymmetric – explain a very useful graph-theoretic view based on Euler's theorem, and describe the classical simple approximation algorithms. In this chapter, we will also introduce basic notation, in particular from graph theory, and some fundamental combinatorial optimization problems.

1.1 Problems and Algorithms

This book is about the traveling salesman problem (TSP), but this is actually more than one problem. Although one could start with the most general variant of the problem, let us begin with the most classical one.

Problem 1.1 (SYMMETRIC TSP WITH TRIANGLE INEQUALITY).

Instance: A finite set V and a distance function $c : V \times V \to \mathbb{R}_{\geq 0}$ such that $c(u, v) = c(v, u)$ for all $u, v \in V$ and $c(u, w) \leq c(u, v) + c(v, w)$ for all $u, v, w \in V$.

Task: Compute a list $v_1, v_2, \ldots, v_n$ that contains every element of V exactly once and minimizes $c(v_n, v_1) + \sum_{i=2}^{n} c(v_{i-1}, v_i)$.

The elements of V are called *cities*, and the number of cities will always be denoted by n in this book. Of course, the distance function c does not necessarily describe geometric distances, but it could also represent driving times or cost. An example with $|V| = 20$ is shown in Figure 1.1.

We will be interested in *algorithms* that accept any instance (V, c) as input and always terminate with a *feasible solution* (an order of the cities) as output. If an algorithm always finds an optimum solution, we speak of an *exact*

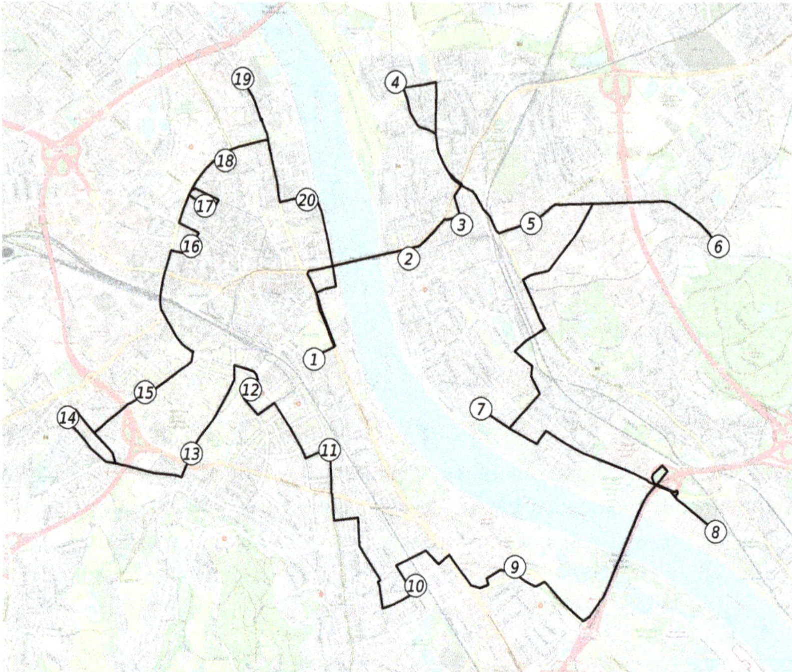

Figure 1.1 A tour visiting 20 locations in Bonn, Germany, by car. The map data is taken from OpenStreetMap (openstreetmap.org/copyright).

algorithm. One such algorithm would simply enumerate all $n!$ permutations of the cities and output the best, but this is too slow already for 20 cities (note that $20! = 2\,432\,902\,008\,176\,640\,000$), and completely hopeless for the 49-city instance that Dantzig, Fulkerson, and Johnson [1954] solved 70 years ago.

To measure the running time of an algorithm, one counts the maximum number of elementary steps that it can take. To avoid technical details, one ignores constant factors and uses the O-notation. For example, an algorithm is said to run in $O(n^3)$ time if there is a constant γ such that the number of elementary steps is never more than $\gamma \cdot n^3$. See, for example, Hougardy and Vygen [2016] for a detailed explanation.

To distinguish algorithms that are, at least asymptotically, much faster than naive enumeration, Edmonds [1965a] suggested the notion of a *polynomial-time algorithm*. For every algorithm that we present in this book, there is a constant k such that the algorithm runs in $O(n^k)$ time.

Karp [1972] showed that the Symmetric TSP with Triangle Inequality is *NP*-hard. This implies that there is no polynomial-time exact algorithm unless

$P = NP$. In fact, there is a polynomial-time exact algorithm if and only if $P = NP$. We assume that the reader is familiar with the notions of P, NP, and NP-hard; otherwise, it is sufficient to know that it is widely believed that $P \neq NP$, which would imply that there is no polynomial-time exact algorithm for any NP-hard problem.

The fastest known exact algorithm is a simple dynamic programming algorithm that computes an optimum solution in $O(n^2 2^n)$ time (Bellman [1962], Held and Karp [1962]; see Exercise 1.1). This time bound has not been improved on for more than 60 years. Since most researchers believe that $P \neq NP$, there is little hope to find a polynomial-time algorithm that always finds an optimum solution. Hence we will study approximation algorithms:

Definition 1.2 (approximation algorithm). An *α-approximation algorithm* (for a minimization problem with nonnegative cost function) is a polynomial-time algorithm that always computes a feasible solution of cost at most α times the optimum.

In the context of the TSP, α can be either a constant or a function of n (the number of cities). For a constant-factor approximation algorithm, we define its *approximation ratio* to be the infimum of all α for which it is an α-approximation algorithm, or, equivalently, the supremum of $\frac{A(I)}{\mathrm{OPT}(I)}$ over all instances I, where $A(I)$ is the cost of the solution computed by the algorithm, $\mathrm{OPT}(I)$ is the cost of an optimum solution, and $\frac{0}{0} := 1$.

Probably the first proof of an approximation ratio for the TSP was due to Rosenkrantz, Stearns, and Lewis [1977]. They proposed algorithms called "nearest insertion" and "cheapest insertion" and showed that they are 2-approximation algorithms for the SYMMETRIC TSP WITH TRIANGLE INEQUALITY. We will see a simpler 2-approximation algorithm in Proposition 1.22.

The *triangle inequality* $c(u, w) \leq c(u, v) + c(v, w)$ for all $u, v, w \in V$ naturally holds in many applications. If we have general nonnegative symmetric distances, not obeying the triangle inequality, we should allow for visiting cities more than once; we will get to this in the next section.

Distances are not always symmetric. Dropping the symmetry assumption yields the following:

Problem 1.3 (ASYMMETRIC TSP WITH TRIANGLE INEQUALITY).

Instance: A finite set V and a distance function $c : V \times V \to \mathbb{R}_{\geq 0}$ such that $c(u, w) \leq c(u, v) + c(v, w)$ for all $u, v, w \in V$.

Task: Compute a list $v_1, v_2, \ldots, v_n$ that contains every element of V exactly once and minimizes $c(v_n, v_1) + \sum_{i=2}^{n} c(v_{i-1}, v_i)$.

This problem seems to be substantially harder; while it is easy to devise a 2-approximation algorithm for the symmetric special case (see Proposition 1.22), no such algorithm is known for the asymmetric version, and no constant-factor approximation algorithm was known at all until 2017.

1.2 Graphs and Euler's Theorem

Often distances are given by a graph $G = (V, E)$ (which can, for example, represent a street network). All our *graphs* are finite; they can be undirected or directed. In both cases, they consist of a finite set V of vertices and a finite set E of edges such that each edge is associated with a pair of distinct vertices. All edge sets in this book can be multi-sets unless specified otherwise, so graphs can have parallel edges (but no loops). Graphs without parallel edges are called *simple*. Directed graphs are also called *digraphs*. For an edge $e \in E$ that goes from v to w, we write $e = \{v, w\}$ in undirected graphs and $e = (v, w)$ in digraphs, and we use this notation even if there are several edges going from v to w. Edges in digraphs are also called *arcs*. If G is a directed graph, the *underlying undirected graph* results from replacing each arc (v, w) with an undirected edge $\{v, w\}$. If H is the underlying undirected graph of a digraph G, then G is called an *orientation* of H. An edge $e = \{v, w\}$ or $e = (v, w)$ is *incident* to v and w, and if such an edge exists, v and w are *neighbors*. A vertex without neighbors is *isolated*. For a graph $G = (V, E)$, we sometimes write $V(G) := V$ and $E(G) := E$.

For a vertex set $U \subseteq V$, we denote by $\delta(U)$ the (multi)set of edges with exactly one endpoint in U. In directed graphs, $\delta^-(U)$ and $\delta^+(U)$ contain the entering and the leaving edges, respectively (so $|\delta(U)| = |\delta^-(U)| + |\delta^+(U)|$). For a single vertex $v \in V$, we write $\delta(v) := \delta(\{v\})$, $\delta^-(v) := \delta^-(\{v\})$, and $\delta^+(v) := \delta^+(\{v\})$. We call $|\delta(v)|$ (the number of edges incident to v) the *degree* of v, and in digraphs, $|\delta^-(v)|$ and $|\delta^+(v)|$ are the *in-degree* and *out-degree* of v, respectively. We add a subscript and, for example, write $\delta_G(U)$ or $\delta_E(U)$ if the graph $G = (V, E)$ is not clear from the context.

Lemma 1.4 (handshake lemma). *In any graph, the number of odd-degree vertices is even.*

Proof. For any graph (V, E), we have $\sum_{v \in V} |\delta(v)| = 2|E|$; hence there is an even number of odd summands on the left-hand side. □

A *walk* (from v_0 to v_k of *length* k) in G is a sequence $v_0, e_1, v_1, e_2, \ldots, v_k$ such that e_i is an edge from vertex v_{i-1} to vertex v_i for all $i = 1, \ldots, k$. If

$v_0 = v_k$, we speak of a *closed walk*. Note that $k = 0$ is possible. The *footprint* of a walk in $G = (V, E)$ is the multi-subset of E that contains r copies of any edge that the walk traverses r times. For a multi-subset F of E and a cost function $c : E \to \mathbb{R}$, we define the *cost* of F by $c(F) := \sum_{e \in F} c(e)$, where the sum counts edges according to their multiplicity in F. Then the cost of a walk $v_0, e_1, v_1, e_2, \ldots, v_k$ with footprint F can be expressed as $c(F) = \sum_{e \in F} c(e) = \sum_{i=1}^{k} c(e_i)$. Sometimes we say *weight* instead of cost.

A walk cannot visit all vertices unless the graph is connected. An undirected graph is *connected* if it contains a walk from v to w for all $v, w \in V$. A directed graph is *connected* if the underlying undirected graph is connected. A directed graph is *strongly connected* if it contains a walk from v to w for all $v, w \in V$.

A *subgraph* of a graph $G = (V, E)$ is a graph $G' = (V', E')$ with $V' \subseteq V$ and $E' \subseteq E$. We also say that G *contains* G' or that G' is *in* G. For a graph $G = (V, E)$ and $\emptyset \neq W \subseteq V$, the graph with vertex set W that contains all edges of E with both endpoints in W is called the subgraph of G *induced by* W and is denoted by $G[W]$; its edge set is denoted by $E[W]$. The maximal connected subgraphs of a graph G are its *connected components*; they are induced subgraphs. A *multi-subgraph* results from a subgraph by possibly adding copies of edges. Sometimes we obtain subgraphs by deleting an edge e, a vertex v (and its incident edges), or a set of vertices X and write $G - e$, $G - v$, and $G - X = G[V(G) \setminus X]$.

Contracting a vertex set W in a graph G means deleting all vertices and edges in $G[W]$, adding a new vertex v_W, and for every edge in $\delta(W)$ replacing the endpoint in W by v_W. We call the result G/W. Contracting an edge means contracting the (two-element) set of its endpoints.

Given a graph G (directed or undirected), we will be looking for a closed walk in G that contains every vertex at least once. Euler [1736] observed that the footprint of such a walk has a simple property:

Definition 1.5 (Eulerian). An undirected graph $G = (V, E)$ (and its edge set E) is called *Eulerian* if every vertex has even degree. A directed graph $G = (V, E)$ (and its edge set E) is called *Eulerian* if for every vertex the in-degree equals the out-degree.

The following characterization is known as Euler's theorem:

Theorem 1.6 (Euler [1736], Hierholzer [1873]). *Let $G = (V, E)$ be a connected graph (directed or undirected). Then G is Eulerian if and only if it contains a closed walk that traverses each edge exactly once. Such a walk can be computed in $O(|E|)$ time.*

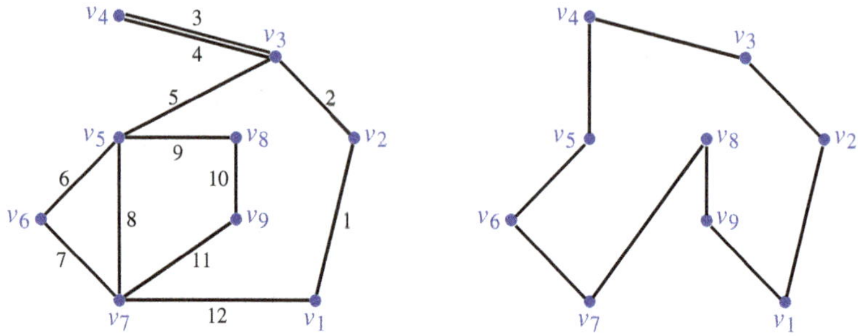

Figure 1.2 The left picture shows a connected Eulerian graph. By Theorem 1.6, this graph contains a closed walk that traverses every edge exactly once. The numbers next to the edges show the order in which the edges appear in one such Eulerian walk. The vertices $v_1, \ldots, v_n$ are numbered in the order of their first appearance in this walk. The right picture shows the solution to the Symmetric TSP with Triangle Inequality that results from taking shortcuts as in the proof of Lemma 1.7.

Proof. Let $G = (V, E)$. If G contains a closed walk $v_0, e_1, v_1, \ldots, e_k, v_k$ that traverses each edge exactly once, then E is the footprint of this walk. If this walk visits a vertex r times (where v_0 and v_k count as one visit), then it enters that vertex r times and leaves it r times; so its degree is $2r$. Hence G is Eulerian.

We prove the converse by induction on $|E|$, the case $E = \emptyset$ being trivial. So let $G = (V, E)$ be Eulerian, and let F be the footprint of a walk $v_0, e_1, v_1 \ldots, e_k, v_k$ in G that contains every edge at most once and is as long as possible. Since E is nonempty, so is F. Moreover, $v_k = v_0$, for otherwise there is an unused edge leaving v_k that we could append to the walk. So we have a closed walk, and by the first part, its footprint F is Eulerian. Hence also $E \setminus F$ is Eulerian. By induction, $E \setminus F$ is the union of footprints of closed walks (one in each connected component), and each of them must contain a vertex of $v_1, \ldots, v_k$ because G is connected. So we can insert the other walks at these positions into the first walk.

This proof easily implies a linear-time algorithm, by greedily extending a walk as long as possible and recursively applying the algorithm to the remainder. □

We use Euler's theorem as follows (cf. Figure 1.2):

Lemma 1.7. *Let (V, c) be an instance of Symmetric TSP with Triangle Inequality, and let (V, F) be a connected Eulerian undirected graph. Then there exists a solution $v_1, \ldots, v_n$ of cost at most $\sum_{e=\{v,w\}\in F} c(v, w)$, and such a solution can be found in $O(|F|)$ time.*

Proof. By Theorem 1.6, one can construct, in $O(|F|)$ time, a closed walk in (V, F) that traverses each edge exactly once. Let $v_1, \ldots, v_n$ be the elements of V in the order in which they appear in that walk for the first time. Then $c(v_n, v_1) + \sum_{i=2}^{n} c(v_{i-1}, v_i) \leq \sum_{e=\{v,w\} \in F} c(v, w)$ because, by the triangle inequality, $c(v_{i-1}, v_i)$ is at most the total cost of the edges in the subwalk from the first appearance of v_{i-1} to the first appearance of v_i (for $i = 1, \ldots, n$, where $v_0 := v_n$). □

The same proof works for the directed version:

Lemma 1.8. *Let (V, c) be an instance of* Asymmetric TSP with Triangle Inequality, *and let (V, F) be a connected Eulerian directed graph. Then there exists a solution $v_1, \ldots, v_n$ of cost at most $\sum_{e=(v,w) \in F} c(v, w)$, and such a solution can be found in $O(|F|)$ time.* □

This motivates the following definition, which plays a central role in this book:

Definition 1.9 (tour)**.** A *tour* in a graph $G = (V, E)$ (directed or undirected) is a multi-subset F of E such that (V, F) is connected and Eulerian.

By Theorem 1.6, an edge set F is a tour if and only if it is the footprint of a closed walk in G that visits every vertex at least once. Using this, we can formulate the TSP in graph-theoretical terms:

Problem 1.10 (Symmetric TSP)**.**

Instance: A simple connected undirected graph $G = (V, E)$ and a cost function $c : E \to \mathbb{R}_{\geq 0}$.

Task: Compute a tour in G with minimum cost.

This has sometimes been called the graphical TSP (see, e.g., Cornuéjols, Fonlupt, and Naddef [1985]). In the asymmetric setting, we can use the same terminology:

Problem 1.11 (Asymmetric TSP)**.**

Instance: A simple strongly connected directed graph $G = (V, E)$ and a cost function $c : E \to \mathbb{R}_{\geq 0}$.

Task: Compute a tour in G with minimum cost.

Now we want to argue that these graph-theoretic versions of the TSP are equivalent to the ones given in the previous section. We say that problem P_1 *reduces to* P_2 if there is a polynomial-time algorithm that computes, for any given instance I_1 of P_1, an instance I_2 of P_2 with the same optimum cost, and for any feasible solution S_2 of I_2, a solution S_1 of I_1 with no larger cost. If P_1 reduces to P_2 and P_2 reduces to P_1, we say that P_1 and P_2 are *equivalent*.

If $v_0, e_1, v_1, e_2, \ldots, v_k$ is a walk without any repetitions of vertices, then the graph with vertex set $\{v_0, \ldots, v_k\}$ and edge set $\{e_1, \ldots, e_k\}$ is called a *path*. If $v_0 = v_k$ but there are no other repetitions, this graph is called a *circuit* (or *cycle*). (A path or circuit can be directed or undirected.) A path or circuit in a graph G is called *Hamiltonian* if it contains all vertices of G. A solution to an instance of the SYMMETRIC TSP WITH TRIANGLE INEQUALITY with $n \geq 3$ can be interpreted as a Hamiltonian circuit in the *complete graph* on vertex set V (whose edge set is $\binom{V}{2}$). We write $c(e) := c(v, w)$ for an edge $e = \{v, w\}$ of this graph. Similarly, a solution to an instance of the ASYMMETRIC TSP WITH TRIANGLE INEQUALITY with $n \geq 2$ can be interpreted as a Hamiltonian circuit in the *complete directed graph* on vertex set V (whose edge set is $\{(v, w) \in V \times V : v \neq w\}$). Again we write $c(e) := c(v, w)$ for an edge $e = (v, w)$ of this digraph.

For an instance (G, c) of the SYMMETRIC TSP or the ASYMMETRIC TSP and two vertices v and w of G, the *distance* from v to w is the minimum total cost of the edges of a walk from v to w. We often denote it by $\mathrm{dist}_{(G,c)}(v, w)$. By *Dijkstra's algorithm*, such a walk can be computed in polynomial time (see Theorem 1.14).

Proposition 1.12. *SYMMETRIC TSP WITH TRIANGLE INEQUALITY (Problem 1.1) and SYMMETRIC TSP (Problem 1.10) are equivalent. ASYMMETRIC TSP WITH TRIANGLE INEQUALITY (Problem 1.3) and ASYMMETRIC TSP (Problem 1.11) are equivalent.*

Proof. We first reduce Problem 1.10 to Problem 1.1. Let $I_1 = (G, c)$ be an instance of Problem 1.10 with $G = (V, E)$. For $v, w \in V$, let $\bar{c}(v, w)$ be the distance from v to w in (G, c), and $I_2 := (V, \bar{c})$. By Lemma 1.7, from any tour F in G, we can construct a solution $v_1, \ldots, v_n$ to I_2 with $\bar{c}(v_n, v_1) + \sum_{i=2}^{n} \bar{c}(v_{i-1}, v_i) \leq \bar{c}(F) \leq c(F)$.

Conversely, for every solution $v_1, \ldots, v_n$ to I_2, we can compute a minimum-cost walk in (G, c) from v_n to v_1 and from v_{i-1} to v_i for $i = 2, \ldots, n$ and append all these walks to obtain a closed walk visiting all vertices. Its footprint is a tour with cost $\bar{c}(v_n, v_1) + \sum_{i=2}^{n} \bar{c}(v_{i-1}, v_i)$. In particular, I_1 and I_2 have the same optimum cost.

The reduction from Problem 1.11 to Problem 1.3 is identical.

To reduce Problem 1.1 to Problem 1.10, let $I_1 = (V, c)$ be an instance of Problem 1.1 and define $I_2 = (G, c)$, where $G = (V, E)$, $E = \binom{V}{2}$, and $c(e) = c(v, w)$ for all $e = \{v, w\} \in E$. Every solution to I_1 corresponds to a Hamiltonian circuit in G with the same cost and vice versa. For every tour F in G, we can construct a Hamiltonian circuit of at most the same cost by Lemma 1.7.

To reduce Problem 1.3 to Problem 1.11, let $I_1 = (V, c)$ be an instance of Problem 1.3 and define $I_2 = (G, c)$, where $G = (V, E)$ has edge set $E = \{(v, w) : v, w \in V,\ v \neq w\}$, and $c(e) = c(v, w)$ for all $e = (v, w) \in E$, and proceed the same way. □

If problem P_1 reduces to problem P_2 and P_2 has an α-approximation algorithm, then so has P_1. We will often work with the graph versions of the TSP (Symmetric TSP and Asymmetric TSP), but sometimes the versions with the triangle inequality are more useful. Proposition 1.12 allows us to switch between the versions and use whichever is more convenient.

If we do not require the triangle inequality but still want to visit every city exactly once, the problem is hopeless, as Sahni and Gonzalez [1976] observed: Any approximation algorithm would imply $P = NP$. This is because any α-approximation algorithm, for any function α, would allow us to decide in polynomial time whether a given graph contains a Hamiltonian circuit (see Problem 1.20): Just define the distance of two cities to be 0 if they are joined by an edge in the graph and 1 otherwise; then any α-approximation algorithm outputs a solution of cost 0 if and only if the given graph contains a Hamiltonian circuit.

1.3 Some Basic Combinatorial Optimization Problems

In this section, we cite three classical combinatorial optimization results without proofs. Proofs can be found in every book on combinatorial optimization, such as Schrijver [2003] or Korte and Vygen [2018].

We already used the fact that a walk from s to t whose footprint has minimum cost can be computed in polynomial time. We may assume that such a walk does not visit any vertex more than once, for otherwise we can omit cycles and obtain a walk with fewer edges that does not cost more. For an instance (G, c) of the Symmetric TSP or the Asymmetric TSP and a subgraph H of G with edge set F, we call $c(F)$ the *cost* of H. So we can formulate the problem of finding a walk from s to t whose footprint has minimum cost as follows:

Problem 1.13 (Shortest Path).

Instance: A simple graph $G = (V, E)$ (directed or undirected), a cost function $c : E \to \mathbb{R}_{\geq 0}$, and two vertices $s, t \in V$.

Task: Compute a path P from s to t in G with minimum cost, or decide that there is no such path.

The classical algorithm of Dijkstra [1959] (see Exercise 1.4) solves this problem efficiently. As before, n denotes the number of vertices in the given graph.

Theorem 1.14 (Dijkstra [1959]). *There is an $O(n^2)$-time algorithm for* Shortest Path.

A faster running time can be obtained for sparse graphs (with $o(n^2)$ edges), but this is not important for the purpose of this book. Note that it is essential that the cost function is nonnegative: The shortest path problem with general weights is *NP*-hard. Using Dijkstra's algorithm, we can compute the distance from s to t for all vertices $s, t \in V$. Since the algorithm in fact computes shortest paths from one vertex s to all vertices $t \in V$, only n applications of Dijkstra's algorithm suffice. The *metric closure* of a pair (G, c), where $G = (V, E)$ is a directed or undirected graph and $c : E \to \mathbb{R}_{\geq 0}$, is the pair $(\bar{G}, \bar{c})$ where $\bar{G}$ has the same vertex set and contains an edge $e = (v, w)$ or $e = \{v, w\}$, respectively, whenever G contains a path from v to w, and $\bar{c}(e)$ is the distance from v to w.

Theorem 1.15. *There is an $O(n^3)$-algorithm that, given a simple graph $G = (V, E)$ and a cost function $c : E \to \mathbb{R}_{\geq 0}$, computes the metric closure of (G, c).*

Another basic problem asks to connect all vertices at minimum cost. A *tree* is a minimal connected graph – that is, the deletion of any edge would destroy connectivity. A subgraph of a graph G is called *spanning* if it contains all vertices of G.

Problem 1.16 (Minimum Spanning Tree).

Instance: A simple undirected graph $G = (V, E)$ and a cost function $c : E \to \mathbb{R}_{\geq 0}$.

Task: Compute a spanning tree (V, S) in G with minimum cost, or decide that G is not connected.

This problem was solved quite early by Borůvka [1926]:

Theorem 1.17 (Borůvka [1926], Jarník [1930], Prim [1957]). *There is an $O(n^2)$-time algorithm for* Minimum Spanning Tree.

In fact, it is well known that the Minimum Spanning Tree problem can be solved by a simple greedy algorithm (see Exercise 1.5 or the proof of Theorem 2.14).

The third problem we cite here is much more difficult to solve. A *perfect matching* in G is a set M of edges such that every vertex of G is incident to exactly one of these edges.

Problem 1.18 (WEIGHTED MATCHING).

Instance: A simple undirected graph $G = (V, E)$ with $|V|$ even and a cost function $c : E \to \mathbb{R}_{\geq 0}$.

Task: Compute a perfect matching in G with minimum cost, or decide that no perfect matching exists.

This problem was solved by Edmonds [1965b]:

Theorem 1.19 (Edmonds [1965b], Gabow [1973]). *There is an $O(n^3)$-time algorithm for WEIGHTED MATCHING.*

For MINIMUM SPANNING TREE and WEIGHTED MATCHING, one could also allow edges with negative cost. Since all spanning trees have $n - 1$ edges and all perfect matchings have $\frac{n}{2}$ edges, adding a constant to all edges costs does not change the set of optimum solutions. This trick does not work for SHORTEST PATH. One can still solve SHORTEST PATH in polynomial time for *conservative weights* (i.e., when there is no circuit of negative total weight), but this is more complicated (see Exercise 1.11).

In contrast to the above three problems, many others have a polynomial-time algorithm only if $P = NP$. We mention one famous example of such an NP-hard problem:

Problem 1.20 (HAMILTONIAN CIRCUIT).

Instance: A undirected graph $G = (V, E)$.

Task: Decide whether G has a Hamiltonian circuit.

Theorem 1.21 (Karp [1972]). *HAMILTONIAN CIRCUIT has a polynomial-time algorithm if and only if $P = NP$.*

A graph with a Hamiltonian circuit is called *Hamiltonian*. Theorem 1.21 easily implies that the shortest path problem with general weights has no polynomial-time algorithm unless $P = NP$ (cf. Exercise 1.7).

1.4 Christofides' Algorithm

Given an instance I of one of our TSP variants, we will denote by $\mathrm{OPT}(I)$ the cost of an optimum solution. For two edge sets A and B, we denote by $A \dot{\cup} B$ (the *disjoint union* of A and B) the multi-set that contains two copies of each edge in $A \cap B$ and one copy of each edge in $(A \cup B) \setminus (A \cap B)$. Our very first approximation algorithm is now almost trivial:

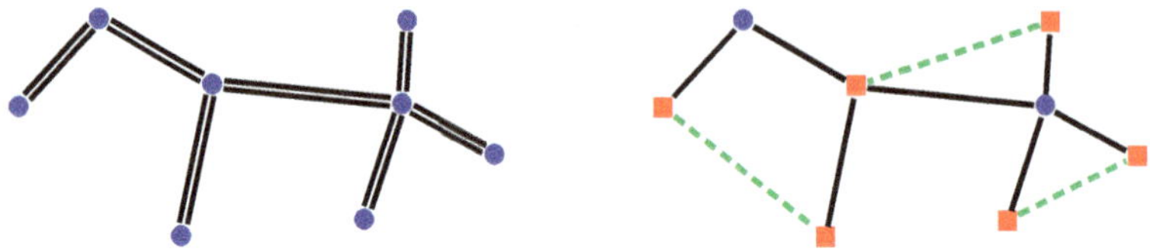

Figure 1.3 Illustrating the double tree algorithm (left) and Christofides' algorithm (right). Both start with a minimum-cost spanning tree. The former then doubles all edges, while the latter adds a minimum-cost perfect matching (green, dashed) among the odd-degree vertices (red squares).

Proposition 1.22 (Rosenkrantz, Stearns, and Lewis [1977]). *There is a 2-approximation algorithm for* Symmetric *TSP.*

Proof. Given an instance (G, c) with $G = (V, E)$, compute a minimum-cost spanning tree (V, S) in (G, c) (cf. Theorem 1.17) and output the tour $S \dot{\cup} S$, which results from S by doubling all edges. Since any tour is connected and thus contains a spanning tree, we have $c(S) \leq \mathrm{OPT}(G, c)$, so the tour $S \dot{\cup} S$ that we compute costs at most $2\,\mathrm{OPT}(G, c)$. □

Rosenkrantz, Stearns, and Lewis [1977] actually proved (already in 1974) that two different algorithms ("nearest insertion" and "cheapest insertion") are 2-approximation algorithms, but the above folklore proof is much simpler. The algorithm in the proof of Proposition 1.22 has often been called the *double tree algorithm.*

Christofides [1976], and independently Serdyukov [1978], showed how to improve on this. (See van Bevern and Slugina [2020] for a historical note.) Christofides' algorithm also begins by computing a minimum-cost spanning tree (V, S). Then, instead of doubling all edges, it finds a potentially cheaper way to make S Eulerian. For a graph (V, F), let $\mathrm{odd}(F) = \{v \in V : |F \cap \delta(v)| \text{ odd}\}$ denote the set of odd-degree vertices. We formulate Christofides' algorithm first for the Symmetric TSP with Triangle Inequality. See Algorithm 1.23, and see Figure 1.3 for an illustration.

Theorem 1.24 (Christofides [1976], Serdyukov [1978]). *Christofides' algorithm (Algorithm 1.23) is a $\frac{3}{2}$-approximation algorithm for* Symmetric *TSP* with Triangle Inequality.

Proof. An optimum solution corresponds to a Hamiltonian circuit H in G of cost $\mathrm{OPT}(V, c)$. We have $c(S) \leq c(H)$ because deleting one edge from H results in a spanning tree. Let $w_1, \ldots, w_k$ be the vertices of W in the order in which they appear in a traversal of H, and let $w_0 := w_k$. Note that

---

Algorithm 1.23: Christofides' Algorithm

Input: an instance (V, c) of SYMMETRIC TSP WITH TRIANGLE INEQUALITY

Output: a solution $v_1, \ldots, v_n$

(1) Let $G = (V, \binom{V}{2})$ be the complete graph on V, and for $e = \{v, w\} \in \binom{V}{2}$, let $c(e) = c(v, w)$.
(2) Compute a minimum-cost spanning tree (V, S) in (G, c).
(3) Let $W = \text{odd}(S)$.
(4) Compute a minimum-cost perfect matching M in $(G[W], c)$.
(5) Apply Lemma 1.7 to the tour $(V, S \dot\cup M)$ and output the resulting solution.

$k = |W|$ is even by the handshake lemma (Lemma 1.4). Then by the triangle inequality, $\sum_{i=1}^{k} c(w_{i-1}, w_i) \leq c(H)$. Hence we have two perfect matchings $M_1 = \{\{w_{i-1}, w_i\} : i \text{ even}\}$ and $M_2 = \{\{w_{i-1}, w_i\} : i \text{ odd}\}$ in $G[W]$ with total cost at most $c(H)$. So

$$c(M) \leq \min \{c(M_1), c(M_2)\} \leq \frac{1}{2}\left(c(M_1) + c(M_2)\right) \leq \frac{1}{2}c(H).$$

By Lemma 1.7, our output has cost $c(S \dot\cup M) = c(S) + c(M) \leq \frac{3}{2}c(H) = \frac{3}{2}\,\text{OPT}(V, c)$. The algorithm can be implemented to run in polynomial time by Theorems 1.17 and 1.19. □

We will now reformulate Christofides' algorithm for the SYMMETRIC TSP. The following notion will be used very often in this book:

Definition 1.25 (T-join)**.** Let V be a finite set and $T \subseteq V$ with $|T|$ even. A *T-join* is a multi-subset J of $\binom{V}{2}$ with $T = \text{odd}(J)$. If $G = (V, E)$ is a graph and $J \subseteq E$, then we say that J is a T-join in G.

We start with a few basic properties. For two sets A and B, their *symmetric difference* $A \triangle B = (A \setminus B) \cup (B \setminus A)$ contains all elements that are in the union of A and B, but not in their intersection.

Proposition 1.26. *Let $G = (V, E)$ be an undirected graph and $T, T' \subseteq V$ with $|T|, |T'|$ even. Let J be a T-join and J' a T'-join. Then $J \triangle J'$ is a $(T \triangle T')$-join.*

Proof. A vertex v has odd degree in $(V, J \triangle J')$ if and only if it has odd degree in $(V, J \dot\cup J')$, and this is the case if and only if it has odd degree in exactly one of (V, J) and (V, J'). □

Proposition 1.27. *Let $G = (V, E)$ be an undirected graph and $T \subseteq V$ with $|T|$ even. Then G contains a T-join if and only if every connected component of G contains an even number of elements of T.*

Proof. Necessity follows from Lemma 1.4. For sufficiency, let $T = \{t_1, \ldots, t_k\}$ such that t_{i-1} and t_i are in the same connected component of G for $i = 2, 4, \ldots, k$. Then take any path P_i from t_{i-1} to t_i for $i = 2, 4, \ldots, k$. The symmetric difference of the edge sets of these paths is a T-join by Proposition 1.26. □

Lemma 1.28. *Let $G = (V, E)$ be an undirected graph and $T \subseteq V$ with $|T|$ even. Let J be a T-join in G. Then there exists a numbering $T = \{t_1, \ldots, t_k\}$ and a path from t_{i-1} to t_i in (V, J) for $i = 2, 4, \ldots, k$ such that these paths are pairwise edge-disjoint.*

Proof. We use induction on $k = |T|$, the case $k = 0$ being trivial. By Proposition 1.27, there are two vertices $t_{k-1}, t_k \in T$ in the same connected component of (V, J), so let P be the edge set of a path from t_{k-1} to t_k in (V, J), and apply the induction hypothesis to $T' = T \setminus \{t_{k-1}, t_k\}$ and the T'-join $J \setminus P$. □

Using an algorithm for WEIGHTED MATCHING, we can compute minimum-cost T-joins in polynomial time:

Theorem 1.29 (Edmonds and Johnson [1973]). *Given a simple undirected graph $G = (V, E)$, $c : E \to \mathbb{R}_{\geq 0}$, and $T \subseteq V$ with $|T|$ even, one can compute a minimum-cost T-join in (G, c) or decide that none exists in $O(n^3)$ time.*

Proof. The existence of a T-join can be decided with Proposition 1.27.

Let $H = \left(T, \binom{T}{2}\right)$ be the complete undirected graph on T, and for $v, w \in T$, let $\bar{c}(\{v, w\})$ be the distance from v to w in (G, c). Note that $\bar{c}$ can be computed in $O(n^3)$ time by Theorem 1.15.

Now compute a minimum-cost perfect matching M in $(H, \bar{c})$, using Theorem 1.19. For $\{v, w\} \in M$, compute a shortest path from v to w in (G, c), and let J be the symmetric difference of these $|M|$ paths. We prove that J is a minimum-cost T-join in (G, c).

Proposition 1.26 implies that J is indeed a T-join. To show that J has minimum cost, let J^* be a minimum-cost T-join. By Lemma 1.28, there exists a numbering $T = \{t_1, \ldots, t_k\}$ and a path P_i from t_{i-1} to t_i in (V, J^*) for $i = 2, 4, \ldots, k$ such that these paths are pairwise edge-disjoint. We conclude

$$c(J) \leq \bar{c}(M) \leq \sum_{i=2,4,\ldots,k} \bar{c}(\{t_{i-1}, t_i\}) \leq \sum_{i=2,4,\ldots,k} c(P_i) \leq c(J^*). \qquad \square$$

The problem can actually be solved for general weights too. We do not need this for now but note it for later use (in Chapter 12):

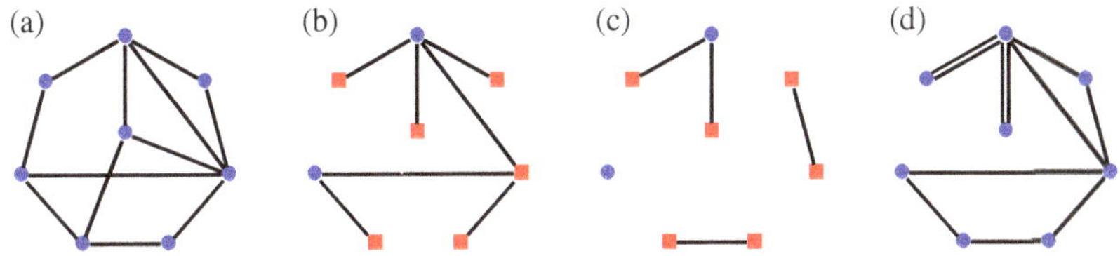

Figure 1.4 Christofides' algorithm illustrated for an unweighted graph instance. (a) An unweighted graph G (i.e., $c(e) = 1$ for all edges e), (b) a spanning tree (V, S) whose odd-degree vertices (elements of $\text{odd}(S)$) are shown as red squares, (c) a minimum $\text{odd}(S)$-join J, and (d) the resulting tour $S \dot{\cup} J$.

Corollary 1.30. *Given a simple undirected graph $G = (V, E)$, $c : E \to \mathbb{R}$, and $T \subseteq V$ with $|T|$ even, one can compute a minimum-cost T-join in (G, c) or decide that none exists in $O(n^3)$ time.*

Proof. Let $E^- = \{e \in E : c(e) < 0\}$ and $c'(e) := |c(e)|$ for all $e \in E$. Then $c'(K \triangle E^-) = c(K) - c(E^-)$ for all $K \subseteq E$. Let $T' := T \triangle \text{odd}(E^-)$. Then J' is a minimum c'-cost T'-join if and only if $J' \triangle E^-$ is a minimum c-cost T-join. Hence the problem reduces to Theorem 1.29. □

With the notion of T-joins, we have an elegant reformulation of Christofides' algorithm: see Algorithm 1.31. Figure 1.4 provides an example.

Algorithm 1.31: Christofides' Algorithm

Input: an instance (G, c) of Symmetric TSP
Output: a tour F

(1) Compute a minimum-cost spanning tree (V, S) in (G, c).
(2) Let $W = \text{odd}(S)$, and let J be a minimum-cost W-join in (G, c).
(3) Output the tour $S \dot{\cup} J$.

The running time of Christofides' algorithm is also $O(n^3)$, dominated by the subroutine to find a minimum-cost $\text{odd}(S)$-join (cf. Theorem 1.29). Let us now prove the approximation guarantee again for this version:

Theorem 1.32 (Christofides [1976], Serdyukov [1978]). *Christofides' algorithm (Algorithm 1.31) is a $\frac{3}{2}$-approximation algorithm for Symmetric TSP.*

Proof. We have $c(S) \le \text{OPT}(G, c)$, like in the proof of Proposition 1.22. Any optimum tour F^* contains a W-join J_1 by Proposition 1.27. Let $J_2 := F^* \setminus J_1$. By Proposition 1.26, $\text{odd}(J_2) = \text{odd}(F^*) \triangle \text{odd}(J_1) = \emptyset \triangle W = W$. After deleting

Table 1.1 *Approximation ratios for* Symmetric *TSP in the order of their discovery. (R) means randomized; this algorithm computes a random tour, and the approximation ratio compares its expected cost to* OPT.

Approximation Ratio	Year	Reference	Chapter
2	1974	Rosenkrantz, Stearns, and Lewis [1977]	–
$\frac{3}{2}$	1976	Christofides [1976]	1.4
$\frac{3}{2}$	1976	Serdyukov [1978]	1.4
$\frac{3}{2} - 10^{-36}$ (R)	2020	Karlin, Klein, and Oveis Gharan [2021]	10–11
$\frac{3}{2} - 10^{-36}$	2022	Karlin, Klein, and Oveis Gharan [2023]	11.6

pairs of parallel edges in J_1 and J_2, we get two W-joins J_1' and J_2' in G with $c(J_1') + c(J_2') \leq c(J_1) + c(J_2) = c(F^*) = \mathrm{OPT}(G, c)$. Hence

$$c(J) \leq \min\{c(J_1'), c(J_2')\} \leq \tfrac{1}{2}\left(c(J_1') + c(J_2')\right) \leq \tfrac{1}{2}\,\mathrm{OPT}(G, c).$$

We conclude $c(S \,\dot\cup\, J) = c(S) + c(J) \leq \frac{3}{2}\,\mathrm{OPT}(G, c)$. □

Adding a matching M in Algorithm 1.23 or a W-join J in Algorithm 1.31 is called *parity correction* because it corrects the parity of every vertex degree (renders it even). Bounding the cost of parity correction will be a central topic in several chapters of this book.

Of course, Theorems 1.24 and 1.32 are equivalent. This bound on the approximation ratio of Christofides' algorithm is tight even for unweighted graph instances: For a complete graph with an even number of vertices, take a spanning tree whose vertices all have odd degree, then we end up with $\frac{3}{2}n - 1$ edges. The special case of Symmetric TSP where $c(e) = 1$ for all $e \in E$ is known as Graph TSP. Today we know a slightly better approximation algorithm for Symmetric TSP (see Table 1.1 and Chapters 10 and 11) and much better approximation algorithms for Graph TSP (see Chapters 12 and 13). However, the following question is still open:

Open Problem 1.33. Find an α-approximation algorithm for Symmetric TSP for some $\alpha \ll \frac{3}{2}$ (say $\alpha \leq 1.49$).

Chekuri and Quanrud [2018] found a $(\frac{3}{2} + \varepsilon)$-approximation algorithm that is faster than Christofides' algorithm, for any $\varepsilon > 0$.

1.5 Cycle Cover Algorithm

For ASYMMETRIC TSP, a constant-factor approximation algorithm is much more difficult to obtain, and indeed such an algorithm was not known until 2017. It is trivial to give an n-approximation algorithm: Given an instance (G, c) with $G = (V, E)$, order the cities arbitrarily (say $V = \{v_1, \ldots, v_n\}$), and take a shortest v_n-v_1-path and a shortest v_{i-1}-v_i-path for $i = 2, \ldots, n$ in (G, c); output the disjoint union of all these paths. Since every tour contains a path from v_{i-1} to v_i for any two cities v_{i-1} and v_i, this algorithm produces a tour at most n times longer than optimal (and this bound is essentially tight; see Exercise 1.12).

The first nontrivial approximation algorithm was found by Frieze, Galbiati, and Maffioli [1982]. It is based on the following concept: A *cycle cover* of a graph $G = (V, E)$ (directed or undirected) is a subset $F \subseteq E$ of edges such that every vertex has degree 2 in (V, F), and in-degree 1 and out-degree 1 in the directed case. In particular, the edge set of a Hamiltonian circuit is a cycle cover, and every cycle cover is Eulerian, but a cycle cover is not necessarily connected.

Lemma 1.34. *Given a simple directed graph $G = (V, E)$ and $c : E \to \mathbb{R}_{\geq 0}$, one can compute a minimum-cost cycle cover in (G, c) or decide that none exists in $O(n^3)$ time.*

Proof. Let G^{12} be the undirected graph that contains two vertices v^1 and v^2 for each $v \in V$ and an edge $\{v^1, w^2\}$ for each $(v, w) \in E$ (with the same cost). There is a one-to-one correspondence between the cycle covers in G and the perfect matchings in G^{12}. Hence the problem reduces to finding a minimum-cost perfect matching in G^{12}, which can be done in $O(n^3)$ time by Theorem 1.19. □

We remark that the graph G^{12} constructed in this proof is *bipartite* (every edge has exactly one endpoint in $\{v^1 : v \in V\}$), and WEIGHTED MATCHING is easier in bipartite graphs, but this is not important here.

The cycle cover algorithm by Frieze, Galbiati, and Maffioli [1982] is best described for the ASYMMETRIC TSP WITH TRIANGLE INEQUALITY (see Algorithm 1.35 and Figure 1.5).

Theorem 1.36 (Frieze, Galbiati, and Maffioli [1982]). *The cycle cover algorithm (Algorithm 1.35) is a $(\log_2 n)$-approximation algorithm for ASYMMETRIC TSP WITH TRIANGLE INEQUALITY.*

Proof. At any stage, F is Eulerian, and the algorithm leaves the while-loop only when F is a tour. The number of connected components decreases by at least a factor of 2 in each iteration of the while-loop. Hence there are at most $\lfloor \log_2 n \rfloor$ iterations, and thus by Lemma 1.34, the algorithm runs in $O(n^3 \log n)$ time.

Algorithm 1.35: Cycle Cover Algorithm

Input: an instance (V, c) of ASYMMETRIC TSP WITH TRIANGLE INEQUALITY

Output: a solution $v_1, \ldots, v_n$

(1) Let $G = (V, \{(v, w) \in V \times V : v \neq w\})$ be the complete directed graph on V. Let $F := \emptyset$.
(2) **while** (V, F) is not connected **do**
(3) Choose one vertex from each connected component of (V, F); let W be the set of these vertices.
(4) Let F_W be a minimum-cost cycle cover in $(G[W], c)$. Set $F := F \dot{\cup} F_W$.
(5) Apply Lemma 1.8 to the tour F and output the resulting solution.

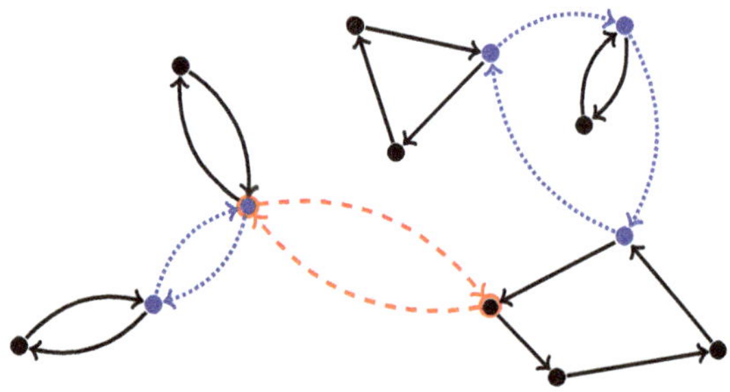

Figure 1.5 Illustrating the cycle cover algorithm. The first iteration chooses a minimum-cost cycle cover (black, solid). The second iteration chooses a representative vertex of each connected component and adds a minimum-cost cycle cover on these (blue, dotted). After two more edges are added in the third iteration (red, dashed), the digraph is connected, and the algorithm terminates.

For any set $W \subseteq V$, the minimum cost of a cycle cover in $G[W]$ is at most $\text{OPT}(V, c)$ because, due to the triangle inequality, we can take shortcuts in any Hamiltonian circuit in G to obtain a Hamiltonian circuit in $G[W]$ without increasing the cost. We conclude that $c(F_W) \leq \text{OPT}(V, c)$ in each iteration, and hence our output has cost at most $\lfloor \log_2 n \rfloor \text{OPT}(V, c)$. □

This bound for the cycle cover algorithm is tight (see Exercise 1.14).

More than 20 years later, the upper bound on the approximation ratio for ASYMMETRIC TSP was improved by a constant factor by Bläser [2008] to $0.99 \log_2 n$, by Kaplan et al. [2005] to $0.842 \log_2 n$, and by Feige and Singh [2007] to $\frac{2}{3} \log_2 n$. We will not present these algorithms here; they are refinements

Table 1.2 *Approximation ratios for* Asymmetric *TSP in the order of their discovery. (R) means randomized; this algorithm computes a random tour, and the approximation ratio compares its expected cost to* OPT. *Moreover, ε stands for an arbitrarily small positive constant.*

Approximation Ratio	Year	Reference	Chapter
$\log_2 n$	1980	Frieze, Galbiati, and Maffioli [1982]	1.5
$0.99 \log_2 n$	2002	Bläser [2008]	–
$0.842 \log_2 n$	2003	Kaplan et al. [2005]	–
$\frac{2}{3} \log_2 n$	2006	Feige and Singh [2007]	–
$O(\frac{\log n}{\log \log n})$ (R)	2009	Asadpour et al. [2017]	5
506	2017	Svensson, Tarnawski, and Végh [2020]	6–8
$22 + \varepsilon$	2019	Traub and Vygen [2022]	6–8
$17 + \varepsilon$	2021	this book	6–8

of the cycle cover algorithm. The first sublogarithmic approximation factor was obtained by Asadpour et al. [2017] and will be presented in Chapter 5. Finally, a constant-factor approximation algorithm was discovered by Svensson, Tarnawski, and Végh [2020]. In Chapters 6–8, we will present an improved version of this algorithm. Table 1.2 summarizes the history.

Exercises

1.1 Show that Asymmetric TSP with Triangle Inequality can be solved exactly in $O(n^2 2^n)$ time.
Hint: Choose an arbitrary vertex s. For every set X with $\{s\} \subsetneq X \subseteq V$ and every vertex $t \in X \setminus \{s\}$, compute a list $s = v_1, v_2, \dots, v_k = t$ that contains every element of X exactly once and minimizes $\sum_{i=2}^{k} c(v_{i-1}, v_i)$.
Note: This technique is called dynamic programming. No faster algorithm is known, even for Symmetric TSP with Triangle Inequality.
(Bellman [1962], Held and Karp [1962])

1.2 Prove that a connected undirected graph contains a walk that traverses each edge exactly once if and only if it has at most two odd-degree vertices.

1.3 Call a tour in a graph *minimal* if no proper subset is a tour in that graph. Prove that a minimal tour in an undirected graph does not contain three parallel edges, and prove that a minimal tour in a directed graph does not contain $n-1$ parallel edges. Show that these bounds are tight: 2 or $n-2$ parallel edges are possible.

1.4 Consider Dijkstra's algorithm to compute the distance from a vertex s to all other vertices in a digraph $G = (V, E)$ with nonnegative weights $c : E \to \mathbb{R}_{\geq 0}$: Initialize $R := \emptyset$, $d(s) := 0$, and $d(v) := \infty$ for all $v \in V \setminus \{s\}$. Then, while $R \neq V$, select $v \in V \setminus R$ with $d(v)$ minimum, add v to R, and set $d(w) := \min\{d(w), d(v)+c(e)\}$ for all $e = (v, w) \in \delta^+(v)$. Prove that this algorithm is correct.
(Dijkstra [1959])

1.5 Consider the following algorithm. Given a connected undirected graph $G = (V, E)$ and edge costs $c : E \to \mathbb{R}$, initialize $F := \emptyset$. As long as there exists an edge $e \in E \setminus F$ such that $(V, F \cup \{e\})$ contains no circuit, choose such an edge with minimum cost and add it to F.
Prove that this algorithm computes an optimum solution to the Minimum Spanning Tree problem.
Hint: Among all optimum spanning trees, consider one that has as many edges as possible in common with the output of the algorithm.
(Kruskal [1956])

1.6 Let $G = (V, E)$ be a graph and $X \subsetneq V$ such that $G[V \setminus X]$ has more than $|X|$ connected components with an odd number of vertices. Show that then G has no perfect matching. (The converse also holds and is known as Tutte's theorem.)

1.7 Deduce from Theorem 1.21 that the shortest path problem with general weights has no polynomial-time algorithm unless $P = NP$.

1.8 Show that for every even $n \geq 4$, there is a Hamiltonian graph G on n vertices in which every vertex has degree 3 and Christofides' algorithm – run on G with unit weights – may compute a tour with $\frac{3}{2}n - 1$ edges.

1.9 The Euclidean TSP is a special case of the Symmetric TSP with Triangle Inequality: Here $V \subsetneq \mathbb{R}^2$, and c is given by the Euclidean distance. Prove that even for Euclidean TSP, Christofides' algorithm is not an α-approximation algorithm for any $\alpha < \frac{3}{2}$.

1.10 In the Rural Postman Problem, we are given a connected undirected graph $G = (V, E)$ with weights $c : E \to \mathbb{R}_{\geq 0}$ and a subset $\bar{E}$ of edges. We ask for a connected (not necessarily spanning) Eulerian multi-subgraph of G that contains at least one copy of every element of $\bar{E}$. Devise a $\frac{3}{2}$-approximation algorithm for the Rural Postman Problem.
(This was first mentioned by Frederickson [1979].)

1.11 Conclude from Corollary 1.30 that there is a polynomial-time algorithm for the Shortest Path problem when the graph G is undirected and the weights c are conservative (i.e., there is no circuit of negative total weight).

1.12 Prove that the trivial algorithm mentioned at the beginning of Section 1.5 is an $(n-1)$-approximation algorithm for ASYMMETRIC TSP, and prove that this bound is tight.

1.13 Show that a minimum-cost cycle cover in an undirected graph can be computed in polynomial time. Note that it is not allowed to take any edge twice.

Hint: Find a reduction to WEIGHTED MATCHING by first replacing every edge by a path of three edges and then duplicating every original vertex.

1.14 Show that whenever n is a power of 2, there are instances with n cities for which the cycle cover algorithm (Algorithm 1.35) can produce a solution that is no better than $\log_2 n$ times the optimum.

1.15 Consider the variant of the cycle cover algorithm (Algorithm 1.35) in which F_W in Step (4) is chosen as the edge set of a cycle C in $G[W]$ with minimum mean weight $\frac{c(E(C))}{|E(C)|}$. Karp [1978] showed that a minimum-mean-weight cycle can be computed in $O(n^3)$ time. Prove (by induction on n) that this variant is a $2(1+\frac{1}{2}+\cdots+\frac{1}{n})$-approximation algorithm. (Bläser [2008] attributed this to Kleinberg and Williamson; see also Williamson and Shmoys [2011].)

2

Linear Programming Relaxations of the Symmetric TSP

For *NP*-hard problems, it is often useful to study relaxations that are easier to solve. In Chapter 1, we already saw two approximation algorithms that started by solving a relaxation: finding a minimum-cost connected spanning subgraph in Christofides' algorithm (Algorithm 1.31) and finding a minimum-cost cycle cover in the cycle cover algorithm (Algorithm 1.35). Another kind of relaxation arises by formulating the problem as an integer linear program and dropping the integrality constraints. In this chapter, we will study such linear programming relaxations for SYMMETRIC TSP WITH TRIANGLE INEQUALITY and SYMMETRIC TSP. These two equivalent versions of the problem give rise to two linear programming relaxations, which turn out to be equivalent as well.

2.1 The Subtour LP

Let (V, c) be an instance of the SYMMETRIC TSP WITH TRIANGLE INEQUALITY, $n = |V| \geq 3$, and $E = \binom{V}{2}$. A solution to this instance can be interpreted as a Hamiltonian circuit in $G = (V, E)$. Recall that we write $c(e) := c(v, w)$ for $e = \{v, w\} \in E$. In order to formulate the problem as a class of integer linear programs, one for each instance, we introduce a variable $x_e \in \{0, 1\}$ for $e \in E$, indicating whether or not e is an edge of this Hamiltonian circuit. Any vector $x \in \{0, 1\}^E$ corresponds to a subset $F \subseteq E$ via $F = \{e \in E : x_e = 1\}$; we say that x is the *incidence vector* (or characteristic vector) of F. More generally, for a multi-subset F of E, the incidence vector of F (with respect to E) is the vector $x \in \mathbb{Z}^E$ with $x_e = k$ if F contains k copies of $e \in E$. We denote this vector by χ^F. For a single edge e, we sometimes write χ^e instead of $\chi^{\{e\}}$.

We can then impose linear constraints to enforce that F is the edge set of a Hamiltonian circuit: For example, by *degree constraints* $\sum_{e\in\delta(v)} x_e = 2$ for all $v \in V$ we enforce that x is the incidence vector of a cycle cover. To enforce connectivity (i.e., avoid *subtours*, which here means non-spanning circuits), we could, for example, write $\sum_{e\in\delta(U)} x_e \geq 1$ for all nonempty proper subsets U of V (such an edge set $\delta(U)$ is called a *cut*). In fact, the incidence vectors of Hamiltonian circuits satisfy $\sum_{e\in\delta(U)} x_e \geq 2$ for $\emptyset \neq U \subsetneq V$, so it does no harm to impose these stronger constraints, often called *subtour elimination constraints*. Then our problem can be written as

$$\begin{aligned}
\min \ & \sum_{e\in E} c(e)x_e \\
\text{subject to} \ & \sum_{e\in\delta(U)} x_e \ \geq \ 2 \qquad (\emptyset \neq U \subsetneq V) \\
& \sum_{e\in\delta(v)} x_e \ = \ 2 \qquad (v \in V) \\
& x_e \ \in \ \{0,1\} \quad (e \in E).
\end{aligned} \tag{2.1}$$

Now the idea is to relax the integrality constraints – that is, replace $x_e \in \{0,1\}$ by $0 \leq x_e \leq 1$. We then arrive at a *linear program* (LP) that is a relaxation of the original instance; we therefore speak of an *LP relaxation*. In the following, we abbreviate the objective function by $c(x) := \sum_{e\in E} c(e)x_e$, and for $F \subseteq E$, we write $x(F) := \sum_{e\in F} x_e$.

We then have the following LP:

$$\begin{aligned}
\min \ & c(x) \\
\text{subject to} \ & x(\delta(U)) \ \geq \ 2 \quad (\emptyset \neq U \subsetneq V) \\
& x(\delta(v)) \ = \ 2 \quad (v \in V) \\
& x_e \ \leq \ 1 \quad (e \in E) \\
& x_e \ \geq \ 0 \quad (e \in E).
\end{aligned} \tag{2.2}$$

This was first formulated by Dantzig, Fulkerson, and Johnson [1954] and has often been called *subtour elimination LP* or simply *subtour LP*. The set of feasible solutions to the subtour LP is called the *subtour polytope*. Appreciating work by Held and Karp [1970] (see Exercise 4.3), the subtour LP has also been called the *Held–Karp relaxation*.

The constraints $x_e \leq 1$ for $e \in E$ are actually redundant:

Proposition 2.1. *The set of feasible solutions to the subtour LP* (2.2) *does not change if the constraints $x_e \leq 1$ for $e \in E$ are omitted.*

Proof. For every edge $e = \{v, w\}$ and every vector $x \in \mathbb{R}^E$ satisfying $x(\delta(\{v,w\}) \geq 2$ and $x(\delta(v)) = x(\delta(w)) = 2$, we have $x_e = \frac{1}{2}\big(x(\delta(v)) + x(\delta(w)) - x(\delta(\{v,w\})\big) \leq 1$. □

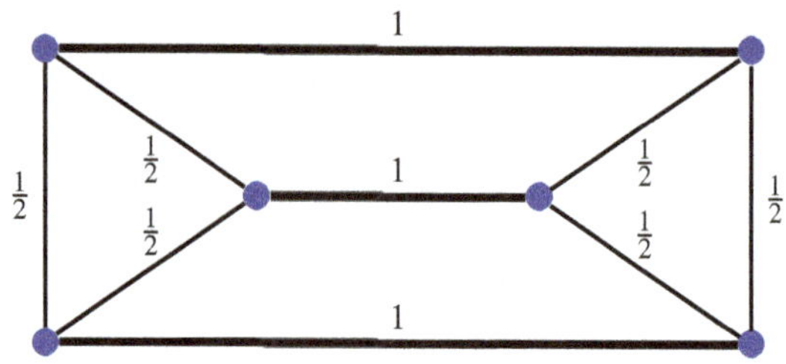

Figure 2.1 A feasible solution to the subtour LP. The value x_e is shown next to each edge e. Edges e with $x_e = 0$ are not shown.

The integral feasible solutions to the subtour LP (2.2) are precisely the incidence vectors of the Hamiltonian circuits, but this LP has other (fractional) feasible solutions (see, e.g., Figure 2.1).

Many approximation algorithms for the TSP (as well as for other *NP*-hard combinatorial optimization problems) begin by solving an LP relaxation and bound the cost of the computed solution by ρ times the LP value for some ρ, implying an approximation ratio of at most ρ. To make this approach viable, we need two properties: We need to be able to solve the LP in polynomial time, and we need that the LP solution is useful; in particular, the LP value should not be much smaller than the optimum cost of a solution for any instance. We will now address these two questions.

2.2 Solving a Linear Program

A *linear program* (or *LP* for short) is an optimization problem of the type

$$\min \left\{ c^\top x : Ax \geq b, x \geq 0 \right\}, \tag{2.3}$$

where $A \in \mathbb{R}^{m \times n}$ is an $(m \times n)$-matrix and $b \in \mathbb{R}^m$ and $c \in \mathbb{R}^n$ are column vectors. The set of feasible solutions $\{x \in \mathbb{R}^n : Ax \geq b, x \geq 0\}$ is a *polyhedron* (the intersection of a finite number of half-spaces). A bounded polyhedron is called a *polytope*.

In general, a linear program can be infeasible (if the set of feasible solutions is empty) or unbounded, but most LPs in this book will have an optimum solution. If x is an optimum solution, then $c^\top x$ is the *value* of the LP.

Khachiyan [1979] showed (using the ellipsoid method) that linear programs can be solved in polynomial time, that is, in a running time bounded by a polynomial in the total number of bits needed to encode all (numerators and denominators of rational) numbers in binary encoding (the *input size*). Tardos [1986] proved a stronger running time guarantee.

Theorem 2.2 (Khachiyan [1979], Tardos [1986]). *A linear program* (2.3) *can be solved in polynomial time. The number of arithmetic operations is bounded by a polynomial in the number of bits that are needed to encode the entries of A.*

However, most of the LPs in this book, including the subtour LP (2.2), have an exponential number of constraints, so the input size is more than 2^n. We can still solve many such LPs, including (2.2), in a running time bounded by a polynomial in n, the number of cities. There are essentially two ways: via an extended formulation or via the equivalence of separation and optimization.

First, we construct an *extended formulation*, that is, an equivalent linear program (extended by additional variables) whose number of variables and constraints is bounded by a polynomial in n. We will replace the constraints $x(\delta(U)) \geq 2$ for all $\emptyset \neq U \subsetneq V$ by flow constraints.

Definition 2.3 (s-t-flow). Let $G = (V, E)$ be a directed graph, $u : E \to \mathbb{R}_{\geq 0}$ (called *capacities*), and $s, t \in V$ (*source* and *sink*). A *flow* in (G, u) is a function $f : E \to \mathbb{R}_{\geq 0}$ with $f(e) \leq u(e)$ for all $e \in E$. An *s-t-flow* in (G, u) is a flow with

- $f(\delta^+(v)) = f(\delta^-(v))$ for all $v \in V \setminus \{s, t\}$ and
- $f(\delta^+(s)) \geq f(\delta^-(s))$.

Then $f(\delta^+(s)) - f(\delta^-(s))$ is the *value* of the s-t-flow f.

Let $E^{\leftrightarrow}$ contain directed edges (v, w) and (w, v) for each $\{v, w\} \in E$ (so $(V, E^{\leftrightarrow})$ is the complete directed graph on V). Choose a city $s \in V$, introduce new variables f_e^t for all $t \in V \setminus \{s\}$ and $e \in E^{\leftrightarrow}$, and replace the constraints $x(\delta(U)) \geq 2$ for all $\emptyset \neq U \subsetneq V$ by

$$
\begin{array}{rcll}
\sum_{e\in\delta^+(s)} f_e^t - \sum_{e\in\delta^-(s)} f_e^t & = & 2 & (t \in V \setminus \{s\}) \\
\sum_{e\in\delta^+(t)} f_e^t - \sum_{e\in\delta^-(t)} f_e^t & = & -2 & (t \in V \setminus \{s\}) \\
\sum_{e\in\delta^+(v)} f_e^t - \sum_{e\in\delta^-(v)} f_e^t & = & 0 & (t \in V \setminus \{s\},\ v \in V \setminus \{s,t\}) \\
f_{(v,w)}^t & \leq & x_{\{v,w\}} & (t \in V \setminus \{s\},\ \{v,w\} \in E) \\
f_e^t & \geq & 0 & (t \in V \setminus \{s\},\ e \in E^{\leftrightarrow}).
\end{array}
\tag{2.4}
$$

These constraints say that one can send two units of flow from s to any other city t while not routing more than x_e units of flow through any edge e. This is indeed equivalent by the max-flow min-cut theorem. For its proof, we need the following basic concept:

Definition 2.4 (residual graph). Let $G = (V, E)$ be a directed graph with capacities $u : E \to \mathbb{R}_{\geq 0}$, and $f : E \to \mathbb{R}_{\geq 0}$ a flow in (G, u).

For an edge $e = (v, w) \in E$, let $e^{\leftarrow}$ be a new edge from w to v and let $E^{\leftarrow} := \{e^{\leftarrow} : e \in E\}$. Then the *residual graph* G_f is the digraph with vertex set V and edge set

$$\{e \in E : f(e) < u(e)\} \mathbin{\dot\cup} \{e^{\leftarrow} : e \in E,\ f(e) > 0\}.$$

For an edge $e \in E$, we define the *residual capacities* of e and $e^{\leftarrow}$ to be $u_f(e) := u(e) - f(e)$ and $u_f(e^{\leftarrow}) := f(e)$.

The residual graph G_f contains precisely those edges from $E \mathbin{\dot\cup} E^{\leftarrow}$ with positive residual capacity. Note that it might contain parallel edges. Now we can prove the max-flow min-cut theorem:

Theorem 2.5 (Ford and Fulkerson [1956], Dantzig and Fulkerson [1956]). *Let $G = (V, E)$ be a directed graph, $u : E \to \mathbb{R}_{\geq 0}$, and $s, t \in V$. Then*

$$\begin{aligned} &\max\{f(\delta^+(s)) - f(\delta^-(s)) : f \text{ is an } s\text{-}t\text{-flow in } (G, u)\} \\ &= \min\{u(\delta^+(R)) : s \in R \subseteq V \setminus \{t\}\}. \end{aligned} \tag{2.5}$$

Proof. Since the left-hand side maximizes a continuous function over a compact set (in fact, it is a linear program), the maximum is attained by some f. The maximum is at most the minimum because $f(\delta^+(s)) - f(\delta^-(s)) = \sum_{v \in R}(f(\delta^+(v)) - f(\delta^-(v))) = f(\delta^+(R)) - f(\delta^-(R))$ for all R with $s \in R \subseteq V \setminus \{t\}$.

To show that equality holds, let $\delta > 0$ such that for each $e \in E$ we have $\delta \leq f(e) \leq u(e) - \delta$ or $f(e) = 0$ or $f(e) = u(e)$. Let R be the set of vertices v for which there is a path from s to v in the residual graph G_f.

If $t \in R$, then let P be such a path from s to t. For each edge $e \in E$ on P, we can increase $f(e)$ by δ, and for each edge $e^{\leftarrow} \in E^{\leftarrow}$ on P, we can decrease $f(e)$ by δ. This increases the value $f(\delta^+(s)) - f(\delta^-(s))$ of the flow, a contradiction.

So $t \notin R$. By the choice of R, we have $f(e) = u(e)$ for every edge $e \in \delta^+_G(R)$ and $f(e) = 0$ for every edge $e \in \delta^-_G(R)$. Thus, $f(\delta^+(s)) - f(\delta^-(s)) = f(\delta^+(R)) - f(\delta^-(R)) = u(\delta^+(R)) - 0$. □

This shows that replacing the constraints $x(\delta(U)) \geq 2$ for all $\emptyset \neq U \subsetneq V$ in the subtour LP (2.2) by the constraints (2.4) leads to an equivalent LP: For every feasible solution (x, f) to the resulting LP, x is a feasible solution to the subtour LP, and every feasible solution x to the subtour LP can be extended to a feasible solution (x, f) to the resulting LP.

Flows feature an important integrality property:

Corollary 2.6. *Let $G = (V, E)$ be a directed graph, $u : E \to \mathbb{Z}_{\geq 0}$, and $s, t \in V$. Then there exists an s-t-flow that has maximum value and is integral.*

Proof. Applying the previous proof to an integral s-t-flow f that has maximum value among all integral s-t-flows allows for choosing $\delta = 1$ and hence yields the assertion. □

Finding an s-t-flow of maximum value is another fundamental combinatorial optimization problem for which efficient algorithms are well-known (and of course it is itself a linear program of polynomial size).

Theorem 2.7 (Edmonds and Karp [1972], Karzanov [1974]). *Given a simple directed graph $G = (V, E)$ with capacities $u : E \to \mathbb{R}_{\geq 0}$ and $s, t \in V$, one can find an s-t-flow in (G, u) with maximum value in $O(n^3)$ time, where $n = |V|$.*

A set $\delta^+(R)$ attaining the minimum in (2.5) is called a minimum-capacity *s-t-cut*. The proof of Theorem 2.5 also reveals that finding a minimum-capacity s-t-cut is easy once we have an s-t-flow of maximum value:

Corollary 2.8. *Given a simple directed graph $G = (V, E)$ with capacities $u : E \to \mathbb{R}_{\geq 0}$ and $s, t \in V$ with $s \neq t$, one can find a set R in $O(n^3)$ time, where $n = |V|$, such that $\delta^+(R)$ is a minimum-capacity s-t-cut in (G, u).*

Proof. Find an s-t-flow f of maximum value by Theorem 2.7. Then output the set R of vertices reachable from s in G_f, which we can compute by initializing $R = \{s\}$ and adding a vertex w to R whenever $\delta^-_{G_f}(w) \cap \delta^+_{G_f}(R) \neq \emptyset$. □

This will be applied later in the following ways:

Corollary 2.9. *Given a simple graph $G = (V, E)$ with $x : E \to \mathbb{R}_{\geq 0}$ and $\emptyset \neq S, T \subsetneq V$ with $S \cap T = \emptyset$, one can do the following in $O(n^4)$ time, where $n = |V|$: find a set U attaining the minimum in* $\min\{x(\delta(U)) : \emptyset \neq U \subsetneq V\}$, *and find a set U attaining the minimum in* $\min\{x(\delta(U)) : S \subseteq U \subseteq V \setminus T\}$ *(if G is undirected) or in* $\min\{x(\delta^+(U)) : S \subseteq U \subseteq V \setminus T\}$ *(if G is directed).*

Proof. If G is an undirected graph, replace each edge $\{v, w\}$ by two edges (v, w) and (w, v), both with capacity $x(\{v, w\})$. Now the first problem reduces to Corollary 2.8 by choosing a vertex s arbitrarily, trying all $t \in V \setminus \{s\}$, computing a minimum-capacity s-t-cut, and taking one with smallest capacity among these. The second problem reduces to Corollary 2.8 by contracting S to a vertex s and contracting T to a vertex t. □

This leads us to the second way to solve the subtour LP (2.2): via the *equivalence of separation and optimization*. Informally, Grötschel, Lovász, and Schrijver [1981] showed (using the ellipsoid method) that one can solve an LP by a polynomial number of calls to a *separation oracle*: Such an oracle takes a vector x as input and decides whether it is a feasible solution and otherwise returns a constraint that x violates. For (2.2), such a separation oracle can be implemented by checking the constraints $\sum_{e\in\delta(v)} x_e = 2$ for all $v \in V$ and $0 \le x_e \le 1$ for all $e \in E$ one by one and then finding a minimum-weight cut – that is, solving $\min\{x(\delta(U)) : \emptyset \neq U \subsetneq V\}$, using Corollary 2.9.

Frank and Tardos [1987] showed that one can even solve (2.2) in *strongly polynomial* time – that is, the number of elementary arithmetic operations is bounded by a polynomial in n, and for rational input, the number of bits needed to encode any number occurring during the algorithm is bounded by a polynomial in the total number of bits in the input. Most algorithms in this book run in strongly polynomial time, although we will usually not mention this explicitly. It is usually either obvious or follows from the following theorem (sometimes in combination with Theorem 2.2):

Theorem 2.10 (Frank and Tardos [1987], Grötschel, Lovász, and Schrijver [1988]). *Let $d, p : \mathbb{Z}_{>0} \to \mathbb{Z}_{>0}$ be any fixed polynomials. For each positive integer n, let $P_n = \{x \in \mathbb{R}^{d(n)} : Ax \ge b, x \ge 0\}$ be a nonempty polyhedron where all entries of the matrix A and the vector b are integers from $\{-2^{p(n)}, \dots, 2^{p(n)}\}$.*

The optimization problem *takes n and a rational vector $c \in \mathbb{Q}^{d(n)}_{\ge 0}$ as input and asks for a solution to* $\min\{c^\top x : x \in P_n\}$.

The separation problem *takes n and a rational vector $x \in \mathbb{Q}^{d(n)}_{\ge 0}$ as input and asks for deciding that $x \in P_n$ or returning a vector $a \in \mathbb{Q}^{d(n)}_{\ge 0}$ with $a^\top x < a^\top y$ for all $y \in P_n$.*

If one of these two problems can be solved in polynomial time, then both can be solved in strongly polynomial time.

Frank and Tardos [1987] proved this by rounding the objective function c to integers whose binary encoding has only a polynomial number of bits and then applying the ellipsoid method.

The sequence of subtour polytopes with n cities and $d(n) = \binom{n}{2}$ satisfies the conditions in Theorem 2.10 with $p(n) = 1$: All coefficients in (2.2) are 0, 1, or 2. We conclude the following:

Corollary 2.11. *The subtour LP* (2.2) *can be solved in polynomial time.*

Proof. Directly from Theorem 2.10, solving the separation problem via Corollary 2.9. □

The resulting algorithm is quite inefficient, in spite of its polynomial running time. In practice, a heuristic approach to iteratively find and add violated subtour elimination constraints works well. Moreover, Held and Karp [1970] and Chekuri and Quanrud [2017] showed how to solve (2.2) fast approximately. Nevertheless, it would be interesting to have a combinatorial algorithm to solve the subtour LP exactly. The term combinatorial algorithm is not always meant the same way, but it never allows using the ellipsoid method. Schrijver [2000] asked (in a different context) for a fully combinatorial algorithm – that is, an algorithm that does not use multiplication or division.

Open Problem 2.12. Find a (fully) combinatorial polynomial-time algorithm to solve the subtour LP (2.2) exactly.

We remark that the subtour LP (2.2) has not only been used in the design of approximation algorithms. Already Dantzig, Fulkerson, and Johnson [1954] used it to solve a 49-city TSP instance optimally. Of course, additional steps – like adding further constraints (*cutting planes*) and case distinctions to fix fractional variables to 0 or 1 (*branching*) – are needed to arrive at an optimum integral solution. Over more than 50 years, this *branch-and-cut* approach was brought to perfection; Applegate et al. [2006] can solve instances with up to 85 900 cities optimally.

2.3 Polyhedral Descriptions of Connectors and T-Joins

The difficulty of the Symmetric TSP lies in the combination of connectivity and parity requirements. In this section, we consider these two aspects separately. Polyhedral descriptions of connectors and T-joins will enable us to bound how much cheaper fractional solutions to the subtour LP can be compared to an optimum integral solution.

If we describe the feasible solutions to a combinatorial optimization problem by incidence vectors (e.g., with respect to the edge set of the given graph), we can try to solve the problem by optimizing over the *convex hull* of these vectors (i.e., the set of convex combinations). The convex hull of a finite set of points is a polytope (Minkowski [1896], Steinitz [1916], Weyl [1935]); hence we can describe it by linear inequalities. If we have a linear objective function, we get a linear program, and one of the incidence vectors is always among the optimum solutions. In many cases, this is not the most efficient way to solve a combinatorial optimization problem, but nevertheless it is a widely used tool in the analysis of approximation algorithms. We need the following simple observation:

Proposition 2.13. *A nonempty polytope $P \subseteq \mathbb{R}^d$ is the convex hull of integral points if and only if for all $c \in \mathbb{R}^d$ there is an integral vector x attaining* $\min\{c^\top x : x \in P\}$.

Proof. Let P_{int} denote the convex hull of the integral points in P. If there is some $x \in P \setminus P_{\text{int}}$, then there is a separating hyperplane (i.e., a vector $c \in \mathbb{R}^d$ with $c^\top x < \min\{c^\top y : y \in P_{\text{int}}\}$).

Conversely, suppose $P = P_{\text{int}}$. Let $c \in \mathbb{R}^d$. The minimum $\min\{c^\top x : x \in P\}$ is attained at some point x^* because we minimize a continuous function over a compact set. Now x^* is a convex combination of integral points in P, and then these also attain the minimum. □

First, we consider connectors. A *connector* in an undirected graph $G = (V, E)$ is a set $F \subseteq E$ of edges such that (V, F) is connected. It is easy to see that the problem of finding a minimum-cost connector in G with respect to edge costs $c : E \to \mathbb{R}$ is equivalent to Minimum Spanning Tree (cf. Exercise 2.4).

To describe the convex hull of incidence vectors of connectors, we need so-called partition constraints. A *partition* of a finite set V is a set of pairwise disjoint nonempty subsets of V whose union is V. For instance, the vertex sets of the connected components of any undirected graph (V, E) form a partition of V. For a partition $\mathcal{W}$ of the vertex set V of a graph, we denote by $\delta(\mathcal{W}) = \bigcup_{W \in \mathcal{W}} \delta(W)$ the set of edges that have the two endpoints in different sets of the partition.

Theorem 2.14. *Let $G = (V, E)$ be an undirected graph. The convex hull of incidence vectors of connectors in G is the set of vectors $x \in \mathbb{R}^E$ with*

$$\begin{array}{rcll} x(\delta(\mathcal{W})) & \geq & |\mathcal{W}| - 1 & (\mathcal{W} \text{ a partition of } V) \\ x_e & \leq & 1 & (e \in E) \\ x_e & \geq & 0 & (e \in E). \end{array} \tag{2.6}$$

Proof. Obviously, the integral vectors that satisfy (2.6) are precisely the incidence vectors of connectors. Let now $c : E \to \mathbb{R}$ and assume that (2.6) has a solution. We show that there is an integral vector x that minimizes $c(x)$ over (2.6). By Proposition 2.13, this proves the theorem.

Let $E = E^- \dot\cup E^+$, where $E^- = \{e \in E : c(e) \leq 0\}$ and $E^+ = \{e_1, \ldots, e_k\}$, sorted so that $c_0 < c_1 \leq \cdots \leq c_k$, where $c_0 = 0$ and $c_i = c(e_i)$ for $i = 1, \ldots, k$. Moreover, for $i = 0, \ldots, k$, let $\mathcal{W}_i$ be the partition of V that is given by the connected components of $(V, E^- \cup \{e_1, \ldots, e_i\})$. Let

$$F := E^- \dot\cup \left\{e_i : i \in \{1, \ldots, k\}, \mathcal{W}_{i-1} \neq \mathcal{W}_i\right\}.$$

We complete the proof by showing that $c(F) = \min\{c(x) : x \text{ satisfies (2.6)}\}$. Indeed, for every $x \in [0,1]^E$ satisfying (2.6), we have

$$\begin{aligned} c(x) &\geq \sum_{e \in E^-} c(e) + \sum_{i=1}^{k} c_i x_{e_i} \\ &= \sum_{e \in E^-} c(e) + \sum_{i=1}^{k} (c_i - c_{i-1})\, x(\{e_i, \dots, e_k\}) \\ &\geq \sum_{e \in E^-} c(e) + \sum_{i=1}^{k} (c_i - c_{i-1})\,(|\mathcal{W}_{i-1}| - 1) \\ &\geq \sum_{e \in E^-} c(e) + \sum_{i=1}^{k} c_i\,(|\mathcal{W}_{i-1}| - |\mathcal{W}_i|) \\ &= c(F). \end{aligned}$$

□

The set of vectors satisfying (2.6) is called the *connector polytope* of G. If we remove the constraints $x_e \leq 1$, we have the *connector polyhedron*:

Corollary 2.15. *Let $G = (V, E)$ be an undirected graph. The convex hull of incidence vectors of multi-subsets F of E such that the graph (V, F) is connected is the set of vectors $x \in \mathbb{R}^E_{\geq 0}$ with*

$$x(\delta(\mathcal{W})) \geq |\mathcal{W}| - 1 \quad \textit{for every partition } \mathcal{W} \textit{ of } V. \tag{2.7}$$

Proof. The incidence vectors of multi-subsets F of E such that the graph (V, F) is connected satisfy (2.7). For the other direction, consider a vector $x \in \mathbb{R}^E_{\geq 0}$ satisfying (2.7). Apply Theorem 2.14 to the graph that arises from G by replacing each edge e by $\lceil x_e \rceil$ parallel copies. □

Theorem 2.14 also implies the following polyhedral description of the set of spanning trees in a graph:

Theorem 2.16 (Edmonds [1970]). *Let $G = (V, E)$ be an undirected graph. The convex hull of incidence vectors of spanning trees in G is the set of vectors $x \in \mathbb{R}^E$ with*

$$\begin{aligned} x(E) &= n - 1 & & \\ x(E[U]) &\leq |U| - 1 & &(\emptyset \neq U \subsetneq V) \\ x_e &\geq 0 & &(e \in E). \end{aligned} \tag{2.8}$$

Proof. Let F be the set of vectors in the connector polytope (cf. (2.6)) that satisfy $x(E) = n-1$. We first show that F is the set of vectors that satisfy (2.8). For "$\subseteq$", let $x \in F$ and $\emptyset \neq U \subsetneq V$. Let $\mathcal{W}$ be the partition $\{U\} \cup \{\{v\} : v \in V \setminus U\}$;

then $x(E[U]) = x(E) - x(\mathcal{W}) \leq n-1-(|\mathcal{W}|-1) = |U|-1$. For "$\supseteq$", let x satisfy (2.8). Then

$$x(\delta(\mathcal{W})) \;=\; x(E) - \sum_{U\in\mathcal{W}} x(E[U]) \;\geq\; n-1-\sum_{U\in\mathcal{W}}(|U|-1) \;=\; |\mathcal{W}|-1$$

for every partition $\mathcal{W}$ of V.

Moreover, every connector has at least $n-1$ edges, and the connectors with $n-1$ edges are the spanning trees. This shows the theorem. □

The set described by (2.8) is called the *spanning tree polytope* of the graph (V,E). We can view the proof of Theorem 2.16 in the following context. If $P \subseteq \mathbb{R}^n$ is a polyhedron and $c \in \mathbb{R}^n$ and $\beta \in \mathbb{R}$ such that $\min\{c^\top x : x \in P\} = \beta$, then $\{x \in P : c^\top x = \beta\}$ is called a *face* of P. We have shown that the spanning tree polytope is a face of the connector polytope. The proof of Theorem 2.14 also shows that a simple greedy algorithm computes a cheapest connector. This generalizes to matroids (see Exercise 2.5).

The following easy observation was made by Held and Karp [1970]. The *support* of a vector $x \in \mathbb{R}^E_{\geq 0}$ is the set of elements $e \in E$ with $x_e > 0$. We call $(V, \{e \in E : x_e > 0\})$ the *support graph* of x.

Corollary 2.17. *If x is a feasible solution to the subtour LP* (2.2)*, then $\frac{n-1}{n}x$ is in the spanning tree polytope of the support graph, where $n = |V| \geq 3$.*

Proof. If x is a feasible solution to (2.2), then $\frac{n-1}{n}x(E) = \frac{n-1}{2n}\sum_{v\in V} x(\delta(v)) = n-1$ as well as

$$x(E[U]) \;=\; \tfrac{1}{2}\left(\sum_{v\in U} x(\delta(v)) - x(\delta(U))\right) \;=\; \tfrac{1}{2}\left(2|U| - x(\delta(U))\right) \;\leq\; |U|-1$$

for every $\emptyset \neq U \subsetneq V$, which implies $\frac{n-1}{n}x(E[U]) \leq \frac{n-1}{n}(|U|-1)$. So $\frac{n-1}{n}x$ satisfies all constraints of (2.8) and thus belongs to the spanning tree polytope by Theorem 2.16. □

In fact, the proof shows that x satisfies all inequalities of (2.8) strictly – that is, $\frac{n-1}{n}x$ is in the relative interior of the spanning tree polytope (Asadpour et al. [2017]).

Now we consider the parity aspect. Let $G = (V,E)$ be an undirected graph and $T \subseteq V$ with $|T|$ even. Recall that a T-join in G is a set $J \subseteq E$ with $\mathrm{odd}(J) = T$. We begin with Guan's lemma:

Lemma 2.18 (Guan [1962]). *Let $G = (V,E)$ be an undirected graph with edge costs $c : E \to \mathbb{R}$. Let $T \subseteq V$ and J a T-join in G. Then J is a minimum-cost T-join if and only if $c(C\cap J) \leq c(C \setminus J)$ for every (edge set of a) circuit C in G.*

Proof. If $c(C \cap J) > c(C \setminus J)$ for some circuit C, then $J \triangle C$ is a cheaper T-join. If J' is a cheaper T-join, then $J \triangle J'$ can be partitioned into edge sets of circuits, and for at least one of these circuits, say C, we have $c(C \cap J) > c(C \cap J') = c(C \setminus J)$. □

The cuts $\delta(U)$ with $|U \cap T|$ odd are called *T-cuts*. Every T-join J has nonempty intersection with every T-cut (by Lemma 1.4 applied to the subgraph of (V, J) induced by U), and this yields a polyhedral description that will play a crucial role many times in this book:

Theorem 2.19 (Edmonds and Johnson [1973]). *Let $G = (V, E)$ be an undirected graph with weights $c : E \to \mathbb{R}_{\geq 0}$ and $T \subseteq V$ with $|T|$ even. Then the minimum weight of a T-join in G equals the value of the LP*

$$
\begin{aligned}
&\min\ c(x) \\
&\textit{subject to}\quad x(\delta(U)) \;\geq\; 1 \quad (U \subseteq V,\ |U \cap T| \ \textit{odd}) \\
&\qquad\qquad\qquad\quad x_e \;\geq\; 0 \quad (e \in E).
\end{aligned}
\tag{2.9}
$$

Proof. Since every T-join has nonempty intersection with every T-cut, the incidence vector of any T-join is a feasible solution to (2.9). This proves "$\geq$".

Note that the inequality "$\leq$" is invariant under multiplying c by a positive constant, and both sides of the inequality are continuous in c. Hence, we may assume that c is integral and $c(C)$ is even for every circuit C. We proceed by induction on $|E| + c(E)$. If $T = \emptyset$, the assertion is trivial. If there is an edge $e = \{v, w\}$ with $c(e) = 0$, we contract e (i.e., identify v and w, remove e, and put the new vertex in T if exactly one of v and w was in T), and apply induction; note that neither side of the inequality changes by the contraction.

Otherwise, $T \neq \emptyset$ and $c(e) \geq 1$ for all $e \in E$. Choose $a, b \in V$ such that the cost of a cheapest $(T \triangle \{a\} \triangle \{b\})$-join is minimum. Note that $a \neq b$, because one possible choice for a and b is the endpoints of an edge in a minimum-cost T-join. Next we prove:

$$|J \cap \delta(a)| = 1 \text{ for every minimum-cost } T\text{-join } J. \tag{2.10}$$

To show this, let J be a minimum-cost T-join and J' a minimum-cost T'-join, where $T' = T \triangle \{a\} \triangle \{b\}$. Then $J \triangle J'$ is an $\{a, b\}$-join (cf. Proposition 1.26) and thus can be partitioned into the edge set P of a path from a to b and edge sets of circuits. By Lemma 2.18, we have $c(C \cap J) = c(C \cap J')$ for each of these circuits. Hence $c(J \triangle P) = c(J')$, and $J \triangle P$ is a minimum-cost T'-join. If $J \triangle P$ contains an edge $e = \{a, a'\}$ incident to a, then $(J \triangle P) \setminus \{e\}$ is a $(T \triangle \{a'\} \triangle \{b\})$-join cheaper than any T'-join, contradicting the choice of a and b. So $(J \triangle P) \cap \delta(a) = \emptyset$, and we conclude $|J \cap \delta(a)| = 1$.

This proves (2.10). In particular, it implies that $\delta(a)$ is a T-cut.

Define a cost function $c^- : E \to \mathbb{R}$ by $c^-(e) := c(e) - 1$ for $e \in \delta(a)$ and $c^-(e) := c(e)$ for $e \in E \setminus \delta(a)$. Note that $c^-(C) \in \{c(C), c(C) - 2\}$ is even for every circuit C. Let J be a minimum-cost T-join in (G, c). Recall that $|J \cap \delta(a)| = 1$. We will show that J is also a minimum-cost T-join in (G, c^-). Then, by the induction hypothesis, any vector x in (2.9) satisfies $c(J) - 1 = c^-(J) \leq c^-(x) = c(x) - x(\delta(a)) \leq c(x) - 1$, completing the proof.

Suppose J is not a minimum-cost T-join in (G, c^-). Then Lemma 2.18 implies that there is a circuit C with $c^-(C \setminus J) < c^-(C \cap J)$, hence $c^-(C \setminus J) \leq c^-(C \cap J) - 2$. On the other hand, $c^-(C \setminus J) + 2 \geq c(C \setminus J) \geq c(C \cap J) \geq c^-(C \cap J)$. So we have equality throughout, and in particular $c(C \setminus J) = c(C \cap J)$, and C contains a but not the edge of J that is incident to a. This means that $J \triangle C$ is a minimum-cost T-join with three edges incident to a, contradicting (2.10). □

This proof is essentially due to Sebő [1987].

In the context of TSP, we are often interested in the set T being the set $\mathrm{odd}(H)$ for some (multi-)edge set H (for example, cf. Algorithm 1.31). For this case, we can give an equivalent characterization of $\mathrm{odd}(H)$-cuts:

Lemma 2.20. *Let (V, H) be an undirected graph and $U \subseteq V$. Then $\delta(U)$ is an $\mathrm{odd}(H)$-cut if and only if $|\delta_H(U)|$ is odd.*

Proof. A cut $\delta(U)$ is an $\mathrm{odd}(H)$-cut if and only if $|\mathrm{odd}(H) \cap U|$ is odd. This is equivalent to $\sum_{v \in U} |\delta_H(v)|$ being odd. We have $\sum_{v \in U} |\delta_H(v)| = 2 \cdot |H[U]| + |\delta_H(U)|$. Hence, the left-hand side is odd if and only if $|\delta_H(U)|$ is odd. □

When some edge weights are negative, the LP (2.9) cannot be used directly. In this case, we need the following:

Theorem 2.21. *Let $G = (V, E)$ be an undirected graph and $T \subseteq V$ with $|T|$ even. Then the convex hull of incidence vectors of T-joins in G is the set of vectors $x \in \mathbb{R}^E$ with*

$$\begin{array}{rcll} |F| - x(F) + x(\delta(U) \setminus F) & \geq & 1 & (U \subseteq V,\ F \subseteq \delta(U), \\ & & & \quad |U \cap T| + |F| \text{ odd}) \\ x_e & \leq & 1 & (e \in E) \\ x_e & \geq & 0 & (e \in E). \end{array} \tag{2.11}$$

Proof. If J is a T-join, then by Lemma 2.20, $|\delta(U) \cap J| + |U \cap T|$ is even for every $U \subseteq V$, so the incidence vector of J satisfies (2.11). This shows "$\subseteq$".

For the reverse direction, let P denote the set of all $x \in [0,1]^E$ satisfying (2.11). Let $c : E \to \mathbb{R}$ and let J be a minimum c-cost T-join. We show that $c(J) \le \min\{c(x) : x \in P\}$; by Proposition 2.13, this will complete the proof.

Let $E^- = \{e \in E : c(e) < 0\}$ and $c'(e) := |c(e)|$ for all $e \in E$. Then $c'(K \triangle E^-) = c(K) - c(E^-)$ for all $K \subseteq E$. Let $T' := T \triangle \text{odd}(E^-)$ and $J' := J \triangle E^-$. Then J' is a minimum c'-cost T'-join.

Now, let $x \in P$. Define $x'_e := x_e$ if $c(e) \ge 0$ and $x'_e := 1 - x_e$ if $c(e) < 0$. Then $x' \ge 0$. Moreover, for every $U \subseteq V$ with $|U \cap T'|$ odd, we have $x'(\delta(U)) \ge 1$. Indeed, let $F = E^- \cap \delta(U)$. Lemma 2.20 with $H = E^-$ implies that $|U \cap T| + |F|$ is odd. Then (2.11) implies $x'(\delta(U)) = x(\delta(U) \setminus F) + |F| - x(F) \ge 1$.

So x' is in the T'-join polyhedron, and by Theorem 2.19, we have $c'(J') \le c'(x')$. We conclude

$$c(J) \;=\; c'(J') + c(E^-) \;\le\; c'(x') + c(E^-) \;=\; c(x). \qquad \square$$

The convex hull of incidence vectors of T-joins in G is called the *T-join polytope* of G, while the set of feasible solutions to (2.9) is called the *T-join polyhedron* (or the *up-hull* of the T-join polytope, where the up-hull – also called *dominant* – results from adding nonnegative vectors).

We call a polyhedron P *integral* if $\min\{c(x) : x \in P\}$ is attained by an integral vector for every vector c for which the minimum is finite. Theorems 2.14, 2.16, 2.19, and 2.21 say that the polyhedra defined by (2.6), (2.8), (2.9), and (2.11) are integral. However, the subtour LP (2.2) is not integral.

Although the combinatorial optimization problems studied in this section can be solved more efficiently by combinatorial algorithms than by linear programming, the polyhedral descriptions will prove very useful. A first and very important application will be shown in the next section.

2.4 Integrality Ratio

Many approximation algorithms can be analyzed by comparing the cost of the computed solution to the LP value (for an appropriate LP relaxation). If the LP value can be arbitrarily smaller than the optimum, this cannot work. Similar to the definition of the approximation ratio, one can define the integrality ratio of a family of linear programs:

Definition 2.22 (integrality ratio)**.** Consider a family of LPs, each of which is of the form $\min\{c^\top x : x \in P\}$ with $P \subseteq \mathbb{R}^n_{\ge 0}$ and $c \in \mathbb{R}^n_{\ge 0}$. Then the *integrality ratio* of this family of LPs is the infimum of all ρ for which $\min\{c^\top x : x \in P \cap \mathbb{Z}^n\} \le \rho \cdot \min\{c^\top x : x \in P\}$ holds for all instances.

Table 2.1 *Upper bounds on the integrality ratio of* (2.2) *for* Symmetric TSP with Triangle Inequality *in the order of their discovery.*

Integrality Ratio	Year	Reference	Chapter
$\frac{3}{2}$	1980	Wolsey [1980]	2.4
$\frac{3}{2} - 10^{-36}$	2021	Karlin, Klein, and Oveis Gharan [2022]	10–11

Equivalently, the integrality ratio (often also called *integrality gap*) is the supremum of

$$\frac{\min\{c(x) : x \in P \cap \mathbb{Z}^n\}}{\min\{c(x) : x \in P\}},$$

where $\frac{0}{0} := 1$. The integrality ratio is 1 if and only if all polyhedra in the family are integral.

What is the integrality ratio of the subtour LP – that is, of the family of the LPs (2.2) for all instances (V, c) of the Symmetric TSP with Triangle Inequality? Surprisingly, we still do not know, although this question has been studied for more than 40 years. Wolsey [1980] proved the following upper bound:

Theorem 2.23 (Wolsey [1980]). *For every instance of the* Symmetric TSP with Triangle Inequality, *Christofides' algorithm (Algorithm 1.23) computes a solution with cost at most* $\frac{3}{2}$ LP, *where* LP *denotes the value of the subtour LP* (2.2). *In particular, the integrality ratio of the subtour LP is at most* $\frac{3}{2}$.

Proof. Let x be a vector in the subtour polytope. By Corollary 2.17, $\frac{n-1}{n}x$ is in the spanning tree polytope, and hence the minimum cost of a spanning tree is at most LP. By Theorem 2.19, $\frac{1}{2}x$ is in the T-join polyhedron for any $T \subseteq V$ with $|T|$ even. In particular, for $T = W$ in Algorithm 1.23, there is a T-join J^* of cost at most $\frac{1}{2}$ LP. By Lemma 1.28 and the triangle inequality, J^* corresponds to a perfect matching M^* in $G[W]$ with cost $c(M^*) \leq c(J^*)$. The matching M that the algorithm computes cannot be more expensive. We conclude that the solution computed by Christofides' algorithm costs at most LP + $\frac{1}{2}$ LP. □

A different proof of Theorem 2.23 was later given by Shmoys and Williamson [1990]. We also say that $\frac{1}{2}x$ is a *parity correction vector* (for the spanning tree S) because it is used to bound the cost for parity correction.

The upper bound in Theorem 2.23 is tight even for Graph TSP instances: as mentioned already in Section 1.4, Christofides' algorithm might end up with

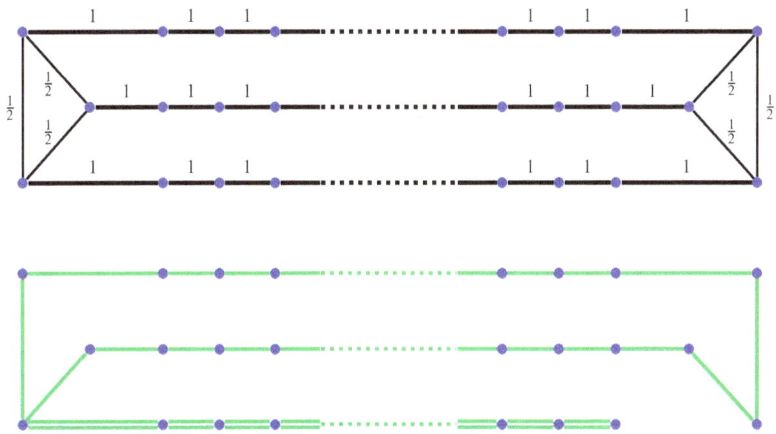

Figure 2.2 Example showing a lower bound of $\frac{4}{3}$ on the integrality ratio of the subtour LP. The top figure shows a graph $G = (V, E)$, which generates the instance and which is also the support graph of an optimum solution to the subtour LP, shown by the numbers x_e $(e \in E)$. The bottom figure shows an optimum tour. This is sometimes called the *envelope example*.

$\frac{3}{2}n - 1$ edges even for an unweighted complete graph with an even number n of vertices (where the subtour LP has value n). Karlin, Klein, and Oveis Gharan [2022] managed to improve on Wolsey's analysis in a sophisticated way (see Chapters 10 and 11 and Table 2.1). Moreover, Wolsey's analysis has been developed further very successfully for more general problems like Path TSP (see Chapter 15).

We now show the best-known lower bound on the integrality ratio of the subtour LP:

Proposition 2.24. *The integrality ratio of the subtour LP* (2.2) *is at least* $\frac{4}{3}$.

Proof. Figure 2.2 shows a well-known example: an infinite family of unweighted undirected graphs $G = (V, E)$, each consisting of two triangles, connected by three vertex-disjoint paths with $\frac{n}{3}$ vertices each (where $n = |V|$). Each such graph $G = (V, E)$ induces an instance (V, c) of the Symmetric TSP with Triangle Inequality by letting $c(v, w)$ be the distance from v to w in G.

Setting $x_e := \frac{1}{2}$ for the six edges of the triangles, $x_e := 1$ for the other edges of G, and $x_{\{v,w\}} := 0$ for $\{v, w\} \in \binom{V}{2} \setminus E$ defines a feasible solution x to (2.2), and we have $c(x) = n$.

We claim that an optimum tour in G has $\frac{4}{3}n - 2$ edges. There is such a tour, as shown in Figure 2.2, but we need to show that there is no tour with fewer edges. Indeed, for each of the three vertex-disjoint paths, any tour contains either all its edges once or all but one of its edges twice (or even more edges), and for at least one of the three paths, the latter alternative applies. This makes $2(\frac{n}{3} - 1) + 2(\frac{n}{3} - 2) = \frac{4n}{3} - 6$ edges and results in four connected components, so we need at least four more edges to connect them to a tour.

Since every integral solution to the subtour LP for this instance is a circuit with vertex set V and can thus be transformed to a tour in G with the same cost, every integral solution has cost at least $\frac{4}{3}n - 2$. □

So we know that the integrality ratio of the subtour LP is between $\frac{4}{3}$ and slightly less than $\frac{3}{2}$, with little progress on these bounds in the last 40 years. Most people believe that $\frac{4}{3}$ is the true answer (this is sometimes called the $\frac{4}{3}$ conjecture).

Open Problem 2.25. Prove or disprove that the integrality ratio of the subtour LP (2.2) is $\frac{4}{3}$.

The $\frac{4}{3}$ conjecture has been verified for $n \leq 12$ by Benoit and Boyd [2008] and Boyd and Elliott-Magwood [2010].

The fact that the worst-known examples (cf. Figure 2.2) are instances of the GRAPH TSP raised interest in this special case (see Chapters 12 and 13).

Goemans [1995] and Carr and Vempala [2004] gave an interesting characterization of the integrality ratio. Before we show this, we note the following:

Proposition 2.26 (Cornuéjols, Fonlupt, and Naddef [1985]). *The convex hull of the incidence vectors of tours is a polyhedron.*

Proof. Let $P_{\min}$ be the convex hull of all incidence vectors of tours with $x_e \leq 2$ for all $e \in E$. Then $P_{\min}$ is a polytope because it is the convex hull of a finite number of vectors. We claim that the convex hull P of all incidence vectors of tours is the polyhedron $P^{\uparrow}_{\min} := \{x + y : x \in P_{\min}, y \geq 0\}$. To see "$\subseteq$", we observe that $P^{\uparrow}_{\min}$ is convex and contains all incidence vectors of tours. Now we prove "$\supseteq$". For every vector $x \in P$ and every edge $e \in E$, the vector $x + 2 \cdot \chi^{\{e\}}$ is contained in P: If x is a convex combination of incidence vectors $x^1, \ldots, x^k$ of tours, then $x + 2\chi^{\{e\}}$ is a convex combination of the vectors $x^1 + 2\chi^{\{e\}}, \ldots, x^k + 2\chi^{\{e\}}$, which are also incidence vectors of tours. Because P is convex, this implies $P = \{x + y : x \in P, y \geq 0\}$. Since $P \supseteq P_{\min}$, this implies $P \supseteq P^{\uparrow}_{\min}$. □

Cornuéjols, Fonlupt, and Naddef [1985] called this the *graphical traveling salesman polyhedron* (see also Goemans [1995]). This is in contrast to the convex hull of incidence vectors of Hamiltonian circuits, which is called the *traveling salesman polytope*.

Theorem 2.27 (Goemans [1995], Carr and Vempala [2004]). *The integrality ratio of the subtour LP* (2.2) *is the smallest number ρ such that for every vector x^* in the subtour polytope, the vector ρx^* is a convex combination of incidence vectors of tours.*

Proof. Let (V, c) be an instance of the SYMMETRIC TSP WITH TRIANGLE INEQUALITY and x^* an optimum solution to the subtour LP. Suppose for some $\rho \geq 1$, we can write $\rho x^* = \sum_{i=1}^{k} \lambda_i y_i$, where $\lambda_1, \ldots, \lambda_k \geq 0$, $\sum_{i=1}^{k} \lambda_i = 1$, and $y_1, \ldots, y_k$ are incidence vectors of tours. Then $\rho c(x^*) = \sum_{i=1}^{k} \lambda_i c(y_i)$, so at least one of the y_i (and thus a tour) costs at most $\rho c(x^*)$. By Lemma 1.7, we conclude that the integrality ratio of the subtour LP is at most ρ.

For the other direction, let x^* be in the subtour polytope, and suppose that $\rho x^* \notin Q$, where Q is the graphical traveling salesman polyhedron (cf. Proposition 2.26).

Then there is a separating hyperplane – that is, a vector w with $w^\top(\rho x^*) < \min\{w^\top q : q \in Q\}$. Note that w is nonnegative because otherwise the right-hand side would be $-\infty$ (we can always add 2 to any component of a vector in Q without leaving Q). We can interpret w as a weight function on the edge set of the complete graph on V. Now, for the metric closure $\bar{w}$ of w,

$$\begin{aligned}
\rho(\bar{w}^\top x^*) &\leq \rho(w^\top x^*) \\
&= w^\top(\rho x^*) \\
&< \min\{w^\top q : q \in Q\} \\
&= \min\{\bar{w}^\top q : q \in Q\} \\
&= \min\{\bar{w}(F) : F \text{ is a tour}\},
\end{aligned}$$

so the integrality ratio is larger than ρ. □

Boyd and Sebő [2021] noted another characterization of the integrality ratio (cf. Exercise 2.12). Call an undirected graph $G = (V, E)$ *k-regular* if every vertex has degree k, and call it *k-edge-connected* if $|\delta_G(U)| \geq k$ for all $\emptyset \neq U \subsetneq V$. Then the integrality ratio is the smallest number ρ for which the following is true: For every positive integer k and every k-regular k-edge-connected graph G, the all-$\frac{2\rho}{k}$ vector is a convex combination of incidence vectors of tours. Haddadan, Newman, and Ravi [2021] showed that the latter statement is true for $k = 3$ and $\rho = \frac{27}{19}$.

Based on earlier work by Carr and Ravi [1998] and Boyd and Carr [2011], Carr and Vempala [2004] showed that in order to prove the $\frac{4}{3}$ conjecture, it is sufficient to consider extreme points x of the subtour polytope for which the edges e with $x_e = 1$ form a perfect matching and the edges e with $0 < x_e < 1$

form a Hamiltonian cycle (so the support graph of x is 3-regular), so-called *fundamental points*.

2.5 Splitting Off

Since we will often work with the SYMMETRIC TSP (without triangle inequality) and ask for a tour rather than a Hamiltonian circuit, an LP that has no degree constraints and allows using edges more than once is more useful in this context. Let $G = (V, E)$ and $c : E \to \mathbb{R}_{\geq 0}$ be an instance of SYMMETRIC TSP. We study the following linear program:

$$\begin{aligned} \min\ & c(x) \\ \text{subject to}\quad x(\delta(U)) &\geq 2 \quad (\emptyset \neq U \subsetneq V) \\ x_e &\geq 0 \quad (e \in E). \end{aligned} \tag{2.12}$$

Lemma 2.28. *Every feasible solution x to* (2.12) *is in the connector polyhedron of G.*

Proof. For every partition $\mathcal{W}$ of V, we have $x(\delta(\mathcal{W})) = \frac{1}{2}\sum_{W\in\mathcal{W}} x(\delta(W)) \geq \frac{1}{2}\sum_{W\in\mathcal{W}} 2 = |\mathcal{W}| > |\mathcal{W}|-1$, so x satisfies (2.7). We are done by Corollary 2.15. □

Wolsey's analysis also works here:

Theorem 2.29 (Wolsey [1980]). *For every instance of the SYMMETRIC TSP, Christofides' algorithm (Algorithm 1.31) computes a tour with cost at most $\frac{3}{2}$* LP, *where* LP *denotes the value of the LP* (2.12).

Proof. Every feasible solution x to (2.12) is in the connector polyhedron of G by Lemma 2.28. The cost of the spanning tree S that Christofides' algorithm (Algorithm 1.31) computes is the minimum cost of a connector (because $c \geq 0$) and hence at most LP. Moreover, by Theorem 2.19, $\frac{1}{2}x$ is in the T-join polyhedron for every $T \subseteq V$ with $|T|$ even, and thus the minimum weight of an odd(S)-join is at most $\frac{1}{2}$ LP. We conclude that the solution computed by Christofides' algorithm costs at most LP + $\frac{1}{2}$ LP. □

We will now show that the values of the two LPs, (2.2) and (2.12), are actually the same whenever the triangle inequality holds. The following useful tool is known as Lovász' [1976] splitting-off technique. *Splitting off* a pair $e = \{v, z\}$ and $f = \{z, w\}$ of distinct edges incident to the same vertex z means replacing e and f by a single edge $\{v, w\}$ (or just remove e and f if $v = w$, i.e., if e and f

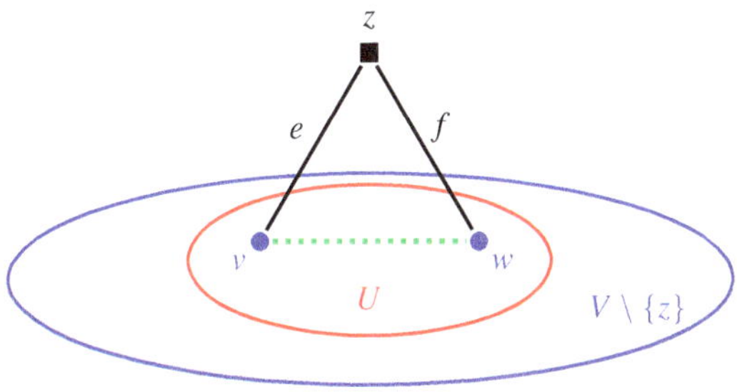

Figure 2.3 Splitting off the pair of edges $e = \{v, z\}$ and $f = \{z, w\}$ means replacing them by the edge $\{v, w\}$ (green, dotted). This reduces $|\delta(U)|$ by 2 for every set $U \subseteq V \setminus \{z\}$ with $v, w \in U$ (one example for such a set U is shown in red).

were parallel edges). See Figure 2.3. We want to do this while keeping the graph sufficiently connected:

Theorem 2.30 (Lovász [1976]). *Let $G = (V, E)$ be an undirected graph and $z \in V$ a vertex whose degree is even. Let $\lambda \geq 2$ be an integer and*

$$|\delta(U)| \geq \lambda \quad \text{for all } \emptyset \neq U \subsetneq V \setminus \{z\}. \tag{2.13}$$

Then for every $e = \{v, z\} \in \delta(z)$, there is an $f = \{z, w\} \in \delta(z) \setminus \{e\}$ such that splitting off e and f preserves (2.13).

Proof. Let $e = \{v, z\} \in \delta(z)$. Call a set U with $v \in U \subsetneq V \setminus \{z\}$ *dangerous* if $|\delta(U)| \leq \lambda + 1$. For any $f = \{z, w\} \in \delta(z) \setminus \{e\}$, splitting off e and f preserves (2.13) unless there is a dangerous set containing (v and) w. So assume that there is a family $\mathcal{F}$ of dangerous sets whose union contains all neighbors of z. Choose $\mathcal{F}$ to be minimal.

Note that for each element $F \in \mathcal{F}$, we have $|\delta(F) \cap \delta(z)| \leq \frac{1}{2}|\delta(z)|$. This is because otherwise $|\delta(V \setminus (F \cup \{z\}))| \leq |\delta(F)| - 2 \leq \lambda - 1$ (in the first inequality, we used that $|\delta(z)|$ is even), contradicting (2.13).

As $e \in \delta(F) \cap \delta(z)$ and $|\delta(F) \cap \delta(z)| \leq \frac{1}{2}|\delta(z)|$ for all $F \in \mathcal{F}$, we conclude $|\mathcal{F}| \geq 3$. Let $A, B, C \in \mathcal{F}$ be distinct. Now

$$\begin{aligned} 3(\lambda + 1) &\geq |\delta(A)| + |\delta(B)| + |\delta(C)| \\ &\geq |\delta(A \cap B \cap C)| + |\delta(A \setminus (B \cup C))| + |\delta(B \setminus (A \cup C))| \\ &\quad + |\delta(C \setminus (A \cup B))| + 2|\delta(A \cap B \cap C) \cap \delta(A \cup B \cup C)| \\ &\geq \lambda + \lambda + \lambda + \lambda + 2, \end{aligned}$$

where we used the minimality of $\mathcal{F}$ in the last inequality. This implies $\lambda \leq 1$, a contradiction. □

This proof is due to Frank [2011]. Note that Lemma 1.7 can be seen as a special case because a tour remains a tour when splitting off at a vertex of degree more than 2. Among the many applications of Theorem 2.30 is the following observation by Cunningham (see Monma, Munson, and Pulleyblank [1990]) and Goemans and Bertsimas [1993]:

Theorem 2.31. *If (V, E) is a complete graph, $|V| \geq 3$, and c obeys the triangle inequality, then the optimum values of* (2.2) *and* (2.12) *are the same.*

Proof. Let x be a rational optimum solution to (2.12). Among all such x, choose one that minimizes $x(E)$. We show that then x is a feasible solution to (2.2).

Choose $K \in \mathbb{N}$ such that Kx_e is an integer for each $e \in E$. If there is a vertex $z \in V$ with $x(\delta(z)) > 2$, choose incident edges $e = \{z, v\}$ and $f = \{z, w\}$ with $x_e > 0$ and $x_f > 0$, reduce x_e and x_f each by $\frac{1}{2K}$ and increase $x_{\{v,w\}}$ by $\frac{1}{2K}$ while maintaining $x(\delta(U)) \geq 2$ for all $\emptyset \neq U \subsetneq V$. The existence of two such edges e, f follows from applying Theorem 2.30 to $\lambda = 4K$ and the Eulerian graph with $2Kx_e$ copies of each edge e. This operation reduces $x(E)$ but does not increase $c(x)$ due to the triangle inequality. This is a contradiction. So $x(\delta(z)) = 2$ for all $z \in V$.

Finally, observe that the constraints $x_e \leq 1$ $(e \in E)$ are satisfied by Proposition 2.1. □

The above proofs suggest an algorithm. It can be implemented as follows:

Theorem 2.32. *Let $G = (V, E)$ be an undirected graph with edge weights $x : E \to \mathbb{R}_{\geq 0}$ and $z \in V$. Let $\lambda > 0$ and*

$$x(\delta(U)) \geq \lambda \quad \text{for all } \emptyset \neq U \subsetneq V \setminus \{z\}. \tag{2.14}$$

Then one can compute in polynomial time a list of triples $(e_i, f_i, \gamma_i) \in \delta(z) \times \delta(z) \times \mathbb{R}_{>0}$, $i = 1, \ldots, k$, such that splitting off all (e_i, f_i, γ_i) maintains (2.14) *and leads to $x(e) = 0$ for all edges e incident to z. Here splitting off (e, f, γ) for $e = \{v, z\}$ and $f = \{w, z\}$ means reducing $x(e)$ and $x(f)$ by γ and adding an edge $\{v, w\}$ with weight γ.*

Proof. We can simply scan the list of all pairs $(e_i, f_i) \in \delta(z) \times \delta(z)$ with $e_i \neq f_i$, greedily set γ_i as large as possible in order to maintain (2.14), and split off (e_i, f_i, γ_i) before continuing with the next pair. The largest feasible γ_i for $e_i = \{v, z\}$ and $f_i = \{w, z\}$ is

$$\gamma_i = \min \left\{ x(e_i),\ x(f_i),\ \tfrac{1}{2} \left(\min \left\{ x(\delta(U)) : v, w \in U \subsetneq V \setminus \{z\} \right\} - \lambda \right) \right\}.$$

Hence, it can be computed by less than n calls to a max-flow algorithm, applying Corollary 2.9 for $S = \{v, w\}$ and $T = \{z, t\}$ for all possibilities of $t \in V \setminus (S \cup \{z\})$. Triples with $\gamma_i = 0$ are omitted from the final list.

If x is rational, Theorem 2.30 (applied to the graph where every edge e is replaced by $2Kx(e)$ parallel edges, where K is a common denominator) implies that at the end, we have $x(e) = 0$ for all edges e incident to z. Since every x is the limit of a monotone sequence of rational functions, this also holds for irrational x. □

Cen, Li, and Panigrahi [2022] devised a faster randomized algorithm.

Although the optimum values of the LPs (2.2) and (2.12) are the same, their integrality ratios might be different because not every integral solution to (2.12) is the incidence vector of a tour. A graph $G = (V, E)$ is *2-edge-connected* if $|\delta_G(U)| \geq 2$ for all $\emptyset \neq U \subsetneq V$. Every tour is the edge set of a 2-edge-connected graph because $|\delta_F(U)|$ is even for every tour F and every vertex set U. The integral solutions to (2.12) are the incidence vectors of the 2-edge-connected spanning multi-subgraphs, not only of tours. Therefore, the integrality ratio of (2.12) might be smaller than the integrality ratio of (2.2). We will return to this question and the problem of finding a minimum-cost 2-edge-connected spanning subgraph in Sections 13.6 and 17.2.

Strengthening the subtour LP by adding further classes of facets of the traveling salesman polytope (the convex hull of incidence vectors of Hamiltonian circuits) has proved very useful in the design of exact branch-and-cut algorithms (see Applegate et al. [2006]), but this has not (yet) been used for approximation algorithms.

Goemans [1995] showed for most of the known classes of facets that adding this class would increase the value of (2.12) by at most a factor of $\frac{4}{3}$, giving further support of the $\frac{4}{3}$ conjecture. Unfortunately, there is no LP relaxation for which we know that it can be solved in polynomial time and has a better integrality ratio than what is known for the subtour LP.

Open Problem 2.33. Prove that there exists a polynomial-time solvable LP relaxation of the SYMMETRIC TSP WITH TRIANGLE INEQUALITY that has a smaller integrality ratio than the subtour LP.

Of course, no complete polyhedral description can be expected for the traveling salesman polytope. In fact, Fiorini et al. [2015] proved that every polyhedron that projects to the TSP polytope (i.e., any *extended formulation*) has $2^{\Omega(n^{1/4})}$ facets. It may not be surprising that the TSP has no compact extended formulation, but this was not known before, and this result is unconditional (i.e., it does not assume $P \neq NP$). The proof reveals an interesting connection to communication complexity.

Exercises

2.1 Prove the undirected version of the max-flow min-cut theorem: If $G = (V, E)$ is an undirected graph, $u : E \to \mathbb{R}_{\geq 0}$, and $s, t \in V$, then the maximum value of an s-t-flow in an orientation of (G, u) equals $\min\{u(\delta(R)) : s \in R \subseteq V \setminus \{t\}\}$.

2.2 Prove (the directed vertex-disjoint version of) Menger's theorem: Let $G = (V, E)$ be a directed graph, $s, t \in V$ such that there is no edge (s, t), and $k \in \mathbb{Z}_{>0}$. Then there are k paths $P_1, \ldots, P_k$ from s to t such that their vertex sets are pairwise disjoint except for the endpoints s and t if and only if for every $X \subseteq V \setminus \{s, t\}$ with $|X| \leq k - 1$ there exists an s-t-path that does not contain any vertex from X.
Hint: Replace every vertex $v \in V \setminus \{s, t\}$ by two vertices v' (inheriting the edges entering v) and v'' (inheriting the edges leaving v) and an edge (v', v''). Apply the max-flow min-cut theorem. For sufficiency, apply Corollary 2.6 and decompose the integral flow into flows along s-t-paths. (Menger [1927])

2.3 Let $G = (V, E)$ be an undirected graph with weights $c : E \to \mathbb{R}_{\geq 0}$. For $s, t \in V$, let $\lambda_{s,t} = \min\{c(\delta(U)) : s \in U \subseteq V \setminus \{t\}\}$ be the minimum weight of a cut separating s and t. Show that for all $u, v, w \in V$, we have $\lambda_{u,w} \geq \min\{\lambda_{u,v}, \lambda_{v,w}\}$.

2.4 Consider the problem of finding a minimum-cost connector in a graph $G = (V, E)$ with edge costs $c : E \to \mathbb{R}$. Show that this problem is equivalent to MINIMUM SPANNING TREE.

2.5 Let E be a finite set and $\mathcal{F} \subseteq 2^E$ with the properties that (i) $\emptyset \in \mathcal{F}$; (ii) if $A \subseteq B \in \mathcal{F}$, then $A \in \mathcal{F}$; and (iii) if $A, B \in \mathcal{F}$ and $|A| < |B|$, then there exists a $b \in B \setminus A$ such that $A \cup \{b\} \in \mathcal{F}$. Then $(E, \mathcal{F})$ is called a *matroid*. A *forest* is an undirected graph without circuits.

- Let $\mathcal{F}$ be the set of edge sets of forests in a connected undirected graph $G = (V, E)$. Prove that $(E, \mathcal{F})$ is a matroid.
- Show (similar to the proof of Theorem 2.14) that one can construct an $F \in \mathcal{F}$ with maximum $c(F)$ in any matroid $(E, \mathcal{F})$ with weights $c : E \to \mathbb{R}_{\geq 0}$ by the greedy algorithm: sort $E = \{e_1, \ldots, e_m\}$ so that $c(e_1) \geq \cdots \geq c(e_m)$, initialize $F = \emptyset$, and successively add e_i to F if this remains in $\mathcal{F}$ ($i = 1, \ldots, m$).

- Prove that the convex hull of incidence vectors of elements of $\mathcal{F}$ is

$$\left\{x \in \mathbb{R}^E : x \geq 0,\ x(A) \leq r(A) \text{ for all } A \subseteq E\right\},$$

where $r(A) := \max\{|F| : F \in \mathcal{F},\ F \subseteq A\}$ is called the *rank* of A.

Hint: Note that the greedy algorithm chooses $r(\{e_1, \ldots, e_i\})$ elements among $e_1, \ldots, e_i$ for each $i = 1, \ldots, m$. Proceed similarly as in the proof of Theorem 2.14.

2.6 Let $G = (V, E)$ be an undirected graph and z a vector in the spanning tree polytope of G. Call a set $U \subseteq V$ *tight* if $z(E[U]) = |U| - 1$. Let A and B be tight sets with $A \cap B \neq \emptyset$. Show that then $A \cap B$ and $A \cup B$ are tight.

2.7 MINIMUM SPANNING TREE can be described by the integer linear program $\min\{c(x) : x(E) = n - 1,\ x(\delta(U)) \geq 1\ (\emptyset \neq U \subsetneq V),\ x_e \in \{0, 1\}\ (e \in E)\}$. Show that replacing $x_e \in \{0, 1\}$ by $0 \leq x_e \leq 1$ leads to an LP with integrality ratio 2.

2.8 Deduce a description of the perfect matching polytope (the convex hull of incidence vectors of perfect matchings in an undirected graph G) from the T-join polyhedron (Theorem 2.19).

2.9 Consider a weaker variant of the subtour LP:

$$\begin{array}{llcll} \min & c(x) & & & \\ \text{subject to} & x(\delta(U)) & \geq & 1 & (\emptyset \neq U \subsetneq V) \\ & x(\delta(v)) & = & 2 & (v \in V) \\ & x_e & \leq & 1 & (e \in E) \\ & x_e & \geq & 0 & (e \in E). \end{array}$$

This LP still has the property that the integral feasible solutions are precisely the incidence vectors of Hamiltonian circuits. Prove, however, that the integrality ratio of this LP is at least $\frac{8}{3}$.

2.10 Prove that the integrality ratio of the LP in Exercise 2.9 is exactly twice the integrality ratio of the subtour LP.

2.11 Show that without the triangle inequality, the integrality ratio of the subtour LP (2.2) would be unbounded.

2.12 Prove that the integrality ratio of the subtour LP (2.2) is the smallest number ρ for which the following is true: For every positive integer k and every k-regular k-edge-connected graph G, the all-$\frac{2\rho}{k}$ vector is a convex combination of incidence vectors of tours.
(Boyd and Sebő [2021])

2.13 A 2-factor (or simple perfect 2-matching, or cycle cover) in a graph is an edge set F that forms a 2-regular spanning subgraph. Show that there are instances of the SYMMETRIC TSP WITH TRIANGLE INEQUALITY in which there is no 2-factor that costs less than $\frac{10}{9}$ times the value of the subtour LP (2.2).
Hint: See the envelope example (Figure 2.2).
Note: Schalekamp, Williamson, and van Zuylen [2014] showed that the worst ratio of an optimum 2-factor over (2.2) is indeed $\frac{10}{9}$, as conjectured by Boyd and Carr [2011].

2.14 Let $G = (V, E)$ be a 2-edge-connected undirected graph and $c(e) = 1$ for all $e \in E$. Show that then the value of (2.12) does not change when adding constraints $x_e \leq 1$ for all $e \in E$.
(Vygen [2012])

2.15 Show the following strengthening of Theorem 2.30. Let G be an undirected graph, $z \in V$ an even-degree vertex, and $\lambda \geq 2$ such that (2.13) holds. Then for every $e = \{v, z\} \in \delta(z)$, there are at least $\frac{1}{2}|\delta(z)| - 1$ edges $f = \{z, w\} \in \delta(z) \setminus \{e\}$ such that splitting off e and f preserves (2.13).
Hint: Consider two dangerous sets A and B whose union contains all neighbors of z that we cannot use for splitting off. Then exploit that $|\delta(V \setminus (A \cup B \cup \{z\}))| \geq \lambda$.
(Bang-Jensen et al. [1999])

3

Linear Programming Relaxations of the Asymmetric TSP

As in the symmetric case, there are two versions of the asymmetric TSP and two corresponding LP relaxations. Again, the two versions are equivalent, and we will present a third equivalent version. We will also study the integrality ratio and show that it is at least 2.

3.1 The Two Basic LP Relaxations of the Asymmetric TSP

First, let (V, c) be an instance of the ASYMMETRIC TSP WITH TRIANGLE INEQUALITY and $E = \{(v, w) : v, w \in V, v \neq w\}$. The following is the natural analogon of the subtour LP (2.2) in this case:

$$\begin{array}{llcll} \min & c(x) & & & \\ \text{subject to} & x(\delta(U)) & \geq & 2 & (\emptyset \neq U \subsetneq V) \\ & x(\delta(v)) & = & 2 & (v \in V) \\ & x(\delta^+(v)) & = & x(\delta^-(v)) & (v \in V) \\ & x_e & \geq & 0 & (e \in E). \end{array} \tag{3.1}$$

Second, let $G = (V, E)$ and $c : E \to \mathbb{R}_{\geq 0}$ be an instance of the ASYMMETRIC TSP. Then the natural LP relaxation is

$$\begin{array}{llcll} \min & c(x) & & & \\ \text{subject to} & x(\delta(U)) & \geq & 2 & (\emptyset \neq U \subsetneq V) \\ & x(\delta^+(v)) & = & x(\delta^-(v)) & (v \in V) \\ & x_e & \geq & 0 & (e \in E). \end{array} \tag{3.2}$$

The integral feasible solutions to (3.1) are the incidence vectors of Hamiltonian circuits, while the integral feasible solutions to (3.2) are the incidence vectors of tours.

Note that $x(\delta^+(v)) = x(\delta^-(v))$ for all $v \in V$ implies $x(\delta^+(U)) = x(\delta^-(U))$ for all $U \subseteq V$, and hence $x(\delta(U)) \geq 2$ is equivalent to $x(\delta^+(U)) \geq 1$ (or $x(\delta^-(U)) \geq 1$).

Again, both LPs can be solved in polynomial time:

Proposition 3.1. *The linear programs* (3.1) *and* (3.2) *can be solved in polynomial time.*

Proof. This can be done either via an extended formulation or by the equivalence of separation and optimization, exactly as in Section 2.2 (cf. Corollary 2.11). □

We will next show that the two LPs are equivalent.

3.2 Directed Splitting Off

The splitting-off technique that we presented in Section 2.5 can also be applied to digraphs, as Mader [1982] showed first. *Splitting off* a pair of edges $e = (v, z)$ and $f = (z, w)$ in a digraph means replacing e and f by a single new edge (v, w) (which can be a loop if $v = w$).

There are several versions of directed splitting-off theorems. The following will contain all we need. The special case $T = V \setminus \{z\}$ is an analogon of Theorem 2.30.

Theorem 3.2 (Jackson [1988], Frank [1989]). *Let $G = (V, E)$ be an Eulerian digraph, $T \subsetneq V$, and $z \in V \setminus T$. Let $\lambda \geq 2$ be an even integer and*

$$|\delta(U)| \geq \lambda \quad \textit{for all } U \subsetneq V \textit{ with } T \cap U \neq \emptyset \textit{ and } T \setminus U \neq \emptyset. \tag{3.3}$$

Then for every edge $e = (v, z) \in \delta^-(z)$, there is an edge $f = (z, w) \in \delta^+(z)$ such that splitting off e and f preserves (3.3).

Proof. Let $e = (v, z) \in \delta^-(z)$. Call a set $U \subsetneq V$ *dangerous* if $T \cap U \neq \emptyset$, $T \setminus U \neq \emptyset$, $e \in \delta^+(U)$, and $|\delta(U)| = \lambda$. Let W be the union of all dangerous sets. For any $f = (z, w) \in \delta^+(z)$, splitting off e and f preserves (3.3) unless $w \in W$ (i.e., unless there is a dangerous set containing w; see Figure 3.1). Hence we are done unless $\delta^+(z) \subseteq \delta^-(W)$.

We next show that W is dangerous, which follows from the following:

$$\text{The union of dangerous sets is dangerous.} \tag{3.4}$$

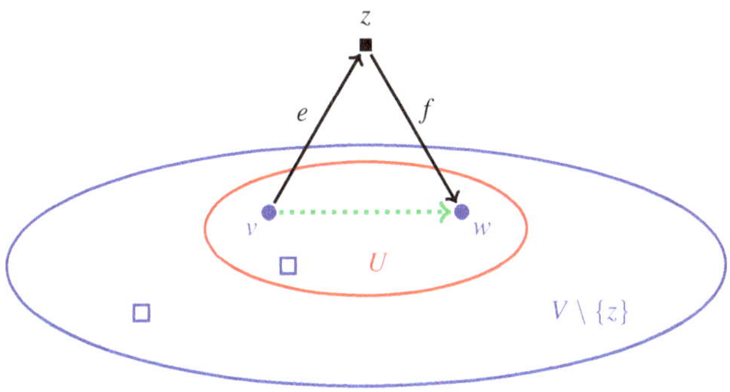

Figure 3.1 Splitting off the pair of edges $e = (v, z)$ and $f = (z, w)$ means replacing them with the edge (v, w) (green, dotted). This reduces $|\delta(U)|$ by 2 for every set $U \subseteq V \setminus \{z\}$ with $v, w \in U$. For sets U with $|\delta(U)| = \lambda$ and $T \cap U \neq \emptyset$ and $T \setminus U \neq \emptyset$, this would lead to violating (3.3). One example for such a set U is shown in red; here, the blue empty squares indicate elements of T.

To prove (3.4), let X and Y be dangerous sets such that neither of them is a subset of the other.

If $T \cap (X \setminus Y) \neq \emptyset$ and $T \cap (Y \setminus X) \neq \emptyset$ (Figure 3.2 (a)), then

$$\begin{aligned} \lambda + \lambda &= |\delta(X)| + |\delta(Y)| \\ &= |\delta(X \setminus Y)| + |\delta(Y \setminus X)| + 2|\delta(X \cap Y) \cap \delta(X \cup Y)| \\ &\geq \lambda + \lambda + 2, \end{aligned}$$

a contradiction (we used (3.3) for $X \setminus Y$ and $Y \setminus X$, and the existence of $e \in \delta^+(X \cap Y) \cap \delta^+(X \cup Y)$).

Therefore we conclude $T \cap (X \cap Y) \neq \emptyset$ and $T \setminus (X \cup Y) \neq \emptyset$ (Figure 3.2 (b)). Then $|\delta(X \cup Y)| = \lambda$ follows from

$$\begin{aligned} \lambda + \lambda &= |\delta(X)| + |\delta(Y)| \\ &\geq |\delta(X \cap Y)| + |\delta(X \cup Y)| \\ &\geq \lambda + \lambda. \end{aligned}$$

So $X \cup Y$ is indeed dangerous, and (3.4) is proved.

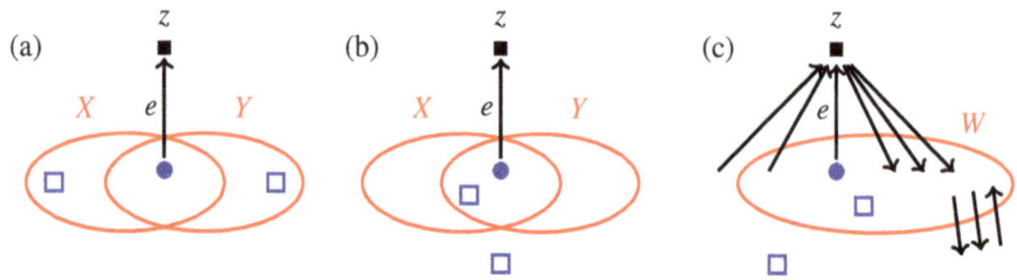

Figure 3.2 (a) and (b): The two cases in the proof of (3.4). (c): The maximal dangerous set W: Assuming $\delta^+(z) \subseteq \delta^-(W)$ leads to a contradiction. The blue empty squares indicate elements of T.

(3.4) says in particular that W is dangerous. But now $\delta^+(z) \subseteq \delta^-(W)$ would imply (cf. Figure 3.2 (c))

$$\begin{aligned}
|\delta(\{z\} \cup W)| &= |\delta(W) \setminus \delta(z)| + |\delta(z) \setminus \delta(W)| \\
&= |\delta(W)| - |\delta^+(z)| - |\delta^+(W) \cap \delta^-(z)| + |\delta^-(z) \setminus \delta^+(W)| \\
&= |\delta(W)| - |\delta^-(z)| - |\delta^+(W) \cap \delta^-(z)| + |\delta^-(z) \setminus \delta^+(W)| \\
&= |\delta(W)| - 2|\delta^+(W) \cap \delta^-(z)| \\
&\leq |\delta(W)| - 2 \\
&= \lambda - 2
\end{aligned}$$

because $e \in \delta^+(W) \cap \delta^-(z)$ and W is dangerous. But then $\{z\} \cup W$ violates (3.3), a contradiction. □

The uncrossing technique that we used in this proof will reappear in the next chapter.

Again, Lemma 1.8 can be seen as a special case (because a tour remains a tour when splitting off at a vertex of degree more than 2 while maintaining connectivity among the other vertices). Very similarly to Theorem 2.32, we obtain an algorithm:

Theorem 3.3. *Let $G = (V, E)$ be a digraph with weights $x : E \to \mathbb{R}_{\geq 0}$ that satisfy $x(\delta^-(v)) = x(\delta^+(v))$ for all $v \in V$. Let $T \subsetneq V$ and $z \in V \setminus T$. Let $\lambda > 0$ and*

$$x(\delta(U)) \geq \lambda \quad \text{for all } U \subsetneq V \text{ with } T \cap U \neq \emptyset \text{ and } T \setminus U \neq \emptyset. \tag{3.5}$$

Then one can compute in polynomial time a list of triples $(e_i, f_i, \gamma_i) \in \delta^-(z) \times \delta^+(z) \times \mathbb{R}_{>0}$, $i = 1, \ldots, k$, such that splitting off all (e_i, f_i, γ_i) maintains (3.5) and leads to $x(e) = 0$ for all edges e incident to z. Here, splitting off (e, f, γ) means reducing $x(e)$ and $x(f)$ by γ and adding an edge with weight γ from the tail of e to the head of f.

Proof. We can simply scan the list of all pairs $(e_i, f_i) \in \delta^-(z)\times\delta^+(z)$, greedily set each γ_i as large as possible in order to maintain (3.5), and split off (e_i, f_i, γ_i) if $\gamma_i > 0$ before continuing with the next pair. The largest feasible γ_i for $e_i = (v, z)$ and $f_i = (z, w)$ is

$$\gamma_i = \min\Big\{x(e_i),\ x(f_i),\\ \frac{1}{2}\Big(\min\big\{x(\delta(U)) : v, w, t \in U \subseteq V \setminus \{z, t'\},\ t, t' \in T\big\} - \lambda\Big)\Big\}$$

(see Figure 3.1), so it can be computed by less than n^2 calls to a max-flow algorithm (cf. Corollary 2.9), one for each possible choice of t and t'.

If x and λ are rational and $K \in \mathbb{N}$ such that Kx_e is an integer for all $e \in E$ and $K\lambda$ is an even integer, Theorem 3.2 (applied to the Eulerian digraph that results from G by replacing every edge e by $Kx(e)$ parallel edges) implies that, at the end, we have $x(e) = 0$ for all edges e incident to z. Since every x, λ is the limit of a monotonely increasing sequence of rational functions, this also holds for irrational x, λ. □

The running time of this algorithm can be improved easily (see Nagamochi, Nichimura, and Ibaraki [1997]). By replacing every undirected edge $\{v, w\}$ by a pair of antiparallel edges (v, w) and (w, v), each with half the weight, and setting $T = V \setminus \{z\}$, we can view Theorem 2.32 as a special case of Theorem 3.3.

The equivalence of the LPs now follows directly (as was noted, for example, by Nagarajan and Ravi [2011]):

Theorem 3.4. *For any instance (V, c) of the* Asymmetric TSP with Triangle Inequality, *the LPs* (3.1) *and* (3.2) *have the same value.*

Proof. Every feasible solution to (3.1) is a feasible solution to (3.2). Conversely, let x be a feasible solution to (3.2). As long as there is a vertex z with $x(\delta(z)) > 2$, apply Theorem 3.3 with $\lambda = 2$ and $T = V \setminus \{z\}$ and apply the resulting splitting-off operations partially, until $x(\delta(z)) = 2$. Remove loops. This yields a feasible solution to (3.1), which by the triangle inequality, has no larger cost. □

Theorem 3.5. *Let (V, c) be an instance of the* Asymmetric TSP with Triangle Inequality, *$E = \{(v, w) : v, w \in V, v \neq w\}$, and $T \subseteq V$. Let $x \in \mathbb{R}^E_{\geq 0}$ satisfy $x(\delta^-(v)) = x(\delta^+(v))$ for all $v \in V$ and $x(\delta(U)) \geq 2$ for all $U \subsetneq V$ with $T \cap U \neq \emptyset$ and $T \setminus U \neq \emptyset$. Then the value of* (3.1) *for the subinstance (T, c) is at most $c(x)$.*

Proof. Apply Theorem 3.3 to all $z \in V \setminus T$ to obtain a solution to (3.2) for (T, c) that has no larger cost. Then apply Theorem 3.4. □

In particular, we get the following monotonicity property: The LP value does not increase if we restrict the instance to a subset of vertices.

Corollary 3.6. *Let (V, c) be an instance of the* Asymmetric TSP with Triangle Inequality *and $T \subseteq V$. Then the value of* (3.2) *for (T, c) is at most the value of* (3.2) *for (V, c).*

Proof. Apply Theorem 3.5 to an optimum solution to (3.2) for (V, c). □

The analogous statement holds for the Symmetric TSP with Triangle Inequality (see Exercise 3.1).

3.3 A Third LP Relaxation of the Asymmetric TSP

It might be interesting to note that one can also obtain a strengthened cut condition. A *circulation* in a digraph (V, E) is a function $f : E \to \mathbb{R}_{\geq 0}$ with $f(\delta^-(v)) = f(\delta^+(v))$ for all $v \in V$. Any circulation can be decomposed into flows along circuits:

Proposition 3.7. *Let $G = (V, E)$ be a digraph and f a circulation in G. Then there are edge sets $C_1, \ldots, C_k$ of circuits in G and $\lambda_1, \ldots, \lambda_k > 0$ such that $f(e) = \sum_{i:e\in C_i} \lambda_i$.*

Proof. Induction on the number of edges e with $f(e) > 0$. The statement is trivial if there are no such edges. If $e_1 = (v_1, v_2)$ is an edge with $f(e_1) > 0$, there must be an edge $e_2 = (v_2, v_3) \in \delta^+(v_2)$ with $f(e_2) > 0$. Iterate until we hit a vertex that we visited before. Then the sequence between the two visits defines the edge set C of a circuit with $\gamma := \min\{f(e) : e \in C\} > 0$. Subtract γ from $f(e)$ for all $e \in C$ and apply induction. □

Of course, any solution to (3.2) is a circulation and hence can be decomposed to flows along circuits. Let $\mathscr{C}$ denote the set of edge sets of all circuits in the given digraph (V, E). Instead of only requiring that the total value of the edges crossing a cut is at least 2, we now require that the total value of circuits crossing any cut is at least 1. This is stronger because circuits crossing a cut several times count only once:

$$
\begin{array}{lrcll}
\min \sum_{C\in\mathscr{C}} c(C)\lambda_C & & & & \\
\text{subject to} \quad \sum_{C\in\mathscr{C}: C\cap\delta^+(U)\neq\emptyset} & \lambda_C & \geq & 1 & (\emptyset \neq U \subsetneq V) \\
& \lambda_C & \geq & 0 & (C \in \mathscr{C})
\end{array}
\tag{3.6}
$$

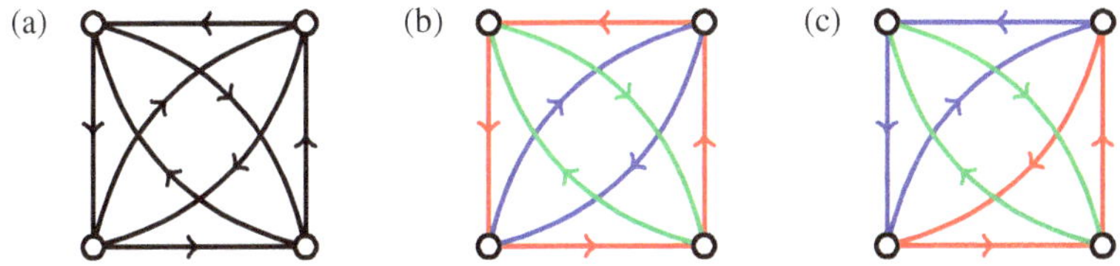

Figure 3.3 Decomposing a solution to (3.1) into circuits in an arbitrary way may not yield a feasible solution to (3.6), as this example (taken from Henke [2018]) shows. In (a) we see the support graph of a solution to (3.1), where each of the eight edges has value $\frac{1}{2}$. The pictures (b) and (c) show two decompositions of this circulation into flows on circuits; each of the circuits has value $\frac{1}{2}$. The decomposition in (c) is a feasible solution to (3.6), but the one in (b) is not.

The example in Figure 3.3 shows that not every decomposition is good: decomposing a solution to (3.2) into circuits in an arbitrary way will not always lead to a feasible solution to (3.6). However, a good decomposition always exists, as the authors showed together with Henke [2018]. The following simpler proof is due to Goemans.

Theorem 3.8. *From any solution x to* (3.2)*, one can compute in polynomial time a solution λ to* (3.6) *with $x_e = \sum_{C\in\mathscr{C}:e\in C} \lambda_C$ (and hence the same objective function value).*

Proof. Take an arbitrary order of the vertices, $V = \{z_1, \ldots, z_n\}$, and apply Theorem 3.3 to $z = z_n$ and $T = \{z_1, \ldots, z_{n-1}\}$, then to $z = z_{n-1}$ and $T = \{z_1, \ldots, z_{n-2}\}$, and so on, until $z_3, \ldots, z_n$ are isolated. Note that after k steps, the subgraph induced by $\{z_1, \ldots, z_{n-k}\}$ contains a feasible solution $x^{(n-k)}$ to (3.2) for this smaller instance.

We show how to compute a feasible solution $\lambda^{(n-k)}$ to (3.6). For $\lambda^{(2)}$, there are only two vertices (z_1 and z_2), and hence any decomposition of $x^{(2)}$ into circuits (with two edges each), which we get greedily like in Proposition 3.7, does the job: the sum of the λ_C will be $\frac{1}{2}x^{(2)}(\delta(v_1)) \geq 1$.

Now, for $k = n-3, n-4, \ldots, 0$ we compute $\lambda^{(n-k)}$ from $\lambda^{(n-k-1)}$. To this end, we note that $x^{(n-k-1)}$ resulted from $x^{(n-k)}$ by splitting off some triples (e_i, f_i, γ_i). We now undo this operation on $\lambda^{(n-k-1)}$: Whenever a circuit C with $\lambda_C^{(n-k-1)} > 0$ contains an edge (v, w) that resulted from splitting off $((v, z_{n-k}), (z_{n-k}, w), \gamma)$, we replace (v, w) in C by the edges (v, z_{n-k}) and (z_{n-k}, w). C remains Eulerian, but z_{n-k} may have in-degree and out-degree more than 1 in C. We partition C into edge sets of circuits and set $\lambda_{C'}^{(n-k)} := \lambda_C^{(n-k-1)}$ for each circuit C' in this partition. The result is a feasible solution to (3.6) on $\{z_1, \ldots, z_{n-k}\}$. □

Theorem 3.8 implies that we can solve the LP (3.6) in polynomial time for any $c : E \to \mathbb{R}_{\geq 0}$ although the LP has exponentially many variables and exponentially many constraints. However, we cannot hope to optimize arbitrary objective function $c : \mathscr{C} \to \mathbb{R}_{\geq 0}$; for example, the LP for $c(C) = 1$ for all circuits C in a given digraph has value 1 if and only if G contains a Hamiltonian circuit (cf. Exercise 3.2).

3.4 Integral and Minimum-Cost Circulations

Feasible solutions to the above LPs are circulations. In this section, we collect some classical facts on circulations for later use. It is often useful to impose lower and upper bounds. Hoffman's circulation theorem tells us when a circulation subject to such bounds exists:

Theorem 3.9 (Hoffman [1960]). *Let $G = (V, E)$ be a digraph with lower and upper bounds $l, u : E \to \mathbb{R}_{\geq 0}$. Then there is a circulation f in G with $l(e) \leq f(e) \leq u(e)$ for all $e \in E$ if and only if $l(e) \leq u(e)$ for all $e \in E$ and $l(\delta^-(U)) \leq u(\delta^+(U))$ for all $U \subseteq V$.*

Proof. Set $u'(e) := u(e) - l(e)$ for all $e \in E$, $b(v) := l(\delta^-(v)) - l(\delta^+(v))$ for all $v \in V$, and construct G' from G by adding two vertices s and t and edges (s, v) with capacity $u'((s, v)) = \max\{0, b(v)\}$ and edges (v, t) with capacity $u'((v, t)) = \max\{0, -b(v)\}$ for all $v \in V$. Then an s-t-flow g of value $u'(\delta^+(s))$ in (G', u') corresponds to a circulation f in G with $l \leq f \leq u$ and vice versa, via $f(e) = g(e) + l(e)$ for $e \in E$. By the max-flow min-cut theorem (Theorem 2.5), such an s-t-flow exists if and only if $u'(\delta^+(\{s\} \cup U)) \geq u'(\delta^+(s))$ for all $U \subseteq V$. Since $u'(\delta^+_{G'}(\{s\} \cup U)) = u'(\delta^+_{G'}(s)) - b(U) + u(\delta^+_G(U)) - l(\delta^+_G(U)) = u'(\delta^+_{G'}(s)) + u(\delta^+_G(U)) - l(\delta^-_G(U))$, this is equivalent to $l(\delta^-(U)) \leq u(\delta^+(U))$ for all $U \subseteq V$. □

Next, we prove the integral circulation theorem:

Theorem 3.10. *Let $G = (V, E)$ be a digraph with lower and upper bounds $l, u : E \to \mathbb{Z}_{\geq 0}$ and costs $c : E \to \mathbb{R}$. Then, among all circulations f in G with $l(e) \leq f(e) \leq u(e)$ for all $e \in E$ that minimize $\sum_{e \in E} c(e) f(e)$, there is an integral one.*

Proof. Call a circulation f with $l(e) \leq f(e) \leq u(e)$ for all $e \in E$ *optimal* if it minimizes $c(f) := \sum_{e \in E} c(e) f(e)$. Among all optimal circulations, let f be one that minimizes $|\{e \in E : f(e) \notin \mathbb{Z}\}|$. We show that f is integral. Suppose not, then there is an edge $e_1 = (v_1, v_2)$ with $f(e_1) \notin \mathbb{Z}$. This cannot be the only edge incident to v_2 with non-integral flow value, so let e_2 be another such edge, and let v_3 be the other endpoint of e_2. Continue with v_3, and iterate until we meet a vertex that we have already visited. Then the sequence between the two visits defines the edge set C of a graph whose underlying undirected graph is a circuit. Let $\delta := \min\{\min\{f(e) - \lfloor f(e) \rfloor, \lceil f(e) \rceil - f(e)\} : e \in C\}$. Then $\delta > 0$, and we define two circulations f^+ and f^- as follows: f^+ arises from f by traversing C and increasing $f(e)$ by δ on forward edges $e \in C$ and decreasing $f(e)$ by δ on backward edges $e \in C$, and f^- is constructed analogously by traversing C in the opposite direction. Then f^+ and f^- satisfy the lower and upper bounds. Moreover, $f^-(e) + f^+(e) = 2f(e)$ for all $e \in E$ and hence $c(f^+) + c(f^-) = 2c(f)$. So both f^+ and f^- are optimal, but at least one of them has at least one edge more with integral flow value, which is a contradiction. □

A polynomial-time algorithm to compute an integral optimum circulation follows from linear programming (Theorem 2.2) and the (algorithmic) proof of Theorem 3.10, but there are faster combinatorial algorithms:

Theorem 3.11 (Tardos [1985], Orlin [1993]). *Let $G = (V, E)$ be a digraph with lower and upper bounds $l, u : E \to \mathbb{R}_{\geq 0}$ and costs $c : E \to \mathbb{R}$. Then one can compute a circulation f in G with $l(e) \leq f(e) \leq u(e)$ for all $e \in E$, minimizing $\sum_{e \in E} c(e)f(e)$, in $O(m^2 \log^2 m)$ time, where $m = |E|$. If l and u are integral, f will be integral.*

If all input numbers are integers bounded by a polynomial in $|V|$, faster randomized algorithms are known (see, e.g., Chen et al. [2022] and van den Brand et al. [2023]). Due to the above theorems, minimum-cost flows (or circulations) are a very powerful tool that we will also use at several places in this book. The following version with bounds on the throughput of every vertex will also be useful:

Corollary 3.12. *Let $G = (V, E)$ be a digraph with lower and upper bounds $l, u : E \cup V \to \mathbb{Z}_{\geq 0}$ and costs $c : E \to \mathbb{R}$. Then, among all circulations f in G with $l(e) \leq f(e) \leq u(e)$ for all $e \in E$ and $l(v) \leq f(\delta^-(v)) \leq u(v)$ for all $v \in V$ that minimize $\sum_{e \in E} c(e)f(e)$, there is an integral one. We can find such an integral optimum circulation in $O(m^2 \log^2 m)$ time, where $m = |E|$.*

Proof. Split each vertex v into two vertices v^- and v^+, and replace each edge (v, w) by (v^+, w^-). For each vertex v, add an edge $e_v = (v^-, v^+)$ with $l(e_v) := l(v)$ and $u(e_v) := u(v)$. Now, apply Theorem 3.11. Finally, contract the added edges. □

Another problem that can be reduced to network flows is bipartite matching. An undirected graph $G = (V, E)$ is called *bipartite* if there exists a set $A \subseteq V$

such that $E = \delta(A)$. Here is a famous theorem, essentially due to Hall [1935]. We formulate it for oriented bipartite graphs:

Theorem 3.13. *Let A and B be disjoint finite sets and $G = (V, E)$ be a digraph with $V = A \cup B$ and $E \subseteq A \times B$. Let $l : A \to \mathbb{R}_{\geq 0}$ and $u : B \to \mathbb{R}_{\geq 0}$. There exists a function $f : E \to \mathbb{R}_{\geq 0}$ with $f(\delta^+(a)) \geq l(a)$ for all $a \in A$ and $f(\delta^-(b)) \leq u(b)$ for all $b \in B$ if and only if*

$$u\left(\left\{b \in B : \delta^+(A') \cap \delta^-(b) \neq \emptyset\right\}\right) \geq l(A') \quad \textit{for all } A' \subseteq A.$$

If l and u are integral, then f can be chosen integral.

Proof. Add a source s and a sink t, and add an edge (s, a) with capacity $l(a)$ for each $a \in A$ and an edge (b, t) with capacity $u(b)$ for each $b \in B$. All original edges have infinite capacity. Then an f as required corresponds to an (integral) s-t-flow of value $l(A)$ subject to the constraints given by the edge capacities. By the max-flow min-cut theorem (Theorem 2.5 and Corollary 2.6), such a flow exists if and only if every s-t-cut $\delta^+(S)$ has capacity at least $l(A)$. If $\delta^+(A \cap S) \not\subseteq \delta^-(B \cap S)$, then the capacity of $\delta^+(S)$ is infinite. For fixed $A \cap S$, the capacity of $\delta^+(S)$ is minimized if $B \cap S = \{b \in B : \delta^+(A \cap S) \cap \delta^-(b) \neq \emptyset\}$. So only such sets S need to be considered, and for those, $\delta^+(S)$ has capacity $l(A \setminus S) + u(B \cap S)$. This is at least $l(A)$ if and only if $u(B \cap S) \geq l(A \cap S)$. □

Hall's [1935] original bipartite matching theorem dealt with the special case $l(a) = 1$ for all $a \in A$ and $u(b) = 1$ for all $b \in B$. By essentially the same construction, one can reduce the minimum-cost perfect matching problem in bipartite graphs (also called the assignment problem) to a minimum-cost circulation problem (see Exercise 3.4).

3.5 Integrality Ratio

We have introduced three classes of LPs, which all have the same integrality ratio:

Proposition 3.14. *The integrality ratios of* (3.1), (3.2), *and* (3.6), *are the same.*

Proof. For any instance (G, c) of the Asymmetric TSP, the LP values of (3.2) and (3.6) are the same by Theorem 3.8, and in both cases, the integral solutions correspond to tours. Hence, the integrality ratios of these two LPs are the same.

Now consider the instance $(V, \bar{c})$ of the Asymmetric TSP with Triangle Inequality that corresponds to the metric closure. Then, by Theorem 3.4, the LP value of (3.1) for $(V, \bar{c})$ is at most the LP value of (3.2) for (G, c).

Moreover, the optimum integral solutions have the same cost (as in the proof of Proposition 1.12). We conclude that the integrality ratio of (3.1) is at least the integrality ratio of (3.2).

For the other inequality, let (V, c) be any instance of the ASYMMETRIC TSP WITH TRIANGLE INEQUALITY, and let (G, c) be the corresponding instance of the ASYMMETRIC TSP (cf. Proposition 1.12). Then the LP value of (3.2) for (G, c) cannot be larger than the LP value of (3.1) for (V, c). The minimum cost of a tour in (G, c) cannot be smaller than the cost of a Hamiltonian circuit by Lemma 1.7. Thus, the integrality ratio of (3.1) is at most the integrality ratio of (3.2). □

Similar to Theorem 2.27, we have the following:

Theorem 3.15 (Goemans [1995], Carr and Vempala [2004]). *The integrality ratio of* (3.2) *is the smallest number ρ such that for every feasible solution x^* to* (3.2), *the vector ρx^* is a convex combination of incidence vectors of tours.*

Proof. Let (G, c) be an instance of the ASYMMETRIC TSP and x^* an optimum solution to (3.2). Suppose, for some $\rho \geq 1$, we can write $\rho x^* = \sum_{i=1}^{k} \lambda_i y_i$, where $\lambda_1, \ldots, \lambda_k > 0$, $\sum_{i=1}^{k} \lambda_i = 1$, and $y_1, \ldots, y_k$ are incidence vectors of tours. Then we have $\rho c(x^*) = \sum_{i=1}^{k} \lambda_i c(y_i)$, and therefore at least one of the vectors y_i (and thus a tour) costs at most $\rho c(x^*)$. We conclude that the integrality ratio is at most ρ.

For the other direction, suppose that the integrality ratio is at most ρ. Let x^* be a feasible solution to (3.2), let Q denote the convex hull of incidence vectors of tours, and let $\mathrm{dom}(Q) := \{y + z : y \in Q,\ z \geq 0\}$.

First, suppose $\rho x^* \notin \mathrm{dom}(Q)$. Then there is a separating hyperplane – that is, a vector w with $w^\top(\rho x^*) < \inf\{w^\top q : q \in \mathrm{dom}(Q)\}$. Note that w is nonnegative because otherwise the right-hand side would be $-\infty$. Now,

$$\begin{aligned}
\rho(w^\top x^*) &= w^\top(\rho x^*) \\
&< \inf\{w^\top q : q \in \mathrm{dom}(Q)\} \\
&= \inf\{w^\top q : q \in Q\} \\
&= \min\{w(F) : F \text{ is a tour}\},
\end{aligned}$$

contradicting the assumption that the integrality ratio is at most ρ.

So $\rho x^* \in \mathrm{dom}(Q)$ – that is, there are vectors $y \in Q$ and $z \geq 0$ with $\rho x^* = y + z$. Since x^* and y are circulations, z is also a circulation and can thus (using Proposition 3.7) be written as $z = \sum_{j=1}^{l} \mu_j z_j$, where $z_1, \ldots, z_l$ are incidence vectors of circuits and $\mu_1, \ldots, \mu_l > 0$. Since $y \in Q$, we can write $y = \sum_{i=1}^{k} \lambda_i y_i$, where $\lambda_1, \ldots, \lambda_k > 0$, $\sum_{i=1}^{k} \lambda_i = 1$, and $y_1, \ldots, y_k$ are incidence vectors of tours.

Let $\Lambda_i := \sum_{j=1}^{i-1} \lambda_j$. We may assume $\mu_j - \lfloor \mu_j \rfloor \in \{\Lambda_i : i = 1, \ldots, k\}$ because we can add duplicates of tours as needed. Then $y_i' = y_i + \sum_{j=1}^{l} \lfloor \mu_j + 1 - \Lambda_i \rfloor z_j$ is the incidence vector of a tour for $i = 1, \ldots, k$, and

$$\sum_{i=1}^{k} \lambda_i y_i' = y + \sum_{j=1}^{l} \sum_{i=1}^{k} \lambda_i \left(\lfloor \mu_j + 1 - \Lambda_i \rfloor \right) z_j = y + \sum_{j=1}^{l} \mu_j z_j = y + z = \rho x^*$$

because $\lfloor \mu_j + 1 - \Lambda_i \rfloor = \lfloor \mu_j \rfloor + 1$ if and only if $\Lambda_i \leq \mu_j - \lfloor \mu_j \rfloor$. We conclude $\rho x^* \in Q$. □

See Exercise 3.9 for an alternative proof.

Analyzing the cycle cover algorithm (Algorithm 1.35) with respect to the LP yields the following upper bound (as noted by Williamson [1990]):

Proposition 3.16. *The integrality ratio of* (3.1) *is at most* $\lfloor \log_2 n \rfloor$, *where* $n = |V|$.

Proof. Let LP denote the value of (3.1). The algorithm computes at most $\lfloor \log_2 n \rfloor$ cycle covers. In each iteration, it considers a subset W of V and computes a cycle cover F_W in $G[W]$. We show $c(F_W) \leq \text{LP}$.

By Corollary 3.6, an optimum solution x' to the LP (3.1) for the subinstance (W, c) costs at most LP. Note that x' is a fractional circulation with $x'(\delta^-(v)) = 1$ for all $v \in W$. A cycle cover in this subinstance corresponds to an integral circulation f with $f(\delta^-(v)) = 1$ for all $v \in W$. By Corollary 3.12, an optimum integral circulation costs no more than an optimum fractional circulation (i.e., at most LP). □

We will show better upper bounds on the integrality ratio in Chapters 6–8. See Table 3.1 for an overview. In the next section, we will consider lower bounds.

3.6 Lower Bound on the Integrality Ratio

A first lower bound on the integrality ratio of (3.1) (which is the same as the integrality ratio of (3.2) and of (3.6) by Proposition 3.14) follows from the LP relaxations of the SYMMETRIC TSP. In particular, we get:

Proposition 3.17. *The integrality ratio of* (3.1) *is at least* $\frac{4}{3}$.

Proof. This follows directly from Proposition 2.24 (cf. Figure 2.2): Any solution x to the subtour LP (2.2) can be transformed to a solution x' to (3.1) by setting $x'_{(v,w)} = x'_{(w,v)} = \frac{1}{2} x_{\{v,w\}}$ for all $\{v, w\} \in \binom{V}{2}$. □

Table 3.1 *Upper bounds on the integrality ratio of* (3.2) *for* Asymmetric TSP *in the order of their discovery. The second and third papers do not mention the integrality ratio explicitly.*

Integrality Ratio	Year	Reference	Chapter
$\log_2 n$	1990	Williamson [1990]	3.5
$0.99 \log_2 n$	2002	Bläser [2008]	–
$0.842 \log_2 n$	2003	Kaplan et al. [2005]	–
$\frac{2}{3} \log_2 n$	2006	Feige and Singh [2007]	–
$O(\frac{\log n}{\log \log n})$	2009	Asadpour et al. [2017]	5
$(\log \log n)^{O(1)}$	2014	Anari and Oveis Gharan [2015]	–
319	2017	Svensson, Tarnawski, and Végh [2020]	6–8
22	2019	Traub and Vygen [2022]	6–8
17	2021	this book	6–8

For many years, it was unclear whether the asymmetric LPs have a larger integrality ratio than the subtour LP (2.2). Then Charikar, Goemans, and Karloff [2006] showed that the integrality ratio of (3.2) is at least 2 (disproving a conjecture of Carr and Vempala [2004]). In contrast to the instances used in Proposition 3.17, the instances constructed by Charikar, Goemans, and Karloff [2006] are not unweighted digraphs. The lower bound for unweighted digraphs was later raised to $\frac{3}{2}$ by Gottschalk [2013]. Based on a different family of examples due to Boyd and Elliott-Magwood [2005] (which is somewhat similar to the one suggested by Charikar, Goemans, and Karloff [2006]), Köhne, Traub, and Vygen [2020] constructed a family of unweighted instances for which the integrality ratio of (3.2) is at least 2. We use this family of examples (with minor simplification) to prove Theorem 3.18.

Theorem 3.18. *The integrality ratio of* (3.2) *is at least 2, even when restricted to unweighted digraphs.*

Proof. Let $l \geq 4$ be an even integer. We will construct a sequence G'_i ($i \in \mathbb{Z}_{\geq 0}$) of directed graphs such that the integrality ratio of the Asymmetric TSP instances with graph G'_i and cost $c(e) = 1$ for all edges e converges to $\frac{2l}{l+1}$ for $i \to \infty$.

Let G_0 be a bidirected path with $l+2$ vertices and $2(l+1)$ edges. Let $v_0 = v'_0$ be one endpoint of this path, and let $w_0 = w'_0$ be the other endpoint.

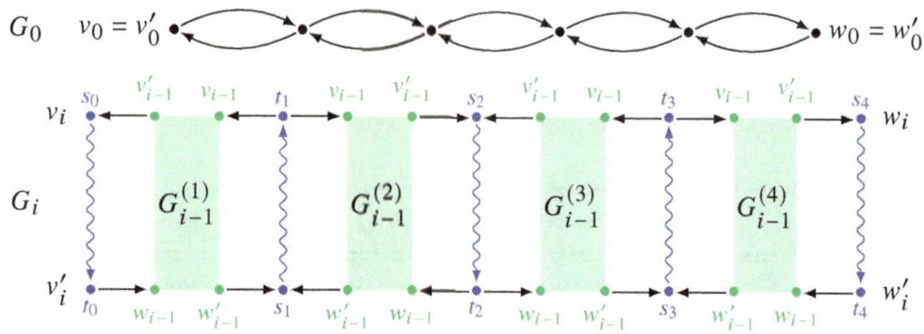

Figure 3.4 Constructing a family of digraphs with integrality ratio arbitrarily close to 2 for Asymmetric TSP with unit weights. For a fixed even integer $l \geq 4$, we define digraphs $G_0, G_1, \ldots$. The digraph G_0 consists of a bidirected path of length $l+1$. Then we construct G_i from G_{i-1} as in the picture. The picture shows the construction for $l = 4$; in general, there are l copies of the graph G_{i-1} (shown in green). The blue wiggly paths indicate the paths $P_i^{(j)}$, each with l^i vertices. Let G'_i be the graph arising from G_i by identifying the blue v_i-v'_i-path with the blue w_i-w'_i-path. Then for $i \to \infty$, the integrality ratio of G'_i converges to $\frac{2l}{l+1}$. This picture is taken from Köhne, Traub, and Vygen [2020] (with permission from Springer Nature).

For $i \geq 1$, let G_i result from G_{i-1} as follows (see Figure 3.4 for an illustration). Let $G_{i-1}^{(1)}, \ldots, G_{i-1}^{(l)}$ be l copies of the graph G_{i-1}. Moreover, let $P_i^{(0)}, \ldots, P_i^{(l)}$ be directed paths with l^i vertices and $l^i - 1$ edges. Denote the first vertex of $P_i^{(j)}$ by s_j and the last vertex by t_j. For every odd number j with $1 \leq j \leq l$, we add an edge from the copy of v'_{i-1} in $G_{i-1}^{(j)}$ to s_{j-1}, an edge from t_j to the copy of v_{i-1} in $G_{i-1}^{(j)}$, an edge from t_{j-1} to the copy of w_{i-1} in $G_{i-1}^{(j)}$, and an edge from the copy of w'_{i-1} in $G_{i-1}^{(j)}$ to s_j. For every even number j with $1 \leq j \leq l$, we add an edge from the copy of v'_{i-1} in $G_{i-1}^{(j)}$ to s_j, an edge from t_{j-1} to the copy of v_{i-1} in $G_{i-1}^{(j)}$, an edge from t_j to the copy of w_{i-1} in $G_{i-1}^{(j)}$, and an edge from the copy of w'_{i-1} in $G_{i-1}^{(j)}$ to s_{j-1}. We define $v_i := s_0$, $v'_i := t_0$, $w_i := s_l$, and $w'_i := t_l$.

Finally, the graph G'_i arises from G_i by identifying the path $P_i^{(0)}$ with the path $P_i^{(l)}$. See Figure 3.6.

Let n_i denote the number of vertices of G_i. We have $n_0 = l + 2$ and $n_i = (l+1)l^i + l n_{i-1}$ for $i \geq 1$. By induction, this yields $n_i = (i+1)l^{i+1} + (i+2)l^i$ for all $i \geq 0$. We conclude that G'_i has $n_i - l^i = (i+1)(l^{i+1} + l^i)$ vertices.

Suppose G_i appears as a subgraph of an instance G'_j for some $j > i$, and consider a tour in G'_j, which can be viewed as a closed walk that visits all vertices. Then the vertices of (this copy of) G_i must be visited by a set of walks,

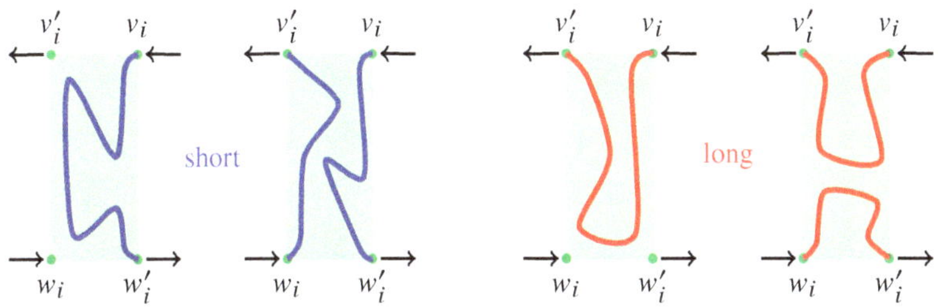

Figure 3.5 Examples of short and long visits of a subgraph G_i.

each starting in v_i or w_i and ending in v'_i or w'_i, such that each vertex is visited by at least one of these walks. We call such a set of walks a *visit* of G_i. We call a visit *short* if it contains a walk from v_i to w'_i or from w_i to v'_i; otherwise, we call it *long*. See Figure 3.5. We denote by S_i the minimum number of edges in a short visit of G_i, and we denote by L_i the minimum total number of edges in a long visit of G_i. We claim:

$$S_i \geq (2i+1)l^{i+1} + l^i \quad \text{and} \quad L_i \geq (2i+2)l^{i+1} \quad \text{for all } i \geq 0. \tag{3.7}$$

This is obvious for $i = 0$. For $i \geq 1$, we distinguish between short and long visits.

First, a short visit of G_i visits all copies of G_{i-1} and traverses each of the blue wiggly paths $P_i^{(j)}$ at least once. For each short visit of a copy of G_{i-1}, we traverse one of these paths once more. Hence the total length is at least $(l+1)l^i + l\min\{L_{i-1}, S_{i-1} + l^i\}$. By the induction hypothesis, this is at least $(2i+1)l^{i+1} + l^i$.

A long visit of G_i consists of walks from v_i to v'_i and/or from w_i to w'_i. The turning point of such a walk (farthest away from start and end) is either a visit of a copy of G_{i-1} or a path $P_i^{(j)}$. Now our long visit of G_i traverses all of the blue wiggly paths $P_i^{(j)}$ at least twice, except one at the turning point of a walk. Moreover, for each short visit of a copy of G_{i-1}, except one at the turning point of a walk, it traverses one of these paths once more. Hence, the total length is at least $(2l+1)l^i + S_{i-1} + (l-1)\min\{L_{i-1}, S_{i-1} + l^i\}$. By the induction hypothesis, this is at least $(2i+2)l^{i+1} + l^{i-1}$.

Having proved (3.7), we can now bound from below the number of edges in a tour in G'_i. We claim:

$$\text{Every tour in } G'_i \text{ has at least } (2i+1)l^{i+1} \text{ edges.} \tag{3.8}$$

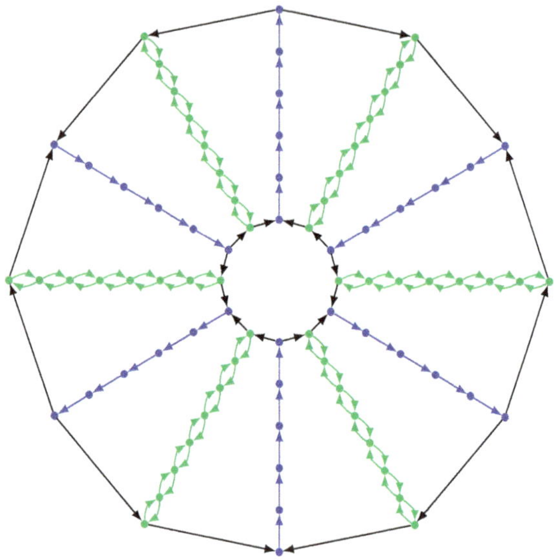

Figure 3.6 The graph G'_1 for $l = 6$. An optimum LP solution has value 1 on the blue edges and value $\frac{1}{2}$ on all other edges, and hence we have LP $= |V(G'_1)|$. This picture is taken from Köhne, Traub, and Vygen [2020] (with permission from Springer Nature).

Let T be a shortest tour in G'_i, which we again view as a closed walk. If T traverses the blue paths of G'_i (cf. Figure 3.6) at least twice on average, then T has at least $2l \cdot l^i + l \cdot S_{i-1} > (2i+1)l^{i+1}$ edges as claimed.

Otherwise, we say that the tour T crosses a blue path P_i if we enter and leave via clockwise edges (cf. Figure 3.6) or if we enter and leave via anti-clockwise edges. Now there is a blue path traversed only once. Either this path is crossed (then call it P_i^*) or every other path is crossed exactly once in clockwise and exactly once in anti-clockwise direction (at least once because the tour visits all vertices, and at most once because we traverse the paths at most twice on average); in the latter case, we choose an arbitrary one of these paths and call it P_i^*.

Without loss of generality, this P_i^* resulted from identifying the left blue path $P_i^{(0)}$ and the right blue path $P_i^{(l)}$ in G_i (cf. Figure 3.4). We map the tour T to a set of walks in G_i in the natural way, mapping a traversal of P_i^* to the left or right blue path according to which side we enter from.

If T crosses P_i^* in both directions, we get an Eulerian edge set (the union of the footprints of two closed walks); otherwise, we get (the footprint of) one walk from v_i' to w_i' or from w_i' to v_i'. In the first case, we add another copy of the left and another copy of the right blue path to obtain a long visit of G_i. In the second case, we add another copy of the left or the right blue path to obtain a short visit of G_i.

We conclude that any tour in G_i' has length at least $\min\{S_i - l^i, L_i - 2l^i\}$, which is at least $(2i+1)l^{i+1}$ by (3.7). This concludes the proof of (3.8).

Finally, we compute the ratio of an optimum tour to the LP value in the digraph $G_i' = (V, E)$. The LP value is equal to the number of vertices: Putting $x_e = 1$ for all edges of the blue paths (on all levels) and $x_e = \frac{1}{2}$ for all other edges constitutes a feasible solution to (3.2), and the value of this solution is $x(E) = \frac{1}{2}\sum_{v \in V} x(\delta(v)) = |V| = (i+1)(l^{i+1} + l^i)$. This shows that the integrality ratio is at least $\frac{(2i+1)l^{i+1}}{(i+1)(l^{i+1}+l^i)} = \frac{(2i+1)l}{(i+1)(l+1)}$. Since we can choose l and i arbitrarily large, the integrality ratio is at least 2. □

Open Problem 3.19. Prove or disprove that the integrality ratio of the LP relaxation (3.2) of the ASYMMETRIC TSP is 2.

Exercises

3.1 Formulate and prove a statement analogous to Corollary 3.6 for the SYMMETRIC TSP WITH TRIANGLE INEQUALITY.

3.2 Let F be the set of feasible solutions of (3.6) that satisfy $\sum_{C \in \mathscr{C}} \lambda_C = 1$. Note that if F is nonempty, it is a face of the polytope of feasible solutions. For $\lambda \in F$, let $x_e^\lambda := \sum_{C \in \mathscr{C}: e \in C} \lambda_C$ for $e \in E$. Show that $\{x^\lambda : \lambda \in F\}$ is the convex hull of incidence vectors of directed Hamiltonian circuits in G.

3.3 Show that for any instance (V, c) of the ASYMMETRIC TSP WITH TRIANGLE INEQUALITY, the value of the LP (3.6) does not change when we add the constraints $\sum_{C \in \mathscr{C}: C \cap \delta^+(v) \neq \emptyset} \lambda_C = 1$ for all $v \in V$.

3.4 Show that finding a minimum-cost perfect matching in a bipartite graph $G = (V, E)$ with weights $c : E \to \mathbb{R}_{\geq 0}$ can be reduced to a circulation problem and thus solved in polynomial time via Theorem 3.11.
Note: This indicates that bipartite matching is easier than general matching. Bipartite matching has many applications. For instance, we used bipartite matching in the proof of Lemma 1.34.

3.5 An *arborescence* (rooted at r) is a digraph whose underlying undirected graph is a tree and such that every vertex except r has in-degree 1. Let $G = (V, E)$ be a digraph with $r \in V$ and edge weights $c : E \to \mathbb{R}_{\geq 0}$. Show that the minimum weight of an arborescence rooted at r in G equals $\max\{\sum_{\emptyset \neq U \subseteq V \setminus \{r\}} y_U : \sum_{U: e \in \delta^-(U)} y_U \leq c(e),\ y \geq 0\}$. Show that if c is

integral, there is an integral vector y attaining the maximum.
Hint: Consider a counterexample (G, r, c) with c integral and $c(E)$ minimum, and let U be the vertex set of a strongly connected component of $(V, \{e \in E : c(e) = 0\})$ that does not contain r and for which $c(e) > 0$ for all $e \in \delta^-(U)$.
Note: This proof can be turned into a polynomial-time algorithm.
(Edmonds [1967a], Bock [1971], Fulkerson [1974])

3.6 For a digraph $G = (V, E)$ with nonnegative edge weights $c : E \to \mathbb{R}_{\geq 0}$, we ask for a minimum-weight strongly connected spanning subgraph. Describe a 2-approximation algorithm for this problem.
Hint: An *anti-arborescence* results from an arborescence by reversing every arc (so every vertex except r has out-degree 1). Compute an arborescence and an anti-arborescence (cf. Exercise 3.5).

3.7 For a digraph $G = (V, E)$ with nonnegative edge weights $c : E \to \mathbb{R}_{\geq 0}$, consider the linear program

$$\min \left\{ c(x) : x(\delta^-(U)) \geq 1 \ (\emptyset \neq U \subsetneq V),\ x \geq 0 \right\},$$

whose integral solutions are the incidence vectors of strongly connected spanning subgraphs. Show that the integrality ratio of this family of LPs is at most 2.
Hint: Use Exercise 3.5 to prove that the algorithm from Exercise 3.6 produces a solution that costs at most twice the LP value.
Note: No better upper bound is known. The integrality ratio is at least $\frac{3}{2}$ (see Exercise 3.12).

3.8 A *feasible potential* in a digraph $G = (V, E)$ with edge weights $c : E \to \mathbb{R}$ is a function $\pi : V \to \mathbb{R}$ such that $c(e) + \pi(v) - \pi(w) \geq 0$ for all $e = (v, w) \in E$. Show that there exists a feasible potential if and only if c is conservative (i.e., there is no circuit with negative total weight).
Hint: Add a vertex r and edges (r, v) for all $v \in V$, and let $\pi(v)$ be the minimum weight of an r-v-path.

3.9 Use Exercise 3.8 to give an alternative proof of Theorem 3.15: let $\rho > 1$ be such that $\rho x^* \notin Q$ for a feasible solution x^* of (3.2). Then there exists a vector c with $c^\top(\rho x^*) < \min\{c^\top q : q \in Q\}$. Note that c is conservative because otherwise the right-hand side would be $-\infty$. Let π be a feasible potential (cf. Exercise 3.8) and $c_\pi(e) := c(e) + \pi(v) - \pi(w)$ for $e = (v, w) \in E$. Conclude that $\rho(c_\pi^\top x^*) = \rho(c^\top x^*) < \inf\{c^\top q : q \in Q\} = \min\{c_\pi(F) : F \text{ is a tour}\}$.

3.10 Prove that the convex hull of incidence vectors of tours in a strongly connected digraph is a polyhedron.

3.11 Show that the digraphs G'_i constructed in the proof of Theorem 3.18 have a tour of length at most twice the number of vertices.

3.12 Prove that the family of linear programs from Exercise 3.7 has integrality ratio at least $\frac{3}{2}$.
Hint: Consider the graph G'_i from the proof of Theorem 3.18 but with different edge weights: In each copy of G_j (for $0 \leq j \leq i$), the blue wiggly paths in Figure 3.4 have weight 0, while the horizontal edges in Figure 3.4 have weight l^j. Call a subset F of edges of G_j a *cover* if $\big(V(G_i), F \cup \{(w'_i, v_i), (v'_i, w_i)\}\big)$ is strongly connected, and call a cover *long* if $(V(G_i), F)$ is strongly connected. Show by induction that a cover has at least $(3i+1)l^{i+1}$ edges and a long cover has at least $(3i+2)l^{i+1}$ edges.
(Laekhanukit, Oveis Gharan, and Singh [2012])

4

Duality, Cuts, and Uncrossing

While many exact and approximation algorithms work with a linear programming formulation (often a relaxation), the dual LP often plays a key role in the algorithms and their analysis. In this chapter, we analyze the structure of optimum dual solutions for the classical LP relaxations of the TSP but also for T-joins, and we deduce properties like laminarity. We obtain optimum primal and dual solutions with linear-size support. Since the primal constraints and dual variables correspond to cuts, enumerating all cuts with a small value is a useful tool in several algorithms.

4.1 LP Duality

For any LP

$$\min\{c^\top x : Ax \geq b,\ x \geq 0\}, \tag{4.1}$$

its *dual* is defined to be the LP

$$\max\{b^\top y : A^\top y \leq c,\ y \geq 0\}. \tag{4.2}$$

In this context, (4.1) is called the *primal LP*. Note that primal constraints correspond to dual variables and vice versa. For any pair (x, y) of primal and dual feasible solutions, we have

$$c^\top x \geq (A^\top y)^\top x = y^\top Ax \geq y^\top b = b^\top y, \tag{4.3}$$

which is called weak duality, and hence $b^\top y$ provides a lower bound on the primal LP value. The duality theorem of linear programming tells us that the two LP values are equal:

Theorem 4.1 (von Neumann [1947], Gale, Kuhn, and Tucker [1951]). *If* (4.1) *and* (4.2) *have feasible solutions, then the two LP values are the same.*

Thus, a pair of feasible solutions x and y to (4.1) and (4.2) are optimum solutions if and only if equality holds in both inequalities of (4.3); this property is known as *complementary slackness*.

Corollary 4.2. *Let x and y be feasible solutions to* (4.1) *and* (4.2), *respectively. Then x and y are optimum solutions if and only if $x^\top(c - A^\top y) = 0$ and $y^\top(Ax - b) = 0$.*

Proof. This follows immediately from (4.3) and Theorem 4.1. □

For example, the dual of (2.12) is

$$\begin{array}{lrcll} \max \displaystyle\sum_{\emptyset \neq U \subsetneq V} 2\, y_U & & & & \\ \text{subject to} \quad \displaystyle\sum_{U: e \in \delta(U)} y_U & \leq & c(e) & (e \in E) & \\ y_U & \geq & 0 & (\emptyset \neq U \subsetneq V). & \end{array} \tag{4.4}$$

For optimum primal and dual solutions x and y, respectively, complementary slackness requires that $\sum_{U:e\in\delta(U)} y_U = c(e)$ whenever $x_e > 0$, and $x(\delta(U)) = 2$ whenever $y_U > 0$.

Some of our LPs have equality constraints. These can of course be described by two inequalities and thus brought into the above form. It is however easier to allow equations, then dual variables corresponding to equality constraints are not restricted to be nonnegative. For example, the dual of (3.2) is

$$\begin{array}{lrcll} \max \displaystyle\sum_{\emptyset \neq U \subsetneq V} 2\, y_U & & & & \\ \text{subject to} \quad a_w - a_v + \displaystyle\sum_{U: e \in \delta(U)} y_U & \leq & c(e) & (e = (v, w) \in E) & \\ y_U & \geq & 0 & (\emptyset \neq U \subsetneq V). & \end{array} \tag{4.5}$$

Here, the variables a_v ($v \in V$) are not restricted to be nonnegative.

The dual LPs (4.4) and (4.5) have exponentially many variables. However, there is always an optimum solution with a linear number of nonzero variables, and such a solution can be found in polynomial time. We first show:

Proposition 4.3. *Given an instance of* Symmetric TSP *or* Asymmetric TSP, *the LP* (4.4) *or* (4.5), *respectively, can be solved in polynomial time.*

Proof. We first solve the primal LP (2.12) or (3.2) by the ellipsoid method (Theorem 2.10) using a separation oracle for the exponentially many cut constraints (cf. Corollary 2.9). While doing so, we record the sets U that the separation oracle returned, producing a constraint $x(\delta(U)) \geq 2$. We set all

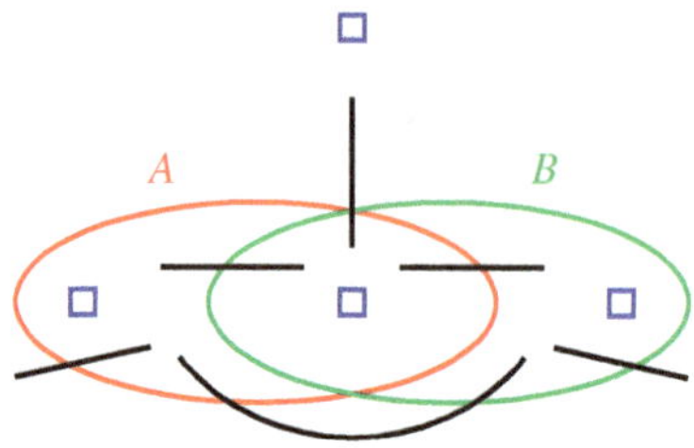

Figure 4.1 Two sets $A, B \subsetneq V$ such that $A \setminus B$, $B \setminus A$, $A \cap B$, and $V \setminus (A \cup B)$ are all nonempty. The black lines indicate the six relevant types of edges.

y-variables of other sets to zero in the dual LP. Then the primal solution is also optimal with respect to this subset of constraints, and hence, by strong duality (Theorem 4.1), restricting the dual LP to these variables does not change the optimum value. But now the restricted dual LP has a polynomial number of variables and constraints, so we can solve it in polynomial time (using Theorem 2.2). □

We will give an alternative proof at the end of Section 4.4.

4.2 Uncrossing

The goal of this section is to obtain a dual solution with a nice structure. The following inequality is fundamental:

Proposition 4.4. *For any undirected graph $G = (V, E)$ and $x \in \mathbb{R}^E_{\geq 0}$ and any $A, B \subseteq V$, we have*

$$x(\delta(A \cap B)) + x(\delta(A \cup B)) \leq x(\delta(A)) + x(\delta(B)). \tag{4.6}$$

Proof. Let first x be a unit vector, so there is an edge $e \in E$ with $x_e = 1$ and $x_{e'} = 0$ for all $e' \in E \setminus \{e\}$. Then $x(\delta(A \cap B)) + x(\delta(A \cup B)) = |\{e\} \cap \delta(A \cap B)| + |\{e\} \cap \delta(A \cup B)| \leq |\{e\} \cap \delta(A)| + |\{e\} \cap \delta(B)| = x(\delta(A)) + x(\delta(B))$; the six relevant cases are illustrated by Figure 4.1. Since every nonnegative vector is a nonnegative combination of unit vectors, the assertion follows by taking a weighted sum of these inequalities. □

We also get $x(\delta(A \setminus B)) + x(\delta(B \setminus A)) \leq x(\delta(A)) + x(\delta(B))$ by applying (4.6) to A and $V \setminus B$.

By complementary slackness, all variables y_U in every optimum solution to the dual LP (4.4) or (4.5) are zero unless $x(\delta(U)) = 2$ for every optimum primal solution x. Given a primal LP solution x, we call a cut $\delta(U)$ *tight* if $x(\delta(U)) = 2$. The following property is very useful:

Proposition 4.5. *Let x be a feasible solution to* (2.12) *or* (3.2) *and $A, B \subsetneq V$ two sets such that $A \setminus B$, $B \setminus A$, $A \cap B$, and $V \setminus (A \cup B)$ are all nonempty. If $\delta(A)$ and $\delta(B)$ are tight cuts, then so are $\delta(A \cup B)$, $\delta(A \cap B)$, $\delta(A \setminus B)$, and $\delta(B \setminus A)$.*

Proof. Using Proposition 4.4 and $x(\delta(U)) \geq 2$ for all $\emptyset \neq U \subsetneq V$, we have $2+2 \leq x(\delta(A \cap B)) + x(\delta(A \cup B)) \leq x(\delta(A)) + x(\delta(B)) = 2+2$, which implies that $\delta(A \cup B)$ and $\delta(A \cap B)$ are tight. To obtain that $\delta(A \setminus B)$ and $\delta(B \setminus A)$ are tight, apply this to A and $V \setminus B$. □

To show that there is always an optimum dual solution with linear-size support, we use an extremely useful technique in combinatorial optimization: uncrossing.

Definition 4.6 (cross-free, laminar)**.** Let V be a finite set and $\mathcal{U}$ a family of subsets of V. Two sets $A, B \in \mathcal{U}$ are *crossing* if the four sets $A \setminus B$, $B \setminus A$, $A \cap B$, and $V \setminus (A \cup B)$ are all nonempty. The family $\mathcal{U}$ is called *cross-free* if no two of its sets are crossing. The family $\mathcal{U}$ is called *laminar* if for any two of its sets that are not disjoint, one is the subset of the other.

The following is immediate from the definition:

Proposition 4.7. *Let V be a finite set and $\mathcal{U}$ a family of subsets of V. If $\mathcal{U}$ is laminar, then $\mathcal{U} \cup \{V \setminus U : U \in \mathcal{U}\}$ is cross-free. If $\mathcal{U}$ is cross-free and $v \in V$, then $\{U \in \mathcal{U} : v \notin U\} \cup \{V \setminus U : U \in \mathcal{U}, v \in U\}$ is laminar.* □

Proposition 4.8. *Let V be a finite set and $\mathcal{L}$ a laminar family of subsets of V. Then $|\mathcal{L}| \leq 2|V|$. If $\emptyset, V \notin \mathcal{L}$, then $|\mathcal{L}| \leq 2|V| - 2$.*

Proof. Suppose $\mathcal{L}$ is a family of laminar subsets of V and $\emptyset, V \notin \mathcal{L}$; it suffices to show $|\mathcal{L}| \leq 2|V| - 2$. This is trivial for $|V| = 1$. Otherwise, let A be a maximal set in $\mathcal{L}$. Then $\mathcal{L}$ consists of A, possibly $V \setminus A$, a laminar family of nonempty proper subsets of A, and a laminar family of nonempty proper subsets of $V \setminus A$. By induction, these families contain at most $2|A| - 2$ and $2|V \setminus A| - 2$ sets, respectively. We obtain $|\mathcal{L}| \leq 2 + (2|A| - 2) + (2|V \setminus A| - 2) = 2|V| - 2$. □

By Proposition 4.7, this also shows that a cross-free family of subsets of V has at most $4|V| - 4$ elements. These bounds are tight (see Exercise 4.8).

Now we return to the dual LPs (4.4) and (4.5) and show that there is always an optimum solution that has laminar support. Let y be an arbitrary optimum solution to (4.4), or let (a, y) be an arbitrary optimum solution to (4.5). First, we can choose an arbitrary vertex v and eliminate the variables y_U for all sets

U that contain v, by adding y_U to $y_{V\setminus U}$ for such sets. This does not change anything since $\delta(U) = \delta(V \setminus U)$.

Next, we will apply uncrossing until the support of y is laminar. This means that we take two crossing sets A and B, set $\varepsilon := \min\{y_A, y_B\}$, reduce y_A and y_B by ε, and increase $y_{A\cap B}$ and $y_{A\cup B}$ by ε. This keeps $\sum_{\emptyset\neq U\subsetneq V} 2y_U$ constant and maintains dual feasibility. To see this for (4.5), note that for any edge $e = (v, w) \in E$, we have $|\{e\} \cap \delta(A \cap B)| + |\{e\} \cap \delta(A \cup B)| \leq |\{e\} \cap \delta(A)| + |\{e\} \cap \delta(B)|$ (this is Proposition 4.4 applied to the unit vector corresponding to e), therefore the left-hand side of the inequality $a_w - a_v + \sum_{U:e\in\delta(U)} y_U \leq c(e)$ does not increase. The analogous argument applies to (4.4).

By the same argument (applied to A and $V \setminus B$), we could also do uncrossing by increasing $y_{A\setminus B}$ and $y_{B\setminus A}$ instead of $y_{A\cap B}$ and $y_{A\cup B}$.

However, it can always happen that the new sets cross other existing sets. It is nontrivial to show that this procedure can be carried out so that it terminates after a polynomial number of steps. We will show this in a slightly more general context so that it can also be used for other LPs with a similar structure. To this end, we define:

Definition 4.9 (uncrossable)**.** Let V be a finite set, $v \in V$, and $\mathcal{U}$ a family of subsets of $V \setminus \{v\}$. We call $\mathcal{U}$ *uncrossable* if for all crossing sets $A, B \in \mathcal{U}$:

- $A \cap B \in \mathcal{U}$ or $A \setminus B \in \mathcal{U}$;
- if $A \cap B \in \mathcal{U}$, then $A \cup B \in \mathcal{U}$; and
- if $A \setminus B \in \mathcal{U}$, then $B \setminus A \in \mathcal{U}$.

In our applications, $\mathcal{U}$ will contain the nonempty subsets of $V \setminus \{v\}$ that correspond to dual variables (all these sets for (4.4) and (4.5)). Following Hurkens et al. [1988], we can view the task of uncrossing (or tidying up) a set family as a game:

Definition 4.10 (uncrossing game)**.** The *uncrossing game* is played by two players, Parent and Child, on an uncrossable family $\mathcal{U}$ of subsets of V and an initial subfamily $\mathcal{F} \subseteq \mathcal{U}$.

In each round, Parent chooses two sets $A, B \in \mathcal{F}$. Moreover, Parent chooses $C = A \cap B$ and $D = A \cup B$ or $C = A \setminus B$ and $D = B \setminus A$, such that $C, D \in \mathcal{U}$. Parent removes A and B from $\mathcal{F}$ and adds C and D to $\mathcal{F}$ (if they are not already in $\mathcal{F}$). Then Child chooses one of A and B and puts it back into $\mathcal{F}$.

The game ends (and Parent wins) as soon as the family $\mathcal{F}$ is laminar at the end of some round.

Hurkens et al. [1988] noted that we need to be careful if we want Parent to win this game. This is illustrated by the following example. Let $V = \{0, 1, 2, 3, 4\}$

and $\mathcal{U} = \{U \subseteq V : 0 \in U, 4 \notin U\}$; consider the initial family

$$\mathcal{F} = \{\{0\}, \{0,1\}, \{0,1,3\}, \{0,2,3\}, \{0,1,2,3\}\}.$$

If we uncross $\{0,1,3\}$ and $\{0,2,3\}$ first, we get the intersection $\{0,3\}$ (the union was already there), and Child might put back $\{0,2,3\}$. Then we have the family

$$\mathcal{F} = \{\{0\}, \{0,1\}, \{0,3\}, \{0,2,3\}, \{0,1,2,3\}\}.$$

If we next uncross $\{0,1\}$ and $\{0,3\}$, we get the union $\{0,1,3\}$ (the intersection was already there), and Child might put back $\{0,1\}$. We get

$$\mathcal{F} = \{\{0\}, \{0,1\}, \{0,1,3\}, \{0,2,3\}, \{0,1,2,3\}\},$$

which was exactly our starting point!

Karzanov [1996], generalizing the work by Hurkens et al. [1988], showed that Parent can guarantee the game to end in polynomial time. Hirai [2016] generalized this result further and improved the running time. The following is a simplified version in a less general setting, which is sufficient for our purposes and for which we obtain an even better running time:

Theorem 4.11. *There is a strategy for Parent to make the uncrossing game end after $4n^2p$ rounds, no matter how Child plays, where $n = |V|$ and p is the initial number of elements of $\mathcal{F}$.*

Proof. To perform their strategy, Parent maintains a coloring of all sets in $\mathcal{F}$. Initially, all sets are red. In the process, some sets will be colored blue and green. The game will end once all sets are green. Throughout this process, we maintain the following invariants:

- The green sets form a laminar family.
- The blue sets form a laminar family.
- For every blue set B, there is a subset S_B such that for every green set A that crosses B, we have $A \cap B \in \{S_B, B \setminus S_B\}$.

By Proposition 4.8, this means in particular that there are less than $2n$ blue and less than $2n$ green sets at any time. Now, Parent plays according to the following strategy:

(i) If there is no blue set but a red set, we color all green sets blue and color an arbitrary red set green. Repeat this until there is a blue set.
(ii) Then take a maximal blue set B^*. Apply Algorithm 4.12 to B^* (see Figure 4.2 for an illustration).

Algorithm 4.12: Uncrossing Algorithm

(1) **while** $B^* \in \mathcal{F}$ and B^* crosses at least one green set **do**
(2) Let $S^* \in \{S_{B^*}, B^* \setminus S_{B^*}\} \cap \mathcal{U}$.
(3) **if** there is a green set A^* that crosses B^* and $A^* \cap B^* = B^* \setminus S^*$, **then**
(4) take the minimal such set A^* and uncross A^* and B^*, replacing one of them by $B^* \setminus A^*$ and $A^* \setminus B^*$,
(5) **else**
(6) take the maximal green set A^* that crosses B^* and uncross A^* and B^*, replacing one of them by $A^* \cap B^*$ and $A^* \cup B^*$.
(7) If new sets arise, color them green.
(8) Finally, if B^* is still in $\mathcal{F}$, color B^* green.

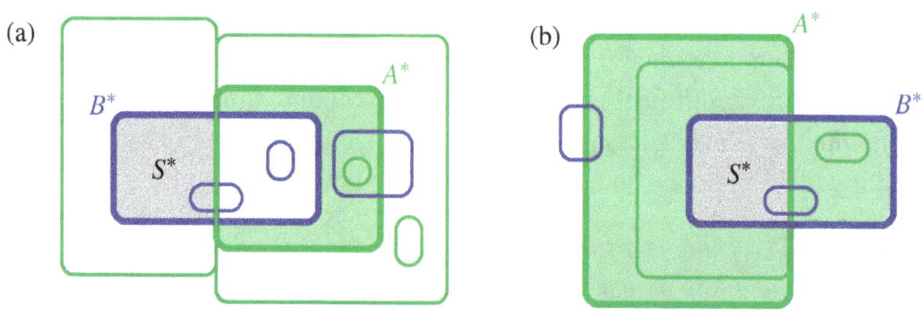

Figure 4.2 Illustration of the uncrossing procedure in the proof of Theorem 4.11. Take a maximal blue set B^*. The gray subset S^* belongs to $\mathcal{U}$. For every green set that crosses B^*, the intersection is either S^* or $B^* \setminus S^*$. (a): In Step (4) of Algorithm 4.12, we take the minimal green set A^* that crosses B^* and with $B^* \setminus A^* = S^*$. We uncross A^* and B^* and obtain $A^* \setminus B^*$ and $B^* \setminus A^*$ (the filled sets). (b): If there is no such green set, we take the maximal green set A^* that crosses B^* in Step (6). We uncross A^* and B^* and obtain $A^* \cup B^*$ and $A^* \cap B^*$. New sets will be colored green.

Step (2) of Algorithm 4.12 can be performed due to the third invariant and since $\mathcal{U}$ is uncrossable. Step (4) is feasible (i.e., $A^* \setminus B^*, B^* \setminus A^* \in \mathcal{U}$), since $B^* \setminus A^* = S^* \in \mathcal{U}$ and $\mathcal{U}$ is uncrossable. Similarly, Step (6) is feasible (i.e., $A^* \cap B^*, A^* \cup B^* \in \mathcal{U}$) because $A^* \cap B^* = S^* \in \mathcal{U}$.

It is obvious that one operation (i) or (ii) can be applied until all sets are green. It is also obvious that operation (i) maintains the invariants. It remains to show that operation (ii) maintains the invariants. Then, since every application of (ii) consists of less than $2n$ uncrossing steps (at most one with each green set that exists at the beginning of (ii)) and reduces the number of blue sets, after at most

$2n$ consecutive applications of (ii), and thus, in less than $4n^2$ uncrossing steps, we have no blue sets anymore. Then we will next apply (i), reducing the number of red sets. Hence, the procedure terminates after less than $4n^2p$ uncrossing steps.

Since no blue sets are added in (ii), the second invariant is obviously maintained. Moreover, Step (8) obviously maintains the invariants. Consider a pair of sets A^* and B^* that we uncross in (4) or (6), and let S^* be the set chosen in (2). We show that this uncrossing step (including (7)) maintains the first and third invariant.

To show that the first invariant is maintained, let A be a green set. If A crosses a new green set, it must cross A^* or B^*. However, A does not cross A^* because the green family was laminar. So A crosses B^*, which by the third invariant implies $A \cap B^* \in \{S^*, B^* \setminus S^*\}$.

If $A \cap B^* = B^* \setminus S^*$, then A^* was chosen in (4), and A is disjoint from $B^* \setminus A^*$ and a superset of A^* (and hence also of $A^* \setminus B^*$) by the minimal choice of A^* in (4).

If $A \cap B^* = S^* \neq A$, then there are two cases: If A^* was chosen in (4), A is disjoint from A^* (and hence also from $A^* \setminus B^*$) and contains $B^* \setminus A^*$; if A^* was chosen in (6), A is a subset of A^* (and hence also of $A^* \cup B^*$) and a superset of $A^* \cap B^*$. In all cases, we see that every new green set is disjoint from A or a subset or superset of A, so the first invariant is maintained.

We finally show that the third invariant is maintained. Let B be a blue set. Due to the maximal choice of B^*, B is not a proper superset of B^*. So it is a subset or it is disjoint.

First, assume that B is a subset of B^*. Then $A^* \setminus B^*$ and $A^* \cup B^*$ do not cross B, and $(B^* \setminus A^*) \cap B = B \setminus A^*$ and $(A^* \cap B^*) \cap B = A^* \cap B$, so the third invariant is also preserved.

Finally, consider the case when B is disjoint from B^*. Then $B^* \setminus A^*$ and $A^* \cap B^*$ do not cross B, and $(A^* \setminus B^*) \cap B = (A^* \cup B^*) \cap B = A^* \cap B$, so the third invariant is also preserved. □

Our main application is to obtain dual solutions with laminar support:

Lemma 4.13. *For every instance (G, c) of the* Asymmetric *TSP, there exists an optimum solution (a, y) to* (4.5) *such that $\{U : y_U > 0\}$ is laminar. Such a solution can be found in polynomial time.*

Proof. We can first find any optimum dual solution (a, y) in polynomial time by Proposition 4.3. Choose an arbitrary $v \in V$, and for all $U \subseteq V \setminus \{v\}$, set $y_U := y_U + y_{V \setminus U}$ and $y_{V \setminus U} := 0$. Then we apply uncrossing: While the support of y contains crossing sets A and B, we *uncross* A and B – that is, we set

$\varepsilon := \min\{y_A, y_B\}$, reduce y_A and y_B by ε, and increase $y_{A\cap B}$ and $y_{A\cup B}$ by ε. This maintains dual feasibility by Proposition 4.4.

By Theorem 4.11 (applied to the uncrossable family $\mathcal{U} = 2^{V\setminus\{v\}} \setminus \emptyset$), these uncrossing operations can be performed in an order so that, in polynomial time, we end up with a laminar family. □

Essentially the same proof works for the SYMMETRIC TSP:

Lemma 4.14. *For every instance (G, c) of the SYMMETRIC TSP, there exists an optimum solution y to (4.4) such that $\{U : y_U > 0\}$ is laminar. Such a solution can be found in polynomial time.* □

We will give an alternative proof of these two results at the end of Section 4.4.

Let us mention another application of the uncrossing game. A similar linear program to (4.4) is the dual of (2.9) (optimizing over the T-join polyhedron for $T \subseteq V$ with $|T|$ even):

$$\begin{array}{lrcll} \max & \displaystyle\sum_{U\subseteq V:|U\cap T| \text{ odd}} y_U & & & \\ \text{subject to} & \displaystyle\sum_{\substack{U\subseteq V:|U\cap T| \text{ odd},\\ e\in\delta(U)}} y_U & \le & c(e) & (e \in E) \\ & y_U & \ge & 0 & (U \subseteq V : |U \cap T| \text{ odd}). \end{array} \tag{4.7}$$

Lemma 4.15. *For every undirected graph $G = (V, E)$, every $T \subseteq V$ with $|T|$ even, and $c : E \to \mathbb{R}_{\ge 0}$, there exists an optimum solution y to (4.7) such that $\{U : y_U > 0\}$ is laminar. Such a solution can be found in polynomial time.*

Proof. As in the proof of Proposition 4.3, we first solve the primal LP (2.9). By Theorem 2.19, the incidence vector of a minimum-cost T-join is an optimum solution, and this can be found in $O(n^3)$ time by Theorem 1.29. Nevertheless, we use the ellipsoid method (Theorem 2.10) for solving (2.9), which we can do because the separation problem has a polynomial-time algorithm, again by Theorem 2.10. We record the sets that the separation oracle returns, set all dual variables corresponding to other sets to zero, and solve the restricted dual LP, which now has only a polynomial number of variables and constraints. Let y be the optimum solution to (4.7) that we obtain.

Finally, we do uncrossing similarly to the proof of Lemma 4.13. Again, choose an arbitrary vertex $v \in V$, and for all $U \subseteq V \setminus \{v\}$, set $y_U := y_U + y_{V\setminus U}$ and $y_{V\setminus U} := 0$ (note that $|(V \setminus U) \cap T|$ is odd if and only if $|U \cap T|$ is odd). Now the support of y is a subset of the family $\mathcal{U} = \{U \subseteq V \setminus \{v\} : |U \cap T| \text{ odd}\}$. This family is uncrossable, so whenever A and B cross, we can reduce y_A and y_B by ε; if $|A \cap B \cap T|$ is odd, increase $y_{A\cap B}$ and $y_{A\cup B}$ by ε; and if $|A \cap B \cap T|$ is even, increase $y_{A\setminus B}$ and $y_{B\setminus A}$ by ε. By Theorem 4.11, these uncrossing operations can be performed in an order so that, in polynomial time, we end up with laminar support. □

The separation problem in the proof of Lemma 4.15 asks for a minimum-capacity T-cut; this problem can be reduced to solving $n-1$ maximum flow problems (Padberg and Rao [1982]). However, a laminar fractional T-cut packing as guaranteed by Lemma 4.15 can also be computed directly by a fast combinatorial algorithm (see, e.g., Edmonds and Johnson [1973] or Sebő [1997]). Moreover, if c is integral and $c(C)$ is even for every edge set C of a circuit in G, then y can be chosen integral (cf. Exercise 4.10).

Uncrossing is also useful to describe the tight constraints of a vector in the spanning tree polytope:

Lemma 4.16. *Let $G = (V, E)$ be an undirected graph and x a vector in the spanning tree polytope (2.8). Then there is a laminar family $\mathcal{L}$ of subsets of V such that*

(i) $x(E[L]) = |L| - 1$ *for all $L \in \mathcal{L}$, and*
(ii) *for every $\emptyset \neq U \subsetneq V$ with $x(E[U]) = |U| - 1$, the incidence vector $\chi^{E[U]}$ is a linear combination of $\chi^{E[L]}$ for $L \in \mathcal{L}$ and $\chi^{\{e\}}$ for $e \in E$ with $x_e = 0$.*

Proof. Let $\mathcal{L}$ be a maximal laminar family with (i). Suppose there is a set U violating (ii); choose U so that it crosses as few sets from $\mathcal{L}$ as possible. By the maximality of $\mathcal{L}$, the set U crosses some $L \in \mathcal{L}$. We have

$$\chi^{E[U\cap L]} + \chi^{E[U\cup L]} = \chi^{E[U]} + \chi^{E[L]} + \chi^{\delta(U\setminus L)\cap\delta(L\setminus U)} \tag{4.8}$$

and hence

$$\begin{aligned}(|U|-1)+(|L|-1) &= (|U\cap L|-1)+(|U\cup L|-1)\\ &\geq x(E[U\cap L]) + x(E[U\cap L])\\ &= x(E[U]) + x(E[L]) + x(\delta(U\setminus L)\cap\delta(L\setminus U))\\ &\geq (|U|-1)+(|L|-1)+0.\end{aligned}$$

We have equality and conclude $x(E[U\cap L]) = |U\cap L| - 1$ and $x(E[U\cup L]) = |U\cup L| - 1$ and $x(\delta(U\setminus L)\cap\delta(L\setminus U)) = 0$. If both $U\cap L$ and $U\cup L$ satisfy (ii), then (4.8) and $x(\delta(U\setminus L)\cap\delta(L\setminus U)) = 0$ imply that also U satisfies (ii), which is a contradiction. So, one of $U\cap L$ and $U\cup L$ violates (ii). Since each of these two sets crosses fewer sets of $\mathcal{L}$ than U, this contradicts the choice of U. □

4.3 Extreme Point Solutions

The set of feasible solutions of a linear program is a polyhedron. We will need some basic polyhedral theory. First, consider the set of optimum solutions to a linear program. Recall that the faces of a polyhedron P are the sets $F = \{x \in P : c^\top x = \gamma\}$, where $\gamma = \min\{c^\top x : x \in P\}$, for all $c \in \mathbb{R}^n$ for which the minimum is bounded.

Definition 4.17 (extreme point)**.** Let $P = \{x \in \mathbb{R}^n : Ax \leq b\}$ be a nonempty polyhedron. If a face of P contains only a single point x, then x is called an *extreme point* of P.

The extreme points of P are sometimes also called the *vertices* of P.

Proposition 4.18. *The faces of $P = \{x \in \mathbb{R}^n : Ax \leq b\}$ are the polyhedra $F = \{x \in \mathbb{R}^n : Ax \leq b,\ A'x = b'\}$ for those subsystems $A'x \leq b'$ of $Ax \leq b$ for which F is nonempty.*

Proof. If $F = \{x \in \mathbb{R}^n : Ax \leq b,\ A'x = b'\} \neq \emptyset$ for a subsystem $A'x \leq b'$ of $Ax \leq b$, let c be the sum of the rows of A' and γ the sum of the entries of b'. Then $c^\top x \leq \gamma$ for all $x \in P$ and $F = \{x \in P : c^\top x = \gamma\}$, so F is a face of P.

Conversely, let $F = \{x \in P : c^\top x = \gamma\}$ be a face of P, where $\gamma = \min\{c^\top x : x \in P\}$. Let $A'x \leq b'$ be the maximal subsystem of $Ax \leq b$ for which $A'x = b'$ for all $x \in F$, and let $A''x \leq b''$ consist of the remaining inequalities. Then $F \subseteq \{x \in \mathbb{R}^n : Ax \leq b,\ A'x = b'\}$. To show the other inclusion, let $y \in P$ with $A'y = b'$. For each inequality of $A''x \leq b''$, there is a vector $x \in F$ that satisfies it strictly, so the arithmetic mean x^* of these vectors satisfies all these inequalities strictly (and belongs to F). Hence, for small enough $\varepsilon > 0$, we have $z := x^* + \varepsilon(x^* - y) \in P$. Then $c^\top z \geq \gamma$, implying $c^\top y = \gamma$. □

This also implies that the faces of the faces of a polyhedron P are faces of P. Moreover, we note:

Proposition 4.19. *Every face of an integral polyhedron is integral.*

Proof. By definition, a polyhedron is integral if and only if every face contains an integral vector. By Proposition 4.18, the faces of a face of a polyhedron P are faces of P. □

Extreme points are described by linear equation systems arising from the description of the polyhedron:

Proposition 4.20 (Hoffman and Kruskal [1956])**.** *Let $P = \{x \in \mathbb{R}^n : Ax \leq b\}$ be a polyhedron and x^* be an extreme point of P. Then there exists a subsystem $A'x \leq b'$ of $Ax \leq b$ such that x^* is the unique solution of $A'x^* = b'$.*

Proof. Let $A'x \le b'$ be a maximal subsystem of $Ax \le b$ such that $\{x^*\} = \{x \in P : A'x = b'\}$ (cf. Proposition 4.18), so x^* satisfies all other inequalities strictly. Suppose there is a vector $y \neq x^*$ with $A'y = b'$. Then $z = x^* + \varepsilon(y - x^*) \in P$ for small enough $\varepsilon > 0$. Since $A'z = b'$ and $z \neq x^*$, this is a contradiction. □

Most polyhedra that we study are nonempty subsets of the nonnegative orthant. If we optimize a linear objective function over such a polyhedron and the optimum is bounded, then it is attained at an extreme point. We call this an *optimum extreme point* of the LP. When we solve a linear program with nonnegativity constraints, we may assume that we get an optimum extreme point:

Corollary 4.21. *If a linear program* $\min\{c^\top x : Ax \ge b,\ x \ge 0\}$ *has an optimum solution, then it has an optimum extreme point. If, for a class of LPs in the setting of Theorem 2.10, one can solve the optimization problem in polynomial time, then one can find an optimum extreme point in polynomial time.*

Proof. We first solve the given LP $\min\{c^\top x : Ax \ge b,\ x \ge 0\}$ and let γ be the LP value. Then we solve $\min\{x_1 : Ax \ge b,\ x \ge 0,\ c^\top x = \gamma\}$, where x_1 is the first entry of x, and let ξ_1 be this LP value. Then we solve $\min\{x_2 : Ax \ge b,\ x \ge 0,\ c^\top x = \gamma,\ x_1 = \xi_1\}$ and so on. We always optimize over a face of the original polyhedron, so we end up with an optimum extreme point.

If we are in the setting of Theorem 2.10, we can solve all these LPs in polynomial time: To solve the separation problem for $\min\{x_i : Ax \ge b,\ x \ge 0,\ c^\top x = \gamma,\ x_1 = \xi_1, \dots, x_{i-1} = \xi_{i-1}\}$, it suffices to solve the separation problem for $\{x : Ax \ge b,\ x \ge 0\}$ (by Theorem 2.10) and check the remaining constraints directly.

The number of bits needed to encode the new right-hand side numbers $\xi_1, \dots, \xi_{n-1}$ is polynomially bounded because the output of the above algorithm, which coincides with $\xi_1, \dots, \xi_{n-1}$ on the first $n-1$ components, is an extreme point; by Proposition 4.20, it is a solution to a linear equation system with polynomially bounded coefficients.

We cannot guarantee that γ is polynomially bounded, but instead of adding the constraint $c^\top x = \gamma$, we can first solve the dual LP as in the proof of Proposition 4.3, let y denote an optimum dual solution, and then add the complementary slackness constraints – that is, instead of the constraint $c^\top x = \gamma$ we add $x_i = 0$ whenever $A_i^\top y < c_i$ and $A_j x = b_j$ whenever $y_j > 0$. Corollary 4.2 says that this is equivalent. □

The *dimension* of a polyhedron P is the minimum dimension of an affine subspace containing P. The following is an algorithmic version of Carathéodory's theorem:

Theorem 4.22 (Carathéodory [1911], Grötschel, Lovász, and Schrijver [1981]). *Consider a class of polytopes in the setting of Theorem 2.10, and let $P \subseteq \mathbb{R}^d$ be a k-dimensional polytope in this class (in particular, we can solve the optimization problem over P in polynomial time). Then, for any given $y \in P$, we can find in polynomial time extreme points $x_1, \ldots, x_{k+1}$ of P and numbers $\lambda_1, \ldots, \lambda_{k+1} \geq 0$ such that $y = \sum_{i=1}^{k+1} \lambda_i x_i$ and $\sum_{i=1}^{k+1} \lambda_i = 1$.*

Proof. By induction on k. First, find an arbitrary extreme point x_{k+1} of P via Corollary 4.21. If $y = x_{k+1}$, we set $\lambda_{k+1} = 1$, and we are done (in particular, this is the case when $k = 0$). Otherwise, by translating P and y, we may assume $x_{k+1} = 0$.

We use the equivalence of optimization and separation (Theorem 2.10) several times. We first compute $\mu' = \max\{\mu : \mu y \in P\}$ using the separation problem for P. Set $y' = \mu' y$. Note that $\mu' \geq 1$. By the maximality of μ', there exists a vector a with $a^\top y' = \max\{a^\top x : x \in P\} = 1$.

Now, let $Q = \{z \in \mathbb{R}^n : z^\top x \leq 1 \text{ for all } x \in P\}$. Note that Q is a polyhedron because it is given by the constraints $z^\top x \leq 1$ for all extreme points x of P. Using the optimization problem for P, we can solve the separation problem for Q and hence the optimization problem for Q. We solve the LP

$$\max\{z^\top y' : z \in Q\}. \tag{4.9}$$

Let $\bar{z}$ be an optimum solution to (4.9). We have $\bar{z}^\top y' \leq 1$ because $\bar{z} \in Q$ and $y' \in P$. Since a is a feasible solution to (4.9), we have $\bar{z}^\top y' = 1$.

Now let $F = \{x \in P : \bar{z}^\top x = 1\}$. We have $y' \in F$, and F is a face of P. Moreover, we can solve the separation problem for F (using the separation problem for P and checking $\bar{z}^\top x = 1$ directly). Since $x_{k+1} = 0 \notin F$, the dimension of F is smaller than the dimension of P. By the induction hypothesis, we can write $y = \sum_{i=1}^{k} \lambda_i x_i$ and $\sum_{i=1}^{k} \lambda_i = 1$ in polynomial time. We conclude $y = \frac{1}{\mu'} y' = \frac{1}{\mu'} y' + (1 - \frac{1}{\mu'}) x_{k+1} = \sum_{i=1}^{k} \frac{\lambda_i}{\mu'} x_i + (1 - \frac{1}{\mu'}) x_{k+1}$.

Note that all induction steps apply to a face of the original polytope P, which is described by linear inequalities with the same coefficients as P by Proposition 4.18, and hence Theorem 2.10 implies an overall polynomial running time. □

Every extreme point solution to any of the LPs (2.2) and (2.12) has at most $2n - 3$ nonzero variables (Cornuéjols, Fonlupt, and Naddef [1985]). In fact, the support graph is "everywhere sparse" – that is, the following stronger property holds:

Theorem 4.23 (Goemans [2006]). *Let x be an extreme point solution to* (2.2) *or* (2.12). *Let $W \subseteq V$ with $|W| \geq 2$. Then the subgraph of the support graph of x induced by W has at most $2|W| - 3$ edges.*

Proof. Let $w \in W$ arbitrary and call a set U with $\emptyset \neq U \subseteq V \setminus \{w\}$ *tight* if $x(\delta(U)) = 2$. Let $F = \{e \in E[W] : x_e > 0\}$. By Proposition 4.20, every extreme point is the unique solution of a linear equation system $Ax = b$, where the columns of A correspond to variables and the rows correspond to constraints satisfied with equality. The restriction of x to F, which we call x', is the unique solution of a system $A'x' = b'$ where the columns of A' correspond to F and the rows of A' are incidence vectors of $\delta(U) \cap F$ (henceforth denoted by r^U) for tight sets U (recall that the constraints $x_e \leq 1$ in (2.2) are redundant by Proposition 2.1). Moreover, A' has rank $|F|$.

We now apply a standard uncrossing argument. Let $\mathcal{B}$ be a maximal family of tight sets such that $\mathcal{B}$ is laminar and the vectors r^B $(B \in \mathcal{B})$ are linearly independent. If $|\mathcal{B}| < |F|$, there is a tight set A such that r^B $(B \in \mathcal{B} \cup \{A\})$ are linearly independent. Choose A such that it crosses as few members of $\mathcal{B}$ as possible. We claim that it crosses none. To show this, suppose that A crosses some $B \in \mathcal{B}$. Then

$$\begin{aligned} 2+2+0 &\leq x(\delta(A \cap B)) + x(\delta(A \cup B)) + 2x(\delta(A \setminus B) \cap \delta(B \setminus A)) \\ &= x(\delta(A)) + x(\delta(B)) \\ &= 2+2, \end{aligned}$$

so $A \cap B$ and $A \cup B$ are tight and $F \cap \delta(A \setminus B) \cap \delta(B \setminus A) = \emptyset$. Hence

$$r^{A \cap B} + r^{A \cup B} = r^A + r^B$$

and therefore r^B $(B \in \mathcal{B} \cup \{A \cap B\})$ or r^B $(B \in \mathcal{B} \cup \{A \cup B\})$ are linearly independent, but $A \cap B$ and $A \cup B$ cross fewer elements of $\mathcal{B}$ than A, a contradiction. So $|\mathcal{B}| = |F|$.

Let $\mathcal{B}' := \{B \cap W : B \in \mathcal{B}\}$; note that this is a laminar family of $|\mathcal{B}|$ distinct nonempty subsets of $W \setminus \{w\}$. Hence $|\mathcal{B}| = |\mathcal{B}'| \leq 2|W| - 3$. □

With the same technique, Goemans [2006] also obtained a similar result for the asymmetric TSP:

Theorem 4.24 (Goemans [2006]). *Let x be an extreme point solution to* (3.1) *or* (3.2). *Let $W \subseteq V$ with $|W| \geq 2$. Then the subgraph of the support graph of x induced by W has at most $3|W| - 4$ edges.*

Proof. We proceed similarly as in the proof of Theorem 4.23. Let again $w \in W$ arbitrary and call a set U with $\emptyset \neq U \subseteq V \setminus \{w\}$ *tight* if $x(\delta(U)) = 2$. Note that the constraint $y(\delta^-(w)) = y(\delta^+(w))$ is redundant. The restriction of x to $F = \{e \in E[W] : x_e > 0\}$, which we call x', is the unique solution of a system $A'x' = b'$ where the columns of A' correspond to F and the rows of A' are vectors r^U for tight sets U and vectors corresponding to constraints $y(\delta^-(v)) = y(\delta^+(v))$ for some $v \in W \setminus \{w\}$. Exactly as in the proof of Theorem 4.23, the matrix A' has at most $2|W| - 3$ linearly independent rows of the first kind. Moreover, it has at most $|W| - 1$ rows of the second kind. Hence, $|F| \leq 3|W| - 4$. □

In particular, any extreme point solution to the LP (3.1) or (3.2) has at most $3n - 4$ nonzero variables.

The extreme points of (3.1) arise as the unique solutions to a linear equation system with a 0-1-matrix, so by Cramer's rule, each of their components can be written as $\frac{a}{b}$ for integers a and $b \leq \frac{(3n-4)!}{2}$. The same holds for (2.2) and (2.12), even with $b \leq \frac{(2n-3)!}{2}$. Boyd and Pulleyblank [1991] found a family of instances with extreme points whose denominators grow linearly in n. Pritchard [2010] found a family of instances where the denominators grow exponentially in n (more precisely, the $\frac{n}{2}$-th Fibonacci number).

4.4 Near-Minimum Cuts

By complementary slackness (Corollary 4.2), in an optimum solution to the dual LP (4.4) or (4.5), all variables are zero except for those corresponding to tight constraints in an optimum solution to the primal LP ((2.12) or (3.2), respectively). These can be viewed as minimum-weight cuts, where an edge $e = \{v, w\}$ has weight x_e (or $x_{(v,w)} + x_{(w,v)}$ for the Asymmetric TSP). For many algorithms, the cuts of (approximately) minimum weight play an important role.

A minimum-weight cut in an undirected graph $G = (V, E)$ with nonnegative weights $c : E \to \mathbb{R}_{\geq 0}$ can be found by $n - 1$ maximum flow computations (cf. Corollary 2.9). For a faster algorithm, see Exercise 4.15. However, we are often interested in *all* minimum-weight cuts.

The number of minimum-weight cuts is at most $\binom{n}{2}$, and these cuts have a nice structure (called cactus) as Dinits, Karzanov, and Lomonosov [1976] showed (see Exercise 10.6). Karger and Stein [1996] gave a simple proof of this bound and generalized it to cuts of near-minimum weight:

Theorem 4.25 (Karger and Stein [1996]). *Let $G = (V, E)$ be an undirected graph, $n = |V|$, and $c : E \to \mathbb{R}_{\geq 0}$. Let $\lambda := \min\{c(\delta(U)) : \emptyset \neq U \subsetneq V\} > 0$*

and $\gamma \geq 1$ be a half-integer. Then there are less than $n^{2\gamma}$ cuts $\delta(U)$ with $c(\delta(U)) \leq \gamma\lambda$.

Proof. Consider the following random contraction algorithm: We randomly choose an edge e, where e is picked with probability proportional to $c(e)$, contract e, and iterate this $n - 2\gamma$ times. At the end, we have 2γ vertices left and output a random cut among the $2^{2\gamma-1} - 1$ possibilities, each with probability $\frac{1}{2^{2\gamma-1}-1}$.

Let $\emptyset \neq U \subsetneq V$ with $c(\delta(U)) \leq \gamma\lambda$. We claim that every such cut is output of this algorithm with probability more than $n^{-2\gamma}$.

After k iterations, the graph has $n - k$ vertices, and the singleton cuts all have value at least λ, showing that the total weight of edges that still exist is at least $\frac{1}{2}(n-k)\lambda$. Hence, the probability that an edge of $\delta(U)$ is contracted is at most $\frac{2\gamma}{n-k}$. Therefore, the probability that we never contract an edge of $\delta(U)$ is at least

$$\prod_{k=0}^{n-2\gamma-1}\left(1 - \frac{2\gamma}{n-k}\right) = \prod_{k=0}^{n-2\gamma-1}\frac{n-k-2\gamma}{n-k} = \frac{1}{\binom{n}{2\gamma}},$$

and hence the probability that we output $\delta(U)$ is at least $\frac{1}{\binom{n}{2\gamma}(2^{2\gamma-1}-1)} > n^{-2\gamma}$. $\square$

This bound is essentially tight for every integer $\gamma \geq 1$ if G is a circuit with unit weights. For $\gamma = 1$, the proof of Theorem 4.25 yields the tight bound $\binom{n}{2}$ on the number of minimum-weight cuts, as well as a randomized algorithm to compute a minimum-weight cut. Nagamochi, Nichimura, and Ibaraki [1997] showed that there are at most $\binom{n}{2}$ cuts of weight less than $\frac{4}{3}$ times the minimum (see also Goemans and Ramakrishnan [1995]). Henzinger and Williamson [1996] showed that there are $O(n^2)$ cuts $\delta(U)$ with $x(\delta(U)) < 3$ in any solution x to the subtour LP. Karger [2000] generalized this by a different randomized algorithm (see Theorem 4.27).

For Karger's [2000] more general bound, we need the following LP relaxation of the global minimum cut problem:

$$\min\left\{c(x) : x(S) \geq 1\ (S \in \mathcal{S}),\ x \in \mathbb{R}^E_{\geq 0}\right\}, \tag{4.10}$$

where $\mathcal{S}$ denotes the set of edge sets of spanning trees in G. We first show that the integrality ratio of this LP is at most 2 (and in fact exactly 2 because the bound is tight for unit-weight circuits).

Lemma 4.26 (Tutte [1961], Nash-Williams [1961]). *Let $G = (V, E)$ be an undirected graph, $n = |V|$, and $c : E \to \mathbb{R}_{\geq 0}$. Let $\lambda := \min\{c(\delta(U)) : \emptyset \neq U \subsetneq V\}$ be the minimum weight of a cut. Then λ is at most $\frac{2(n-1)}{n}$ times the value of the LP* (4.10).

Proof. First, we show that we may assume that there is an optimum LP solution with $x_e > 0$ for all $e \in E$. Indeed, if there is an edge e with $x_e = 0$, we contract it and arrive at a smaller instance. In this instance, the minimum weight of a cut cannot be smaller, and we now show that the LP value cannot be larger. This follows from the fact that x is still feasible in the smaller instance because, for every spanning tree S in the smaller instance, $S \cup \{e\}$ is a spanning tree in the original instance and $x(S) = x(S \cup \{e\}) \geq 1$.

Now, assuming $x_e > 0$ for all $e \in E$, let y be an optimum solution to the dual LP

$$\max\left\{y(\mathcal{S}) : \sum_{S \in \mathcal{S}: e \in S} y_S \leq c(e)\ (e \in E),\ y \geq 0\right\}. \tag{4.11}$$

The singleton cuts have total weight $2c(E)$, so $\lambda \leq \frac{2c(E)}{n}$. On the other hand, complementary slackness (Corollary 4.2) yields

$$c(E) = \sum_{e \in E} \sum_{S \in \mathcal{S}: e \in S} y_S = \sum_{S \in \mathcal{S}} y_S |S| = y(\mathcal{S})(n-1).$$

Hence $\lambda \leq \frac{2(n-1)}{n} y(\mathcal{S})$. □

Here is Karger's [2000] more general bound:

Theorem 4.27 (Karger [2000]). *Let $\gamma \geq 1$ be a constant. Let $G = (V, E)$ be an undirected graph, $n = |V|$, and $c : E \to \mathbb{R}_{\geq 0}$. Let $\lambda := \min\{c(\delta(U)) : \emptyset \neq U \subsetneq V\} > 0$. Then there are $O(n^{\lfloor 2\gamma \rfloor})$ cuts $\delta(U)$ in G with $c(\delta(U)) \leq \gamma\lambda$.*

Proof. Let $\delta(U)$ be a fixed cut with $c(\delta(U)) \leq \gamma\lambda$. Let again y be an optimum solution to (4.11). Then

$$\sum_{S \in \mathcal{S}} y_S |S \cap \delta(U)| = \sum_{e \in \delta(U)} \sum_{S \in \mathcal{S}: e \in S} y_S \leq \sum_{e \in \delta(U)} c(e) \leq \gamma\lambda \tag{4.12}$$

and $y(\mathcal{S}) > \frac{\lambda}{2}$ by Lemma 4.26. Combining these inequalities, we get

$$\frac{1}{y(\mathcal{S})} \sum_{S \in \mathcal{S}} y_S |S \cap \delta(U)| < 2\gamma. \tag{4.13}$$

Call a tree $S \in \mathcal{S}$ *good* (for $\delta(U)$) if $|S \cap \delta(U)| \leq \lfloor 2\gamma \rfloor$.

Let $p := \frac{1}{y(\mathcal{S})} \sum_{S \in \mathcal{S}:\, S \text{ good}} y_S$ be the probability that a random tree (where the probability of picking S is proportional to y_S) is good. Then

$$\frac{1}{y(\mathcal{S})} \sum_{S \in \mathcal{S}} y_S |S \cap \delta(U)| \geq p + (1-p)(\lfloor 2\gamma \rfloor + 1),$$

which together with (4.13) yields

$$p\lfloor 2\gamma \rfloor > \lfloor 2\gamma \rfloor + 1 - 2\gamma,$$

so p is bounded from below by a positive constant (depending on γ only).

For a good tree $S \in \mathcal{S}$ and $F = S \cap \delta(U)$, we have $|F| \leq \lfloor 2\gamma \rfloor$. From S and F, we can recover $\delta(U)$ by contracting the edges $S \setminus F$ in (V, S), coloring the vertices of the resulting tree red and blue so that the endpoints of each edge have different colors, and letting U contain the vertices of G that result from uncontracting the red vertices.

Hence, choosing a random spanning tree (where the probability of picking S is proportional to y_S) and a random subset $F \subseteq S$ with $|F| \leq \lfloor 2\gamma \rfloor$ (with uniform distribution) produces $\delta(U)$ with probability at least $pn^{-\lfloor 2\gamma \rfloor}$. Since this holds for every cut $\delta(U)$ with $c(\delta(U)) \leq \gamma\lambda$, we conclude that there are at most $\frac{1}{p} n^{\lfloor 2\gamma \rfloor}$ of those. □

See also Chekuri, Quanrud, and Xu [2020] for extensions of this idea. Both algorithms underlying Theorems 4.25 and 4.27 are randomized. Nagamochi, Nichimura, and Ibaraki [1997] showed how to enumerate all small cuts in deterministic polynomial time:

Theorem 4.28 (Nagamochi, Nichimura, and Ibaraki [1997]). *Let $\gamma \geq 1$ be a constant. Then there is a deterministic polynomial-time algorithm that, given an undirected graph $G = (V, E)$ and $c : E \to \mathbb{R}_{\geq 0}$ with $\lambda := \min\{c(\delta(U)) : \emptyset \neq U \subsetneq V\} > 0$, computes all sets $\emptyset \neq U \subsetneq V$ with $c(\delta(U)) \leq \gamma\lambda$.*

Proof. Let $V = \{v_1, \ldots, v_n\}$. Without loss of generality, the graph is complete (some edges e may have $c(e) = 0$).

Let $c_n := c$, and for $k = n, n-1, \ldots, 3$, obtain c_{k-1} from c_k by applying Theorem 2.32 (splitting off) to $z = v_k$ so that $c_{k-1}(\delta(v_k)) = 0$ and

$$c_{k-1}(\delta(U)) \geq \lambda \quad \text{for all } \emptyset \neq U \subsetneq \{v_1, \ldots, v_{k-1}\}. \tag{4.14}$$

Now define $\mathcal{U}_k := \{U \subsetneq \{v_1, \ldots, v_k\} : v_1 \in U,\ c_k(\delta(U)) \leq \gamma\lambda\}$ for $k = 2, \ldots, n$. We want to compute $\mathcal{U}_n$. We have $\mathcal{U}_2 \subseteq \{\{v_1\}\}$. To compute $\mathcal{U}_k$ from $\mathcal{U}_{k-1}$ (for $k = 3, \ldots, n$), observe that for every set $U \subsetneq \{v_1, \ldots, v_k\}$ with $v_1 \in U$ and $c_k(\delta(U)) \leq \gamma\lambda$, we have either $U = \{v_1, \ldots, v_{k-1}\}$ or $U \setminus \{v_k\} \in \mathcal{U}_{k-1}$ because $c_{k-1}(\delta(U \setminus \{v_k\})) \leq c_k(\delta(U)) \leq \gamma\lambda$. So we can compute $\mathcal{U}_k$ from $\mathcal{U}_{k-1}$ in $O(|\mathcal{U}_{k-1}| \cdot n^2)$ time. By Theorem 4.27 and (4.14), we have $|\mathcal{U}_{k-1}| = O\left((k-1)^{\lfloor 2\gamma \rfloor}\right)$. Hence, the total running time is $O\left(n^{3+\lfloor 2\gamma \rfloor}\right)$. □

We remark that this result also follows from combining Theorem 4.25 with the enumeration algorithm by Vazirani and Yannakakis [1992]. Another proof was suggested by Beideman, Chandrasekaran, and Wang [2023].

Theorem 4.28 also yields an alternative proof of Lemma 4.13 (and analogously of Lemma 4.14 and hence also of Proposition 4.3): After having solved the primal LP (3.2), let x be an optimum solution. Consider the complete undirected graph on V and, for $\{v,w\} \in \binom{V}{2}$, let $x'(\{v,w\}) := x_{(v,w)} + x_{(w,v)}$. Enumerate all sets $\emptyset \neq S \subsetneq V$ with $x'(\delta(S)) = 2$; these are the minimum-weight cuts. By complementary slackness (Corollary 4.2), dual variables that correspond to other sets must be zero in every optimum dual solution. Therefore, restricting the dual LP to the at most $\binom{n}{2}$ subsets of $V \setminus \{w\}$ (for some arbitrary vertex w) inducing minimum cuts does not change the dual. Since this LP now has polynomially many variables and constraints, we can solve it in polynomial time. To obtain a solution with laminar support, we add a constraint $\sum_{\emptyset \neq U \subsetneq V} 2y_U = \text{LP}$, where LP again denotes the LP value, and minimize $\sum_{\emptyset \neq U \subsetneq V} |U| y_U$ (using Theorem 2.2). The resulting dual solution will be optimum and laminar because otherwise uncrossing a pair of sets (reducing y_A and y_B and increasing $y_{A \setminus B}$ and $y_{B \setminus A}$; note that these variables exist by Proposition 4.5) would decrease the objective function value.

The same proof works for Lemma 4.14.

Exercises

4.1 Show that the dual of the dual LP is equivalent to the primal LP.

4.2 Describe the convex hull of incidence vectors of spanning arborescences rooted at r in a digraph $G = (V, E)$ with $r \in V$.
Hint: Consider the dual of the LP in Exercise 3.5 and add a constraint that the sum of all variables is $|V| - 1$.

4.3 Let (V, c) be an instance of the SYMMETRIC TSP WITH TRIANGLE INEQUALITY and $v_1 \in V$ arbitrary. A *1-tree* is a tree on $V \setminus \{v_1\}$ plus two edges incident to v_1 – that is, a graph T with $V(T) = V$ and $|\delta(v_1)| = 2$ such that $T[V \setminus \{v_1\}]$ is a tree. Show that the value of the subtour LP (2.2) equals

$$\max\left\{\min\left\{c(E(T)) + \sum_{v \in V}(|\delta_T(v)| - 2)\lambda_v : T \text{ is a 1-tree}\right\} : \lambda \in \mathbb{R}^V\right\}.$$

Hint: Derive a polyhedral description of 1-trees from Theorem 2.16 and apply LP duality (Theorem 4.1) twice: first to the inner minimization problem and then to the maximization problem.
(Held and Karp [1970])

4.4 Let $G = (V, E)$ be an undirected graph and $c : E \to \mathbb{R}_{\geq 0}$. Let $\lambda := \min\{c(\delta(U)) : \emptyset \neq U \subsetneq V\}$ and $\varepsilon \geq 0$. Let $A, B \subsetneq V$ be crossing with $c(\delta(A)) \leq \lambda + \varepsilon$ and $c(\delta(B)) \leq \lambda + \varepsilon$. Show that then $c(\delta(A \cap B)) \leq \lambda + 2\varepsilon$ and $c(\delta(A \cup B)) \leq \lambda + 2\varepsilon$ and $c(\delta(A \setminus B) \cap \delta(B \setminus A)) \leq 2\varepsilon$.

4.5 Let V be a finite set and $f : 2^V \to \mathbb{R}$. If

$$f(A \cap B) + f(A \cup B) \;\leq\; f(A) + f(B)$$

for all $A, B \subseteq V$, then f is called *submodular* (and $-f$ is called *supermodular*). See (4.6) for an example. Let $G = (V, E)$ be a digraph, $u : E \to \mathbb{R}_{\geq 0}$, and $f(A) = u(\delta^-(A))$ for $A \subseteq V$. Show that f is submodular.

4.6 Show that the rank function $r : 2^E \to \mathbb{Z}_{\geq 0}$ of a matroid $(E, \mathcal{F})$ (cf. Exercise 2.5) is submodular.

4.7 The submodular function minimization problem takes as input a finite set V and has access to a submodular function $f : 2^V \to \mathbb{R}$ via an oracle that computes $f(A)$ for any given $A \subseteq V$; the task is to compute a set $A \subseteq V$ with $f(A)$ minimum. Show that computing a minimum-capacity s-t-cut (cf. Corollary 2.8) reduces to submodular function minimization.
Note: (Strongly) polynomial-time algorithms for submodular function minimization have been designed by Grötschel, Lovász, and Schrijver [1988], Schrijver [2000], and Iwata, Fleischer, and Fujishige [2001]. The currently fastest algorithms are due to Lee, Sidford, and Wong [2015] and Dadush, Végh, and Zambelli [2021].

4.8 Show that for every $n \in \mathbb{N}$ with $n \geq 2$, there is a cross-free family of subsets of $\{1, \ldots, n\}$ with $4n - 4$ sets, and show that no cross-free family contains more sets.

4.9 Prove that the LP in Exercise 3.5 (cf. Exercise 4.2) has an optimum solution with laminar support and that one can compute such a solution in polynomial time.

4.10 Let $G = (V, E)$ be an undirected graph, $T \subseteq V$ with $|T|$ even, and $c : E \to \mathbb{Z}_{\geq 0}$ such that $c(C)$ is even for every edge set C of a circuit. Show that then there exists an optimum solution y to the T-cut packing LP (4.7) such that y is integral and $\{U : y_U > 0\}$ is laminar.
Hint: Review the proof of Theorem 2.19.

4.11 Let P be a polyhedron and $x \in P$. Show that x is an extreme point of P if and only if there are no $y, z \in P \setminus \{x\}$ with $x = \frac{1}{2}(y + z)$.

4.12 Let x be an extreme point of the subtour polytope (cf. (2.2), for n cities). Call a city $v \in V$ *fractional* if $x_e < 1$ for all $e \in \delta(v)$. Suppose the support graph G_x of x contains a set U of three fractional degree-3 vertices such that $G_x[U]$ is a circuit and $x(\delta(U)) = 2$ (such a set is called a *triangle configuration*). Let x' result from x by contracting U. Show that x' is again an extreme point of the subtour polytope (for $n - 2$ cities).
Hint: Use Exercise 4.11.
(Boyd and Pulleyblank [1991])

4.13 Let x^* be an extreme point of the subtour polytope (cf. (2.2)). Let $G = (V, E)$ be the support graph of x^*.

(a) Assume there is no triangle configuration (cf. Exercise 4.12). Let V_3^f denote the set of degree-3 vertices of G that have no incident edge $e \in \delta_G(v)$ with $x_e^* = 1$. Let $\mathcal{B}$ be a laminar family of tight sets such that x^* (restricted to E) is the unique element of $\{x \in \mathbb{R}^E : x(\delta(U)) = 2 \text{ for all } U \in \mathcal{B}\}$ (cf. Theorem 4.23). Show that there is a family $\mathcal{C}$ of $\lceil \frac{1}{2}|V_3^f| \rceil$ nonempty proper subsets of V such that $\mathcal{B} \cap \mathcal{C} = \emptyset$ and $\mathcal{B} \cup \mathcal{C}$ is laminar.
Hint: For each non-singleton set $B \in \mathcal{B}$, let $\hat{B}$ denote the set of vertices of B that are not in any non-singleton set $B' \in \mathcal{B}$ that is a proper subset of B. If $|\hat{B}| = |\hat{B} \cap V_3^f| = 1$, add $B \setminus \hat{B}$ (which is not tight!) to $\mathcal{C}$. If $|\hat{B}| > 1$, add a maximal laminar family of non-singleton subsets of $\hat{B}$ to $\mathcal{C}$ (but do not add B if $\hat{B} = B$).

(b) Show that if there is no triangle configuration, then $|E| \leq 2n-2-\frac{1}{2}|V_3^f|$.
Hint: Use (a) and recall the proof of Theorem 4.23.

(c) Prove that there are two edges $e \in E$ with $x_e^* = 1$.
Hint: Use Exercise 4.12 and (b).

Note: By refining this proof, one can obtain three edges $e \in E$ with $x_e^* = 1$.
(Boyd and Pulleyblank [1991])

4.14 Let $G = (V, E)$ be an undirected graph and $c : E \to \mathbb{R}_{\geq 0}$. For $i = 1, \ldots, n$, let v_i be a vertex maximizing $c(\delta(\{v_1, \ldots, v_{i-1}\}) \cap \delta(v_i))$. Prove that then $\delta(v_n)$ is a minimum-weight cut separating v_{n-1} and v_n.
Hint: Induction on n, use the induction hypothesis for $G - v_n$ and $G - v_{n-1}$ and Exercise 2.3.
(Stoer and Wagner [1997], Frank [1994])

4.15 Given an undirected graph $G = (V, E)$ and weights $c : E \to \mathbb{R}_{\geq 0}$, show how to find a minimum-weight cut in $O(n^3)$ time, exploiting Exercise 4.14.
(Nagamochi and Ibaraki [1992], Stoer and Wagner [1997])

5

Thin Trees and Random Trees

After the $O(\log n)$-approximation algorithms for the Asymmetric TSP by Frieze, Galbiati, and Maffioli [1982], Bläser [2008], Kaplan et al. [2005], and Feige and Singh [2007] (cf. Section 1.5), the first algorithm to beat the cycle cover algorithm by more than a constant factor was found in 2009 by Asadpour, Goemans, Mądry, Oveis Gharan, and Saberi [2017]. Their approach is based on finding a "thin" (oriented) spanning tree and then adding edges to obtain a tour. A major open question is how thin trees are guaranteed to exist.

Asadpour et al. [2017] sampled a random spanning tree from the maximum entropy distribution. To show how this works, we discuss interesting connections between random spanning trees and electrical networks. Some results of this chapter will be used again in Chapters 10 and 11 when we discuss the maximum entropy sampling for the Symmetric TSP.

5.1 Completing Connected Digraphs to Tours

Asadpour, Goemans, Mądry, Oveis Gharan, and Saberi [2017] were the first to obtain an $o(\log n)$-approximation algorithm for the Asymmetric TSP, albeit their algorithm is randomized.

There are two definitions of randomized approximation algorithms. For us, a *randomized α-approximation algorithm* (for a minimization problem with nonnegative cost function) is a polynomial-time algorithm that uses random bits in addition to the input and always computes a feasible solution, such that the expected cost of this solution is at most α times the optimum. In a different definition, rather than bounding the expected cost, one demands that with probability at least $\frac{1}{2}$ the algorithm produces a solution of cost at most α times the optimum. These two definitions are almost equivalent (see Exercise 5.1).

A tour is connected and Eulerian, and it is quite natural to first compute a connected subgraph and then add a minimum-cost multi-set of edges in order to make the graph Eulerian. This idea (underlying Christofides' algorithm) works also in the directed case, but bounding the cost is more difficult. The following was observed by Goemans et al. [2009] and Asadpour et al. [2017] (see Figure 5.1 for an example).

Lemma 5.1. *Let $G = (V, E)$ be a digraph and $c : E \to \mathbb{R}_{\geq 0}$. Let (V, R) be a connected spanning subgraph of G, and let $x \in \mathbb{R}^E_{\geq 0}$. Then one can find a tour F in G in polynomial time such that $c(F) \leq c(R) + \alpha c(x)$ for every $\alpha > 0$ with $|R \cap \delta^-(U)| \leq \alpha x(\delta^+(U))$ for all $U \subseteq V$.*

Proof. Let $l(e) := 1$ for $e \in R$ and $l(e) := 0$ for $e \in E \setminus R$. Any integral circulation f in (V, E) with $f \geq l$ corresponds to a tour. We compute an integral minimum-cost circulation $f^* \geq l$ (using Theorem 3.11) and note that the resulting tour has cost $c(f^*)$.

To prove that the cost of f^* is at most $c(R) + \alpha c(x)$, we show that there exists a fractional circulation g with $g \geq l$ and cost at most $c(R) + \alpha c(x)$. This is sufficient by Theorem 3.10.

To prove that such a circulation g exists, we define upper bounds $u(e) := \max\{l(e), \alpha x_e\}$ for all $e \in E$ and observe that a circulation g with $l \leq g \leq u$ exists; then $c(f^*) \leq c(g) \leq \sum_{e \in E} c(e)u(e) \leq c(R) + \alpha c(x)$.

The existence of g follows from Hoffman's circulation theorem (Theorem 3.9): We have $l \leq u$ and $l(\delta^-(U)) = |R \cap \delta^-(U)| \leq \alpha x(\delta^+(U)) \leq u(\delta^+(U))$ for all $U \subseteq V$. □

However, a minimum-cost oriented spanning tree does often not give the best overall result (see Exercise 5.3). A cheap and "thin" tree, containing not too many edges in any cut, is better. It is useful to define thinness in an undirected setting:

Definition 5.2 (thin tree). Let $G = (V, E)$ be an undirected graph, $\alpha > 0$, and (V, S) a spanning tree in G. Then (V, S) is *α-thin* with respect to G if $|S \cap \delta(U)| \leq \alpha|\delta(U)|$ for all $U \subseteq V$. For $z \in \mathbb{R}^E_{\geq 0}$, we say that (V, S) is *α-thin* with respect to z if $|S \cap \delta(U)| \leq \alpha z(\delta(U))$ for all $U \subseteq V$.

The main ingredient of the randomized approximation algorithm of Asadpour et al. [2017] is the following result:

Theorem 5.3 (Asadpour et al. [2017]). *There is a randomized polynomial-time algorithm that, given a feasible solution z to the subtour LP* (2.2) *for any $n := |V| \geq 3$, computes a spanning tree (V, S) such that each edge e belongs to S with probability at most z_e, and (V, S) is α-thin with respect to z with probability at least $\frac{9}{10}$, where $\alpha = \frac{3 \ln n}{\ln \ln n}$.*

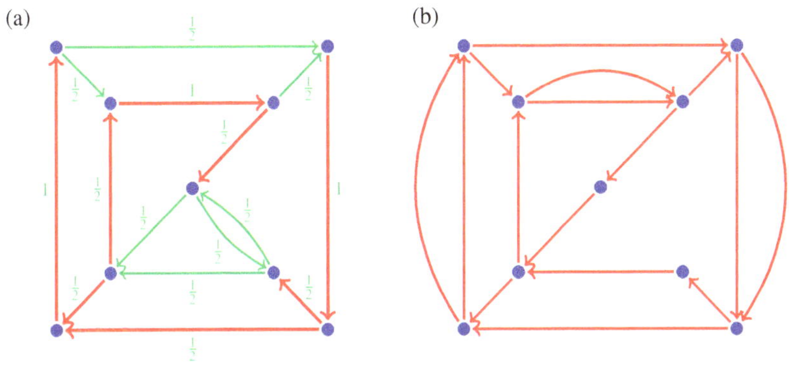

Figure 5.1 (a): A digraph $G = (V, E)$ with a vector $x \in \mathbb{R}^E_{\geq 0}$. The (weakly) connected subgraph (V, R) (shown in bold red) satisfies $|R \cap \delta^-(U)| \leq 2x(\delta^+(U))$ for all $U \subseteq V$. (b): Therefore, for any cost function c, it can be extended to a tour in G that costs at most $2c(x)$ more than R (see Lemma 5.1). If we ignore all orientations in (a), the red tree (V, R) is 2-thin with respect to the symmetrized vector z (resulting from x as in Step (2) of Algorithm 5.4).

Assuming Theorem 5.3, the algorithm of Asadpour et al. [2017] and its analysis can be described easily. Without loss of generality, $n \geq 3$. By Proposition 1.12, we can work in the metric closure, so we assume that c satisfies the triangle inequality.

Algorithm 5.4: Thin Tree Completion

Input: an instance (V, c) of the ASYMMETRIC TSP WITH TRIANGLE INEQUALITY

Output: a tour in the complete digraph on V

(1) Let x be an optimum solution to the LP (3.1).
(2) Set $z_{\{v,w\}} := x_{(v,w)} + x_{(w,v)}$ for all $\{v, w\} \in \binom{V}{2}$.
(3) Apply Theorem 5.3 to z to obtain a spanning tree (V, S).
(4) Orient each edge of S in the cheaper direction and apply Lemma 5.1 to the result.
(5) Repeat Steps (3) and (4) n times and output the best of the resulting tours.

Theorem 5.5 (Asadpour et al. [2017]). *Algorithm 5.4 (thin tree completion) is a randomized $O(\frac{\log n}{\log \log n})$-approximation algorithm for the ASYMMETRIC TSP WITH TRIANGLE INEQUALITY.*

Proof. Observe that z as constructed in Step (2) is a feasible solution to (2.2) by Proposition 2.1. By Theorem 5.3, the expected cost of S (with respect to symmetrized cost $c'(\{v,w\}) := \min\{c(v,w), c(w,v)\}$) is at most $c'(z)$. By Markov's inequality, $c'(S) \leq 2c'(z)$ with probability at least $\frac{1}{2}$. Orienting the edges of this tree (by replacing each $\{v,w\} \in S$ by the cheaper one of (v,w) and (w,v)) yields an arc set R. With probability $\frac{1}{2}$, this set R satisfies $c(R) = c'(S) \leq 2c'(z) \leq 2c(x)$, and with probability at least $\frac{9}{10}$, it satisfies

$$\begin{aligned}|R \cap \delta^-(U)| &\leq |S \cap \delta(U)| \\ &\leq \alpha z(\delta(U)) \\ &= \alpha(x(\delta^-(U)) + x(\delta^+(U))) \\ &= 2\alpha x(\delta^+(U))\end{aligned}$$

for all $U \subseteq V$, where $\alpha = \frac{3 \ln n}{\ln \ln n}$. We apply Lemma 5.1 to R and x. With probability at least $\frac{4}{10}$, we obtain a tour of cost at most $(2\alpha + 2)c(x)$. Repeating this process n times increases the success probability to at least $1 - (\frac{6}{10})^n$.

Let $p \leq (\frac{6}{10})^n$ be the probability of failure. In this case, we analyze the tour that we obtain from the first spanning tree (V, S_1). In the event of failure, the expected cost of S_1 is at most $\frac{1}{p}c(x)$ because in Step (3) we sample a spanning tree with expected cost at most $c(x)$. Every spanning tree, in particular (V, S_1), is $\frac{n-1}{2}$-thin with respect to z, and therefore $|S_1 \cap \delta(U)| \leq \frac{n-1}{2}x(\delta(U)) = (n-1)x(\delta^+(U))$ for all $U \subseteq V$. By Lemma 5.1, the resulting tour cannot cost more than $\frac{1}{p}c(x) + (n-1)c(x)$.

The expected cost of the final solution is at most

$$\begin{aligned}(1-p)(2\alpha+2) \cdot c(x) + p\left(\tfrac{1}{p} + n - 1\right) \cdot c(x) &\leq (2\alpha + 3 + np) \cdot c(x) \\ &\leq (2\alpha + 4) \cdot c(x).\end{aligned}$$

□

It remains to prove Theorem 5.3. This will take much of the rest of this chapter.

5.2 Random Spanning Trees with Negative Correlation

Let us see how Asadpour et al. [2017] proved Theorem 5.3. First, if z is a feasible solution to (2.2), then $\frac{n-1}{n}z$ is in the relative interior of the spanning tree polytope of the support graph (V, E) (cf. Corollary 2.17). So $\frac{n-1}{n}z$ is a convex combination of incidence vectors of spanning trees – that is, $\frac{n-1}{n}z_e = \sum_{S \in \mathcal{S}: e \in S} \mu(S)$ for all $e \in E$, where $\mathcal{S}$ is the set of (edge sets of) spanning trees, $\mu(S) \geq 0$ for all $S \in \mathcal{S}$, and $\sum_{S \in \mathcal{S}} \mu(S) = 1$. One could obtain such an explicit convex combination in polynomial time (cf. Theorem 4.22 or Theorem 15.4), but here we will only

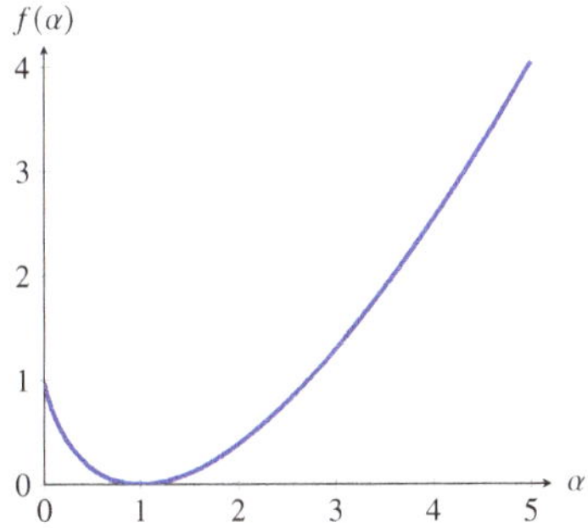

Figure 5.2 The function f in Lemma 5.6.

sample from an implicitly given distribution. If we pick each tree $S \in \mathcal{S}$ with probability $\mu(S)$, the expected cost is $\frac{n-1}{n} \sum_{e \in E} c'(e) z_e$, where c' is again the symmetrized cost function.

The difficulty is that such a random spanning tree will in general not be thin enough. Therefore, Asadpour et al. [2017] chose the probability distribution carefully. It turns out that it is sufficient to have negative correlation, which we will show in Theorem 5.7. Let $\mathbb{P}[\cdot]$ denote the probability of an event, and let $\mathbb{E}[\cdot]$ denote the expectation of a random variable. A set $X_1, \ldots, X_k$ of nonnegative real random variables is called *negatively correlated* if $\mathbb{E}[\prod_{i \in I} X_i] \leq \prod_{i \in I} \mathbb{E}[X_i]$ for all $I \subseteq \{1, \ldots, k\}$. It is well-known that the sum of negatively correlated random variables is unlikely to deviate much from the expectation:

Lemma 5.6 (Chernoff [1952], Panconesi and Srinivasan [1997]). *Let $X_1, \ldots, X_k$ be random variables in $[0, 1]$ that are negatively correlated. Let s be the sum of their expectations, and let $\alpha \geq 1$. Then the probability that $X_1 + \cdots + X_k > \alpha s$ is at most $e^{-sf(\alpha)}$, where $f(\alpha) := 1 - \alpha + \alpha \ln \alpha$ (cf. Figure 5.2).*

Proof. The assertion is trivial for $\alpha = 1$. For $\alpha > 1$, we compute

$$
\begin{aligned}
&\mathbb{P}[X_1 + \cdots + X_k > \alpha s] \\
&= \mathbb{P}\left[\frac{\prod_{i=1}^{k} \alpha^{X_i}}{\alpha^{\alpha s}} > 1\right] \\
&\leq \mathbb{P}\left[\frac{\prod_{i=1}^{k} (1 + (\alpha - 1)X_i)}{\alpha^{\alpha s}} > 1\right] && (\alpha^x \leq 1 + (\alpha - 1)x \text{ for } 0 \leq x \leq 1) \\
&\leq \mathbb{E}\left[\frac{\prod_{i=1}^{k} (1 + (\alpha - 1)X_i)}{\alpha^{\alpha s}}\right] && (\text{Markov's inequality})
\end{aligned}
$$

$$
\begin{aligned}
&= \mathbb{E}\left[\frac{1}{\alpha^{\alpha s}} \sum_{I \subseteq \{1,\dots,k\}} (\alpha-1)^{|I|} \prod_{i \in I} X_i\right] \\
&= \frac{1}{\alpha^{\alpha s}} \sum_{I \subseteq \{1,\dots,k\}} (\alpha-1)^{|I|} \mathbb{E}\left[\prod_{i \in I} X_i\right] && \text{(linearity of expectation)} \\
&\le \frac{1}{\alpha^{\alpha s}} \sum_{I \subseteq \{1,\dots,k\}} (\alpha-1)^{|I|} \prod_{i \in I} \mathbb{E}\left[X_i\right] && \text{(negative correlation)} \\
&= \frac{\prod_{i=1}^{k}(1+(\alpha-1)\mathbb{E}[X_i])}{\alpha^{\alpha s}} \\
&\le \frac{\prod_{i=1}^{k} e^{(\alpha-1)\mathbb{E}[X_i]}}{\alpha^{\alpha s}} && (1+x \le e^x \text{ for } x \in \mathbb{R}) \\
&= \frac{e^{(\alpha-1)s}}{\alpha^{\alpha s}} \\
&= e^{-sf(\alpha)}.
\end{aligned}
$$

□

For a probability distribution over the spanning trees of a graph $G = (V, E)$ and $e \in E$, let q_e be the probability that the random spanning tree contains e. Then the values $(q_e)_{e \in E}$ are called the *marginals* of that probability distribution. Now, we can prove:

Theorem 5.7 (Asadpour et al. [2017]). *Let $G = (V, E)$ be an undirected graph with $n = |V| \ge 3$ vertices, and let z be a feasible solution to the LP* (2.2). *Let (V, S) be a spanning tree of G sampled according to a probability distribution with marginals at most z_e ($e \in E$) such that the 0-1 random variables $|\{e\} \cap S|$ (for $e \in E$) are negatively correlated. Then (V, S) is $\frac{3 \ln n}{\ln \ln n}$-thin with respect to z with probability at least $\frac{9}{10}$.*

Proof. Applying Lemma 5.6 to $\alpha = \frac{3 \ln n}{\ln \ln n}$ yields

$$
f(\alpha) = 1 + \tfrac{3 \ln n}{\ln \ln n}\left(\ln \ln n + \ln 3 - \ln \ln \ln n - 1\right) > 1 + 2 \ln n
$$

for all $n \ge 3$.

Now, thinness is implied by the fact that there are only few small cuts (see Theorem 4.25). The union bound yields that the probability that a random tree is not α-thin is at most $\sum_{k=3}^{\infty} n^k e^{-(k-1)f(\alpha)}$, where the k-th summand takes care of the cuts $\delta(U)$ with $k - 1 \le z(\delta(U)) < k$. Using $f(\alpha) > 1 + 2 \ln n$ for

$\alpha = \frac{3 \ln n}{\ln \ln n}$, the tree is α-thin with probability at least

$$\begin{aligned} 1 - \sum_{k=3}^{\infty} n^k e^{-(k-1)(1+2\ln n)} &= 1 - \sum_{k=3}^{\infty} e^{-(k-1)} n^{2-k} \\ &\geq 1 - e^{-2} \sum_{k=3}^{\infty} n^{2-k} \\ &= 1 - \frac{e^{-2}}{n-1} \\ &> \frac{9}{10}. \end{aligned}$$

□

Now the proof of Theorem 5.3 reduces to sampling a random spanning tree S according to a probability distribution with marginals at most z_e $(e \in E)$ such that the 0-1 random variables $|\{e\} \cap S|$ (for $e \in E$) are negatively correlated. This is the subject of the next two sections.

5.3 Electrical Networks

Still following Asadpour et al. [2017], we will exploit a nice connection of random spanning trees to electrical networks, discovered by Kirchhoff [1847].

Given a connected undirected graph $G = (V, E)$, positive edge weights λ_e $(e \in E)$, and two distinct vertices $s, t \in V$, we interpret every edge e as a wire with resistance $\frac{1}{\lambda_e}$ and send one unit of electrical flow from s to t. We call (G, λ, s, t) an electrical network. It is convenient to fix an arbitrary orientation $G^{\rightarrow} = (V, E^{\rightarrow})$ of G and interpret a negative flow along an edge $(v, w) \in E^{\rightarrow}$ as a flow in the reverse direction (from w to v). Then $f \in \mathbb{R}^{E^{\rightarrow}}$ is an s-t-flow of value 1 if $f(\delta^+(v)) - f(\delta^-(v))$ is 0 for all $v \in V \setminus \{s, t\}$ and is 1 for $v = s$ and -1 for $v = t$. Note that we now allow negative flow along an oriented edge. Sometimes, we consider $E^{\leftrightarrow}$ (with edges oriented both ways) and write $f_{e^{\leftarrow}} := -f_e$ for every edge $e \in E^{\rightarrow}$ and its reverse edge $e^{\leftarrow}$. We write $G^{\leftrightarrow} := (V, E^{\leftrightarrow})$ and $\lambda_{(v,w)} := \lambda_{\{v,w\}}$ for $(v, w) \in E^{\leftrightarrow}$.

Definition 5.8 (electrical flow). Let $G = (V, E)$ be a connected undirected graph with an orientation $E^{\rightarrow}$, $\lambda \in \mathbb{R}^E_{>0}$, and $s, t \in V$. The *electrical s-t-flow of value 1 in* (G, λ), or simply the *electrical flow*, is the s-t-flow $f \in \mathbb{R}^{E^{\rightarrow}}$ of value 1 that minimizes the *energy*

$$\mathcal{E}(f) := \sum_{e \in E^{\rightarrow}} \frac{(f_e)^2}{\lambda_e}. \tag{5.1}$$

Since $\mathcal{E}$ is a strictly convex function and the set of s-t-flows of value 1 is convex, the electrical flow is indeed unique. Note that it is irrelevant how we

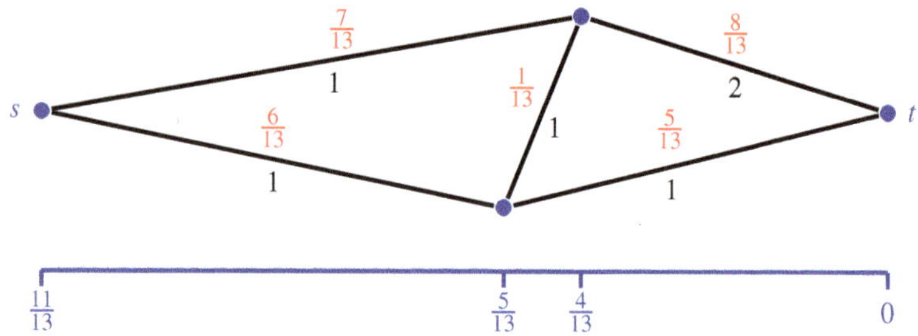

Figure 5.3 A small electrical network (with λ_e shown in black below every edge e). The electrical s-t-flow of value 1 is shown in red above the edges, flowing from left to right. A corresponding potential π is shown in blue below.

oriented the edges of G. Here are two alternative definitions (cf. Figure 5.3 for a small example):

Theorem 5.9. *Let $G = (V, E)$ be a connected undirected graph, $\lambda_e > 0$ for $e \in E$, and $s, t \in V$. Let $f \in \mathbb{R}^{E^\rightarrow}$ be an s-t-flow of value 1. Then the following statements are equivalent:*

(i) *f is the electrical flow.*
(ii) *There is a potential $\pi \in \mathbb{R}^V$ such that $f_e = \lambda_e(\pi_v - \pi_w)$ for every edge $e = (v, w) \in E^\rightarrow$ (Ohm's law).*
(iii) *$\sum_{e \in C} \frac{f_e}{\lambda_e} = 0$ for every directed circuit C in $G^\leftrightarrow$, where $f_{e^\leftarrow} := -f_e$ for every edge $e \in E^\rightarrow$ (Kirchhoff's potential law).*

Proof. (i)$\Rightarrow$(iii): Let f be the electrical flow, and suppose $\sum_{e \in C} \frac{f_e}{\lambda_e} \neq 0$ for some directed circuit C. Then let $\varepsilon := \left(\sum_{e \in C} \frac{f_e}{\lambda_e}\right) / \left(\sum_{e \in C} \frac{1}{\lambda_e}\right)$, and let $g \in \mathbb{R}^{E^\rightarrow}$ be defined by $g_e = f_e - \varepsilon$ for $e \in C$ and $g_e = f_e$ elsewhere. Then g is an s-t-flow of value 1 and

$$
\begin{aligned}
\mathcal{E}(g) - \mathcal{E}(f) &= \sum_{e \in E^\rightarrow} \frac{1}{\lambda_e} \left((g_e)^2 - (f_e)^2 \right) \\
&= \sum_{e \in C} \left(\frac{\varepsilon^2}{\lambda_e} - 2\varepsilon \frac{f_e}{\lambda_e} \right) \\
&= -\varepsilon \sum_{e \in C} \frac{f_e}{\lambda_e} \\
&< 0,
\end{aligned}
$$

contradicting the assumption that f is the electrical flow.

(iii)⇒(ii): Let (V, S) be any anti-arborescence in $G^{\leftrightarrow}$ rooted at t. Let π_v be the distance from v to t in (V, S) with respect to the edge weights $\frac{f_e}{\lambda_e}$ (for $e \in S$). Then for all $e = (v, w) \in S$, we have $\pi_v - \pi_w = \frac{f_e}{\lambda_e}$. Now let $e = (v, w) \in E^{\rightarrow} \setminus S$. Then $\pi_v - \pi_w = \mathrm{dist}_{(S,f/\lambda)}(v, t) - \mathrm{dist}_{(S,f/\lambda)}(w, t) = \frac{f_e}{\lambda_e}$, where the last equation follows from (iii) applied to the unique directed circuit in $S^{\leftrightarrow} \cup \{e\}$ that contains e. Hence, this potential function π satisfies (ii) (and is independent of the choice of S).

(ii)⇒(i): Let π be a potential as in (ii). For any s-t-flow g of value 1 with $g \neq f$, we have

$$
\begin{aligned}
&\mathcal{E}(g) - \mathcal{E}(f) \\
&\quad = \sum_{e \in E^{\rightarrow}} \tfrac{1}{\lambda_e}\left((g_e)^2 - (f_e)^2\right) \\
&\quad = \sum_{e \in E^{\rightarrow}} \tfrac{1}{\lambda_e}(g_e - f_e)^2 + \sum_{e \in E^{\rightarrow}} \tfrac{2}{\lambda_e} f_e (g_e - f_e) \\
&\quad = \sum_{e \in E^{\rightarrow}} \tfrac{1}{\lambda_e}(g_e - f_e)^2 + \sum_{e=(v,w) \in E^{\rightarrow}} 2(\pi_v - \pi_w)(g_e - f_e) \\
&\quad = \sum_{e \in E^{\rightarrow}} \tfrac{1}{\lambda_e}(g_e - f_e)^2 \\
&\qquad\quad + \sum_{v \in V} 2\pi_v \left(\left(g(\delta^+(v)) - g(\delta^-(v))\right) - \left(f(\delta^+(v)) - f(\delta^-(v))\right)\right) \\
&\quad = \sum_{e \in E^{\rightarrow}} \tfrac{1}{\lambda_e}(g_e - f_e)^2 \\
&\quad > 0,
\end{aligned}
$$

where we used in the last equation that both f and g are s-t-flows of value 1. □

The electrical flow can be computed easily by linear algebra, using the Laplacian matrix:

Definition 5.10 (Laplacian). Given an undirected graph $G = (V, E)$ and $\lambda_e > 0$ for $e \in E$, the weighted Laplacian of (G, λ) is the matrix $L = A \,\mathrm{diag}(\lambda) A^{\top}$ where $\mathrm{diag}(\lambda)$ is the diagonal matrix with entries λ_e ($e \in E$) and A is the vertex-edge incidence matrix of any orientation of G – that is, the matrix whose column with index $e = \{v, w\} \in E$ has entries $a_{v,e} = 1$ and $a_{w,e} = -1$ if e is oriented from v to w, and $a_{u,e} = 0$ for all $u \in V \setminus \{v, w\}$.

Note that this is well-defined: The matrix L is independent of the orientation of G. The weighted Laplacian is always symmetric and positive semi-definite: For every $\pi \in \mathbb{R}^V$, we have $\pi^{\top} L \pi = \sum_{e=\{v,w\} \in E} \lambda_e (\pi_v - \pi_w)^2 \geq 0$. Now we can compute the electrical flow as follows:

Corollary 5.11. *Let $G = (V, E)$ be a connected undirected graph, $\lambda_e > 0$ for $e \in E$, and $s, t \in V$. Consider the linear equation system $L\pi = b$, where L is the weighted Laplacian of (G, λ) and $b = \chi^{\{s\}} - \chi^{\{t\}}$ is the vector $b \in \{-1, 0, 1\}^V$ with $b_s = 1$, $b_t = -1$, and $b_v = 0$ for $v \in V \setminus \{s, t\}$. Then $L\pi = b$ has a solution, and for every solution $\pi \in \mathbb{R}^V$, we obtain the electrical flow f by $f_e = \lambda_e(\pi_v - \pi_w)$ for all $e = (v, w) \in E^{\rightarrow}$. Moreover, $\mathcal{E}(f) = \pi_s - \pi_t$.*

Proof. Let f be the electrical flow and π as in Theorem 5.9 (ii). Let A be as in Definition 5.10. Then $f = \mathrm{diag}(\lambda)A^\top\pi$ and hence $L\pi = Af = b$.

Next, let π be a solution to $L\pi = b$ and $f_e = \lambda_e(\pi_v - \pi_w)$ for all $e = (v, w) \in E^{\rightarrow}$. Then f is an s-t-flow of value 1 because $f = \mathrm{diag}(\lambda)A^\top\pi$ implies $Af = L\pi = b$. By Theorem 5.9 (ii)⇒(i), f is the electrical flow. Finally, $\mathcal{E}(f) = f^\top\mathrm{diag}(\lambda)^{-1}f = (\mathrm{diag}(\lambda)A^\top\pi)^\top\mathrm{diag}(\lambda)^{-1}\mathrm{diag}(\lambda)A^\top\pi = \pi^\top A\,\mathrm{diag}(\lambda)A^\top\pi = \pi^\top L\pi = \pi^\top b = \pi_s - \pi_t$. □

Corollary 5.11 also shows that for connected graphs, the solution π to $L\pi = b$ is unique up to adding a constant to all entries (because the electrical flow is unique), and hence L has rank $n - 1$. The linear equation system $L\pi = b$ can be solved by Gaussian elimination in polynomial time (Edmonds [1967b]). In fact, an approximate solution can be computed in (randomized) near-linear time, as shown by Spielman and Teng [2014].

Since $\mathcal{E}(f) = \pi_s - \pi_t$, the electrical network behaves like a single resistor between s and t with resistance $\pi_s - \pi_t$; hence, the following useful notation:

Definition 5.12 (effective resistance)**.** Let $G = (V, E)$ be a connected undirected graph and $\lambda_e > 0$ for $e \in E$. Let $s, t \in V$. The *effective resistance* between s and t is defined as

$$\mathrm{R_{eff}}(s, t) := \pi_s - \pi_t,$$

where π is a solution to $L\pi = b$ for the weighted Laplacian L of (G, λ) and $b = \chi^{\{s\}} - \chi^{\{t\}}$.

An important property (that will lead us to negative correlation) is Rayleigh's monotonicity law: If resistances increase, effective resistances cannot decrease.

Lemma 5.13 (Rayleigh [1871])**.** *Let (G, λ, s, t) be an electrical network and let $0 < \lambda'_e \le \lambda_e$ for all $e \in E$. Then $\mathrm{R_{eff}}(s, t)$ is not smaller in (G, λ', s, t) than in (G, λ, s, t).*

Proof. Let f be the electrical flow in (G, λ, s, t) and g be the electrical flow in (G, λ', s, t). In the original network, f has minimum energy, so

$$\sum_{e \in E^{\rightarrow}} \frac{(f_e)^2}{\lambda_e} \le \sum_{e \in E^{\rightarrow}} \frac{(g_e)^2}{\lambda_e} \le \sum_{e \in E^{\rightarrow}} \frac{(g_e)^2}{\lambda'_e}.$$

But (by the last sentence of Corollary 5.11) the left-hand side is the effective resistance between s and t in (G, λ), and the right-hand side is the effective resistance between s and t in (G, λ'). □

We now return to random spanning trees. Let $G = (V, E)$ be a connected undirected graph and again $\mathcal{S}$ the set of edge sets of spanning trees in G. For any positive edge weights $\lambda_e > 0$ for $e \in E$, we consider the probability distribution μ^λ with $\mu^\lambda(S) = \frac{1}{\Lambda} \prod_{e \in S} \lambda_e$ for $S \in \mathcal{S}$, where $\Lambda = \sum_{S \in \mathcal{S}} \prod_{e \in S} \lambda_e$. Such a distribution is called λ-uniform.

A useful property of λ-uniform distributions is that we can easily condition on the event that an edge belongs or does not belong to the spanning tree:

Proposition 5.14. *Let $G = (V, E)$ be a connected undirected graph, $\lambda \in \mathbb{R}^E_{>0}$, and S be a random spanning tree sampled according to the λ-uniform distribution. Let $e \in E$. Then for all spanning trees (V, S^+) and (V, S^-) with $e \in S^+ \setminus S^-$, we have*

$$\begin{aligned} \mathbb{P}[S = S^+ \mid e \in S] &= \mu_{G/e}(S^+/e) \text{ and} \\ \mathbb{P}[S = S^- \mid e \notin S] &= \mu_{G-e}(S^-), \end{aligned}$$

where $\mu_{G/e}$ and μ_{G-e} denote the $\lambda|_{E(G/e)}$-uniform spanning tree distribution on G/e and the $\lambda|_{E(G-e)}$-uniform spanning tree distribution on $G - e$, respectively.

Proof. We have

$$\begin{aligned} \mathbb{P}[S = S^+ | e \in S] &= \frac{\prod_{f \in S^+} \lambda_f}{\sum_{S \in \mathcal{S}: e \in S} \prod_{f \in S} \lambda_f} \\ &= \frac{\prod_{f \in S^+/e} \lambda_f}{\sum_{S \in \mathcal{S}[G/e]} \prod_{f \in S} \lambda_f} \\ &= \mu_{G/e}(S^+/e) \end{aligned}$$

and

$$\mathbb{P}[S = S^- \mid e \notin S] = \frac{\prod_{f \in S^-} \lambda_f}{\sum_{S \in \mathcal{S}: e \notin S} \prod_{f \in S} \lambda_f} = \mu_{G-e}(S^-). \qquad □$$

We will prove that the random variables $|\{e\} \cap S|$ $(e \in E)$ are negatively correlated if $S \in \mathcal{S}$ is chosen according to the distribution μ^λ. This will be easy by exploiting the connection to electrical flows. In fact, as we show now, the probability that an edge e belongs to a λ-uniform spanning tree is λ_e times the effective resistance of e (see Figure 5.4 for examples). Recall that we always work with an arbitrary orientation of G and write $f_{e^\leftarrow} := -f_e$ for every edge $e \in E^\rightarrow$.

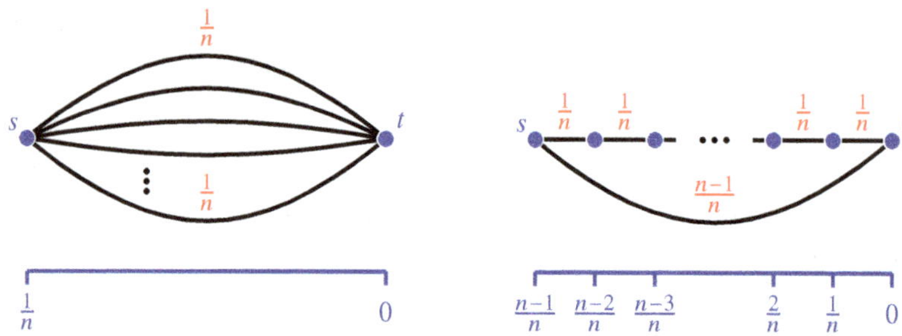

Figure 5.4 Two graphs $G = (V, E)$, both with n edges and with $\lambda_e = 1$ for all $e \in E$. The electrical s-t-flow of value 1 is shown in red above the edges, flowing from left to right. A corresponding potential π is shown in blue below. The effective resistance of every edge is $\frac{1}{n}$ on the left and $\frac{n-1}{n}$ on the right; this is the probability that an edge belongs to a random spanning tree in the uniform distribution.

Theorem 5.15 (Kirchhoff [1847]). *Let $G = (V, E)$ be a connected undirected graph and $\lambda_e > 0$ for $e \in E$. Let $\bar{e} = \{s, t\} \in E$, and f the electrical s-t-flow of value 1 in (G, λ). Then $\mathbb{P}[\bar{e} \in S] = f_{\vec{e}}$, where $\mathbb{P}$ refers to the probability distribution μ^λ and $\vec{e} = (s, t)$.*

Proof. Let again A denote the vertex-edge-incidence matrix of the orientation $G^{\rightarrow} = (V, E^{\rightarrow})$ of G and $b = \chi^{\{s\}} - \chi^{\{t\}}$.

Let $\hat{f} := \frac{1}{\Lambda} \sum_{S \in \mathcal{S}} \prod_{e \in S} \lambda_e f^S$, where $f^S \in \{-1, 0, 1\}^{E^{\rightarrow}}$ denotes the unique s-t-flow of value 1 in $S^{\rightarrow}$ (i.e., $Af^S = b$ and $f^S_e = 0$ for $e \in E^{\rightarrow} \setminus S^{\rightarrow}$). Then, looking at the component with index $\vec{e} = (s, t)$, we get

$$\hat{f}_{\vec{e}} = \frac{1}{\Lambda} \sum_{S \in \mathcal{S}} \prod_{e \in S} \lambda_e f^S_{\vec{e}} = \frac{1}{\Lambda} \sum_{S \in \mathcal{S} : \bar{e} \in S} \prod_{e \in S} \lambda_e = \mathbb{P}[\bar{e} \in S].$$

Since $\hat{f}$ is a convex combination of the f^S, we have $A\hat{f} = b$. We claim that $\hat{f}$ is the electrical flow. To this end, we check Kirchhoff's potential law (i.e., (iii) of Theorem 5.9).

Let $\mathcal{U} = \{U \subsetneq V : s \in U, t \notin U, G[U] \text{ connected}, G[V \setminus U] \text{ connected}\}$. Note that $|f^S_e| = 1$ if and only if there is a set $U \in \mathcal{U}$ such that e is the only edge in $S \cap \delta(U)$, and the sign depends on whether e leaves or enters U. If U exists, it is unique. Each such spanning tree S consists of e, a spanning tree $S' \in \mathcal{S}[U]$, and a spanning tree $S'' \in \mathcal{S}[V \setminus U]$, where $\mathcal{S}[U]$ denotes the set of (edge sets of) spanning trees in $G[U]$.

Let C be a directed circuit in $G^{\leftrightarrow}$. Using the above observation, we have

$$\begin{aligned}
&\Lambda \sum_{e \in C} \frac{\hat{f}_e}{\lambda_e} \\
&= \sum_{e \in C} \frac{1}{\lambda_e} \sum_{S \in \mathcal{S}} \prod_{e' \in S} \lambda_{e'} f_e^S \\
&= \sum_{e \in C} \sum_{S \in \mathcal{S}} \prod_{e' \in S \setminus \{e\}} \lambda_{e'} f_e^S \\
&= \sum_{e \in C} \sum_{S \in \mathcal{S}} \prod_{e' \in S \setminus \{e\}} \lambda_{e'} \sum_{U \in \mathcal{U}: S \cap \delta(U) = \{e\}} \left(|\delta^+(U) \cap \{e\}| - |\delta^-(U) \cap \{e\}|\right) \\
&= \sum_{e \in C} \sum_{U \in \mathcal{U}} \sum_{S \in \mathcal{S}: S \cap \delta(U) = \{e\}} \prod_{e' \in S \setminus \{e\}} \lambda_{e'} \left(|\delta^+(U) \cap \{e\}| - |\delta^-(U) \cap \{e\}|\right) \\
&= \sum_{U \in \mathcal{U}} \sum_{S' \in \mathcal{S}[U]} \sum_{S'' \in \mathcal{S}[V \setminus U]} \prod_{e' \in S' \cup S''} \lambda_{e'} \left(|\delta^+(U) \cap C| - |\delta^-(U) \cap C|\right) \\
&= 0
\end{aligned}$$

because for every $U \subseteq V$, every directed circuit C enters and leaves U the same number of times. By Theorem 5.9 (iii)⇒(i), $\hat{f}$ is the electrical flow. □

This proof is taken from Williamson [2019]. The special case $\lambda_e = 1$ for all $e \in E$ says that the fraction of spanning trees containing an edge $\bar{e}$ equals the effective resistance of $\bar{e}$.

Now it is easy to show that λ-uniform distributions have negative correlation (for any λ):

Theorem 5.16 (Brooks et al. [1940], Feder and Mihail [1992]). *Let $G = (V, E)$ be a connected undirected graph and $\lambda_e > 0$ for $e \in E$. Let $\mathcal{S}$ be the set of edge sets of spanning trees in G, and let $S \in \mathcal{S}$ be chosen randomly according to the λ-uniform probability distribution μ^λ. Then the 0-1 random variables $|\{e\} \cap S|$ (for $e \in E$) are negatively correlated.*

Proof. We show $\mathbb{P}[F \subseteq S] \leq \prod_{e \in F} \mathbb{P}[e \in S]$ for any $F \subseteq E$, by induction on $|F|$, where S is a random variable distributed according to μ^λ.

For $|F| = 2$, let $F = \{e', e''\}$. Negative correlation is obvious if e' and e'' are parallel edges, so assume that this is not the case. Consider the effective resistance of $e' = \{s, t\}$ as a function of $\lambda_{e''}$ (keeping all other λ-values fixed). If π is a solution to $L\pi = b$ as in Corollary 5.11 and f is the electrical flow, we have, using Kirchhoff's Theorem 5.15,

$$\mathrm{R_{eff}}(s,t) = \pi_s - \pi_t = \frac{f_{(s,t)}}{\lambda_{e'}} = \frac{\mathbb{P}[e' \in S]}{\lambda_{e'}} = \frac{\sum_{S \in \mathcal{S}[G/e']} \prod_{e \in S} \lambda_e}{\sum_{S \in \mathcal{S}} \prod_{e \in S} \lambda_e}.$$

Then by Rayleigh's monotonicity law (Lemma 5.13),

$$0 \geq \frac{\partial \, \mathrm{R_{eff}}(s,t)}{\partial \lambda_{e''}}$$
$$= \frac{\left(\sum_{S \in \mathcal{S}[G/e'/e'']} \prod_{e \in S} \lambda_e\right)\left(\sum_{S \in \mathcal{S}} \prod_{e \in S} \lambda_e\right) - \left(\sum_{S \in \mathcal{S}[G/e']} \prod_{e \in S} \lambda_e\right)\left(\sum_{S \in \mathcal{S}[G/e'']} \prod_{e \in S} \lambda_e\right)}{\left(\sum_{S \in \mathcal{S}} \prod_{e \in S} \lambda_e\right)^2}.$$

Multiplying this inequality by $\lambda_{e'}\lambda_{e''}$ and setting $\Lambda = \sum_{S \in \mathcal{S}} \prod_{e \in S} \lambda_e$ yields

$$\frac{1}{\Lambda} \sum_{S \in \mathcal{S}: e', e'' \in S} \prod_{e \in S} \lambda_e \leq \left(\frac{1}{\Lambda} \sum_{S \in \mathcal{S}: e' \in S} \prod_{e \in S} \lambda_e\right)\left(\frac{1}{\Lambda} \sum_{S \in \mathcal{S}: e'' \in S} \prod_{e \in S} \lambda_e\right),$$

which means $\mathbb{P}[e', e'' \in S] \leq \mathbb{P}[e' \in S] \cdot \mathbb{P}[e'' \in S]$ and hence pairwise negative correlation.

If $|F| > 2$, let $e' \in F$. Applying the induction hypothesis first to $(G/e', F \setminus \{e'\})$ and then to $(G, \{e', e''\})$ for all $e'' \in F \setminus \{e'\}$ (using Proposition 5.14), we obtain

$$\begin{aligned}
\mathbb{P}[F \subseteq S] &= \mathbb{P}[e' \in S] \cdot \mathbb{P}[F \subseteq S \mid e' \in S] \\
&\leq \mathbb{P}[e' \in S] \cdot \prod_{e'' \in F \setminus \{e'\}} \mathbb{P}[e'' \in S \mid e' \in S] \\
&\leq \mathbb{P}[e' \in S] \cdot \prod_{e'' \in F \setminus \{e'\}} \mathbb{P}[e'' \in S] \\
&= \prod_{e \in F} \mathbb{P}[e \in S]
\end{aligned}$$

as required. □

This is sufficient for the purpose of this chapter, and the remaining results of this section will only be used in Chapter 11.

The negative correlation property from Theorem 5.16 can be generalized, as Feder and Mihail [1992] showed. In Chapter 11, we will need to condition on the event that S induces a tree on a subset U of vertices:

Corollary 5.17 (Feder and Mihail [1992]). *Let $G = (V, E)$ be a connected undirected graph. Sample the edge set S of a spanning tree from a λ-uniform distribution. Let $\emptyset \neq U \subseteq V$, let τ denote the event that $(U, S[U])$ is a tree, and let $A \subseteq E[U]$ and $B \subseteq E \setminus E[U]$. Suppose τ has positive probability. Then*

$$\mathbb{E}[|A \cap S| \mid \tau] \ \geq\ \mathbb{E}[|A \cap S|] \quad \textit{and} \quad \mathbb{E}[|B \cap S| \mid \tau] \ \leq\ \mathbb{E}[|B \cap S|].$$

Proof. Let $n = |V|$. For an edge e, we denote the events that $e \in S$ and that $e \notin S$ simply by e and $\bar{e}$, respectively. We claim

$$\mathbb{P}[\tau \mid e] \geq^* \mathbb{P}[\tau] \quad \text{for all } e \in A. \tag{5.2}$$

(For $e \in B$, all inequalities marked with an asterisk will be reversed.) This is equivalent to $\mathbb{P}[e \mid \tau] \geq^* \mathbb{P}[e]$ for all $e \in A$ and implies

$$\mathbb{E}[|A \cap S| \mid \tau] \ =\ \sum_{e \in A} \mathbb{P}[e \mid \tau] \ \geq^*\ \sum_{e \in A} \mathbb{P}[e] \ =\ \mathbb{E}[|A \cap S|]$$

as required.

We prove the claim (5.2) by induction on $|E|$. The assertion is trivial if $n \leq 2$. Let now $n \geq 3$ and $e \in A$. If E contains a bridge (an edge that belongs to every spanning tree), we contract it and apply the induction hypothesis. Now assume that E contains no bridge.

Because $\sum_{f \in E \setminus \{e\}} \mathbb{P}[f \mid \tau, e] = n - 2 = \sum_{f \in E \setminus \{e\}} \mathbb{P}[f \mid e]$, there exists an edge f with $\mathbb{P}[f \mid \tau, e] \leq^* \mathbb{P}[f \mid e]$ and $\mathbb{P}[f \mid e] > 0$. This implies $\mathbb{P}[\tau \mid f, e] \leq^* \mathbb{P}[\tau \mid e]$ and hence $\mathbb{P}[\tau \mid f, e] \leq^* \mathbb{P}[\tau \mid \bar{f}, e]$. Using this and $\mathbb{P}[f \mid e] \leq \mathbb{P}[f]$ (negative correlation, cf. Theorem 5.16), we compute

$$\begin{aligned}
\mathbb{P}[\tau \mid e] &= \mathbb{P}[f \mid e] \cdot \mathbb{P}[\tau \mid f, e] + \mathbb{P}[\bar{f} \mid e] \cdot \mathbb{P}[\tau \mid \bar{f}, e] \\
&= \mathbb{P}[f \mid e] \cdot \big(\mathbb{P}[\tau \mid f, e] - \mathbb{P}[\tau \mid \bar{f}, e]\big) + \mathbb{P}[\tau \mid \bar{f}, e] \\
&\geq^* \mathbb{P}[f] \cdot \big(\mathbb{P}[\tau \mid f, e] - \mathbb{P}[\tau \mid \bar{f}, e]\big) + \mathbb{P}[\tau \mid \bar{f}, e] \\
&= \mathbb{P}[f] \cdot \mathbb{P}[\tau \mid f, e] + \mathbb{P}[\bar{f}] \cdot \mathbb{P}[\tau \mid \bar{f}, e].
\end{aligned}$$

By applying the induction hypothesis to G/f and to $G - f$ (and their induced λ-uniform distributions for the appropriate restrictions of λ; cf. Proposition 5.14), we obtain $\mathbb{P}[\tau \mid f, e] \geq^* \mathbb{P}[\tau \mid f]$ and $\mathbb{P}[\tau \mid \bar{f}, e] \geq^* \mathbb{P}[\tau \mid \bar{f}]$, respectively. We conclude $\mathbb{P}[\tau \mid e] \geq^* \mathbb{P}[f] \cdot \mathbb{P}[\tau \mid f] + \mathbb{P}[\bar{f}] \cdot \mathbb{P}[\tau \mid \bar{f}] = \mathbb{P}[\tau]$ as claimed. □

As we show next, conditioning on the event that a subset U induces a subtree, the distributions inside U and outside U are independent. When writing $S \sim \mu$, we mean that S is sampled from the probability distribution μ.

Lemma 5.18. *Let $G = (V, E)$ be a connected undirected graph, $\lambda \in \mathbb{R}^E_{>0}$, and μ the λ-uniform spanning tree distribution on G. Let $\emptyset \neq U \subseteq V$ such that $G[U]$*

is connected, and let $\tilde{\mu}$ be the distribution we obtain from μ by conditioning on the event that $(U, S[U])$ is a tree. Then, for every spanning tree T of G,

$$\tilde{\mu}(T) = \mu_{G[U]}(T[U]) \cdot \mu_{G/U}(T/U),$$

where $\mu_{G[U]}$ and $\mu_{G/U}$ denote the $\lambda|_{E[U]}$-uniform spanning tree distribution on $G[U]$ and the $\lambda|_{E(G/U)}$-uniform spanning tree distribution on G/U, respectively.

Proof. We may assume that $(U, T[U])$ is a tree because otherwise the assertion is trivial. Using that S is a spanning tree in G such that $(U, S[U])$ is a tree if and only if $S[U]$ is a spanning tree of $G[U]$ and S/U is a spanning tree of G/U, we get

$$\begin{aligned}
&\mathbb{P}_{S\sim\mu}\Big[S = T \,\Big|\, (U, S[U]) \text{ is a tree}\Big] \\
&= \frac{\prod_{e\in T}\lambda_e}{\sum_{S\in\mathcal{S}:(U,S[U]) \text{ is a tree}} \prod_{e\in S}\lambda_e} \\
&= \frac{\prod_{e\in T[U]}\lambda_e \cdot \prod_{e\in T/U}\lambda_e}{\left(\sum_{S \text{ spanning tree in } G[U]} \prod_{e\in S}\lambda_e\right)\cdot\left(\sum_{S \text{ spanning tree in } G/U} \prod_{e\in S}\lambda_e\right)} \\
&= \mu_{G[U]}(T[U]) \cdot \mu_{G/U}(T/U).
\end{aligned}$$

□

In Chapter 11, we will also need another version of Theorem 5.15, known as Kirchhoff's matrix tree theorem:

Theorem 5.19 (Kirchhoff [1847]). *Let $G = (V, E)$ be an undirected graph with $n := |V| \geq 2$. Let $\mathcal{S}$ denote the set of edge sets of spanning trees and $\lambda \in \mathbb{R}^E_{>0}$. Then $\sum_{S\in\mathcal{S}} \prod_{e\in S} \lambda_e$ is equal to the cofactor* $\det L^{1..n-1}$ *of the weighted Laplacian L of (G, λ), where $L^{1..n-1}$ arises from L by deleting the n-th row and the n-th column.*

Proof. If G is disconnected, then it contains no spanning tree. For a connected component that does not contain vertex n, its weighted Laplacian L' is singular (all column sums are zero) and part of $L^{1..n-1}$ – that is, after permutation of rows and columns $L^{1..n-1} = \begin{pmatrix} L' & 0 \\ 0 & L'' \end{pmatrix}$ for some matrix L''. Hence, $\det L^{1..n-1} = (\det L')(\det L'') = 0 \cdot \det L'' = 0$.

Now assume that G is connected. We use induction on $|V|$. The base case $|V| = 2$ is trivial because every edge forms a spanning tree and $\sum_{e\in E}\lambda_e$ is the only entry of $L^{1..n-1}$. Let now $|V| \geq 3$.

Let $\bar{e} = \{s, t\} \in E$ such that t corresponds to the n-th row and column of L, and without loss of generality, s corresponds to the row and column $n-1$. By Corollary 5.11, the system $L\pi = (0, \ldots, 0, 1, -1)^\top$, $\pi_n = 0$ has a unique solution, which we obtain by solving $L^{1..n-1}\pi = (0, \dots, 0, 1)^\top$ because the sum of all rows of L is the all-zero vector. By Cramer's rule, $\pi_{n-1} = \frac{\det L^{1..n-2}}{\det L^{1..n-1}}$.

Let $\Lambda(G) := \sum_{S \in \mathcal{S}} \prod_{e \in S} \lambda_e$. By Theorem 5.15,

$$\frac{\lambda_{\bar{e}} \Lambda(G/\bar{e})}{\Lambda(G)} = \mathbb{P}[\bar{e} \in S] = f_{(s,t)} = \lambda_{\bar{e}}(\pi_{n-1} - \pi_n) = \lambda_{\bar{e}} \frac{\det L^{1..n-2}}{\det L^{1..n-1}}.$$

By the induction hypothesis, $\Lambda(G/\bar{e}) = \det L^{1..n-2}$, and we conclude $\Lambda(G) = \det L^{1..n-1}$. □

5.4 How to Sample Spanning Trees

In this section, we complete the proof of Theorem 5.3 and thus the analysis of the randomized $O(\log n / \log \log n)$-approximation algorithm for the Asymmetric TSP. By Theorem 5.16 and Theorem 5.7, it suffices to sample a spanning tree from a λ-uniform distribution such that $\mathbb{P}[e \in S] \le z_e$, where z is given by Step (2) of Algorithm 5.4. We set $q_e = \frac{n-1}{n} z_e$ for $e \in E$, recall that q is in (the relative interior of) the spanning tree polytope, and aim at $\mathbb{P}[e \in S] = q_e$ (such a distribution will be called *marginal-preserving*).

Asadpour et al. [2017] showed that a λ-uniform marginal-preserving distribution exists and that it is the unique distribution μ that maximizes the *entropy* $\sum_{S \in \mathcal{S}: \mu(S) > 0} \mu(S) \ln \frac{1}{\mu(S)}$ among all marginal-preserving distributions (cf. Exercise 5.12). The entropy is a measure of how even a distribution is. For example, the entropy is 0 if $\mu(S) = 1$ for some $S \in \mathcal{S}$ (no randomness), and the entropy is maximized for the uniform distribution ($\mu(S) = \frac{1}{|\mathcal{S}|}$ for all $S \in \mathcal{S}$); then the entropy is $\ln |\mathcal{S}|$. For a simpler notation, we will write $\mu(S) \ln \frac{1}{\mu(S)} = 0$ if $\mu(S) = 0$. Moreover, we again abbreviate $\gamma(S) := \sum_{e \in S} \gamma_e$.

Theorem 5.20 (Asadpour et al. [2017]). *Let $G = (V, E)$ be a graph and q in the relative interior of the spanning tree polytope. Then*

$$\sup \left\{ \sum_{S \in \mathcal{S}} \mu(S) \ln \frac{1}{\mu(S)} : \mu : \mathcal{S} \to \mathbb{R}_{\ge 0}, \sum_{S \in \mathcal{S}: e \in S} \mu(S) = q_e \ (e \in E) \right\}$$
$$= \inf \left\{ \sum_{S \in \mathcal{S}} e^{\gamma(S)} - 1 - \sum_{e \in E} \gamma_e q_e : \gamma \in \mathbb{R}^E \right\},$$

there exist unique optimum solutions μ^* and γ^*, and we have $\mu^*(S) = e^{\gamma^*(S)}$ for all $S \in \mathcal{S}$.

Summing over the marginal-preserving constraints $\sum_{S \in \mathcal{S}: e \in S} \mu(S) = q_e$ for all $e \in E$, we get $(n-1)\mu(\mathcal{S}) = q(E) = n-1$ because q is in the spanning tree polytope, and thus μ is indeed a probability distribution. Theorem 5.20 is a strong duality theorem and can be shown by nonlinear programming theory. For our purposes, we only need the inequality "$\le$" (weak duality), and we do not need that the infimum is attained. This is easy to show:

Proposition 5.21. *Let $G = (V, E)$ be a graph and q in the spanning tree polytope. Then*

$$\sup\left\{\sum_{S\in\mathcal{S}} \mu(S)\ln\tfrac{1}{\mu(S)} : \mu : \mathcal{S} \to \mathbb{R}_{\ge 0},\ \sum_{S\in\mathcal{S}:e\in S} \mu(S) = q_e\ (e \in E)\right\}$$
$$\le \inf\left\{\sum_{S\in\mathcal{S}} e^{\gamma(S)} - 1 - \sum_{e\in E} \gamma_e q_e : \gamma \in \mathbb{R}^E\right\}.$$

Proof. In the supremum, we maximize a strictly concave function over a compact convex set, which is nonempty because q is in the spanning tree polytope. Hence the supremum is attained by a unique distribution μ^*. The supremum is at most the Lagrangian dual

$$\inf\left\{\sup\left\{\sum_{S\in\mathcal{S}} \mu(S)\ln\tfrac{1}{\mu(S)} + \sum_{e\in E}\kappa_e\left(\sum_{S\in\mathcal{S}:e\in S}\mu(S) - q_e\right) : \mu : \mathcal{S}\to\mathbb{R}_{\ge 0}\right\} : \kappa\in\mathbb{R}^E\right\} \tag{5.3}$$

because μ^* is a feasible solution to every inner optimization problem. For fixed κ, the inner supremum in (5.3) is easily obtained by optimizing each $\mu(S)$ independently, setting $\mu(S) = e^{\kappa(S)-1}$. Hence, (5.3) simplifies to

$$\inf\left\{\sum_{S\in\mathcal{S}} e^{\kappa(S)-1}(1-\kappa(S)) + \sum_{e\in E}\kappa_e\left(\sum_{S\in\mathcal{S}:e\in S} e^{\kappa(S)-1} - q_e\right) : \kappa\in\mathbb{R}^E\right\}.$$

Cancelling terms and substituting $\gamma_e := \kappa_e - \frac{1}{n-1}$ yields that the Lagrangian dual equals

$$\inf\left\{\sum_{S\in\mathcal{S}} e^{\gamma(S)} - \sum_{e\in E}\gamma_e q_e - 1 : \gamma\in\mathbb{R}^E\right\}, \tag{5.4}$$

where we used $\sum_{e\in E} q_e = n-1$ because q is in the spanning tree polytope. □

Note that the Lagrangian dual (5.4) is strictly convex and bounded from below (e.g., by 0; cf. Proposition 5.21). One can evaluate $\sum_{S\in\mathcal{S}} e^{\gamma(S)}$ for any vector γ in polynomial time using Kirchhoff's matrix tree theorem (Theorem 5.19). Alternatively, we can take an arbitrary tree $\bar{S} = \{e_1, \ldots, e_{n-1}\} \in \mathcal{S}$, compute

$\mathbb{P}[e_i \in S \mid e_1, \ldots, e_{i-1} \in S]$ for $i = 1, \ldots, n-1$ by Proposition 5.14 and Theorem 5.15 and Corollary 5.11, and note that

$$\prod_{i=1}^{n-1} \mathbb{P}\left[e_i \in S \mid e_1, \ldots, e_{i-1} \in S\right] = \mathbb{P}\left[S = \bar{S}\right] = \frac{e^{\gamma(\bar{S})}}{\sum_{S \in \mathcal{S}} e^{\gamma(S)}}.$$

Asadpour et al. [2017] showed how to compute a near-optimal solution γ to the Lagrangian dual (5.4) by convex programming techniques, noting that one can restrict the variables to a polynomially bounded range. See also Singh and Vishnoi [2014] for details and generalizations.

If γ is a near-optimal solution to (5.4), letting the probability of $S \in \mathcal{S}$ be proportional to $e^{\gamma(S)}$ yields a λ-uniform distribution (for $\lambda_f = e^{\gamma_f}$ for $f \in E$) and in fact an almost marginal-preserving distribution:

Theorem 5.22. *Let $G = (V, E)$ be a graph, $\mathcal{S}$ the set of edge sets of spanning trees, and q in the spanning tree polytope of G with $q_e > 0$ for all $e \in E$. Let $0 < \varepsilon \leq 1$ and $\eta = \frac{\varepsilon^2}{36} \min_{e \in E} q_e$. Let $\gamma \in \mathbb{R}^E$ such that*

$$\sum_{S \in \mathcal{S}} e^{\gamma(S)} - \sum_{f \in E} \gamma_f q_f - 1 \leq \sum_{S \in \mathcal{S}} e^{\delta(S)} - \sum_{f \in E} \delta_f q_f - 1 + \eta \tag{5.5}$$

for every $\delta \in \mathbb{R}^E$. Let $\mu(S) := e^{\gamma(S)} / \sum_{S' \in \mathcal{S}} e^{\gamma(S')}$ for $S \in \mathcal{S}$. Then

$$\sum_{S \in \mathcal{S}: e \in S} \mu(S) \leq (1 + \varepsilon) q_e$$

for all $e \in E$.

Proof. Let $e \in E$. Abbreviate $\hat{\varepsilon} := \frac{\varepsilon}{6}$. Define $\gamma'_e := \gamma_e - \hat{\varepsilon}$ and $\gamma'_f := \gamma_f$ for all $f \in E \setminus \{e\}$. Using (5.5) for $\delta = \gamma'$ yields

$$(1 - e^{-\hat{\varepsilon}}) \sum_{S \in \mathcal{S}: e \in S} e^{\gamma(S)} = \sum_{S \in \mathcal{S}} e^{\gamma(S)} - \sum_{S \in \mathcal{S}} e^{\gamma'(S)} \leq \eta + \hat{\varepsilon} q_e. \tag{5.6}$$

Now define $\gamma''_f := \gamma_f + \frac{\hat{\varepsilon}}{n-1}$ for all $f \in E$, where $n = |V|$. Note that $\gamma''(S) = \gamma(S) + \hat{\varepsilon}$ for all $S \in \mathcal{S}$. Using (5.5) for $\delta = \gamma''$ yields

$$(e^{\hat{\varepsilon}} - 1) \sum_{S \in \mathcal{S}} e^{\gamma(S)} = \sum_{S \in \mathcal{S}} e^{\gamma''(S)} - \sum_{S \in \mathcal{S}} e^{\gamma(S)} \geq \sum_{f \in E} \frac{\hat{\varepsilon}}{n-1} q_f - \eta = \hat{\varepsilon} - \eta \tag{5.7}$$

because $q(E) = n-1$. Using $e^x - 1 \leq x + x^2$ for all $x \leq 1$ and $\eta = \hat{\varepsilon}^2 \min_{f \in E} q_f \leq \hat{\varepsilon}^2 q_e \leq \hat{\varepsilon}^2$, we get from (5.6) and (5.7):

$$(\hat{\varepsilon} - \hat{\varepsilon}^2) \sum_{S \in \mathcal{S}: e \in S} e^{\gamma(S)} \leq \hat{\varepsilon}(1 + \hat{\varepsilon}) q_e$$

and

$$(\hat{\varepsilon} + \hat{\varepsilon}^2) \sum_{S \in \mathcal{S}} e^{\gamma(S)} \geq \hat{\varepsilon}(1 - \hat{\varepsilon}).$$

Putting these inequalities together yields

$$\sum_{S\in\mathcal{S}:e\in S}\mu(S) = \frac{\sum_{S\in\mathcal{S}:e\in S}e^{\gamma(S)}}{\sum_{S\in\mathcal{S}}e^{\gamma(S)}} \le \frac{(1+\hat{\varepsilon})q_e/(1-\hat{\varepsilon})}{(1-\hat{\varepsilon})/(1+\hat{\varepsilon})} = \frac{(1+\hat{\varepsilon})^2}{(1-\hat{\varepsilon})^2}q_e \le (1+\varepsilon)q_e,$$

where the last inequality follows from $6\hat{\varepsilon} = \varepsilon \le 1$. □

Hence, we get an almost marginal-preserving λ-uniform distribution (with $\lambda_f = e^{\gamma_f}$ for all $f \in E$):

Corollary 5.23. *Given an undirected graph $G = (V, E)$, a number $\varepsilon > 0$, and a vector $q \in \mathbb{R}^E_{>0}$ in the spanning tree polytope of G, we can compute in polynomial time $\lambda_e > 0$ for $e \in E$ such that the λ-uniform distribution satisfies $\mathbb{P}[e \in S] \le (1+\varepsilon)q_e$ for all $e \in E$.*

Proof. Set $\eta = \frac{\varepsilon^2}{36}\min_{e\in E} q_e$. Compute a near-optimal solution γ to (5.4), satisfying (5.5) (see Singh and Vishnoi [2014] for details). Set $\lambda_f = e^{\gamma_f}$ for all $f \in E$ and apply Theorem 5.22. □

Asadpour et al. [2017] also suggested an alternative combinatorial approach using multiplicative weight updates (however, with a running time that depends polynomially on $\frac{1}{\varepsilon}$ instead of $\log\frac{1}{\varepsilon}$). The total difference that results from preserving the marginals only approximately can be bounded as follows:

Theorem 5.24 (Straszak and Vishnoi [2019]). *Let $G = (V, E)$ be an undirected graph with n vertices and $q \in \mathbb{R}^E_{>0}$ a vector in the spanning tree polytope of G. Let $\mathcal{S}$ be the set of edge sets of spanning trees, and μ the maximum entropy distribution on $\mathcal{S}$ with marginals q_e ($e \in E$). Let $\varepsilon_\mu > 0$ and μ^λ a λ-uniform distribution on $\mathcal{S}$ with $\mathbb{P}_{S\sim\mu^\lambda}[e \in S] \le \left(1 + \frac{\varepsilon_\mu^3}{n^4}\right)q_e$ for all $e \in E$. Then $\sum_{S\in\mathcal{S}}|\mu^\lambda(S) - \mu(S)| \le \varepsilon_\mu$.*

We omit the proof here, and we will need this bound only in Chapter 10. Once we have the vector λ, sampling a tree from the λ-uniform distribution is easy:

Theorem 5.25. *Given a connected graph $G = (V, E)$ and $\lambda_e > 0$ for all $e \in E$, we can sample a tree from the λ-uniform distribution in polynomial time.*

Proof. We can just scan the edges one by one and pick an edge with the appropriate probability (computed by Proposition 5.14 and Theorem 5.15 and Corollary 5.11). If we pick an edge, we contract it; otherwise, we delete it. □

A faster method was proposed by Anari et al. [2021]. Corollary 5.23 applied to $q = \frac{n-1}{n}z$ and Theorem 5.25 imply Theorem 5.3. Indeed, by Theorem 5.16, the random variables $|\{e\} \cap S|$ are negatively correlated, and hence by Theorem 5.7, the tree is $\frac{3\ln n}{\ln\ln n}$-thin with respect to z with probability at least $\frac{9}{10}$. This shows Theorem 5.3.

For applying Theorem 5.7, one does not need a λ-uniform distribution. Negative correlation is sufficient. Sampling a spanning tree according to a different marginal-preserving distribution with negative correlation can be easier. Two such approaches are known. Let again $q = \frac{n-1}{n}z$.

First, the swap rounding approach of Chekuri, Vondrák, and Zenklusen [2010] starts by writing q as an arbitrary convex combination of spanning trees (which can be done in polynomial time by Theorem 4.22; see also Theorem 15.4). Then it iteratively takes two (edge sets of) spanning trees S_1, S_2 in the support and merges them: Merging S_1 and S_2 means choosing $e_1 \in S_1 \setminus S_2$ and $e_2 \in S_2 \setminus S_1$ such that $(V, S_1 \setminus \{e_1\} \cup \{e_2\})$ and $(V, S_2 \setminus \{e_2\} \cup \{e_1\})$ are trees (cf. Exercise 5.13), exchanging e_1 for e_2 in S_1 (with probability α) or e_2 for e_1 in S_2 (with probability $1 - \alpha$), until the two trees are the same. Here α is chosen so that the marginals remain constant. It is obvious that swap rounding runs in polynomial time. Chekuri, Quanrud, and Torres [2021] showed how to implement swap rounding in near-linear time.

Second, the pipage rounding approach of Ageev and Sviridenko [2004] and Calinescu et al. [2011] starts with q, then iteratively takes a minimal face F of the spanning tree polytope such that $q \in F$, as well as two edges e and f such that $q' = q + \delta(\chi^{\{e\}} - \chi^{\{f\}}) \in F$ whenever $|\delta| > 0$ is small enough. Choosing δ as large as possible (with probability α) or as small as possible (with probability $1 - \alpha$) so that $q' \in F$, we land on a lower-dimensional face, so after at most $n - 1$ steps, we end at an extreme point. Again, α is chosen so that the marginals remain constant. One can see that finding such an F and the maximum and minimum δ so that $q' \in F$ reduces to submodular function minimization (cf. Exercise 5.14).

By Exercise 5.15, both swap rounding and pipage rounding lead to probability distributions over spanning trees such that the marginals are preserved and the random variables $|\{e\} \cap S|$ are negatively correlated. In general, neither of these distributions is identical to the maximum entropy distribution. The maximum entropy distribution will be used again in Chapter 10.

5.5 Thin Trees Suffice

Goddyn [undated] conjectured that there exists a constant α such that all connected undirected graphs G have a spanning tree (V, S) with $|\delta_S(U)| \leq \frac{2\alpha}{k}|\delta_G(U)|$ for all $U \subseteq V$, where k is the edge-connectivity of G. If such a tree could be found in polynomial time, this would imply a constant-factor approximation algorithm for ASYMMETRIC TSP, as we will show in Theorem 5.27. The following version of Goddyn's thin tree conjecture is equivalent (but the constant may differ by at most a factor 2):

Open Problem 5.26. Does there exist a constant α such that for every connected undirected graph $H = (V, F)$ and any point y in the spanning tree polytope of H, there exists a spanning tree S with $|\delta_S(U)| \leq \alpha \cdot y(\delta(U))$ for all $U \subseteq V$?

If the answer is yes, the constant α must be at least 2, as the complete graph with $y_e = \frac{2}{n}$ for all edges e shows.

Oveis Gharan and Saberi [2011] showed that such a constant α exists for bounded-genus graphs (graphs that can be embedded on any fixed orientable surface).

For general graphs, Theorem 5.7, Theorem 5.16, and Corollary 5.23 imply that a spanning tree as in Open Problem 5.26 exists for $\alpha = \frac{3 \ln n}{\ln \ln n}$. Anari and Oveis Gharan [2015] even showed the upper bound $\alpha = (\log \log n)^{O(1)}$, but it is not known how to find such a thin tree in polynomial time. Nevertheless, this implies that the integrality ratio of (3.2) is at most $(\log \log n)^{O(1)}$ by Theorem 5.27. This was the best-known bound before Svensson, Tarnawski, and Végh [2020] showed a constant upper bound on this integrality ratio.

It is *NP*-hard to compute how thin a given tree is (see Exercise 5.16).

Klein and Olver [2023] showed that there is a constant α such that for every laminar family $\mathcal{L}$, there exists a spanning tree S for which $|\delta_S(U)| \leq \alpha \cdot y(\delta(U))$ for all $U \in \mathcal{L}$.

Oveis Gharan and Saberi [2011] and Goemans [2012] observed that a resolution of the thin tree conjecture would have immediate consequences for the ASYMMETRIC TSP:

Theorem 5.27. *Let $\alpha : \mathbb{N} \to \mathbb{R}_{>0}$. Suppose for every connected undirected graph $H = (V, F)$ and any point y in the spanning tree polytope of H, there exists a spanning tree S with $|\delta_S(U)| \leq \alpha(n) \cdot y(\delta(U))$ for all $U \subseteq V$, where $n = |V|$. Then the integrality ratio of* (3.2) *is at most $\frac{20}{3}\alpha(n)$. If S can be found in polynomial time, then there is a $\frac{20}{3}\alpha(n)$-approximation algorithm for the ASYMMETRIC TSP.*

Proof. Since every spanning tree is $(n-1)$-thin with repect to y, we may assume $\alpha(n) \leq n$ for $n \in \mathbb{N}$. Moreover, we may assume $n \geq 4$.

Take an optimum solution x to the LP (3.2) for an instance (G, c), and let $G = (V, E)$. We may assume that G is simple. Let $N \geq 50n^2$. Let $H = (V, F)$ be the undirected graph with vertex set V that contains $\lfloor N(x_{(v,w)} + x_{(w,v)}) \rfloor$ copies of edge e for all $e = \{v, w\} \in \binom{V}{2}$, where we set $x_{(v,w)} = 0$ if $(v, w) \notin E$.

Since $|\delta_H(U)| \geq \frac{6N}{5}$ for all $\emptyset \neq U \subsetneq V$ (actually initially more than $(2-\frac{1}{200})N$ because $x(\delta(U)) \geq 2$ and $|\delta_G(U)| \leq \frac{n^2}{4} \leq \frac{N}{200}$), the vector $y \in \mathbb{R}^F_{\geq 0}$ with all components equal to $\frac{5}{3N}$ is in the connector polytope of H by Lemma 2.28, hence in the up-hull of the spanning tree polytope of H. By our assumption, H has a thin tree S_1, with

$$
\begin{aligned}
|\delta_{S_1}(U)| &\leq \alpha(n) \cdot y(\delta(U)) \\
&= \frac{5\alpha(n)}{3N} \cdot |\delta_H(U)| \\
&\leq \frac{5\alpha(n)}{3} \cdot x(\delta(U)) \\
&= \frac{10\alpha(n)}{3} \cdot x(\delta^+(U)).
\end{aligned}
\tag{5.8}
$$

Delete the edges of S_1 from H. By the first line of (5.8), this removes at most a $\frac{5\alpha(n)}{3N}$ fraction of the edges from each cut. Then find a thin tree S_2 in the remaining graph and iterate this $\lceil \frac{3N}{10\alpha(n)} \rceil$ times. Before the last iteration, there is still at least a fraction of

$$
\left(\left(1 - \frac{5\alpha(n)}{3N} \right)^{\frac{3N}{5\alpha(n)}} \right)^{\frac{1}{2}} \geq 0.605 \tag{5.9}
$$

left from the (originally more than $(2 - \frac{1}{200})N$) edges in any cut $\delta_H(U)$, in particular more than $\frac{6N}{5}$ edges. In the inequality (5.9), we used $\alpha(n) \leq n$.

Hence (5.8) holds for all the trees that we get, which are $\lceil \frac{3N}{10\alpha(n)} \rceil$ edge-disjoint spanning trees in $H = (V, F)$. We can orient all edges of H in the cheaper direction; then $c(F) \leq Nc(x)$. The cheapest one among our oriented thin trees has cost at most $\frac{10\alpha(n)}{3N} c(F) \leq \frac{10\alpha(n)}{3} c(x)$. Now apply Lemma 5.1 to this thin tree. We get a tour with total cost at most $\frac{10\alpha(n)}{3} c(x) + \frac{10\alpha(n)}{3} c(x) = \frac{20}{3}\alpha(n) \cdot c(x)$. □

The constant $\frac{20}{3}$ can be improved slightly (see Exercise 5.17).

If the thin tree conjecture (Open Problem 5.26) holds, this would also imply a constant-factor approximation algorithm for the ASYMMETRIC BOTTLENECK TSP (An, Kleinberg, and Shmoys [2021]; see Exercise 5.5); such an algorithm is not known.

Exercises

5.1 Let $\varepsilon > 0$ and consider a minimization problem. We assume that the cost of any given feasible solution can be computed in polynomial time.

(a) Suppose there is a randomized polynomial-time algorithm such that the expected cost of the computed solution is at most $\alpha \cdot \text{OPT}$. Show that then there is a randomized polynomial-time algorithm that computes a solution with cost at most $(\alpha + \varepsilon) \cdot \mathrm{OPT}$ with probability at least $\frac{1}{2}$.

(b) Suppose there is a randomized polynomial-time algorithm that computes a solution with cost at most $\alpha \cdot \mathrm{OPT}$ with probability at least $\frac{1}{2}$, and always computes a solution with cost at most $2^n \cdot \mathrm{OPT}$ (where n is the size of the instance). Show that then there is a randomized polynomial-time algorithm such that the expected cost of the computed solution is at most $(\alpha + \varepsilon) \cdot \mathrm{OPT}$.

5.2 Show the following strengthening of Lemma 5.1. Let (V, R) be a connected spanning subgraph of a digraph $G = (V, E)$ with weights $c \in \mathbb{R}^E_{\geq 0}$, and let $x \in \mathbb{R}^E_{\geq 0}$ and $\alpha > 0$ such that $|R \cap \delta^-(U)| - |R \cap \delta^+(U)| \leq \alpha x(\delta^+(U))$ for all $U \subseteq V$. Then one can find a tour F in G with $c(F) \leq c(R) + \alpha c(x)$ in polynomial time.

5.3 Consider an algorithm that computes a minimum-cost oriented spanning tree and adds a cheapest multi-set of edges to obtain a tour. Show that the resulting tour can cost $n - 1$ times more than an optimum tour.

5.4 Let (G, c) be an instance of the Asymmetric TSP with $G = (V, E)$, and let $x \in \mathbb{R}^E_{\geq 0}$ be a feasible solution to the LP (3.2). Let (V, R) be a connected spanning subgraph of G, and let $\alpha > 0$ such that $|R \cap \delta(U)| \leq \alpha x(\delta(U))$ for all $U \subseteq V$. Prove that one can find in polynomial time a tour F in G with $|F \cap \delta(v)| \leq 4\lceil \alpha \cdot x(\delta(v)) \rceil$ for all $v \in V$.

5.5 Given an instance (V, c) of the Asymmetric TSP with Triangle Inequality, where $E = \{(v, w) \in V \times V : v \neq w\}$, the Asymmetric Bottleneck TSP asks for a Hamiltonian circuit (V, C) in (V, E) minimizing $\max_{e \in C} c(e)$.

(a) Let $k \in \mathbb{N}$. Let (V, F) be a tour in (V, E) where $|F \cap \delta(v)| \leq k$ for every $v \in V$. Prove that one can find a Hamiltonian circuit C in (V, E) such that every edge in C has length at most $k \cdot \max_{e \in F} c(e)$ in polynomial time.

(b) Let $\alpha \geq 1$. Suppose that for every feasible solution x to the LP (2.2), we can compute in polynomial time an α-thin spanning tree with respect to x in the support graph of x. Prove that then there is a $8\lceil \alpha \rceil$-approximation algorithm for the Asymmetric Bottleneck TSP.

Hint: For (a), partition an Eulerian walk into segments containing at most $\lceil \frac{k}{2} \rceil$ vertices each (with at most one segment containing fewer than $\lceil \frac{k}{2} \rceil$ vertices) and use Hall's bipartite matching theorem (cf. Theorem 3.13). For (b), apply Exercise 5.4 to a suitable subgraph (V, E') of (V, E) and then use (a).

(An, Kleinberg, and Shmoys [2021])

5.6 Consider the following randomized algorithm for the ASYMMETRIC TSP: Starting with a solution x to (3.1), make $100 \log n$ copies of each edge e, then sample every copy of every edge e independently with probability x_e. Finally, apply Lemma 5.1 to the resulting edge set R (if (V, R) is connected, otherwise output an arbitrary tour). Show that this is a randomized $O(\log n)$-approximation algorithm.
Hint: Show that (V, R) is very likely connected and thin, using Theorem 4.25 similarly as in the proof of Theorem 5.7.
(Goemans et al. [2009])

5.7 Let $G = (V, E)$ be a connected undirected graph and $\lambda \in \mathbb{R}^E_{>0}$. Show that $\mathrm{R_{eff}}(s, u) \le \mathrm{R_{eff}}(s, t) + \mathrm{R_{eff}}(t, u)$ for all $s, t, u \in V$.
Hint: Use the max-flow min-cut theorem and Exercise 2.3.

5.8 Let $G = (V, E)$ be an undirected circuit and μ a probability distribution of spanning trees in G such that every spanning tree has positive probability. Prove that then μ is a λ-uniform distribution for some $\lambda : E \to \mathbb{R}_{>0}$.

5.9 Describe an undirected graph G and a probability distribution μ of spanning trees in G such that every spanning tree has positive probability and the random variables $|\{e\} \cap S|$ are negatively correlated, but μ is not a λ-uniform distribution for any $\lambda : E \to \mathbb{R}_{>0}$.

5.10 Let $G = (V, E)$ be a connected undirected graph and (V, S) a spanning tree sampled according to a λ-uniform distribution. Let $F \subseteq E$. Prove that $\mathbb{P}[F \cap S = \emptyset] \le \prod_{f \in F} \mathbb{P}[f \notin S]$.
Hint: Let $X \subseteq E$, and let τ denote the event that $X \cap S = \emptyset$. Prove that $\mathbb{P}[\tau \mid \bar{e}] \le \mathbb{P}[\tau]$ for every $e \in E \setminus X$ that is not a bridge, similarly to the proof of Corollary 5.17.

5.11 Give a direct proof of Theorem 5.19, not using Theorem 5.15. Use induction and $\Lambda(G) = \lambda_e \Lambda(G/e) + \Lambda(G - e)$. Show that Theorem 5.19 implies Theorem 5.15.
Hint: Review the proof of Theorem 5.19.

5.12 (a) Let $G = (V, E)$ be an undirected graph and q be a vector in the spanning tree polytope of G. Let $\gamma \in \mathbb{R}^E$ and $\mu(S) := e^{\gamma(S)}$ for all $S \in \mathcal{S}$. Suppose $\sum_{S \in \mathcal{S}: e \in S} \mu(S) = q_e$ for all $e \in E$. Show that then μ is a probability distribution and maximizes the entropy over all probability distributions $\mu' : \mathcal{S} \to [0, 1]$ with $\sum_{S \in \mathcal{S}: e \in S} \mu'(S) = q_e$ for all $e \in E$.

(b) Describe an undirected graph $G = (V, E)$ and a vector q in the spanning tree polytope of G with $0 < q_e < 1$ for all $e \in E$ such that there exists no $\gamma \in \mathbb{R}^E$ and $\mu(S) := e^{\gamma(S)}$ for all $S \in \mathcal{S}$ for which $\sum_{S \in \mathcal{S}: e \in S} \mu(S) = q_e$ for all $e \in E$.
Hint: Such a γ is only guaranteed to exist if q is in the relative interior of

the spanning tree polytope – that is, all constraints $x(E[U]) \leq |U| - 1$ for $2 \leq |U| < n$ are satisfied strictly.

5.13 Let (V, S_1) and (V, S_2) be two different trees on the same vertex set. Show that there are edges $e_1 \in S_1 \setminus S_2$ and $e_2 \in S_2 \setminus S_1$ such that $(V, (S_1 \setminus \{e_1\}) \cup \{e_2\})$ and $(V, (S_2 \setminus \{e_2\}) \cup \{e_1\})$ are trees. (See also Exercise 13.10.)

5.14 Let $G = (V, E)$ be an undirected graph, and let q be a vector in the spanning tree polytope P of G. Call a set $W \subseteq V$ *tight* if $q(E[W]) = |W| - 1$.

(a) Let $e, f \in E$ such that q_e and q_f are strictly between 0 and 1. Show that finding a number $\delta \geq 0$ and a set $X \subsetneq V$ with $e \in E[X]$ and $f \notin E[X]$ such that $q' := q + \delta(\chi^{\{e\}} - \chi^{\{f\}}) \in P$ and X is tight for q' reduces to submodular function minimization (cf. Exercise 4.7).

(b) Let $W \subseteq V$ be a tight set such that there is an edge (and thus two edges) $e \in E[W]$ with $0 < q_e < 1$. Show that (a) (and an algorithm for submodular function minimization) can be used to compute a vector $q' \in P$ such that one of the following holds:

- $q'_e = 1$.
- $q'_f = 0$ for some edge f in the support of q.
- We find a proper subset $W' \subsetneq W$ with $e \in E[W']$ and $q'(E[W']) = |W'| - 1$.

Hint: Use Exercise 2.6.

(c) Using that there is a polynomial-time algorithm for submodular function minimization, conclude that pipage rounding can be implemented to run in polynomial time.

5.15 Suppose we have a probability distribution μ over points in the spanning tree polytope P of a graph, and we obtain another probability distribution by reducing the probability for some point $x \in P$ and increasing it for $x + \delta(\chi^{\{e\}} - \chi^{\{f\}}) \in P$ and $x - \delta(\chi^{\{e\}} - \chi^{\{f\}}) \in P$ so that the marginals remain the same.

(a) Show that then, for any $F \subseteq E$, the expectation of $\sum_{x \in P} \mu(x) \prod_{e' \in F} x_{e'}$ does not increase.

(b) Conclude that if we start with $\mu(x) = 1$ for some $x \in P$, apply such operations, and end with a probability distribution of (the incidence vectors of) spanning trees, then the random variables $|\{e\} \cap S|$ $(e \in E)$ are negatively correlated, where S is a random spanning tree sampled from the final distribution.

(Chekuri, Vondrák, and Zenklusen [2010])

5.16 Suppose we can compute in polynomial time $\max\left\{\frac{|\delta_S(U)|}{|\delta_G(U)|} : \emptyset \neq U \subsetneq V\right\}$ for a given undirected graph $G = (V, E)$ and a spanning tree S in G. Show that then we can solve the SPARSEST CUT PROBLEM in polynomial time: Given an undirected graph (V, E), compute $\min\left\{\frac{|\delta(U)|}{|U|\cdot|V\setminus U|} : \emptyset \neq U \subsetneq V\right\}$.
Hint: Let $G = (V, E)$ be a graph for which we want to compute a sparsest cut. If $k_{v,w}$ denotes the number of parallel edges between v and w and $K = 1 + \max\{k_{v,w} : v, w \in V,\ v \neq w\}$, let $\bar{G}$ denote the graph that has $K - k_{v,w}$ parallel edges between v and w. Now we define a new graph H. For sufficiently large N, make N copies of every element of V. Connect any two copies of the same vertex by many parallel edges. For each pair $\{v, w\} \in \binom{V}{2}$, add a matching of N edges between the copies of v and the copies of w. Call this graph H. Define a spanning tree S in H such that after contracting the copies of each vertex, S becomes $\bar{G}$.
Note: The SPARSEST CUT PROBLEM is *NP*-hard (Matula and Shahrokhi [1990]).

5.17 Improve the constant in the integrality gap upper bound of Theorem 5.27 from $\frac{20}{3}$ to less than 5.
Hint: In the i-th iteration, set $y_e = \frac{1}{N-(i-1)\alpha}$ for all $e \in H_{i-1}$, where $H_0 = H$ and H_i results from H_{i-1} by removing the edges of the tree S_i, which is α-thin with respect to y. Show that $|\delta_{H_i}(U)| \geq 2(N - i\alpha)$ for all $\emptyset \neq U \subsetneq V$ and all i. Assuming that the tour resulting from tree S_i costs at least $r\alpha c(x)$, deduce a lower bound on the total cost of $S_1, \ldots, S_{\frac{1}{\alpha}(1-\frac{2}{r})N}$.

6

Asymmetric Graph TSP

A major step towards the first constant-factor approximation algorithm for the ASYMMETRIC TSP was made by Svensson [2015]. He devised a constant-factor approximation algorithm for the ASYMMETRIC GRAPH TSP, which is the special case of the ASYMMETRIC TSP with $c(e) = 1$ for all $e \in E$.

In this chapter, we present Svensson's [2015] algorithm for the ASYMMETRIC GRAPH TSP. We also incorporate some improvements, mostly from Traub and Vygen [2022] and Traub [2020a], who gave a variant of Svensson's algorithm with improved approximation ratio. Moreover, we present an improved algorithm for finding a graph subtour cover, which is the main subroutine of Svensson's algorithm. Overall, we will obtain an approximation ratio of $8 + \varepsilon$ for every $\varepsilon > 0$. Almost all the techniques presented in this chapter will be used again in Chapters 7 and 8 for the general ASYMMETRIC TSP.

6.1 Preliminaries on Asymmetric Graph TSP

Let us formally define the problem that we deal with in this chapter.

Problem 6.1 (ASYMMETRIC GRAPH TSP).

Instance: A strongly connected directed graph $G = (V, E)$.

Task: Compute a tour in G with a minimum number of edges.

We will compare the result of the algorithm to the value of the linear programming relaxation (3.2), which we restate here for the special case of ASYMMETRIC GRAPH TSP (where $c(e) = 1$ for all $e \in E$):

$$
\begin{array}{llcll}
\min & x(E) & & & \\
\text{subject to} & x(\delta(U)) & \geq & 2 & (\emptyset \neq U \subsetneq V) \\
& x(\delta^+(v)) & = & x(\delta^-(v)) & (v \in V) \\
& x_e & \geq & 0 & (e \in E)
\end{array}
\tag{6.1}
$$

Theorem 3.18 says that this linear program has integrality ratio at least 2, and no worse example is known even for the general ASYMMETRIC TSP.

Table 6.1 summarizes the history of the ASYMMETRIC GRAPH TSP. In fact, Svensson [2015] considered *node-weighted* instances, but these are essentially equivalent (see Exercise 6.2).

Let us start with a very simple observation.

Proposition 6.2. *For every instance $G = (V, E)$ of the ASYMMETRIC GRAPH TSP with n vertices, we have $n \leq \mathrm{LP} \leq \mathrm{OPT} \leq n^2$, where* LP *denotes the value of the linear program* (6.1).

Proof. For any instance $G = (V, E)$ and an optimum solution x to the LP (6.1), we have $n \leq \frac{1}{2}\sum_{v\in V} x(\delta(v)) = x(E) = \mathrm{LP} \leq \mathrm{OPT} \leq n(n-1)$, where the last inequality follows since OPT is the minimum cost of a Hamiltonian circuit in the metric closure (cf. Proposition 1.12), and in a strongly connected unweighted digraph, the distance of any pair of vertices is at most $n - 1$. □

This upper bound can be improved by a constant factor (see Exercise 6.3), but we will not need this.

Instances of the ASYMMETRIC GRAPH TSP are strongly connected, but we will also deal with digraphs that are not. The *strongly connected components* of a directed graph $G = (V, E)$ are its maximal strongly connected (induced) subgraphs. Their vertex sets form a partition of V. The following simple fact will be used several times:

Proposition 6.3. *Let G be a digraph and $G[U]$ a strongly connected component of G. Let (W, F) be a strongly connected subgraph of G. Then $W \subseteq U$ or $W \cap U = \emptyset$.*

Proof. Suppose there exist vertices $v \in W \cap U$ and $w \in W \setminus U$. Then (W, F) and thus also G contains a path from v to w and a path from w to v. Hence, v and w are in the same strongly connected component of G, contradicting $v \in U$ and $w \notin U$. □

Contracting each strongly connected component results in an *acyclic* digraph: a digraph that contains no (directed) circuit. The following is well-known:

Table 6.1 *Approximation ratios and upper bounds on the integrality ratio of* (6.1) *for* Asymmetric Graph TSP *in the order of their discovery. Only results that are (or were) better than for general* Asymmetric TSP *are shown. Again, ε stands for an arbitrarily small positive constant.*

Approximation Ratio	Integrality Ratio	Year	Reference	Chapter
$27 + \varepsilon$	13	2015	Svensson [2015]	6
$13 + \varepsilon$	13	2019	Traub [2020a]	6
$8 + \varepsilon$	8	2021	this book	6

Proposition 6.4. *A digraph $G = (V, E)$ is acyclic if and only if it has a* topological order *– that is, $V = \{v_1, \ldots, v_n\}$ such that $i < j$ for all $(v_i, v_j) \in E$. Every digraph G has a strongly connected component $G[U]$ such that $\delta^-(U) = \emptyset$, and such a set U can be found in polynomial time.*

Proof. Clearly, a digraph with a topological order is acyclic. We prove the converse by induction on $n = |V|$, the case $n = 1$ being trivial. So, let G be an acyclic digraph with $n \geq 2$. Start at any vertex and follow an outgoing arc as long as there is one. We never return to any vertex because G is acyclic, so this terminates at a vertex u with $\delta^+(u) = \emptyset$. Let $v_n := u$ and append this to a topological order of $G - u$, which exists by induction.

For the second statement, note that contracting the vertex set of each strongly connected component of G yields an acyclic digraph G', which has a topological order $w_1, \ldots, w_{n'}$ by the first part. Then the strongly connected component of G that corresponds to w_1 has no entering arc. □

In fact, the strongly connected components and a topological order of the digraph that results from contracting them can be found in linear time. We call this order a topological order of the strongly connected components.

6.2 Covering Subtours for Asymmetric Graph TSP

Like the cycle cover algorithm (Algorithm 1.35) by Frieze, Galbiati, and Maffioli [1982], Svensson's [2015] algorithm always maintains an Eulerian multi-set H of edges. It repeatedly computes another Eulerian multi-set F of edges such that F enters and leaves every connected component of (V, H) at least once.

Definition 6.5 (graph subtour cover). Given a digraph $G = (V, E)$ and an Eulerian multi-subset H of E, a *graph subtour cover* of H in G is an Eulerian multi-subset F of E such that $F \cap \delta(W) \neq \emptyset$ for all vertex sets W of connected components of (V, H).

An essential difference of Svensson's algorithm is that, in contrast to the cycle cover algorithm, it adds only a carefully selected subset of F to H. To analyze Svensson's algorithm, it is therefore not sufficient to have a bound on the total cardinality of F. Instead, F needs to fulfill certain "local" bounds. The following was proved by Svensson [2015] with the constant 3 instead of 2.

Theorem 6.6. *Given a directed graph $G = (V, E)$, a solution x to the LP (6.1), and an Eulerian multi-subset H of E, we can compute in polynomial time a graph subtour cover F of H such that for every connected component D of (V, F), we have*

$$|E(D)| \leq \sum_{v \in V(D)} 2 \cdot x(\delta^-(v)).$$

Proof. Let $W_1, \ldots, W_k$ be the vertex sets of the connected components of (V, H). First, we make the support of x inside each set W_i acyclic: While there is a circuit C in $G[W_i]$ with $\gamma := \min\{x_e : e \in E(C)\} > 0$, reduce x_e by γ for all $e \in C$. Let the resulting circulation be $\tilde{x}$. We have $\tilde{x}(\delta^-(W_i)) = x(\delta^-(W_i)) \geq 1$ for $i = 1, \ldots, k$.

The idea is to round $\tilde{x}$ to an integral circulation while guaranteeing that we use at least one edge of $\delta^-(W_i)$ for each $i = 1, \ldots, k$. We say that $w \in V$ is a *high-throughput vertex* if $\tilde{x}(\delta^-(w)) \geq 1$. If W_i contains a high-throughput vertex w, we will enforce routing at least one unit of flow through w and hence through W_i; here, we use that the support of $\tilde{x}$ inside W_i is acyclic. If W_i contains no high-throughput vertex, we enforce a unit of flow through W_i by an auxiliary vertex.

To this end, we construct an auxiliary digraph $\bar{G}$ from G and transform $\tilde{x}$ to a circulation $\bar{x}$ in $\bar{G}$ (cf. Figure 6.1). Start with $\bar{G} = G$ and $\bar{x} = \tilde{x}$. Let $i \in \{1, \ldots, k\}$ such that W_i contains no high-throughput vertex. We add an auxiliary vertex a_i and add an edge $e' = (v, a_i)$ for each $e = (v, w) \in \delta^-(W_i)$ and an edge $e' = (a_i, v)$ for each $e = (w, v) \in \delta^+(W_i)$. We take one unit of flow that goes through W_i and reroute it through a_i, scaling down the flow inside W_i. More precisely, we set $\alpha := \frac{1}{\bar{x}(\delta^-(W_i))}$, define $\bar{x}_{e'} = \alpha \cdot \bar{x}_e$ for all edges $e \in \delta(W_i)$, and then $\bar{x}_e = (1 - \alpha) \cdot \bar{x}_e$ for all edges e with at least one endpoint in W_i.

After doing this transformation successively for all $i \in \{1, \ldots, k\}$ for which W_i contains no high-throughput vertex, we end up with a circulation $\bar{x}$ in $\bar{G}$

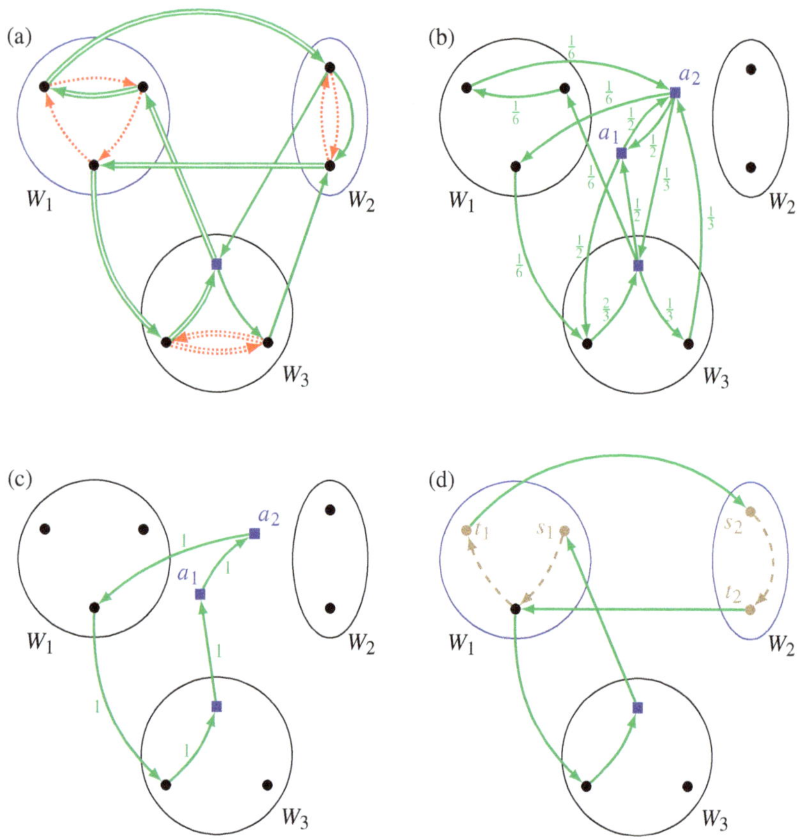

Figure 6.1 An example of the construction of a graph subtour cover (proof of Theorem 6.6). Let W_1, W_2, and W_3 be the vertex sets of the connected components of (V, H); note that H itself is not shown. (a): An LP solution x. Single edges e have $x_e = \frac{1}{3}$, double edges e have $x_e = \frac{2}{3}$. After deleting the red dotted circulation, the remaining circulation $\tilde{x}$ (green, solid) is acyclic within each W_i. The set W_3 contains a vertex w with $\tilde{x}(\delta^-(w)) \geq 1$ (the blue square). (b): For the other sets, W_1 and W_2, we reroute one unit of flow through the auxiliary vertices a_1 and a_2, respectively. This yields the circulation $\bar{x}$ in $\bar{G}$. (c): A possible integer circulation $\bar{x}^*$ in $\bar{G}$. (d): Mapping $\bar{x}^*$ back to the original edges, we get the solid green edges. Together with the dashed brown paths inside W_1 and W_2, this yields a graph subtour cover F.

that satisfies the following throughput bounds: First, $\bar{x}(\delta^-(a_i)) = 1$ for each $i \in \{1, \ldots, k\}$ for which the auxiliary vertex a_i was created; second, $\lfloor \tilde{x}(\delta^-(v)) \rfloor \leq \bar{x}(\delta^-(v)) \leq \lceil \tilde{x}(\delta^-(v)) \rceil$ for all $v \in V$ (note that the lower bound is zero unless v is a high-throughput vertex). By Corollary 3.12, there is an integral circulation $\bar{x}^*$ in the support graph of $\bar{x}$ satisfying the same throughput bounds, and such a circulation can be computed in polynomial time.

We now construct F. First, taking $\bar{x}^*_e$ copies of every edge yields an Eulerian multi-set $\bar{F}$ in $\bar{G}$. Since we route one unit of flow through every high-throughput vertex, and all the flow is routed in the support graph of $\bar{x}$, which is acyclic within each W_i, we have $\bar{F} \cap \delta^-(W_i) \neq \emptyset$ for all i for which no vertex a_i was added. For each new vertex a_i, we replace the two edges incident to a_i by their corresponding original edges in $\delta(W_i)$ and obtain a multi-set F of edges in G. For all $i = 1, \ldots, k$, we have $|F \cap \delta^-(W_i)| = |F \cap \delta^+(W_i)| \geq 1$. Moreover, $|F \cap \delta^-(v)| = |F \cap \delta^+(v)|$ for all $v \in W_i$, except possibly for one pair of vertices $s_i, t_i \in W_i$ with $|F \cap \delta^-(s_i)| = |F \cap \delta^+(s_i)| + 1$ and $|F \cap \delta^-(t_i)| = |F \cap \delta^+(t_i)| - 1$, in which case we take an s_i-t_i-path in $G[W_i]$ and add it to F. (Note that such a path exists because $G[W_i]$ is strongly connected since W_i induces a connected component of the Eulerian digraph (V, H).) We end up with a graph subtour cover F of H in G.

For all i for which W_i contains no high-throughput vertex, we rerouted the unit of flow that went through a_i to enter and leave W_i, and we have, for all $v \in W_i$,

$$\begin{aligned} |F \cap \delta^-(v)| &\leq |\bar{F} \cap \delta^-(v)| + 1 \\ &= \bar{x}^*(\delta^-(v)) + 1 \leq \lceil \tilde{x}(\delta^-(v)) \rceil + 1 \leq 2 \leq 2 \cdot x(\delta^-(v)). \end{aligned}$$

For all i for which W_i contains a high-throughput vertex w, we enforced at least one unit of flow through w and needed no rerouting. Hence, for every $v \in W_i$,

$$\begin{aligned} |F \cap \delta^-(v)| &= |\bar{F} \cap \delta^-(v)| \\ &= \bar{x}^*(\delta^-(v)) \leq \lceil \tilde{x}(\delta^-(v)) \rceil \leq \lceil x(\delta^-(v)) \rceil < 2 \cdot x(\delta^-(v)), \end{aligned}$$

where the last inequality follows from $x(\delta^-(v)) \geq 1$. □

The algorithm in this proof will be generalized in Chapter 8.

6.3 Outline of Svensson's Algorithm

In this section, we outline Svensson's algorithm. It maintains an Eulerian multi-set H of edges and stops once H is a tour. In each iteration, the algorithm computes a graph subtour cover F of H and adds a carefully selected subset of F to H. Theorem 6.6 says that for every connected component D of (V, F), we have $|E(D)| \leq \sum_{v \in V(D)} 2 \cdot x(\delta^-(v))$; such a graph D (and also such an edge set F) will be called *2x-light* in this outline section.

We will start Svensson's algorithm with an Eulerian multi-set $\tilde{H}$ of edges that is $4x$-light (such an $\tilde{H}$ will be called an *initialization*). So if $\tilde{W}_1, \ldots, \tilde{W}_k$ are the vertex sets of the connected components of $(V, \tilde{H})$, then we have $|\tilde{H}[\tilde{W}_i]| \leq \sum_{v \in \tilde{W}_i} 4 \cdot x(\delta^-(v))$ for $i = 1, \ldots, k$. The empty set would be a feasible initialization, but we aim at a better one, already forming large connected components. We order the connected components so that $\sum_{v \in \tilde{W}_1} x(\delta^-(v)) \geq \cdots \geq \sum_{v \in \tilde{W}_k} x(\delta^-(v))$. Call an initialization $\tilde{H}$ *optimal* if (among all possible $4x$-light Eulerian multi-sets) $\sum_{v \in \tilde{W}_1} x(\delta^-(v))$ is as large as possible, then among all those $\sum_{v \in \tilde{W}_2} x(\delta^-(v))$ is as large as possible, and so on.

We do not know how to find an optimal initialization in polynomial time, but let us assume for now that we can. For proving an upper bound on the integrality ratio, polynomial running time is irrelevant anyway. For a fixed initialization, we say that a graph has *index i* if it contains a vertex of $\tilde{W}_i$ but no vertex of $\tilde{W}_1 \cup \cdots \cup \tilde{W}_{i-1}$.

Recall that we compute a $2x$-light graph subtour cover F for a superset H of $\tilde{H}$ in each iteration of the algorithm. Given F, let us denote by F_i the union of the connected components of (V, F) with index i. A key property of $\tilde{H}$ and F is the following:

Proposition 6.7. $|E(F_i)| \leq 4 \sum_{v \in \tilde{W}_i} x(\delta^-(v))$.

Proof. Suppose this is not true. Then (since F is $2x$-light)

$$2 \sum_{v \in V(F_i)} x(\delta^-(v)) \geq |E(F_i)| > 4 \sum_{v \in \tilde{W}_i} x(\delta^-(v)) \geq |\tilde{H}[\tilde{W}_i]|,$$

and therefore $|\tilde{H}[\tilde{W}_i]| + |E(F_i)| \leq 4 \sum_{v \in V(F_i)} x(\delta^-(v))$. We conclude that $\tilde{H}[\tilde{W}_i] \cup E(F_i)$ is $4x$-light. But then $E(F_i) \cup \tilde{H}\left[V \setminus (\tilde{W}_{i+1} \cup \cdots \cup \tilde{W}_k)\right]$ is a better initialization because $\sum_{v \in \tilde{W}_i \cup V(F_i)} x(\delta^-(v)) > \sum_{v \in \tilde{W}_i} x(\delta^-(v))$, contradicting our choice of the initialization $\tilde{H}$. □

Now assume for a moment the following lucky property: For all subtour covers F that we compute for intermediate Eulerian multi-sets H throughout the algorithm, $H \cup E(F_i)$ connects all of $\tilde{W}_i \cup \cdots \cup \tilde{W}_k$ (but H alone does not), for the largest i for which F_i is nonempty. Then we will buy F_i (add its edges to H) and make $\tilde{W}_i$ pay for F_i (recall the bound in Proposition 6.7). Note that later the same set $\tilde{W}_i$ will never pay again for anything, so the total number of edges that we added is at most $\sum_{i=1}^{k-1} 4 \sum_{v \in \tilde{W}_i} x(\delta^-(v)) \leq 4 \sum_{v \in V} x(\delta^-(v)) = 4\,\mathrm{LP}$.

In addition, we have the edges in the initialization, but this was $4x$-light, so it cannot have more than 4 LP edges. Our tour then has at most 8 LP edges.

Unfortunately, we cannot expect to be so lucky. Nevertheless, we consider a connected component of $(V, H \cup F)$ with maximum index, say index i. If there is a "cheap" circuit C that connects this component to some other component, we tentatively add the edges of C to the pool F from which we will eventually buy edges. This circuit will also be paid for by $\tilde{W}_i$ if we end up buying this new larger connected component of (V, F) that contains C. Since this new component must have smaller index, only one circuit C will be paid for by $\tilde{W}_i$.

We iterate until no such cheap circuit exists for the connected component with largest index (in the graph consisting of edges from $H \cup F$ and possibly additional circuits). Only then do we buy the edges of this (and only this) component (add them to H). Now we reduced the number of connected components of (V, H), and we iterate the entire procedure until (V, H) is connected.

One can argue that every $\tilde{W}_i$ still pays at most once for some F_i, as in the lucky case: If at a later stage F_i is nonempty, it must contain a cheap circuit that we would have added to the pool (as we will show in Lemma 6.15).

The assumption that we start with an optimal initialization only works if we do not care about the running time (i.e., when only proving an upper bound on the integrality ratio). However, one can simply start with an arbitrary initialization (for example, the empty set) and assume that it is optimal. Either the algorithm works essentially as described above, or we find a "significantly better" initialization. In the latter case, we can simply restart.

We will now describe all this in detail. In order to arrive at essentially the same bound of 8 LP as in the lucky case, we will pay the circuits C from slack that we have because $\tilde{H}$ may contain fewer edges than is allowed for $4x$-light sets. This will require a slightly different definition of "optimal initialization," but the overall idea remains the same.

6.4 Initializing Svensson's Algorithm

Svensson's algorithm is initialized with an Eulerian multi-set $\tilde{H}$ of E and then computes either a "better" initialization $\tilde{H}'$ or extends $\tilde{H}$ to a tour H. The overall algorithm for Asymmetric Graph TSP starts with $\tilde{H} = \emptyset$ and repeatedly applies Svensson's algorithm until it outputs a tour H. In this section, we discuss how to find a better initialization $\tilde{H}'$ in certain cases.

The initialization $\tilde{H}$ of the algorithm will always be *light* (see Definition 6.8). To define what a light edge set is, we fix a function $\ell : V \to \mathbb{R}_{\geq 0}$. We will make a particular choice for the function ℓ later (in (6.9)), but at the moment, we don't make any further assumptions on ℓ. This will be useful when we adapt Svensson's algorithm for the general Asymmetric TSP in Chapter 7. There, we will apply some results from this section (in particular Lemma 6.9) for a different choice of ℓ. For this reason, we also work with a cost function $c : E \to \mathbb{R}_{\geq 0}$ here, although for Asymmetric Graph TSP, we have $c(e) = 1$ for all $e \in E$.

Definition 6.8 (light). Let $\tilde{H}$ be an Eulerian multi-subset of E. We call $\tilde{H}$ *light* if $c(E(D)) \leq \ell(V(D))$ for every connected component D of $(V, \tilde{H})$.

To measure what a "better" initialization for Svensson's algorithm is, we introduce a potential function Φ. For a subset $\tilde{V}$ of V and a multi-subset $\tilde{E}$ of E, we write

$$\text{slack}(\tilde{V}, \tilde{E}) := \ell(\tilde{V}) - c(\tilde{E}[\tilde{V}]).$$

When we apply this notation, $\tilde{V}$ will normally be the vertex set of a connected component of $(V, \tilde{E})$.

For a multi-subset $\tilde{H}$ of E such that the connected components of $(V, \tilde{H})$ have vertex sets $\tilde{W}_1, \ldots, \tilde{W}_k$, we write

$$\Phi(\tilde{H}) := \sum_{i=1}^{k} \text{slack}(\tilde{W}_i, \tilde{H})^{1+p},$$

where $p := \log_{1+\varepsilon'}(\frac{1+\varepsilon'}{\varepsilon'})$ and $0 < \varepsilon' \leq \frac{1}{4}$ is an arbitrary but fixed constant. Since $\tilde{H}$ will always be light,

$$0 \leq \Phi(\tilde{H}) \leq \ell(V)^{1+p}.$$

We will choose ℓ such that $\ell(V)$ is $O(\text{LP})$, where LP denotes the value of (3.2) (which is (6.1) for Asymmetric Graph TSP). Then $c(\tilde{H}) = O(\text{LP})$, again because $\tilde{H}$ is light.

We sort the connected components $\tilde{W}_1, \ldots, \tilde{W}_k$ of $(V, \tilde{H})$ such that

$$\text{slack}(\tilde{W}_1, \tilde{H}) \geq \cdots \geq \text{slack}(\tilde{W}_k, \tilde{H}).$$

In the limit for $\varepsilon' \to 0$ and thus $p \to \infty$, we then have $\Phi(\tilde{H}') > \Phi(\tilde{H})$ whenever the sorted slack vector $\left(\text{slack}(\tilde{W}_1, \tilde{H}), \ldots, \text{slack}(\tilde{W}_k, \tilde{H})\right)$ is lexicographically larger for $\tilde{H}'$ than for $\tilde{H}$. Working with this limit potential would be sufficient for showing the integrality ratio upper bound (as Svensson [2015] showed; cf. Exercise 6.4) but not for a polynomial-time algorithm.

We will start with the initialization $\tilde{H} = \emptyset$. When Svensson's algorithm computes an improved initialization, $\Phi(\tilde{H})$ increases significantly. This will imply that after a polynomial number of steps, no further significant improvement is possible.

Let $\tilde{H}$ be a light Eulerian multi-edge set, which we fix for the rest of this section. Let again $\tilde{W}_1, \ldots, \tilde{W}_k$ be the vertex sets of the connected components of $(V, \tilde{H})$, ordered so that $\text{slack}(\tilde{W}_1, \tilde{H}) \geq \cdots \geq \text{slack}(\tilde{W}_k, \tilde{H})$. For a connected multi-subgraph D of G, we define the *index* of D to be

$$\text{ind}(D) := \min\{j \in \{1, \ldots, k\} : V(D) \cap \tilde{W}_j \neq \emptyset\}.$$

We now prove the main lemma that we will use to find a better initialization $\tilde{H}'$. We are able to find such a better initialization whenever we have a connected Eulerian multi-subgraph D of G that has significantly larger slack than every connected component of $(V, \tilde{H})$ it intersects:

Lemma 6.9. *Let D be a connected Eulerian multi-subgraph of $G = (V, E)$ such that*

$$\text{slack}(V(D), E(D)) > \text{slack}(\tilde{W}_{\text{ind}(D)}, \tilde{H}) + \tfrac{\varepsilon'}{1+\varepsilon'} \cdot \ell(V(D)). \qquad (6.2)$$

Then we can compute in polynomial time a light Eulerian multi-set $\tilde{H}' \subseteq \tilde{H} \,\dot{\cup}\, E(D)$ such that

$$\Phi(\tilde{H}') - \Phi(\tilde{H}) > \min\left\{\ell(v)^{1+p} : v \in \tilde{W}_{\text{ind}(D)}\right\}. \qquad (6.3)$$

Proof. If $\text{slack}(\tilde{W}_{\text{ind}(D)}, \tilde{H}) < \min\{\ell(v) : v \in \tilde{W}_{\text{ind}(D)}\}$, we simply set $\tilde{H}' := \tilde{H} \setminus \tilde{H}[\tilde{W}_{\text{ind}(D)}]$, which does the job because we replace the connected component induced by $\tilde{W}_{\text{ind}(D)}$ by at least two singletons, each of which has slack at least $\min\{\ell(v) : v \in \tilde{W}_{\text{ind}(D)}\}$. Henceforth, we assume $\text{slack}(\tilde{W}_{\text{ind}(D)}, \tilde{H}) \geq \min\{\ell(v) : v \in \tilde{W}_{\text{ind}(D)}\}$.

Let $I := \{j \in \{1, \ldots, k\} : V(D) \cap \tilde{W}_j \neq \emptyset\}$ and $i := \min I = \text{ind}(D)$. We will compute a subset J of I and replace the components $\tilde{H}[\tilde{W}_j]$ for $j \in I$ by one new component that is the union of $E(D)$ and all $\tilde{H}[\tilde{W}_j]$ with $j \in J$, as well as possibly some singleton components. More precisely, we set

$$\tilde{H}' := \bigcup_{h \in \{1,\ldots,k\} \setminus I} \tilde{H}[\tilde{W}_h] \,\dot{\cup}\, E(D) \,\dot{\cup}\, \bigcup_{j \in J} \tilde{H}[\tilde{W}_j].$$

See Figure 6.2. Let D^* be the connected component of $(V, \tilde{H}')$ with edge set

$$E(D) \,\dot{\cup} \bigcup_{j \in J} \tilde{H}[\tilde{W}_j].$$

Choose

$$J := \left\{j \in I : \ell(\tilde{W}_j \cap V(D)) \leq \text{slack}(\tilde{W}_j, \tilde{H})\right\}. \qquad (6.4)$$

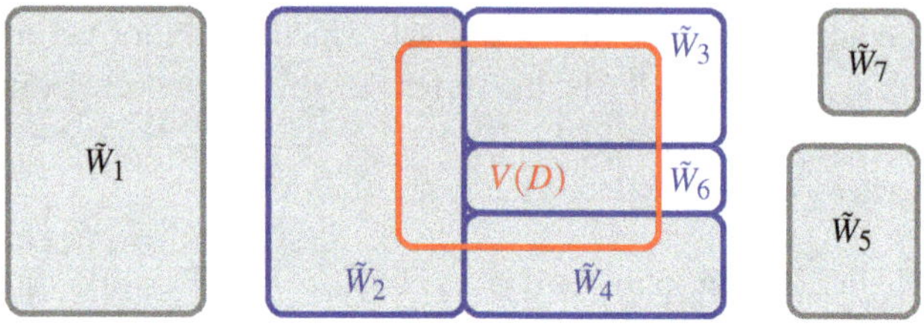

Figure 6.2 Illustration of the proof of Lemma 6.9. The gray and blue rectangles show the partition of V into $\tilde{W}_1, \ldots, \tilde{W}_7$. In red, we see the vertex set $V(D)$ of the given connected graph D with $\mathrm{ind}(D) = 2$. The rectangles with blue boundaries show the sets $\tilde{W}_i$ with $i \in I$. In this example, $I = \{2, 3, 4, 6\}$. The filled areas show vertex sets of connected components of $(V, \tilde{H}')$. In this example, we have $J = \{2, 4\}$. The connected components $\tilde{H}[\tilde{W}_1]$, $\tilde{H}[\tilde{W}_5]$, and $\tilde{H}[\tilde{W}_7]$ remain unchanged, and we get a new component D^* with vertex set $V(D) \cup \tilde{W}_2 \cup \tilde{W}_4$; we also get singleton components (without edges) for all vertices in $\tilde{W}_3 \setminus V(D)$ and $\tilde{W}_6 \setminus V(D)$. This picture is taken from Traub and Vygen [2022].

Then

$$\begin{aligned} \mathrm{slack}(V(D^*), E(D^*)) &= \mathrm{slack}(V(D), E(D)) \\ &\quad + \sum_{j \in J} \left(\mathrm{slack}(\tilde{W}_j, \tilde{H}) - \ell(\tilde{W}_j \cap V(D))\right) \\ &\geq \mathrm{slack}(V(D), E(D)), \end{aligned} \tag{6.5}$$

so $E(D^*)$ is light, and hence $\tilde{H}'$ is also light.

We will now show (6.3). To this end, we will first prove two lower bounds on $\mathrm{slack}(V(D^*), E(D^*))$. Using the definition of J and (6.2), we get

$$\begin{aligned} &\mathrm{slack}(V(D^*), E(D^*)) \\ &\quad = \mathrm{slack}(V(D), E(D)) + \sum_{j \in J} \left(\mathrm{slack}(\tilde{W}_j, \tilde{H}) - \ell(\tilde{W}_j \cap V(D))\right) \\ &\quad \geq \mathrm{slack}(V(D), E(D)) + \tfrac{\varepsilon'}{1+\varepsilon'} \sum_{j \in J} \left(\mathrm{slack}(\tilde{W}_j, \tilde{H}) - \ell(\tilde{W}_j \cap V(D))\right) \\ &\quad \geq \mathrm{slack}(V(D), E(D)) + \tfrac{\varepsilon'}{1+\varepsilon'} \sum_{j \in I} \left(\mathrm{slack}(\tilde{W}_j, \tilde{H}) - \ell(\tilde{W}_j \cap V(D))\right) \\ &\quad > \mathrm{slack}(\tilde{W}_i, \tilde{H}) + \tfrac{\varepsilon'}{1+\varepsilon'} \cdot \ell(V(D)) \\ &\qquad\qquad + \tfrac{\varepsilon'}{1+\varepsilon'} \sum_{j \in I} \left(\mathrm{slack}(\tilde{W}_j, \tilde{H}) - \ell(\tilde{W}_j \cap V(D))\right) \\ &\quad = \mathrm{slack}(\tilde{W}_i, \tilde{H}) + \tfrac{\varepsilon'}{1+\varepsilon'} \sum_{j \in I} \mathrm{slack}(\tilde{W}_j, \tilde{H}). \end{aligned} \tag{6.6}$$

For the second lower bound on $\text{slack}(V(D^*), E(D^*))$, we first observe that from (6.2) we also get

$$\begin{aligned} \text{slack}(\tilde{W}_i, \tilde{H}) &< \text{slack}(V(D), E(D)) - \tfrac{\varepsilon'}{1+\varepsilon'} \cdot \ell(V(D)) \\ &\leq \tfrac{1}{1+\varepsilon'} \ell(V(D)). \end{aligned} \tag{6.7}$$

From (6.5) and (6.2) and (6.7), we obtain

$$\begin{aligned} \text{slack}(V(D^*), E(D^*)) &\geq \text{slack}(V(D), E(D)) \\ &> \text{slack}(\tilde{W}_i, \tilde{H}) + \tfrac{\varepsilon'}{1+\varepsilon'} \cdot \ell(V(D)) \\ &> (1+\varepsilon') \cdot \text{slack}(\tilde{W}_i, \tilde{H}). \end{aligned} \tag{6.8}$$

Combining the two lower bounds (6.6) and (6.8) for $\text{slack}(V(D^*), E(D^*))$ and the definition of $p = \log_{1+\varepsilon'}(\frac{1+\varepsilon'}{\varepsilon'})$, we get

$$\begin{aligned} &\text{slack}(V(D^*), E(D^*))^{1+p} \\ &> (1+\varepsilon')^p \cdot \text{slack}(\tilde{W}_i, \tilde{H})^p \Big(\text{slack}(\tilde{W}_i, \tilde{H}) + \tfrac{\varepsilon'}{1+\varepsilon'} \sum_{j \in I} \text{slack}(\tilde{W}_j, \tilde{H}) \Big) \\ &= \tfrac{1+\varepsilon'}{\varepsilon'} \Big(\text{slack}(\tilde{W}_i, \tilde{H})^{1+p} + \text{slack}(\tilde{W}_i, \tilde{H})^p \cdot \tfrac{\varepsilon'}{1+\varepsilon'} \sum_{j \in I} \text{slack}(\tilde{W}_j, \tilde{H}) \Big) \\ &\geq \text{slack}(\tilde{W}_i, \tilde{H})^{1+p} + \sum_{j \in I} \text{slack}(\tilde{W}_j, \tilde{H})^{1+p}, \end{aligned}$$

where we used $\text{slack}(\tilde{W}_i, \tilde{H}) \geq \text{slack}(\tilde{W}_j, \tilde{H})$ for all $j \in I$.

We conclude

$$\begin{aligned} \Phi(\tilde{H}') - \Phi(\tilde{H}) &\geq \text{slack}(V(D^*), E(D^*))^{1+p} - \sum_{j \in I} \text{slack}(\tilde{W}_j, \tilde{H})^{1+p} \\ &> \text{slack}(\tilde{W}_i, \tilde{H})^{1+p}, \end{aligned}$$

which is sufficient since $\text{slack}(\tilde{W}_i, \tilde{H}) \geq \min\{\ell(v) : v \in \tilde{W}_{\text{ind}(D)}\}$. □

For the rest of this chapter, we will deal with ASYMMETRIC GRAPH TSP only. Here, we set

$$\ell(v) := 4(1+\varepsilon') \cdot x(\delta^-(v)) \tag{6.9}$$

for all $v \in V$, where x is an optimum solution to the LP (6.1). Then we have $\ell(v) \geq 4(1+\varepsilon')$ for all $v \in V$. Moreover,

$$\ell(V) = 4(1+\varepsilon') \cdot x(E) = 4(1+\varepsilon') \cdot \text{LP}, \tag{6.10}$$

where LP denotes the value of (6.1). By Proposition 6.2, this also implies $\ell(V) \leq 4(1+\varepsilon') \cdot n^2$. Thus, Lemma 6.9 guarantees

$$\begin{aligned} \Phi(\tilde{H}') - \Phi(\tilde{H}) &> \min\left\{\ell(v)^{1+p} : v \in V\right\} \\ &\geq (4(1+\varepsilon'))^{1+p} \\ &\geq \left(\frac{\ell(V)}{n^2}\right)^{1+p}. \end{aligned} \tag{6.11}$$

Since $0 \leq \Phi(\tilde{H}) \leq \ell(V)^{1+p}$, this shows that we can apply Lemma 6.9 at most $n^{2(1+p)}$ times until there exists no connected Eulerian multi-subgraph D with (6.2). For such an initialization $\tilde{H}$, Svensson's algorithm will complete $\tilde{H}$ to a tour.

Let us now explain how we obtain the graphs D to which we apply Lemma 6.9 during Svensson's algorithm. First, we will apply Theorem 6.6 to obtain a graph subtour cover F of H. For $i \in \{1, \ldots, k\}$, we denote by F_i the union of the connected components D' of (V, F) with $\mathrm{ind}(D') = i$. We will find a better initialization for Svensson's algorithm whenever F violates any of the following two conditions:

$$|E(F_i)| \leq \ell(\tilde{W}_i) \text{ for all } i \in \{1, \ldots, k\}; \tag{6.12}$$

$$\mathrm{slack}(V(D), E(D)) \leq \mathrm{slack}(\tilde{W}_{\mathrm{ind}(D)}, \tilde{H}) + \tfrac{\varepsilon'}{1+\varepsilon'} \cdot \ell(V(D)) \quad \text{for every connected component } D \text{ of } (V, F). \tag{6.13}$$

The fact that we can find an improved initialization and thus make progress whenever any of (6.12) and (6.13) is violated will allow us to assume that we always obtain a set F fulfilling both (6.12) and (6.13).

Figure 6.3 illustrates the two different ways of obtaining D as in Lemma 6.9 whenever (6.12) or (6.13) is violated. This is trivial if (6.13) is violated and follows from Lemma 6.10 if (6.12) is violated.

Lemma 6.10. *Let F be an Eulerian multi-edge set such that $|E(D')| \leq \sum_{v \in V(D')} 2 \cdot x(\delta^-(v))$ for every connected component D' of (V, F). For $i \in \{1, \ldots, k\}$, let the graph F_i be the union of the connected components D' of (V, F) with $\mathrm{ind}(D') = i$.*

Suppose we have $|E(F_i)| > \ell(\tilde{W}_i)$ for some $i \in \{1, \ldots, k\}$. Then

$$D := \left(\tilde{W}_i \cup V(F_i),\ \tilde{H}[\tilde{W}_i] \,\dot{\cup}\, E(F_i)\right)$$

fulfills the conditions of Lemma 6.9 – that is, D is a connected Eulerian multi-subgraph of G with (6.2).

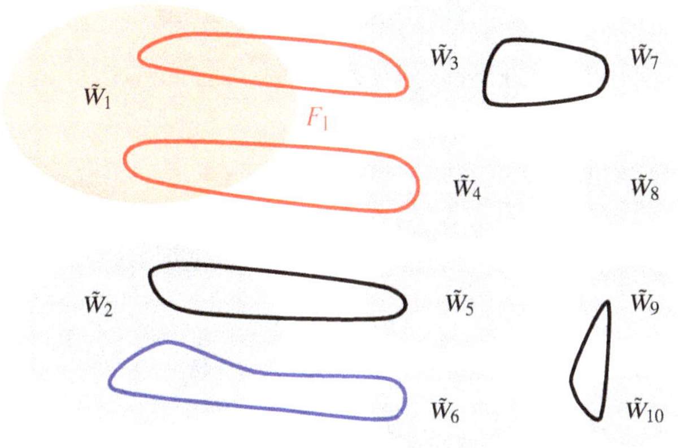

Figure 6.3 Illustration of the two possibilities for obtaining a better initialization via Lemma 6.9 when (6.12) or (6.13) is violated. Here, the filled ellipses show the partition $\tilde{W}_1, \dots, \tilde{W}_{10}$ of V (the connected components of $(V, \tilde{H})$). The curves depict an Eulerian edge set F resulting from Theorem 6.6 for some H. (The set H is not shown here.) In blue, we see a subgraph D as in (6.13): Here D is a single connected component of (V, F), and in this example, we have $\text{ind}(D) = 2$. In red, we see a subgraph D as in Lemma 6.10: Here the red curves constitute the graph F_1, and D is the disjoint union of $E(F_1)$ and $\tilde{H}[\tilde{W}_1]$. This picture is taken from Traub and Vygen [2022].

Proof. We have

$$\begin{aligned} \tfrac{1}{2+2\varepsilon'} \cdot \ell(V(D)) &\geq \tfrac{1}{2+2\varepsilon'} \cdot \ell(V(F_i)) \\ &= \sum_{v \in V(F_i)} 2 \cdot x(\delta^-(v)) \\ &\geq |E(F_i)| \\ &> \ell(\tilde{W}_i). \end{aligned} \tag{6.14}$$

We get

$$\begin{aligned} |E(D)| &= |\tilde{H}[\tilde{W}_i]| + |E(F_i)| \\ &= \ell(\tilde{W}_i) + |E(F_i)| - \text{slack}(\tilde{W}_i, \tilde{H}) \\ &< \tfrac{1}{2+2\varepsilon'} \cdot \ell(V(D)) + \tfrac{1}{2+2\varepsilon'} \cdot \ell(V(D)) - \text{slack}(\tilde{W}_i, \tilde{H}) \\ &= \ell(V(D)) - \tfrac{\varepsilon'}{1+\varepsilon'} \cdot \ell(V(D)) - \text{slack}(\tilde{W}_i, \tilde{H}), \end{aligned}$$

where we used (6.14) twice (once to bound $|E(F_i)|$ and once to bound $\ell(\tilde{W}_i)$). This proves (6.2). □

6.5 Svensson's Algorithm

In this section, we present Svensson's algorithm. The following lemma states its result:

Lemma 6.11. *Given an instance $G = (V, E)$ of* Asymmetric Graph TSP *and a light Eulerian multi-set $\tilde{H} \subseteq E$, we can compute in polynomial time*

(i) *a tour H in G with*

$$|H| \leq 2 \cdot \ell(V) \tag{6.15}$$

or

(ii) *a light Eulerian multi-subset $\tilde{H}'$ of E such that*

$$\Phi(\tilde{H}') - \Phi(\tilde{H}) > \left(\tfrac{1}{n^2} \cdot \ell(V)\right)^{1+p}. \tag{6.16}$$

From Lemma 6.11, we can derive the main result of this chapter.

Theorem 6.12. *Let $\varepsilon > 0$. Then there is a polynomial-time algorithm that computes a solution of cardinality at most $(8+\varepsilon)$ times the value of the LP (6.1) for every instance of* Asymmetric Graph TSP.

Proof. We may assume $\varepsilon \leq 2$. Let $\varepsilon' := \frac{\varepsilon}{8}$ and define p, ℓ, and Φ as in Section 6.4. We start with $\tilde{H} = \emptyset$ and apply Lemma 6.11. If we obtain a set $\tilde{H}'$ as in Lemma 6.11 (ii), we set $\tilde{H} := \tilde{H}'$ and iterate – that is, we apply Lemma 6.11 again until we finally obtain a tour H as in Lemma 6.11 (i).

Since $0 \leq \Phi(\tilde{H}) \leq \ell(V)^{1+p}$, we need at most $n^{2(1+p)}$ iterations. At the end, the algorithm guaranteed by Lemma 6.11 returns a solution H to the instance G such that

$$|H| \leq 2 \cdot \ell(V) = 2 \cdot 4(1+\varepsilon') \cdot \mathrm{LP} = (8+\varepsilon) \cdot \mathrm{LP},$$

where we used (6.10) in the first equation. □

In the remaining part of this chapter, we prove Lemma 6.11. To this end, we consider Algorithm 6.13. We maintain an Eulerian edge set H, which is initialized with $H = \tilde{H}$. We iterate the following steps. First, we apply Theorem 6.6 to obtain a graph subtour cover F. Then we try to find an improved initialization $\tilde{H}'$ as discussed in Section 6.4, and finally, if we could not find a better initialization, we extend the set H. The careful update of H in Step (6) of Algorithm 6.13 is illustrated in Figure 6.4 (see also the outline in Section 6.3).

Notice that Step (4b) is just a cleanup step that removes unnecessary parts; it maintains a graph subtour cover with $|E(D)| \leq \sum_{v \in V(D)} 2 \cdot x(\delta^-(v))$ for every connected component D of (V, F'). (This cleanup step is actually not

Algorithm 6.13: Svensson's Algorithm

Input: a directed graph G with $G = (V, E)$,
a light Eulerian multi-subset $\tilde{H}$ of E

Output: either H as in Lemma 6.11 (i) or $\tilde{H}'$ as in Lemma 6.11 (ii)

(1) Let $\tilde{W}_1, \ldots, \tilde{W}_k$ be the vertex sets of the connected components of $(V, \tilde{H})$ such that $\text{slack}(\tilde{W}_1, \tilde{H}) \geq \text{slack}(\tilde{W}_2, \tilde{H}) \geq \cdots \geq \text{slack}(\tilde{W}_k, \tilde{H})$.

(2) Set $H := \tilde{H}$.

(3) **while** (V, H) is not connected **do**

(4) **Compute a graph subtour cover F:**

(4a) Apply Theorem 6.6 to H to obtain a graph subtour cover F' with $|E(D)| \leq \sum_{v \in V(D)} 2 \cdot x(\delta^-(v))$ for every connected component D of (V, F').

(4b) Let F result from F' by deleting all edges of connected components of (V, F') whose vertex sets are contained in a connected component of (V, H).

(5) **Try to find a better initialization $\tilde{H}'$:**
For $i \in \{1, \ldots, k\}$, let the graph F_i be the union of the connected components D' of (V, F) with $\text{ind}(D') = i$.

(5a) If for some $i \in \{1, \ldots, k\}$ we have $|E(F_i)| > \ell(\tilde{W}_i)$, apply Lemma 6.9 to $D = \big(\tilde{W}_i \cup V(F_i),\ \tilde{H}[\tilde{W}_i] \dot{\cup} E(F_i)\big)$ to obtain a better initialization $\tilde{H}'$. Then return $\tilde{H}'$.

(5b) If (V, F) has a connected component D with $\text{slack}(V(D), E(D)) > \text{slack}(\tilde{W}_{\text{ind}(D)}, \tilde{H}) + \frac{\varepsilon'}{1+\varepsilon'} \cdot \ell(V(D))$, apply Lemma 6.9 to obtain a better initialization $\tilde{H}'$. Then return $\tilde{H}'$.

(6) **Extend H:**

(6a) Set $X := \emptyset$.

(6b) Select the connected component Z of $(V, H \dot{\cup} F \dot{\cup} X)$ for which $\text{ind}(Z)$ is largest.

(6c) If there is a circuit C with
- $E(C) \cap \delta(V(Z)) \neq \emptyset$ and
- $|E(C)| \leq \text{slack}(\tilde{W}_{\text{ind}(Z)}, \tilde{H})$,

then add $E(C)$ to X and go to Step (6b).

(6d) Add the edges $(F \dot{\cup} X)[V(Z)]$ to H.

(7) Return H.

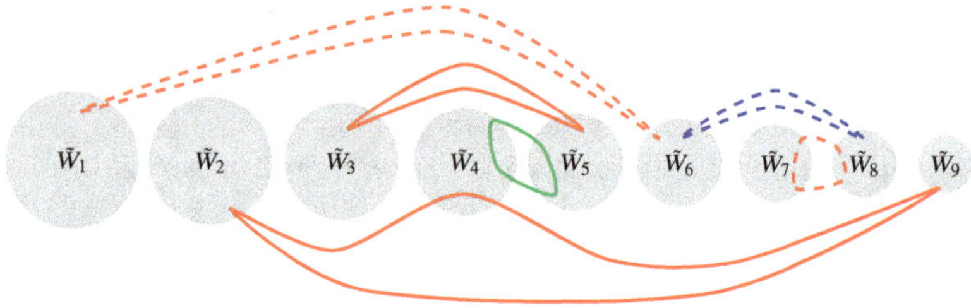

Figure 6.4 An illustration of Step (6) in the first iteration of Svensson's algorithm. The edge set F is shown in red (solid and dashed). First, the component Z with vertex set $\tilde{W}_7 \cup \tilde{W}_8$ and $\text{ind}(Z) = 7$ is considered. We may find the blue circuit C with $|E(C)| \leq \text{slack}(\tilde{W}_7, \tilde{H})$. After adding $E(C)$ to X, the component Z with vertex set $\tilde{W}_3 \cup \tilde{W}_5$ and $\text{ind}(Z) = 3$ is considered next. Then we may find the green circuit C' with $|E(C')| \leq \text{slack}(\tilde{W}_3, \tilde{H})$. Then $E(C')$ is added to X, and now $(V, H \,\dot{\cup}\, F \,\dot{\cup}\, X)$ has two connected components. The component Z with vertex set $\tilde{W}_2 \cup \tilde{W}_3 \cup \tilde{W}_4 \cup \tilde{W}_5 \cup \tilde{W}_9$ and $\text{ind}(Z) = 2$ is considered next. Suppose there is no circuit C'' connecting it to the rest and with $|E(C'')| \leq \text{slack}(\tilde{W}_2, \tilde{H})$. Then the edges drawn as solid curves are added to H, concluding the first iteration. This picture is taken from Traub and Vygen [2022].

necessary when using the graph subtour cover algorithm from Section 6.2; see Exercise 6.5.)

To show that in Step (5a) the application of Lemma 6.9 is indeed possible, we apply Lemma 6.10. To implement Step (6c), consider each edge $e = (v, w) \in \delta^+(V(Z))$ and compute a shortest w-v-path P in G (cf. Theorem 1.14) and check whether $1 + |E(P)| \leq \text{slack}(\tilde{W}_{\text{ind}(Z)}, \tilde{H})$.

Note that adding $E(C)$ to X in Step (6c) decreases the number of connected components of $(V, H \,\dot{\cup}\, F \,\dot{\cup}\, X)$, and adding edges to H in Step (6d) decreases the number of connected components of (V, H). Thus, the procedure terminates after a polynomial number of steps.

If Algorithm 6.13 returns a multi-set $\tilde{H}'$ in Step (5), we have $\Phi(\tilde{H}') - \Phi(\tilde{H}) > \left(\frac{1}{n^2} \cdot \ell(V)\right)^{1+p}$ as we observed in (6.11). Therefore, in this case, $\tilde{H}'$ is a multi-set as in Lemma 6.11 (ii).

Now suppose the algorithm does not terminate in Step (5). Since H remains Eulerian throughout the algorithm and (V, H) is connected at the end of Algorithm 6.13, the returned edge set H is a tour. It remains to show the upper bound (6.15) on the number of edges in H. Initially, we have $|H| = |\tilde{H}| \leq \ell(V)$. We bound the number of X-edges and the number of F-edges added to H separately.

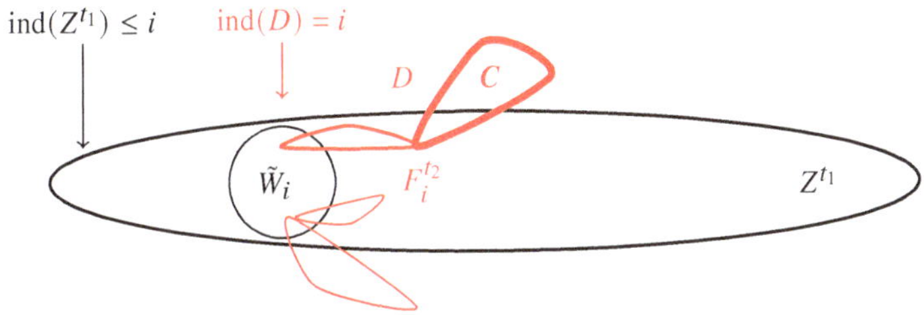

Figure 6.5 Proof of Lemma 6.15: An example of the graph $F_i^{t_2}$ is shown in red. The circuit C contains an edge of $\delta(V(Z^{t_1}))$, and D is the connected component containing C. This picture is taken from Traub and Vygen [2022].

Lemma 6.14. *The total number of all X-edges that are added to H is at most $\ell(V) - |\tilde{H}|$.*

Proof. A circuit C that is selected in Step (6c) and will later be added to H connects Z to another connected component Y with $\text{ind}(Y) < \text{ind}(Z)$. We say that it marks $\text{ind}(Z)$. It has at most $\text{slack}(\tilde{W}_{\text{ind}(Z)}, \tilde{H})$ edges. No circuit added later can mark $\text{ind}(Z)$ because the new connected component of $(V, H \,\dot\cup\, F \,\dot\cup\, X)$ containing both Y and Z will have smaller index by the choice of Z. Hence, the total number of edges in the added circuits is at most $\sum_{i=1}^{k} \text{slack}(\tilde{W}_i, \tilde{H}) = \ell(V) - |\tilde{H}|$. □

Lemma 6.15. *The total number of all F-edges that are added to H is at most $\ell(V)$.*

Proof. Let Z^t denote Z at the end of iteration t of the while-loop. Let F_i^t be the graph F_i in iteration t if the set of edges of F_i is nonempty and is added to H at the end of this iteration, and let $F_i^t = \emptyset$ otherwise.

If F_i^t is nonempty for some i and t, then $|E(F_i^t)| \leq \ell(\tilde{W}_i)$ by Step (5a). We claim that for any i, at most one of the F_i^t is nonempty. Then summing over all i and t concludes the proof.

Suppose there are $t_1 < t_2$ such that $F_i^{t_1} \neq \emptyset$ and $F_i^{t_2} \neq \emptyset$. Then $V(F_i^{t_1}) \subseteq V(Z^{t_1})$ and thus $\tilde{W}_i \subseteq V(Z^{t_1})$. Moreover, $F_i^{t_2}$ contains a vertex of $\tilde{W}_i$ and is not completely contained in Z^{t_1} by Step (4b) of the algorithm. Thus, $F_i^{t_2}$ contains a circuit C with $E(C) \cap \delta(V(Z^{t_1})) \neq \emptyset$.

If $|E(C)| \leq \text{slack}(\tilde{W}_{\text{ind}(Z^{t_1})}, \tilde{H})$, due to Step (6c), this is a contradiction to reaching Step (6d) in iteration t_1 with the component Z^{t_1}. Otherwise, let D be the connected component of $F_i^{t_2}$ containing C (cf. Figure 6.5). Note that $\text{ind}(D) = i \geq \text{ind}(Z^{t_1})$. We get

$$|V(C)| = |E(C)| > \text{slack}(\tilde{W}_{\text{ind}(Z^{t_1})}, \tilde{H}) \geq \text{slack}(\tilde{W}_{\text{ind}(D)}, \tilde{H}).$$

Moreover, using the facts that $|E(D)| \leq \sum_{v \in V(D)} 2x(\delta^-(v)) < \frac{1}{2}\ell(V(D))$, that D contains C, that $\varepsilon' \leq \frac{1}{4}$, and that $l(v) \geq 4(1+\varepsilon')$ for all $v \in V$, we get

$$\begin{aligned}\text{slack}(V(D), E(D)) &> \tfrac{1}{2} \cdot \ell(V(D)) \\ &> \tfrac{1}{4} \cdot \ell(V(C)) + \tfrac{\varepsilon'}{1+\varepsilon'} \cdot \ell(V(D)) \\ &> |V(C)| + \tfrac{\varepsilon'}{1+\varepsilon'} \cdot \ell(V(D)).\end{aligned}$$

Together, we obtain

$$\text{slack}(V(D), E(D)) > \text{slack}(\tilde{W}_{\text{ind}(D)}, \tilde{H}) + \tfrac{\varepsilon'}{1+\varepsilon'} \cdot \ell(V(D)).$$

Due to Step (5b), this is a contradiction to reaching Step (6) in iteration t_2 and adding $F_i^{t_2}$ there. □

Using Lemmas 6.14 and 6.15, we conclude that the returned edge set H has cardinality at most $2 \cdot \ell(V)$. This concludes the proof of Lemma 6.11.

This $(8+\varepsilon)$-approximation algorithm is the best known today. However, there is no reason to believe that a factor 2 is impossible (this is the known lower bound on the integrality ratio by Theorem 3.18). Note that the factor 2 also appeared in Theorem 6.6. We leave the following as a challenging open problem:

Open Problem 6.16. Devise a 2-approximation algorithm for the Asymmetric Graph TSP.

Exercises

6.1 Let $\varepsilon > 0$. Let (G, c) be an Asymmetric TSP instance with n vertices, such that there are nonnegative node weights $c_v \geq 0$ $(v \in V)$ determining the edge weights via $c(e) = c_v + c_w$ for each edge $e = (v, w)$. Such instances are called *node-weighted*. Show that given a node-weighted Asymmetric TSP instance with n vertices, we can find in polynomial time a number $M > 0$ and an unweighted digraph G' with $O(\frac{n^2}{\varepsilon})$ vertices such that

- $\text{LP}(G, c) \leq M \cdot \text{LP}(G') \leq (1+\varepsilon) \cdot \text{LP}(G, c)$,
- $\text{OPT}(G, c) \leq M \cdot \text{OPT}(G') \leq (1+\varepsilon) \cdot \text{OPT}(G, c)$, and
- for every tour F' in the unweighted digraph G', there is a corresponding tour F in G such that $c(F) \leq M \cdot |F'|$, and F can be obtained from F' in polynomial time.

Here, $\mathrm{LP}(\mathcal{I})$ refers to the value of the LP (3.2) or (6.1), respectively, and $\mathrm{OPT}(\mathcal{I})$ to the cost of an optimum tour for an instance $\mathcal{I}$.
(Köhne, Traub, and Vygen [2020])

6.2 Conclude from Exercise 6.1 that the integrality ratio of (3.2) is the same whether we restrict it to unweighted or to node-weighted instances. Moreover, every α-approximation algorithm for unweighted instances implies an $(\alpha + \varepsilon)$-approximation algorithm for node-weighted instances, for every $\varepsilon > 0$.
(Köhne, Traub, and Vygen [2020])

6.3 The bounds in Proposition 6.2 are tight up to a constant factor: Show that $\mathrm{OPT} \le \frac{(n+1)^2}{4}$, and for every odd $n \ge 3$, there is an instance with $\mathrm{LP} = \frac{(n+1)^2}{4}$.

6.4 Show that in the situation of Lemma 6.9, one immediately gets $\Phi(\tilde{H}') > \Phi(\tilde{H})$ for the limit potential Φ (for $p \to \infty$). More precisely, show that the sorted slack vector $(\mathrm{slack}(\tilde{W}_1, \tilde{H}), \ldots, \mathrm{slack}(\tilde{W}_k, \tilde{H}))$ increases lexicographically. Deduce that to prove the upper bound on the integrality ratio, one can simply start Svensson's algorithm with a light Eulerian multi-subset $\tilde{H}$ of E for which the sorted slack vector is lexicographically maximal.

6.5 Show that Step (4b) in Algorithm 6.13 is not needed (we can set $F = F'$) if we compute the graph subtour cover in Step (4a) as in the proof of Theorem 6.6.

6.6 Show that in Step (6c) in Algorithm 6.13, one can replace the inequality $|E(C)| \le \mathrm{slack}(\tilde{W}_{\mathrm{ind}(Z)}, \tilde{H})$ by $|E(C)| \le \frac{1}{2}\mathrm{slack}(\tilde{W}_{\mathrm{ind}(Z)}, \tilde{H})$ and still obtain the same bound on the number of F-edges as in Lemma 6.15; the bound on the number of X-edges from Lemma 6.14 improves by a factor 2.

7

Constant-Factor Approximation for the Asymmetric TSP

In this and the following chapter, we describe a constant-factor approximation algorithm for the ASYMMETRIC TSP. Such an algorithm was first devised by Svensson, Tarnawski, and Végh [2020]. We present the improved version by Traub and Vygen [2022], with an additional improvement that has not been published before.

The overall algorithm consists of four main components, three of which we will present in this chapter. First, we show that we can restrict attention to instances whose cost function is given by a solution to the dual LP with laminar support and an additional strong connectivity property. Second, we reduce such instances to so-called vertebrate pairs. Third, we will adapt Svensson's algorithm from Chapter 6 to deal with vertebrate pairs.

7.1 Outline of the Asymmetric TSP Algorithm

An obvious idea would be to try to reduce the problem to the ASYMMETRIC GRAPH TSP and apply the result of the previous chapter. However, such a reduction is not known; the structure of general ASYMMETRIC TSP instances is more complicated. Instead, we show a reduction to what Svensson, Tarnawski, and Végh [2020] called *vertebrate pairs*. We will see that the ASYMMETRIC GRAPH TSP techniques can be extended to work also for vertebrate pairs.

The first step reduces ASYMMETRIC TSP to so-called *strongly laminar instances* by exploiting properties of a dual LP solution and complementary slackness. In particular, we will work with a dual solution with laminar support as in Lemma 4.13. We will strengthen this further and obtain that the sets in the support induce strongly connected subgraphs.

Then we define vertebrate pairs, which consist of a strongly laminar instance and a subtour visiting all non-singleton sets in the support of the dual LP solution.

Table 7.1 *Approximation ratios and upper bounds on the integrality ratio of* (3.2) *for* Asymmetric *TSP in the order of their discovery. The third and fourth papers do not mention the integrality ratio explicitly. (R) means randomized; this algorithm computes a random tour, and the approximation ratio compares its expected cost to* OPT. *Moreover, ε is an arbitrarily small positive constant.*

Approximation Ratio	Integrality Ratio	Year	Reference	Chapter
$\log_2 n$	–	1980	Frieze, Galbiati, and Maffioli [1982]	1.5
–	$\log_2 n$	1990	Williamson [1990]	3.5
$0.99 \log_2 n$	$0.99 \log_2 n$	2002	Bläser [2008]	–
$0.842 \log_2 n$	$0.842 \log_2 n$	2003	Kaplan et al. [2005]	–
$\frac{2}{3} \log_2 n$	$\frac{2}{3} \log_2 n$	2006	Feige and Singh [2007]	–
$O(\frac{\log n}{\log \log n})$ (R)	$O(\frac{\log n}{\log \log n})$	2009	Asadpour et al. [2017]	5
–	$(\log \log n)^{O(1)}$	2014	Anari and Oveis Gharan [2015]	–
506	319	2017	Svensson, Tarnawski, and Végh [2020]	6–8
$22 + \varepsilon$	22	2019	Traub and Vygen [2022]	6–8
$17 + \varepsilon$	17	2021	this book	6–8

By a *subtour*, we mean the edge set of a connected Eulerian multi-subgraph, which however does not necessarily contain all vertices. In the reduction to vertebrate pairs, we will repeatedly contract sets that are not visited so far, apply an algorithm for vertebrate pairs, and iterate until we have a tour. See Figure 7.1 for the overall structure of our algorithm for Asymmetric TSP.

Overall, we will present a $(17 + \varepsilon)$-approximation algorithm for the Asymmetric TSP. Most of our presentation is based on Traub and Vygen [2022]. We again summarize the history in Table 7.1.

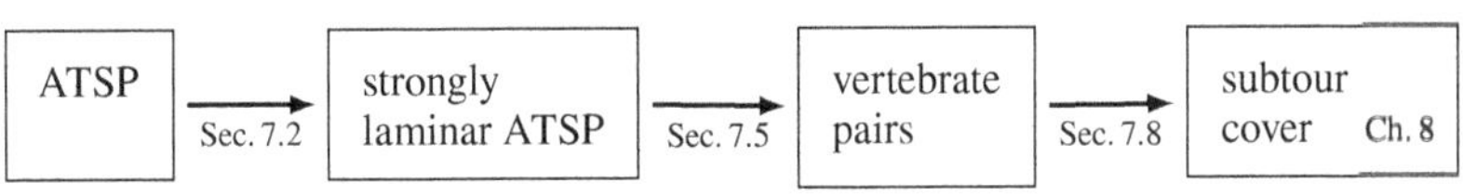

Figure 7.1 The sequence of reductions of our Asymmetric TSP algorithm.

7.2 Reducing to Strongly Laminar Instances

In this section, we show how to reduce the Asymmetric TSP to a special case where the instance has a particular structure. We already showed in Lemma 4.13 that we can assume a dual LP solution with laminar support. This will play a key role in this section. We strengthen this property as follows:

Definition 7.1 (strongly laminar support). Let (G, c) be an instance of Asymmetric TSP. Moreover, let (a, y) be a dual LP solution (i.e., a solution to (4.5)). We say that y or (a, y) has *strongly laminar support* if

- $\mathcal{L} := \{U : y_U > 0\}$ is a laminar family, and
- for every set $U \in \mathcal{L}$, the digraph $G[U]$ is strongly connected.

The following lemma allows us to assume that our optimum dual solution has strongly laminar support:

Lemma 7.2. *Let (G, c) be an instance of the Asymmetric TSP. Moreover, let (a, y) be an optimum solution to (4.5) with laminar support. Then we can compute in polynomial time (a', y') such that*

- *(a', y') is an optimum solution to (4.5), and*
- *(a', y') has strongly laminar support.*

Proof. As long as there is a set U with $y_U > 0$, but $G[U]$ is not strongly connected, we do the following. Let U be a minimal set with $y_U > 0$ such that $G[U]$ is not strongly connected. Moreover, let $G[S]$ be a strongly connected component of $G[U]$ with $\delta_G^-(S) \subseteq \delta_G^-(U)$; note that S exists and can be computed by Proposition 6.4.

Define a dual solution (a', y') as follows (see Figure 7.2). We set $y'_U := 0$, $y'_S := y_S + y_U$, and $y'_W := y_W$ for all other sets W. Moreover, $a'_v := a_v - y_U$ for $v \in U \setminus S$ and $a'_v := a_v$ for all other vertices v. The only edges $e = (v, w)$ for which

$$a'_w - a'_v + \sum_{W: e \in \delta(W)} y'_W > a_w - a_v + \sum_{W: e \in \delta(W)} y_W$$

are edges from $U \setminus S$ to S. However, such edges do not exist by choice of S. Hence, (a', y') is a feasible dual solution. Since $\sum_{\emptyset \neq W \subsetneq V} 2y'_W = \sum_{\emptyset \neq W \subsetneq V} 2y_W$, it is also optimum.

We now show that the support of y' is laminar. Suppose there is a set W in the support of y' that crosses S. Then W must be in the support of y and hence a subset of U because the support of y is laminar. By the minimal choice of U, the digraph $G[W]$ is strongly connected. But this implies that G contains an edge from $W \setminus S$ to $W \cap S$, contradicting $\delta_G^-(S) \subseteq \delta_G^-(U)$.

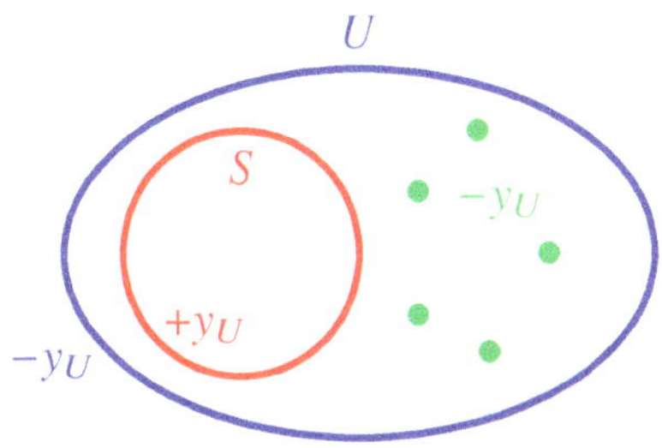

Figure 7.2 Illustration of the proof of Lemma 7.2. A nonempty set $U \subsetneq V$ is shown in blue. The vertex set S of a strongly connected component of $G[U]$ with $\delta^-(S) \subseteq \delta^-(U)$ is shown in red. We modify our dual solution by increasing the dual variable corresponding to S, decreasing the dual variable corresponding to U (blue), and decreasing the variables corresponding to the vertices in $U \setminus S$ (green).

We have now decreased the number of sets U in the support for which $G[U]$ is not strongly connected. Because the support of y is laminar, it has at most $2|V|$ elements by Proposition 4.8. Therefore, after iterating the above procedure at most $2|V|$ times, the dual solution has the desired properties. □

Definition 7.3 (strongly laminar Asymmetric TSP instance). A *strongly laminar Asymmetric TSP instance* is a quadruple $(G, \mathcal{L}, x, y)$, where

- $G = (V, E)$ is a strongly connected digraph,
- $\mathcal{L}$ is a laminar family of subsets of V such that $G[U]$ is strongly connected for all $U \in \mathcal{L}$,
- x is a feasible solution to the linear program (3.2) such that $x(\delta(U)) = 2$ for all $U \in \mathcal{L}$ and $x_e > 0$ for all $e \in E$, and
- $y : \mathcal{L} \to \mathbb{R}_{\geq 0}$.

This induces the Asymmetric TSP instance (G, c^y), where c^y is the *induced cost function* defined by $c^y(e) := \sum_{U \in \mathcal{L}: e \in \delta(U)} y_U$ for all $e \in E$.

See Figures 7.3 and 7.4. In the following, c^y will always denote the induced cost function of the given strongly laminar Asymmetric TSP instance. The cost $c^y(F)$ of a tour F in G is $\sum_{U \in \mathcal{L}} |\delta_F(U)| \cdot y_U$. By complementary slackness (Corollary 4.2), x and $(0, y)$ are optimum solutions to the LPs (3.2) and (4.5) for (G, c^y). For a strongly laminar instance $\mathcal{I}$, we denote by $\text{LP}(\mathcal{I}) = c^y(x) = \sum_{U \in \mathcal{L}} 2 \cdot y_U$ the value of these LPs.

We now prove that for Asymmetric TSP, it is sufficient to consider strongly laminar Asymmetric TSP instances.

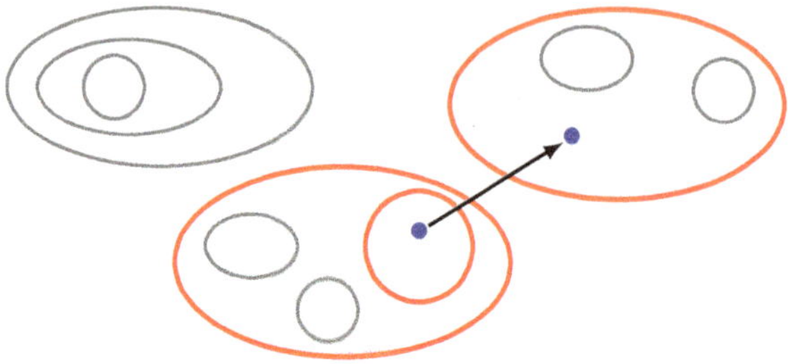

Figure 7.3 Structure of the cost function c^y induced by a strongly laminar Asymmetric TSP instance. The cost of the edge e shown in black is the sum of the values y_U for the sets $U \in \mathcal{L}$ with $e \in \delta(U)$; these elements of the laminar family $\mathcal{L}$ are shown in red.

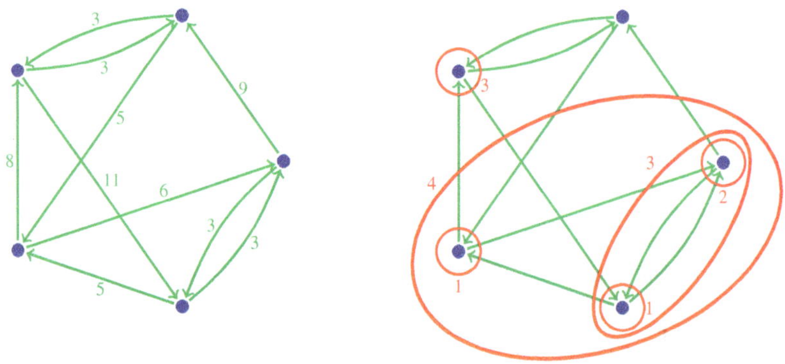

Figure 7.4 An example of a cost function (left) induced by a vector y whose support is a laminar family of vertex sets (right).

Theorem 7.4. *Let* $\alpha \geq 1$. *If there is a polynomial-time algorithm that computes for every* strongly laminar Asymmetric TSP instance $(G, \mathcal{L}, x, y)$ *a solution of cost at most* $\alpha \cdot c^y(x)$, *then there is a polynomial-time algorithm that computes for every Asymmetric TSP instance a solution of cost at most* α *times the value of the linear program* (3.2).

Proof. Let (G, c) be an arbitrary instance of the Asymmetric TSP. We compute an optimum solution x to (3.2) and an optimum solution (a, y) to (4.5) such that the support of y is a laminar family $\mathcal{L}$. This is possible by Proposition 3.1 and Lemma 4.13.

Now let E' be the support of x and define $G' := (V, E')$. Let x' be the vector x restricted to its support E'. Then apply Lemma 7.2 to G' and (a, y). We obtain an optimum dual solution (a', y') to (4.5) with strongly laminar support $\mathcal{L}$. By complementary slackness (Corollary 4.2), we have $x'(\delta(U)) = 2$ for all $U \in \mathcal{L}$.

Then the induced weight function of the strongly laminar ASYMMETRIC TSP instance $(G', \mathcal{L}, x', y')$ is given by $c^{y'}(e) = \sum_{U \in \mathcal{L}: e \in \delta(U)} y'_U = c(e) + a'_v - a'_w$ for all $e = (v, w) \in E'$ (by complementary slackness). Because every tour in G' is Eulerian, it has the same cost with respect to c and with respect to $c^{y'}$. Moreover, $c(x) = c^{y'}(x')$, and thus the theorem follows. □

7.3 Nice Paths

One advantage of the strongly laminar structure is that it implies the existence of *nice paths* (see Figure 7.5):

Lemma 7.5. *Let $G = (V, E)$ be a strongly connected directed graph and let $\mathcal{L}$ be a laminar family such that $G[U]$ is strongly connected for every $U \in \mathcal{L}$. Let $v, w \in V$ and let $\tilde{U}$ be the minimal set in $\mathcal{L} \cup \{V\}$ with $v, w \in \tilde{U}$.*

Then there is a v-w-path in $G[\tilde{U}]$ that enters every set $U \in \mathcal{L}$ at most once and leaves every set $U \in \mathcal{L}$ at most once; we will call such a path nice. *A nice v-w-path can be found in polynomial time.*

Proof. Let P be a path from v to w in $G[\tilde{U}]$. Now repeat the following until P enters and leaves every set in $\mathcal{L}$ at most once. Let U be a maximal set with $U \in \mathcal{L}$ that P enters or leaves more than once. Let v' be the first vertex that P visits in U, and let w' be the last vertex that P visits in U. Since $G[U]$ is strongly connected, we can replace the v'-w'-subpath of P by a path in $G[U]$. After at most $|\mathcal{L}|$ iterations, P is a nice v-w-path. Because $\mathcal{L}$ is laminar, we have $|\mathcal{L}| < 2|V|$ by Proposition 4.8. □

Let $(G, \mathcal{L}, x, y)$ be a strongly laminar ASYMMETRIC TSP instance and c^y the induced cost function. In the following, we fix for every $u, v \in V$ a nice u-v-path $P_{u,v}$. Such paths can be computed in polynomial time by Lemma 7.5. The fact that the path $P_{u,v}$ is nice allows us to express its cost as follows:

Lemma 7.6. *Let $W \in \mathcal{L} \cup \{V\}$ and let $u, v \in W$. Then*

$$c^y(E(P_{u,v})) = \sum_{L \in \mathcal{L}:\, L \subsetneq W, L \cap V(P_{u,v}) \neq \emptyset} 2y_L - \sum_{L \in \mathcal{L}:\, u \in L \subsetneq W} y_L - \sum_{L \in \mathcal{L}:\, v \in L \subsetneq W} y_L.$$

Proof. Since the path $P_{u,v}$ is nice, it is contained in $G[W]$. Moreover, it leaves every set $L \in \mathcal{L}$ at most once and enters every set $L \in \mathcal{L}$ at most once. A set $L \in \mathcal{L}$ with $u \in L$ is never entered by $P_{u,v}$, and a set $L \in \mathcal{L}$ with $v \in L$ is never left by $P_{u,v}$. □

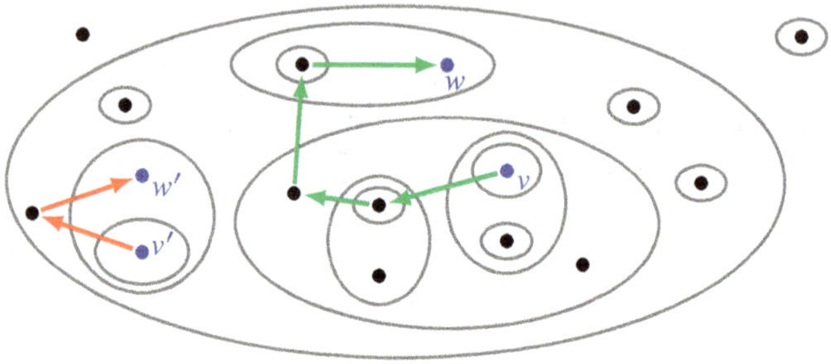

Figure 7.5 A nice path from v to w is shown in green. The laminar family $\mathcal{L}$ is shown in gray. The edges of G are not shown. The red path from v' to w' is not nice because it re-enters a set from $\mathcal{L}$ that it has previously left.

We define

$$\text{value}(W) := \sum_{L \in \mathcal{L}: L \subsetneq W} 2y_L$$

and

$$D_W(u,v) := \sum_{L \in \mathcal{L}: u \in L \subsetneq W} y_L + \sum_{L \in \mathcal{L}: v \in L \subsetneq W} y_L + c^y(E(P_{u,v})) \tag{7.1}$$

for $u, v \in W$. Intuitively, $D_W(u,v)$ is the cost for entering and leaving proper subsets of W, when entering W by an edge from $V \setminus W$ to u, following the path $P_{u,v}$, and leaving W by an edge from v to $V \setminus W$. Note that $D_W(u,v) \le \text{value}(W)$ by Lemma 7.6. We write

$$D_W := \max\{D_W(u,v) : u, v \in W\}.$$

On the one hand, it can be useful if D_W is small: Then we can enter the set W at some vertex $s \in W$ and leave it at some other vertex $t \in W$, following a cheap s-t-path inside $G[W]$ to get from s to t. On the other hand, we can also gain from D_W being large: We can find a nice path inside W that visits many sets $L \in \mathcal{L}$ (more precisely, sets of high total weight in the dual solution y) without paying more than necessary. Such paths will be useful (for constructing so-called backbones) in the following section.

7.4 Vertebrate Pairs

If the laminar family $\mathcal{L}$ of our given strongly laminar Asymmetric TSP instance consists of singletons only, the instance is a node-weighted instance, which is essentially equivalent to an Asymmetric Graph TSP instance (see Exercise 6.2). However, this assumption is too strong.

A natural idea is to contract the maximal non-singleton sets in $\mathcal{L}$ and solve the resulting node-weighted instance. After uncontracting, our tour will often enter a set at a vertex u and leave it at another vertex v; then we need to add a u-v-path to obtain an Eulerian edge set again. One could use a cheapest u-v-path for this purpose, which will not cost more than a nice path. We can define the node weights of the vertices corresponding to contracted sets so that visiting these vertices incurs a cost that is sufficient to pay for the u-v-path after uncontracting.

We will normally still not visit all the vertices in an uncontracted set, so we will apply the algorithm recursively to every such set. This algorithm will not lead to a constant approximation guarantee unless the laminar family has constant depth (Exercise 7.4). However, the approach can be adapted, leading to a reduction to vertebrate pairs, which is a generalization of the node-weighted asymmetric TSP that we will introduce shortly. For this reduction, we exploit that a nice u-v-path in U is either cheap or visits many subsets of U in the laminar family. In the latter case, we can avoid contracting many of the subsets of U when applying the algorithm to U recursively. More precisely, we will not contract any subset that is already visited by a given "backbone."

For a strongly laminar Asymmetric TSP instance $(G, \mathcal{L}, x, y)$, let $\mathcal{L}_{\geq 2} := \{L \in \mathcal{L} : |L| \geq 2\}$ be the family of all non-singleton elements of $\mathcal{L}$. We show how to reduce Asymmetric TSP to the case where we have already a given subtour B, called *backbone*, that visits all elements of $\mathcal{L}_{\geq 2}$ (see Figure 7.6). We call a strongly laminar Asymmetric TSP instance together with a given backbone a *vertebrate pair*.

Definition 7.7 (vertebrate pair)**.** A *vertebrate pair* consists of

- a strongly laminar Asymmetric TSP instance $\mathcal{I} = (G, \mathcal{L}, x, y)$ and
- a connected Eulerian multi-subgraph B of G (the *backbone*) such that $V(B) \cap L \neq \emptyset$ for all $L \in \mathcal{L}_{\geq 2}$.

A *solution* of the vertebrate pair $(\mathcal{I}, B)$ is a multi-set F of edges such that $E(B) \dot{\cup} F$ is a tour. Let $\kappa, \eta \geq 0$. A *(κ, η)-algorithm for vertebrate pairs* is a polynomial-time algorithm that computes, for any given vertebrate pair $(\mathcal{I}, B)$, a solution F such that

$$c^y(F) \leq \kappa \cdot \mathrm{LP}(\mathcal{I}) + \eta \cdot \sum_{v \in V \setminus V(B) : \{v\} \in \mathcal{L}} 2y_{\{v\}}.$$

In the following, we will show that a (κ, η)-algorithm for vertebrate pairs (for any constants κ and η) implies a $(3\kappa + \eta + 2)$-approximation algorithm for Asymmetric TSP.

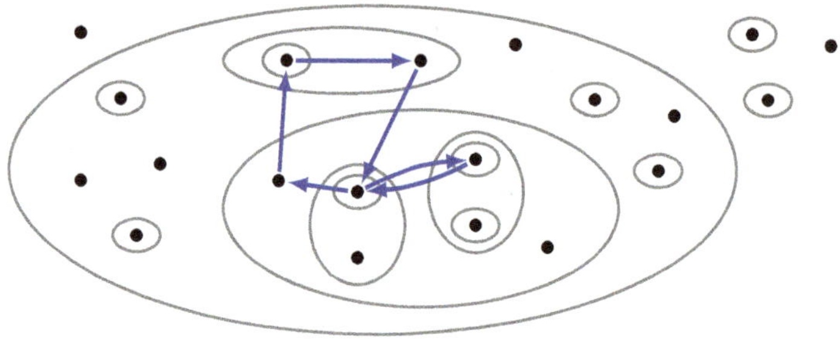

Figure 7.6 Example of a vertebrate pair $(\mathcal{I}, B)$. The laminar family $\mathcal{L}$ of $\mathcal{I}$ is shown in gray, and a backbone B is shown in blue. The edges of G are not shown. This picture is taken from Traub and Vygen [2022].

7.5 Reducing to Vertebrate Pairs

Algorithm 7.8 is our reduction to vertebrate pairs. It is a recursive algorithm that, given a set $W \in \mathcal{L} \cup \{V\}$, constructs a tour in $G[W]$. See Figures 7.7 and 7.8 for an illustration. First, we observe that the algorithm indeed returns a tour in $G[W]$.

Lemma 7.9. *Let $\kappa, \eta \geq 0$. Suppose we have a (κ, η)-algorithm $\mathcal{A}$ for vertebrate pairs. Then Algorithm 7.8 has polynomial running time and returns a tour in $G[W]$ for any given strongly laminar* Asymmetric TSP *instance $\mathcal{I} = (G, \mathcal{L}, x, y)$ and any given $W \in \mathcal{L} \cup \{V\}$.*

Proof. We apply induction on $|W|$. For $|W| = 1$, the algorithm returns $F = \emptyset$. Now let $|W| > 1$. Note that Steps (1) and (2) indeed construct a strongly laminar Asymmetric TSP instance $\mathcal{I}'$; in particular, x is a feasible primal LP solution (since it results from the original x by contracting subsets of V). At the end of Step (3), we have that F' is Eulerian and $F' \dot\cup E(B)$ is a tour in the instance $\mathcal{I}'$. In Step (4), the set F' remains Eulerian, and $F' \dot\cup E(B)$ remains connected. Moreover, the subtour $F' \dot\cup E(B)$ visits all sets in $\mathcal{L}_{\bar{B}}$ (i.e., we have $F' \cap \delta(L) \neq \emptyset$ for all $L \in \mathcal{L}_{\bar{B}}$). The subtour $F' \dot\cup E(B)$ also visits all vertices in W that are not contained in any set $L \in \mathcal{L}_{\bar{B}}$ (i.e., for these vertices v, we have $\delta(v) \cap (F' \cup E(B)) \neq \emptyset$). After Step (4), we have $F' \subseteq E[W]$. We conclude that the graph $(W, F'' \dot\cup E(B))$ is connected and Eulerian; here we applied the induction hypothesis to the sets $L \in \mathcal{L}_{\bar{B}}$ and use that F' visits every $L \in \mathcal{L}_{\bar{B}}$.

To see that the runtime of the algorithm is polynomially bounded, we observe that by Proposition 4.8, there are in total at most $|\mathcal{L}| + 1 < 2|V|$ recursive calls of the algorithm because $\mathcal{L}$ is a laminar family with $\emptyset, V \notin \mathcal{L}$. $\square$

Algorithm 7.8: Recursive Algorithm to Reduce to Vertebrate Pairs

Input: a strongly laminar Asymmetric TSP instance $\mathcal{I} = (G, \mathcal{L}, x, y)$ with $G = (V, E)$,
a set $W \in \mathcal{L} \cup \{V\}$, and
a (κ, η)-algorithm $\mathcal{A}$ for vertebrate pairs (for some constants $\kappa, \eta \geq 0$)

Output: a tour F in $G[W]$

(1) **Contract $V \setminus W$:** If $W \neq V$, contract $V \setminus W$ into a single vertex $v_{\bar{W}}$ and redefine $y_W := \frac{D_W}{2}$.

(2) **Construct a vertebrate pair:** Let $u^*, v^* \in W$ such that $D_W(u^*, v^*) = D_W$. Let B be the multi-graph corresponding to the closed walk that results from appending P_{u^*,v^*} and P_{v^*,u^*}.

Let $\mathcal{L}_{\bar{B}}$ denote the set of all maximal sets $L \in \mathcal{L}$ with $L \subsetneq W$ and $L \cap V(B) = \emptyset$. Contract each set $L \in \mathcal{L}_{\bar{B}}$ to a single vertex v_L and let G' be the resulting graph.

Let $\mathcal{L}'$ be the laminar family of subsets of $V(G')$ that contains singletons $\{v_L\}$ for $L \in \mathcal{L}_{\bar{B}}$ and all the sets arising from $L \in \mathcal{L}$ with $L \subseteq W$ and $L \cap V(B) \neq \emptyset$.

Set $y_{\{v_L\}} := y_L + \frac{D_L}{2}$ for all $L \in \mathcal{L}_{\bar{B}}$.

Let $\mathcal{I}' = (G', \mathcal{L}', x, y)$ be the resulting strongly laminar instance.

(3) **Compute a solution for the vertebrate pair:** Apply the given algorithm $\mathcal{A}$ to the vertebrate pair $(\mathcal{I}', B)$. Let F' be the resulting Eulerian edge set.

(4) **Lift the solution to a subtour:** Fix an Eulerian walk in every connected component of F'. Now uncontract every $L \in \mathcal{L}_{\bar{B}}$. Whenever an Eulerian walk passes through v_L, we get two edges $(u', u) \in \delta^-(L)$ and $(v, v') \in \delta^+(L)$. To connect u and v within L, add the path $P_{u,v}$.

Moreover, if $W \neq V$, do the following. Whenever an Eulerian walk passes through $v_{\bar{W}}$ using the edges $(u, v_{\bar{W}})$ and $(v_{\bar{W}}, v)$, replace them by the path $P_{u,v}$.

(5) **Recurse to complete to a tour of the original instance:** For every set $L \in \mathcal{L}_{\bar{B}}$, apply Algorithm 7.8 recursively to obtain a tour F_L in $G[L]$. Let F'' be the disjoint union of F' and all these tours F_L for $L \in \mathcal{L}_{\bar{B}}$.

(6) Return $F := F'' \,\dot{\cup}\, E(B)$.

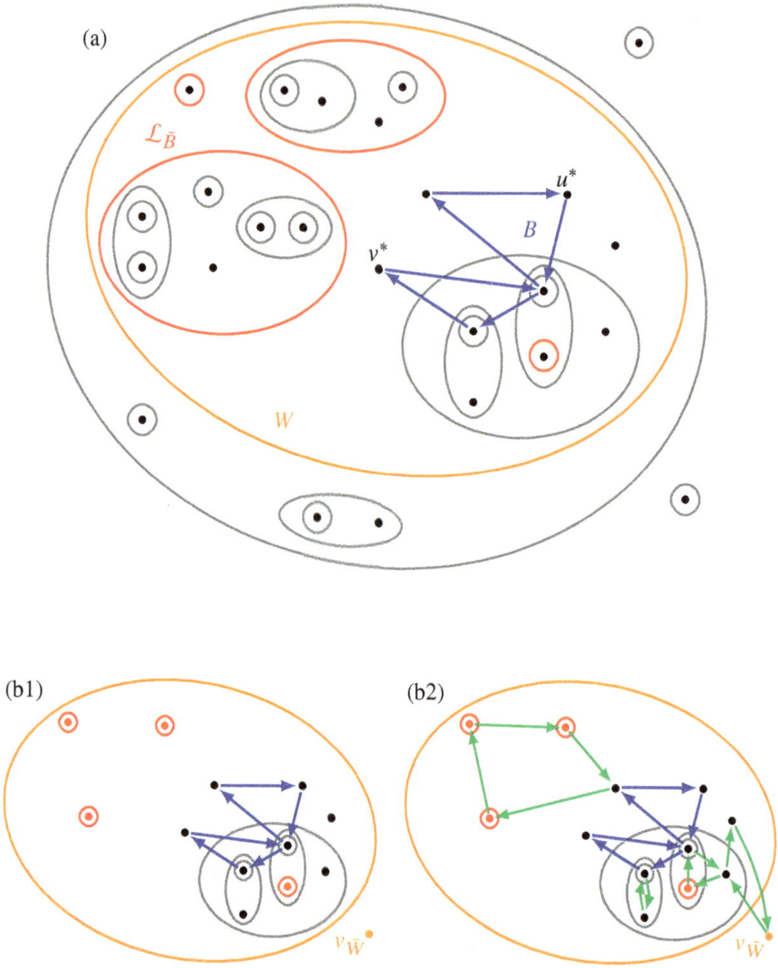

Figure 7.7 Illustration of Algorithm 7.8. The ellipses show the laminar family $\mathcal{L}$. Picture (a) shows the set W (orange), the subtour B (blue), and the elements of $\mathcal{L}_{\bar{B}}$ (red). The subtour B is the union of the paths P_{u^*,v^*} and P_{v^*,u^*}. Picture (b1) shows the resulting vertebrate pair as constructed in Step (2) of Algorithm 7.8. The vertices resulting from the contraction of elements of $\mathcal{L}_{\bar{B}}$ are shown in red, and the vertex $v_{\bar{W}}$ that results from the contraction of $V \setminus W$ is shown in orange. Picture (b2) shows in green a possible solution to this vertebrate pair as computed in Step (3). Step (4) of Algorithm 7.8 is illustrated by Figure 7.8. This picture is taken from Traub and Vygen [2022].

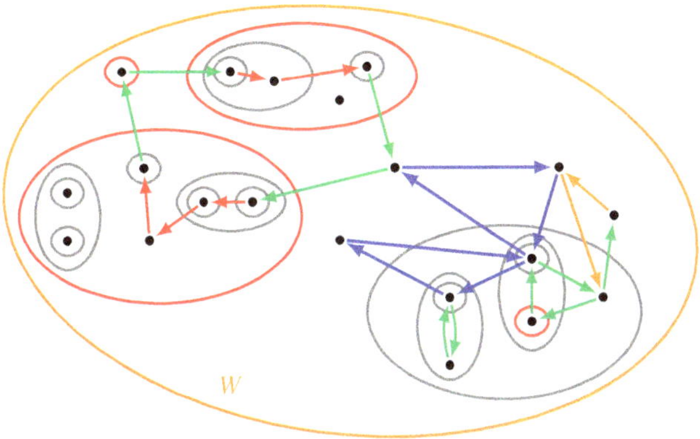

Figure 7.8 Illustration of Step (4) of Algorithm 7.8, continuing Figure 7.7. The green edges are those that arise from the vertebrate pair solution from Figure 7.7 (b2) by undoing the contraction of the sets in $\mathcal{L}_{\bar{B}}$. The red edges are the paths that we add to connect within $L \in \mathcal{L}_{\bar{B}}$ when uncontracting L. The orange edges show the u-v-path in $G[W]$ that we add to replace the edges $(u, v_{\bar{W}})$ and $(v_{\bar{W}}, v)$ in the vertebrate pair solution from Figure 7.7 (b2). This picture is taken from Traub and Vygen [2022].

Next, we observe that our backbone B visits many sets $L \in \mathcal{L}$ inside W if D_W is large.

Lemma 7.10. *Let $\mathcal{I} = (G, \mathcal{L}, x, y)$ be a strongly laminar* Asymmetric *TSP instance, and let $W \in \mathcal{L} \cup \{V\}$. Moreover, let B be as in Step* (2) *of Algorithm 7.8. Then*

$$\sum_{L \in \mathcal{L}_{\bar{B}}} (2y_L + \text{value}(L)) \;\leq\; \text{value}(W) - D_W. \tag{7.2}$$

Proof. By the choice of u^* and v^* in Step (2) and Lemma 7.6, we have

$$\begin{aligned} D_W &= D_W(u^*, v^*) \\ &= c^y(E(P_{u^*,v^*})) + \sum_{L \in \mathcal{L}:\, u^* \in L \subsetneq W} y_L + \sum_{L \in \mathcal{L}:\, v^* \in L \subsetneq W} y_L \\ &= \sum_{L \in \mathcal{L}:\, L \subsetneq W,\, L \cap V(P_{u^*,v^*}) \neq \emptyset} 2y_L \\ &\leq \sum_{L \in \mathcal{L}:\, L \subsetneq W,\, L \cap V(B) \neq \emptyset} 2y_L \\ &= \text{value}(W) - \sum_{L \in \mathcal{L}_{\bar{B}}} (2y_L + \text{value}(L)). \end{aligned}$$

□

Now, we analyze the cost of the tour F in $G[W]$ computed by Algorithm 7.8. Note that the bound in the following lemma becomes smaller the larger D_W is. Hence, the more expensive it can be to visit W (entering W at some vertex s, following the nice path $P_{s,t}$, and leaving it at t), the cheaper is the tour in $G[W]$ that we compute.

Lemma 7.11. *Let $\kappa, \eta \geq 0$. Suppose we have a (κ, η)-algorithm $\mathcal{A}$ for vertebrate pairs. Let $\mathcal{I} = (G, \mathcal{L}, x, y)$ be a strongly laminar* ASYMMETRIC *TSP instance, c^y the induced cost function, and $W \in \mathcal{L} \cup \{V\}$. Then the tour F in $G[W]$ returned by Algorithm 7.8 has cost at most*

$$c^y(F) \leq (2\kappa + 2) \cdot \text{value}(W) + (\kappa + \eta) \cdot (\text{value}(W) - D_W).$$

Proof. By induction on $|W|$. The statement is trivial for $|W| = 1$ since then $c^y(F) = 0$ (because $F \subseteq E[W] = \emptyset$). Let now $|W| \geq 2$. By definition of D_W, we have

$$c^y(E(B)) = c^y(E(P_{u^*,v^*})) + c^y(E(P_{v^*,u^*})) \leq 2 \cdot D_W. \tag{7.3}$$

We now analyze the cost of F' in Step (3) of Algorithm 7.8. Since F' is the output of a (κ, η)-algorithm applied to the vertebrate pair $(\mathcal{I}', B)$, we have $c^y(F') \leq \kappa \cdot \text{LP}(\mathcal{I}') + \eta \cdot \sum_{L \in \mathcal{L}_{\bar{B}}} 2y_{\{v_L\}}$. Using $\sum_{L \in \mathcal{L}_{\bar{B}}} 2y_{\{v_L\}} = \sum_{L \in \mathcal{L}_{\bar{B}}} (2y_L + D_L)$ and

$$\text{LP}(\mathcal{I}') \leq D_W + \sum_{\substack{L \in \mathcal{L}, L \subsetneq W, \\ L \cap V(B) \neq \emptyset}} 2y_L + \sum_{L \in \mathcal{L}_{\bar{B}}} 2y_{\{v_L\}},$$

this implies

$$c^y(F') \leq \kappa \cdot D_W + \sum_{\substack{L \in \mathcal{L}, L \subsetneq W, \\ L \cap V(B) \neq \emptyset}} \kappa \cdot 2y_L + \sum_{L \in \mathcal{L}_{\bar{B}}} (\kappa + \eta) \cdot (2y_L + D_L) \tag{7.4}$$

at the end of Step (3). The lifting and all the amendments of F' in Step (4) do not increase the cost of F' by Lemma 7.6 and the choices of the values $y_{\{v_L\}}$ in Step (2) and y_W in Step (1). (Here we use that whenever an Eulerian walk passes through $v_{\bar{W}}$, we leave and enter W.)

To bound the cost increase in Step (5), we apply the induction hypothesis. Adding the edges resulting from a single recursive call of Algorithm 7.8 in Step (5) for some $L \in \mathcal{L}_{\bar{B}}$ increases the cost by at most

$$c^y(F_L) \leq (2\kappa + 2) \cdot \text{value}(L) + (\kappa + \eta) \cdot (\text{value}(L) - D_L).$$

Using (7.4), we obtain the following bound:

$$\begin{aligned}
c^y(F'') &\le \kappa \cdot D_W + \sum_{L\in\mathcal{L}, L\subsetneq W, L\cap V(B)\neq\emptyset} \kappa \cdot 2y_L \\
&\quad + \sum_{L\in\mathcal{L}_{\bar{B}}} \big((2\kappa+2)\mathrm{value}(L) + (\kappa+\eta)(2y_L + \mathrm{value}(L))\big) \\
&\le \kappa \cdot D_W + \kappa \cdot \mathrm{value}(W) \\
&\quad + \sum_{L\in\mathcal{L}_{\bar{B}}} \big((\kappa+2)\mathrm{value}(L) + (\kappa+\eta)(2y_L + \mathrm{value}(L))\big) \\
&\le \kappa \cdot D_W + \kappa \cdot \mathrm{value}(W) \\
&\quad + (\kappa+2)(\mathrm{value}(W) - D_W) + (\kappa+\eta)(\mathrm{value}(W) - D_W) \\
&= (2\kappa+2)\mathrm{value}(W) - 2\,D_W + (\kappa+\eta)(\mathrm{value}(W) - D_W),
\end{aligned} \tag{7.5}$$

where we used the definition of $\mathcal{L}_{\bar{B}}$ for the second inequality and Lemma 7.10 for the third inequality. Together with (7.3), this implies the claimed bound on $c^y(F)$. □

Now we prove the main result of this section.

Theorem 7.12. *Let $\kappa, \eta \ge 0$. Suppose we have a (κ, η)-algorithm for vertebrate pairs. Then there is a polynomial-time algorithm that computes a solution of cost at most*

$$3\kappa + \eta + 2$$

times the value of the LP (3.2) for any given Asymmetric *TSP instance.*

Proof. By Theorem 7.4, it suffices to show that there is a polynomial-time algorithm that computes a solution of cost at most $(3\kappa + \eta + 2) \cdot \mathrm{LP}(\mathcal{I})$ for any given strongly laminar Asymmetric TSP instance $\mathcal{I} = (G, \mathcal{L}, x, y)$. Given such an instance, we apply Algorithm 7.8 to $W = V$, where V is the vertex set of G. By Lemma 7.9 and Lemma 7.11, this algorithm computes in polynomial time a tour of cost at most

$$\begin{aligned}
c^y(F) &\le (2\kappa+2) \cdot \mathrm{value}(V) + (\kappa+\eta) \cdot (\mathrm{value}(V) - D_V) \\
&= (2\kappa+2) \cdot \mathrm{LP}(\mathcal{I}) + (\kappa+\eta) \cdot (\mathrm{LP}(\mathcal{I}) - D_V) \\
&\le (3\kappa+\eta+2) \cdot \mathrm{LP}(\mathcal{I}).
\end{aligned}$$

□

See Exercises 7.5–7.7 for a small improvement of this bound. In the following, we will present a $(2, 9+\varepsilon)$-algorithm for vertebrate pairs for any $\varepsilon > 0$.

In Chapter 9, we will apply a slight modification of Algorithm 7.8 to a more general problem, the Asymmetric Path TSP. This modification works with a given B instead of constructing it in Step (2). We will need:

Lemma 7.13. *Let $\kappa, \eta \geq 0$. Suppose we have a (κ, η)-algorithm $\mathcal{A}$ for vertebrate pairs. Let $\mathcal{I} = (G, \mathcal{L}, x, y)$ be a strongly laminar* Asymmetric TSP *instance, $W \in \mathcal{L} \cup \{V\}$, and B a connected Eulerian multi-subgraph of G with $V(B) \cap W \neq \emptyset$. Let $F'' \dot{\cup} E(B)$ be the tour in $G[W \cup V(B)]$ that is computed by the variant of Algorithm 7.8 that does not compute B in Step (2) but works with the given B. Then F'' has cost at most*

$$c^y(F'') \leq \kappa \cdot D_W + \kappa \cdot \text{value}(W) + (2\kappa + \eta + 2) \cdot \sum_{L \in \mathcal{L}: L \subsetneq W, V(B) \cap L = \emptyset} 2y_L.$$

Proof. Directly from the bound (7.5) in the proof of Lemma 7.11. □

7.6 Subtour Cover

In the first part of this chapter, we showed that the Asymmetric TSP reduces to vertebrate pairs. Svensson, Tarnawski, and Végh [2020] showed that an algorithm for vertebrate pairs can be designed similarly to the algorithm for Asymmetric Graph TSP. To this end, we will adapt Algorithm 6.13 (Svensson's algorithm for Asymmetric Graph TSP) to vertebrate pairs.

In Chapter 6, we used an algorithm to find a graph subtour cover as subroutine. Here, we need a more general subroutine: an algorithm to find a *subtour cover*. Recall that we had an Eulerian multi-subset H of the edge set E of our digraph $G = (V, E)$, and a graph subtour cover for H is an Eulerian multi-subset set F of E that enters and leaves every connected component of (V, H). Now, we have a strongly laminar Asymmetric TSP instance $(G, \mathcal{L}, x, y)$ with a backbone B, and H is a *local* Eulerian multi-subset H of E, where local means that H does not contain any edge from $\delta(L)$ for any $L \in \mathcal{L}_{\geq 2}$ (recall that $\mathcal{L}_{\geq 2} = \{L \in \mathcal{L} : |L| \geq 2\}$ is the family of all non-singleton elements of $\mathcal{L}$). For a subtour cover, we require in addition that every connected component of (V, F) is local or connected to the backbone.

Definition 7.14 (local, subtour cover)**.** Let $(\mathcal{I}, B)$ be a vertebrate pair with $\mathcal{I} = (G, \mathcal{L}, x, y)$ and $G = (V, E)$. We call a multi-subset H of E (and a graph with edge set H) *local* if $H \cap \delta(L) = \emptyset$ for all $L \in \mathcal{L}_{\geq 2}$.

A *subtour cover instance* $(\mathcal{I}, B, H)$ consists of a vertebrate pair $(\mathcal{I}, B)$ and a local Eulerian multi-subset H of $E[V \setminus V(B)]$. A *subtour cover* for $(\mathcal{I}, B, H)$ is an Eulerian multi-subset F of E satisfying the following two conditions:

(i) $F \cap \delta(W) \neq \emptyset$ for all vertex sets W of connected components of the graph $(V \setminus V(B), H)$.

(ii) For every connected component D of (V, F), the edge set $E(D)$ is local or $V(D) \cap V(B) \neq \emptyset$.

See Figure 7.9 for an example. In the following, we write $y_v := y_{\{v\}}$ if $\{v\} \in \mathcal{L}$ and $y_v := 0$ otherwise. In Chapter 8, we will give a polynomial-time algorithm that computes, for any given subtour cover instance $(\mathcal{I}, B, H)$, a subtour cover F such that

$$c^y(F) \leq 2 \cdot \mathrm{LP}(\mathcal{I}) + \sum_{v \in V \setminus V(B)} 2y_v, \tag{7.6}$$

and for every connected component D of (V, F) with $V(D) \cap V(B) = \emptyset$, we have

$$c^y(E(D)) \leq 2 \cdot \sum_{v \in V(D)} 2y_v. \tag{7.7}$$

Again, c^y denotes the induced cost function of the strongly laminar Asymmetric TSP instance $\mathcal{I}$.

While adapting Svensson's algorithm to vertebrate pairs requires only relatively small changes compared to the algorithm for Asymmetric Graph TSP, generalizing the algorithm for graph subtour cover from Section 6.2 to subtour cover requires substantially new techniques.

We remark that requiring property (ii) would not be necessary to obtain a constant-factor approximation algorithm for Asymmetric TSP. However, an inequality like (7.7) is crucial. Since we will prove (7.7) using property (ii) anyway, and requiring this property explicitly yields a better approximation ratio, we defined subtour covers as in Definition 7.14.

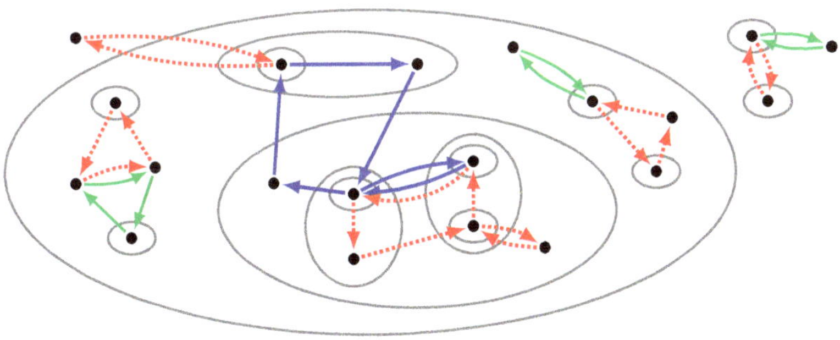

Figure 7.9 Example of a vertebrate pair $(\mathcal{I}, B)$, a local Eulerian edge set H, and a subtour cover for $(\mathcal{I}, B, H)$. The laminar family $\mathcal{L}$ of $\mathcal{I}$ is shown in gray, and the backbone B is shown in blue. The edge set H is shown in green, and a subtour cover consists of the dotted red edges.

In order to exhibit the dependence of the approximation guarantee of Svensson's algorithm for vertebrate pairs on the subtour cover algorithm, we introduce the notion of an (α, κ, β)-*algorithm* for subtour cover. In Chapter 8, we will describe a $(2, 2, 1)$-algorithm for subtour cover.

Definition 7.15 (algorithm for subtour cover). Let $\alpha, \kappa, \beta \geq 0$ be constants. An (α, κ, β)-*algorithm for subtour cover* is a polynomial-time algorithm that computes a subtour cover F for any given subtour cover instance $(\mathcal{I}, B, H)$ such that

$$c^y(F) \leq \kappa \cdot \mathrm{LP}(\mathcal{I}) + \beta \cdot \sum_{v \in V \setminus V(B)} 2y_v, \tag{7.8}$$

and for every connected component D of (V, F) with $V(D) \cap V(B) = \emptyset$, we have

$$c^y(E(D)) \leq \alpha \cdot \sum_{v \in V(D)} 2y_v. \tag{7.9}$$

Let $\alpha > 1$ and $\kappa, \beta \geq 0$ be constants such that there is an (α, κ, β)-algorithm $\mathcal{A}$ for subtour cover, and let $\varepsilon > 0$ be a fixed constant. The goal of the rest of this chapter is to show that then there is a $(\kappa, 4\alpha + \beta + \varepsilon)$-algorithm for vertebrate pairs.

To this end, we adapt Svensson's algorithm from Chapter 6 to vertebrate pairs. While most of the algorithm and analysis are the same, there are a few essential differences. Most importantly, there is a backbone, and a subtour cover can connect to it, but we will never buy edges that connect to the backbone except for the very last iteration. Until then, we will only consider local edges (here, we exploit property (ii) of Definition 7.14). The cost of any local edge $e = (v, w)$ is $y_v + y_w$. Hence, as long as we buy only local edges, the algorithm behaves like for node-weighted instances, which are essentially equivalent to unweighted instances by Exercise 6.1.

7.7 Initializing Svensson's Algorithm for Vertebrate Pairs

Let $(\mathcal{I}, B)$ be a vertebrate pair. Svensson's algorithm for vertebrate pairs is initialized with a local Eulerian multi-set $\tilde{H} \subseteq E[V \setminus V(B)]$ and then computes either a "better" initialization $\tilde{H}'$ or extends $\tilde{H}$ to a solution H of the given vertebrate pair $(\mathcal{I}, B)$.

As in Svensson's algorithm for Asymmetric Graph TSP, the initialization $\tilde{H}$ of the algorithm will always be *light*. However, we use a different function

$\ell : V \to \mathbb{R}_{\geq 0}$ for defining what a light edge set is. For $v \in V$, we set

$$\ell(v) := \begin{cases} (1+\varepsilon') \cdot 2\alpha \cdot 2y_v + \frac{\varepsilon'}{n} \cdot \sum_{u \in V \setminus V(B)} 2y_u & \text{if } v \in V \setminus V(B) \\ \frac{\kappa \cdot \mathrm{LP}(\mathcal{I}) + \beta \cdot \sum_{u \in V \setminus V(B)} 2y_u}{|V(B)|} & \text{if } v \in V(B), \end{cases} \tag{7.10}$$

where $\varepsilon' := \min\{\frac{\varepsilon}{2+4\alpha}, 1 - \frac{1}{\alpha}\}$ and $n := |V|$.

Recall that an Eulerian multi-subset $\tilde{H}$ of E is *light* if $c^y(E(D)) \leq \ell(V(D))$ for every connected component D of $(V, \tilde{H})$.

Note that for $v \in V \setminus V(B)$, the first term of the definition of $\ell(v)$ is proportional to the corresponding dual variable y_v. We need the small additional term $\frac{\varepsilon'}{n} \cdot \sum_{u \in V \setminus V(B)} 2y_u$ to guarantee that the numbers $\ell(v)$ $(v \in V \setminus V(B))$ cannot differ by a too large factor; this will allow us to bound the progress when we find a better initialization as in Lemma 6.9. For vertices in $V(B)$, we will only need that $\ell(V(B)) = \kappa \cdot \mathrm{LP}(\mathcal{I}) + \beta \cdot \sum_{u \in V \setminus V(B)} 2y_u$.

As in Asymmetric Graph TSP, we use a potential function Φ to measure what a "better" initialization for Svensson's algorithm is. For a light Eulerian multi-subset $\tilde{H}$ of $E[V \setminus V(B)]$ such that the connected components of $(V \setminus V(B), \tilde{H})$ have vertex sets $\tilde{W}_1, \ldots, \tilde{W}_k$, we write

$$\Phi(\tilde{H}) = \sum_{i=1}^{k} \mathrm{slack}(\tilde{W}_i, \tilde{H})^{1+p},$$

where again $p := \log_{1+\varepsilon'}(\frac{1+\varepsilon'}{\varepsilon'})$ and $\mathrm{slack}(\tilde{V}, \tilde{E}) = \ell(\tilde{V}) - c^y(\tilde{E}[\tilde{V}])$.

As for Asymmetric Graph TSP, we will use Lemma 6.9 to find improved initializations for Svensson's algorithm. We will apply it to connected Eulerian multi-subgraphs D of $G[V \setminus V(B)]$.

We again order the connected components of $(V \setminus V(B), \tilde{H})$ such that $\mathrm{slack}(\tilde{W}_1, \tilde{H}) \geq \cdots \geq \mathrm{slack}(\tilde{W}_k, \tilde{H})$. Moreover, we define $\tilde{W}_0 := V(B)$. Then, for a connected multi-subgraph D of G, we again define its *index* by $\mathrm{ind}(D) := \min\{j \in \{0, \ldots, k\} : V(D) \cap \tilde{W}_j \neq \emptyset\}$.

During Svensson's algorithm for vertebrate pairs, we obtain D from a subtour cover F. For $i \in \{0, 1, \ldots, k\}$, we denote by F_i the union of the connected components D' of (V, F) with $\mathrm{ind}(D') = i$. We will find a better initialization for Svensson's algorithm for vertebrate pairs whenever F violates any of the following two conditions:

$$c^y(E(F_i)) \leq \ell(\tilde{W}_i) \text{ for all } i \in \{1, \ldots, k\}; \tag{7.11}$$

$$\begin{aligned} \mathrm{slack}(V(D), E(D)) \leq \mathrm{slack}(\tilde{W}_{\mathrm{ind}(D)}, \tilde{H}) + \tfrac{\varepsilon'}{1+\varepsilon'} \cdot \ell(V(D)) \\ \text{for every connected component } D \text{ of } (V, F) \text{ with } \mathrm{ind}(D) > 0. \end{aligned} \tag{7.12}$$

Next, we show that we can indeed find an improved initialization if (7.11) or (7.12) is violated. We show that we can then find a connected Eulerian multi-subgraph D of $G[V \setminus V(B)]$ with (6.2). This allows for applying Lemma 6.9 to D. While this is trivial if (7.12) is violated, we need a lemma similar to Lemma 6.10 for the case when (7.11) is violated:

Lemma 7.16. *Let F be an Eulerian multi-edge set such that $c^y(E(D')) \leq \alpha \cdot \sum_{v \in V(D')} 2y_v$ for every connected component D' of (V, F) with $V(D') \cap V(B) = \emptyset$. For $i \in \{0, \ldots, k\}$, let the graph F_i be the union of the connected components D' of (V, F) with $\text{ind}(D') = i$.*

Suppose we have $c^y(E(F_i)) > \ell(\tilde{W}_i)$ for some $i \in \{1, \ldots, k\}$. Then

$$D := \left(\tilde{W}_i \cup V(F_i),\ \tilde{H}[\tilde{W}_i] \dot{\cup} E(F_i)\right)$$

is a connected Eulerian multi-subgraph of $G[V \setminus V(B)]$ with (6.2).

Proof. Let $i \in \{1, \ldots, k\}$ such that $c^y(E(F_i)) > \ell(\tilde{W}_i)$. Then

$$\tfrac{1}{2(1+\varepsilon')} \cdot \ell(V(D)) \geq \tfrac{1}{2(1+\varepsilon')} \cdot \ell(V(F_i)) \geq c^y(E(F_i)) > \ell(\tilde{W}_i), \qquad (7.13)$$

where the second inequality follows from $c^y(E(F_i)) \leq \alpha \cdot \sum_{v \in V(F_i)} 2y_v$ and the definition of ℓ (see (7.10)). We get

$$\begin{aligned} c^y(E(D)) &= c^y(\tilde{H}[\tilde{W}_i]) + c^y(E(F_i)) \\ &= \ell(\tilde{W}_i) + c^y(E(F_i)) - \text{slack}(\tilde{W}_i, \tilde{H}) \\ &< \tfrac{1}{2(1+\varepsilon')} \cdot \ell(V(D)) + \tfrac{1}{2(1+\varepsilon')} \cdot \ell(V(D)) - \text{slack}(\tilde{W}_i, \tilde{H}) \\ &= \ell(V(D)) - \tfrac{\varepsilon'}{1+\varepsilon'} \cdot \ell(V(D)) - \text{slack}(\tilde{W}_i, \tilde{H}), \end{aligned}$$

where we used (7.13) twice (once to bound $c^y(E(F_i))$ and once to bound $\ell(\tilde{W}_i)$). This proves (6.2). □

7.8 Adapting Svensson's Algorithm to Vertebrate Pairs

The following lemma states the result of Svensson's algorithm for vertebrate pairs:

Lemma 7.17. *Let $\alpha > 1$, $\kappa, \beta \geq 0$ such that there is an (α, κ, β)-algorithm for subtour cover, and let ε' be a fixed constant with $0 < \varepsilon' \leq 1 - \frac{1}{\alpha}$. Given a vertebrate pair $(\mathcal{I}, B)$ with $\mathcal{I} = (G, \mathcal{L}, x, y)$ and $G = (V, E)$ and a light local Eulerian multi-subset $\tilde{H}$ of $E[V \setminus V(B)]$, we can compute in polynomial time*

(i) *a solution H for the vertebrate pair $(\mathcal{I}, B)$ such that*

$$c^y(H) \leq \ell(V(B)) + 2 \cdot \ell(V \setminus V(B)) \tag{7.14}$$

or

(ii) *a light local Eulerian multi-subset $\tilde{H}'$ of $E[V \setminus V(B)]$ such that*

$$\Phi(\tilde{H}') - \Phi(\tilde{H}) > \min\{\ell(v)^{1+p} : v \in V \setminus V(B)\}. \tag{7.15}$$

From Lemma 7.17, we can derive the main result of this section.

Theorem 7.18. *Let $\alpha > 1$, $\kappa, \beta \geq 0$ such that there is an (α, κ, β)-algorithm for subtour cover, and let $\varepsilon > 0$ be a fixed constant.*

Then there is a $(\kappa, 4\alpha + \beta + \varepsilon)$-algorithm for vertebrate pairs.

Proof. Define $\varepsilon' := \min\{\frac{\varepsilon}{2+4\alpha}, 1 - \frac{1}{\alpha}\}$, p, ℓ, and Φ as earlier. We start with $\tilde{H} = \emptyset$ and apply Lemma 7.17. If we obtain a set $\tilde{H}'$ as in Lemma 7.17 (ii), we set $\tilde{H} := \tilde{H}'$ and restart – that is, we apply Lemma 7.17 again until we finally obtain a set H as in Lemma 7.17 (i).

To bound the number of restarts, note that

$$\begin{aligned}\ell(V \setminus V(B)) &\leq (1 + \varepsilon') \cdot 2\alpha \cdot \sum_{u \in V \setminus V(B)} 2y_u + \varepsilon' \cdot \sum_{u \in V \setminus V(B)} 2y_u \\ &= ((1 + \varepsilon') \cdot 2\alpha + \varepsilon') \cdot \sum_{u \in V \setminus V(B)} 2y_u,\end{aligned}$$

which implies that for every vertex $v \in V \setminus V(B)$, we have

$$\ell(v) \geq \frac{\varepsilon'}{n} \cdot \sum_{u \in V \setminus V(B)} 2y_u \geq \frac{\varepsilon'}{((1 + \varepsilon') \cdot 2\alpha + \varepsilon') \cdot n} \cdot \ell(V \setminus V(B)).$$

Since $0 \leq \Phi(\tilde{H}) \leq \ell(V \setminus V(B))^{1+p}$, we need at most

$$\left(\frac{((1 + \varepsilon') \cdot 2\alpha + \varepsilon') \cdot n}{\varepsilon'}\right)^{1+p}$$

restarts, which is bounded by a polynomial in n as p, α, and ε' are constants.

At the end, the algorithm guaranteed by Lemma 7.17 returns a solution H for the vertebrate pair $(\mathcal{I}, B)$ such that

$$\begin{aligned}c^y(H) &\leq \ell(V(B)) + 2 \cdot \ell(V \setminus V(B)) \\ &\leq \kappa \cdot \mathrm{LP}(\mathcal{I}) + (\beta + 2 \cdot ((1 + \varepsilon') \cdot 2\alpha + \varepsilon')) \cdot \sum_{v \in V \setminus V(B)} 2y_v \\ &\leq \kappa \cdot \mathrm{LP}(\mathcal{I}) + (4\alpha + \beta + \varepsilon) \cdot \sum_{v \in V \setminus V(B)} 2y_v,\end{aligned}$$

where we used $\varepsilon' \leq \frac{\varepsilon}{2+4\alpha}$ in the last inequality. □

In the remaining part of this section, we prove Lemma 7.17. To this end, we consider Algorithm 7.19, which is a variant of Algorithm 6.13.

To implement Step (6c), consider each local edge $e = (v, w) \in \delta^+(V(Z))$ and compute a shortest w-v-path P in $\big(V \setminus V(B), E[V \setminus V(B)] \setminus \big(\cup_{L \in \mathcal{L}_{\geq 2}} \delta(L)\big)\big)$ (cf. Theorem 1.14) and check whether $c^y(e) + c^y(E(P)) \leq \text{slack}(\tilde{W}_{\text{ind}(Z)}, \tilde{H})$.

Note that adding $E(C)$ to X in Step (6c) decreases the number of connected components of $(V, H \dot{\cup} F \dot{\cup} X)$, and adding edges to H in Step (6d) decreases the number of connected components of (V, H). Thus, the procedure terminates after a polynomial number of steps.

The following observation implies that $(\mathcal{I}, B, H)$ is indeed a subtour cover instance in Step (4a) of Algorithm 7.19.

Lemma 7.20. *As long as $(V, E(B) \cup H)$ is not connected in Algorithm 7.19, H is a local Eulerian multi-subset of $E[V \setminus V(B)]$.*

Proof. In Step (2) of Algorithm 7.19, we set $H := \tilde{H}$ and thus H is a local Eulerian multi-subset of $E[V \setminus V(B)]$. The circuits that we find in Step (6c) neither contain a vertex from the backbone nor an edge in $\delta(L)$ for any $L \in \mathcal{L}_{\geq 2}$ by construction. Moreover, by (ii) of the definition of a subtour cover (Definition 7.14), F_i is local for every $i \in \{1, \ldots, k\}$. By the choice of Z in Step (6b) of Algorithm 7.19, the component Z contains edges from F_0 only if $(V, E(B) \dot{\cup} H \dot{\cup} F \dot{\cup} X)$ is connected, and in this case, $(V, E(B) \dot{\cup} H)$ becomes connected when the edges in $F \dot{\cup} X$ are added to H. □

Also notice that Step (4b) maintains a subtour cover with (7.8) and (7.9). Therefore, we can apply Lemma 7.16 for Step (5a) to show that the application of Lemma 6.9 is indeed possible (in Step (5b), this is trivial). Moreover, $E(D)$ is local in Steps (5a) and (5b) because F_i is local for all $i > 0$ and $\tilde{H}$ is local. Hence, if Algorithm 7.19 returns a multi-set $\tilde{H}'$ in Step (5), then $\tilde{H}'$ is local and thus it is a multi-set as in Lemma 7.17 (ii).

Otherwise, the returned edge set H is a solution for the vertebrate pair $(\mathcal{I}, B)$. To show the upper bound (7.14) on the cost of H, we proceed analogously to Chapter 6. Initially, we have $c^y(H) = c^y(\tilde{H})$. Then we bound the cost of the X-edges and the cost of the F-edges added to H separately.

We need the following observation:

Lemma 7.21. *For every local circuit C in $G[V \setminus V(B)]$, we have*

$$c^y(E(C)) \leq \frac{1}{(1 + \varepsilon') \cdot 2\alpha} \cdot \ell(V(C)).$$

Algorithm 7.19: Svensson's Algorithm for Vertebrate Pairs

Input: a vertebrate pair $(\mathcal{I}, B)$ with $\mathcal{I} = (G, \mathcal{L}, x, y)$ and $G = (V, E)$, a light local Eulerian multi-subset $\tilde{H} \subseteq E[V \setminus V(B)]$, an (α, κ, β)-algorithm $\mathcal{A}$ for subtour cover

Output: either H as in Lemma 7.17 (i) or $\tilde{H}'$ as in Lemma 7.17 (ii)

(1) Let $\tilde{W}_0 := V(B)$ and let $\tilde{W}_1, \ldots, \tilde{W}_k$ be the vertex sets of the connected components of $(V \setminus V(B), \tilde{H})$ such that $\text{slack}(\tilde{W}_1, \tilde{H}) \geq \cdots \geq \text{slack}(\tilde{W}_k, \tilde{H})$.

(2) Set $H := \tilde{H}$.

(3) **while** $(V, E(B) \dot{\cup} H)$ is not connected **do**

(4) **Compute a subtour cover F:**

(4a) Apply $\mathcal{A}$ to the subtour cover instance $(\mathcal{I}, B, H)$ to obtain a subtour cover F'.

(4b) Let F result from F' by deleting all edges of connected components of (V, F') whose vertex sets are contained in a connected component of $(V, E(B) \cup H)$.

(5) **Try to find a better initialization $\tilde{H}'$:**
For $i \in \{0, \ldots, k\}$, let the graph F_i be the union of the connected components D' of (V, F) with $\text{ind}(D') = i$.

(5a) If for some $i \in \{1, \ldots, k\}$ we have $c^y(E(F_i)) > \ell(\tilde{W}_i)$, apply Lemma 6.9 to $D = \big(\tilde{W}_i \cup V(F_i),\ \tilde{H}[\tilde{W}_i] \dot{\cup} E(F_i)\big)$ to obtain a better initialization $\tilde{H}'$. Then return $\tilde{H}'$.

(5b) If (V, F) has a connected component D with $\text{ind}(D) > 0$ and $\text{slack}(V(D), E(D)) > \text{slack}(\tilde{W}_{\text{ind}(D)}, \tilde{H}) + \frac{\varepsilon'}{1+\varepsilon'} \cdot \ell(V(D))$, apply Lemma 6.9 to obtain a better initialization $\tilde{H}'$. Then return $\tilde{H}'$.

(6) **Extend H:**

(6a) Set $X := \emptyset$.

(6b) Select the connected component Z of $(V, E(B) \dot{\cup} H \dot{\cup} F \dot{\cup} X)$ for which $\text{ind}(Z)$ is largest.

(6c) If there is a circuit C in $G[V \setminus V(B)]$ with
- $E(C) \cap \delta(V(Z)) \neq \emptyset$,
- C is local, and
- $c^y(E(C)) \leq \text{slack}(\tilde{W}_{\text{ind}(Z)}, \tilde{H})$,

then add $E(C)$ to X and go to Step (6b).

(6d) Add the edges $(F \dot{\cup} X)[V(Z)]$ to H.

(7) Return H.

Proof. Let C be a circuit with $E(C) \cap \delta(L) = \emptyset$ for all $L \in \mathcal{L}_{\geq 2}$. Then we have $c^y(E(C)) = \sum_{v \in V(C)} 2y_v \leq \frac{1}{(1+\varepsilon')\cdot 2\alpha} \cdot \ell(V(C))$, using the definition of ℓ (see (7.10)). □

The following proofs are very similar to the proofs of Lemmas 6.14 and 6.15.

Lemma 7.22. *The total cost of all X-edges that are added to H is at most $\ell(V \setminus V(B)) - c^y(\tilde{H})$.*

Proof. A circuit C that is selected in Step (6c) and will later be added to H connects Z to another connected component Y with $\text{ind}(Y) < \text{ind}(Z)$. We say that it marks $\text{ind}(Z)$. It has cost at most $\text{slack}(\tilde{W}_{\text{ind}(Z)}, \tilde{H})$. No circuit added later can mark $\text{ind}(Z)$ because the new connected component of $(V, E(B) \dot\cup H \dot\cup F \dot\cup X)$ containing $Y \cup Z$ will have a smaller index by the choice of Z. Hence, the total cost of the added circuits is at most $\sum_{i=1}^{k} \text{slack}(\tilde{W}_i, \tilde{H}) = \ell(V \setminus V(B)) - c^y(\tilde{H})$. □

Lemma 7.23. *The total cost of all F-edges that are added to H is at most $\ell(V)$.*

Proof. Let Z^t denote Z at the end of iteration t of the while-loop. Let F_i^t be the graph F_i in iteration t if the set of edges of F_i is nonempty and is added to H at the end of this iteration, and let $F_i^t = \emptyset$ otherwise.

For $i = 1, \ldots, k$, the total cost of F_i^t is $c^y(E(F_i^t)) \leq \ell(\tilde{W}_i)$ by Step (5a). Moreover,

$$c^y(E(F_0^t)) \leq c^y(F) \leq \kappa \cdot \text{LP}(\mathcal{I}) + \beta \cdot \sum_{v \in V \setminus V(B)} 2y_v = \ell(\tilde{W}_0)$$

because $\mathcal{A}$ is an (α, κ, β)-algorithm for subtour cover and thus F satisfies (7.8).

We claim that for any i, at most one of the F_i^t is nonempty. Then summing over all i and t concludes the proof.

Suppose there are $t_1 < t_2$ such that $F_i^{t_1} \neq \emptyset$ and $F_i^{t_2} \neq \emptyset$. We have $i > 0$ because otherwise the algorithm would terminate after iteration t_1 by the choice of Z^{t_1}. Then $V(F_i^{t_1}) \subseteq V(Z^{t_1})$ and thus $\tilde{W}_i \subseteq V(Z^{t_1})$. Moreover, $F_i^{t_2}$ contains a vertex of $\tilde{W}_i$ and is not completely contained in Z^{t_1} by Step (4b) of the algorithm. Thus, $F_i^{t_2}$ contains a circuit C with $E(C) \cap \delta(V(Z^{t_1})) \neq \emptyset$. Note that $V(F_i^{t_2}) \cap V(B) = \emptyset$ (since $i > 0$). Hence, C is a circuit in $G[V \setminus V(B)]$, and moreover C is local because it is a subset of $E(F_i^{t_2})$ and F is a subtour cover (cf. item (ii) of Definition 7.14).

If $c^y(E(C)) \leq \text{slack}(\tilde{W}_{\text{ind}(Z^{t_1})}, \tilde{H})$, due to Step (6c), this is a contradiction to reaching Step (6d) in iteration t_1 with the component Z^{t_1}. Otherwise, let D be the connected component of $F_i^{t_2}$ containing C. Note that $\text{ind}(D) = i \geq \text{ind}(Z^{t_1})$ (cf. Figure 6.5).

Since C is a circuit in $G[V \setminus V(B)]$ and C is local, we can apply Lemma 7.21 to obtain

$$\tfrac{1}{(1+\varepsilon')\cdot 2\alpha} \cdot \ell(V(C)) \geq c^y(E(C)) > \operatorname{slack}(\tilde{W}_{\operatorname{ind}(Z^{t_1})}, \tilde{H}) \\ \geq \operatorname{slack}(\tilde{W}_{\operatorname{ind}(D)}, \tilde{H}).$$

Moreover, using the facts that $c^y(E(D)) \leq \alpha \sum_{v\in V(D)} 2y_v \leq \frac{1}{2}\ell(V(D))$, that D contains C, and that $\frac{1}{\alpha} \leq 1-\varepsilon'$ (by the choice of ε'),

$$\operatorname{slack}(V(D), E(D)) \geq \tfrac{1}{2} \cdot \ell(V(D)) \geq \tfrac{1}{(1+\varepsilon')\cdot 2\alpha} \cdot \ell(V(C)) + \tfrac{\varepsilon'}{1+\varepsilon'} \cdot \ell(V(D)).$$

Together, we obtain

$$\operatorname{slack}(V(D), E(D)) > \operatorname{slack}(\tilde{W}_{\operatorname{ind}(D)}, \tilde{H}) + \tfrac{\varepsilon'}{1+\varepsilon'} \cdot \ell(V(D)).$$

Due to Step (5b), this is a contradiction to reaching Step (6) in iteration t_2 and adding $F_i^{t_2}$ there. □

Using Lemmas 7.22 and 7.23, we conclude that the cost of the returned edge set H is at most $\ell(V(B)) + 2 \cdot \ell(V \setminus V(B))$. This concludes the proof of Lemma 7.17.

Exercises

7.1 Show that the set S in the proof of Lemma 7.2 is unique.

7.2 Define strongly laminar Symmetric TSP instances analogously to Definition 7.3, but replacing "strongly connected" with "connected." Conclude an analogous statement to Theorem 7.4.

7.3 Let $(G, \mathcal{L}, x, y)$ be a strongly laminar Asymmetric TSP instance with $G = (V, E)$, and let $W \in \mathcal{L}$. Show that the value of the LP (3.2) on the instance $(G[W], c^y)$ is at most $\sum_{e\in E[W]} c^y(e)x_e + D_W$.
Hint: Use splitting off (Theorem 3.3).

7.4 Let $\varepsilon > 0$ be a constant. Given a strongly laminar Asymmetric TSP instance $\mathcal{I} = (G, \mathcal{L}, x, y)$ such that no vertex is contained in more than k sets of $\mathcal{L}$, show how to compute a solution of cost at most $(16k+\varepsilon)\cdot \mathrm{LP}(\mathcal{I})$ in polynomial time.
Hint: Proceed as indicated at the beginning of Section 7.4. Contract the elements of $\mathcal{L}_{\geq 2}$. Use that if $\mathcal{L}$ contains only singletons, one can compute a solution of cost at most $(8+\varepsilon)\cdot \mathrm{LP}(\mathcal{I})$ in polynomial time by Exercise 6.1 and Theorem 6.12. When recursing, also contract the outside as in Step (1) of Algorithm 7.8.

7.5 In this and the following two exercises, we will show how to obtain a slightly better approximation ratio for Asymmetric TSP. Let $0 < \delta < 1$. Modify Algorithm 7.8 as follows. As in Lemma 7.13, it also takes a connected Eulerian multi-subgraph B with $V(B) \cap W \neq \emptyset$ as input, and uses this as the backbone instead of computing a backbone in Step (2). Call a set $L \in \mathcal{L}_{\bar{B}}$ *busy* if $D_L(u, v) \geq (1 - \delta)D_L$ for at least one of the paths $P_{u,v}$ that we add within this set L in Step (4). For the recursive calls, use F' as the backbone for L whenever L is busy; otherwise, construct a new backbone for L by appending P_{u^*,v^*} and P_{v^*,u^*}, where u^* and v^* are chosen such that $D_L(u^*, v^*) = D_L$, and add this new backbone to F'. In the end, return only F''.

(a) Show that, for any input $(\mathcal{I}, W, B)$, the set F'' returned by the modified algorithm, together with the given backbone B, forms a tour in the subgraph $G[W \cup V(B)]$.

(b) Show that for any input $(\mathcal{I}, W, B)$ to the modified algorithm, we have $\sum_{L \in \mathcal{L}_{\bar{B}}} (2y_L + \text{value}(L)) \leq \text{value}(W) - D_W(B)$, where $D_W(B) := \sum_{L \in \mathcal{L}:\, L \subsetneq W,\, L \cap V(B) \neq \emptyset} 2y_L$.

7.6 Let $0 < \delta < 1$ with $(2\kappa + \eta + 1 - \delta)\delta \leq 2 - \delta$. Show that then, for any input $(\mathcal{I}, W, B)$, the cost of the output F'' of the modified algorithm described in Exercise 7.5 can be bounded by

$$c^y(F'') \leq \kappa(\text{value}(W) + D_W) + (2\kappa + \eta + 2 - \delta)(\text{value}(W) - D_W(B)),$$

where $D_W(B) = \sum_{L \in \mathcal{L}:\, L \subsetneq W,\, L \cap V(B) \neq \emptyset} 2y_L$ as defined in Exercise 7.5 (b). *Hint*: In the induction step, bound the cost that we pay for each $L \in \mathcal{L}_{\bar{B}}$ in addition to (7.4). We possibly save when uncontracting and lifting, we may have to pay a new backbone, and we pay for the recursive call. Distinguish the cases when L is busy or not busy and obtain a bound of $(3\kappa + \eta + 2 - \delta)\text{value}(L) - (\kappa + \eta)D_L$ in both cases. Then use Exercise 7.5 (b).

7.7 Conclude from Exercise 7.6 that any (κ, η)-algorithm for vertebrate pairs implies a $(3\kappa + \eta + 2 - \delta)$-approximation algorithm for Asymmetric TSP, where $\delta = \frac{1}{2}\left(2\kappa + \eta + 2 - \sqrt{(2\kappa + \eta + 2)^2 - 8}\right)$. For $\kappa = 2$ and $\eta = 9$, this saves $\delta \approx 0.13454$.

7.8 Assume $\alpha \geq 2$. Suppose the third condition in Step (6c) of Algorithm 7.19 is strengthened to $c^y(E(C)) \leq \frac{1}{2}\text{slack}(\tilde{W}_{\text{ind}(Z)}, \tilde{H})$.

(a) Show that then Lemma 7.23 would still hold.

(b) Conclude that for $\alpha = 2$, the bound in (7.14) could be strengthened by $\frac{1}{2} \cdot \text{slack}(V \setminus V(B), \tilde{H})$.

8

Algorithms for Subtour Cover

In this chapter, we will present an algorithm for the subtour cover problem, which we defined in Chapter 7. This will complete the constant-factor approximation algorithm for the Asymmetric TSP. The subtour cover problem was introduced (in a slightly different form) by Svensson, Tarnawski, and Végh [2020], who gave a $(4, 2, 1)$-algorithm for subtour cover. Traub and Vygen [2022] strengthened this to a $(3, 2, 1)$-algorithm. Based on this work, we further improve this here to a $(2, 2, 1)$-algorithm. Our subtour cover algorithm builds on the algorithm for the graph subtour cover problem that we presented in Section 6.2.

We will first describe a $(3, 2, 1)$-algorithm (Sections 8.1–8.5). In Section 8.6, we then strengthen this to a $(2, 2, 1)$-algorithm. As a consequence, we obtain a $(17 + \varepsilon)$-approximation for the Asymmetric TSP for any fixed $\varepsilon > 0$.

8.1 The Split Graph

The most important difference to a graph subtour cover is property (ii) of the definition of a subtour cover F (Definition 7.14), which requires that every connected component of (V, F) must be local or connected to the backbone B. Due to the structure of the cost function c^y, this will also be crucial to obtain the following property of solutions returned by an (α, κ, β)-algorithm: we must have

$$c^y(E(D)) \leq \alpha \cdot \sum_{v \in V(D)} 2y_v \tag{8.1}$$

for every connected component D of (V, F) with $V(D) \cap V(B) = \emptyset$.

To achieve that each connected component of (V, F) is local or connected to the backbone, the key tool is the *split graph*. The concept of the split graph was first suggested by Svensson, Tarnawski, and Végh [2018], albeit in a slightly

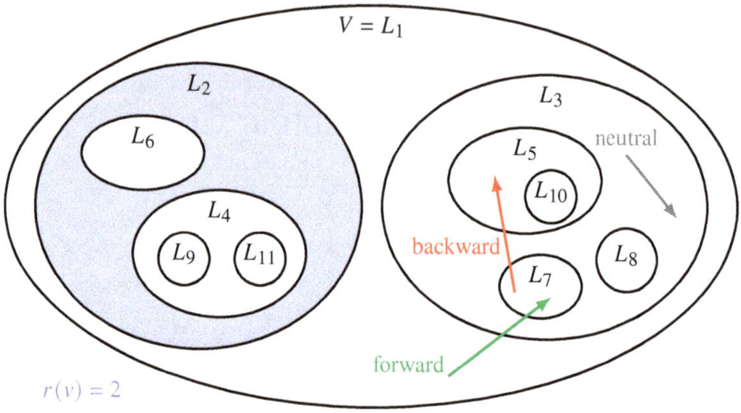

Figure 8.1 The laminar family $\mathcal{L}_{\geq 2} \cup \{V\} = \{L_1, \ldots, L_{11}\}$. In this example, the set $L_2 \setminus (L_6 \cup L_4)$ is the set of all vertices v with $r(v) = 2$; it is shown in blue. On the right, we see examples of a forward edge (green), a backward edge (red), and a neutral edge (gray). This picture is adapted from Traub and Vygen [2022].

different context. To define the split graph, we number the non-singleton elements of our laminar family $\mathcal{L}$ as follows. Number $\mathcal{L}_{\geq 2} \cup \{V\} = \{L_1, \ldots, L_{r_{\max}}\}$ such that $|V| = |L_1| \geq \cdots \geq |L_{r_{\max}}| \geq 2$. For $v \in V$, let $r(v) := \max\{i : v \in L_i\}$, and call an edge $e = (v, w) \in E$ *forward* if $r(v) < r(w)$, *backward* if $r(v) > r(w)$, and *neutral* if $r(v) = r(w)$. See Figure 8.1. An edge e is neutral if and only if for all $L \in \mathcal{L}_{\geq 2}$, we have $e \notin \delta(L)$.

We will need the following simple observation:

Lemma 8.1. *Let C be the edge set of a cycle. If C is not local, then C contains a forward edge and a backward edge.*

Proof. If C is not local, there exists an edge $e = (v, w) \in C \cap \delta^+(L)$ for some $L \in \mathcal{L}_{\geq 2}$. By the choice of the numbering $L_1, \ldots, L_{r_{\max}}$, we have $L_{r(v)} \subseteq L$ and hence $w \notin L_{r(v)}$. Therefore, the cycle with edge set C contains vertices v, w with $r(v) \neq r(w)$. Hence, C contains both a forward and a backward edge. □

Next, we define the *split graph* G^{01} of G (see Figure 8.2).

- For every vertex $v \in V$, the split graph G^{01} contains two vertices v^0 and v^1 (on the lower and upper level).
- For every $v \in V$, it contains an edge $e_v^{\downarrow} = (v^1, v^0)$ with $c^y(e_v^{\downarrow}) = 0$.
- For every $v \in V(B)$, it also contains an edge $e_v^{\uparrow} = (v^0, v^1)$ with $c^y(e_v^{\uparrow}) = 0$.
- For every forward edge $e = (v, w) \in E$, it contains an edge $e^0 = (v^0, w^0)$ with $c^y(e^0) = c^y(e)$.

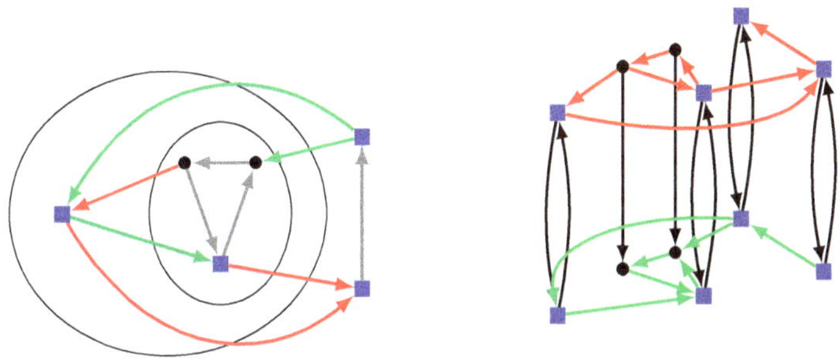

Figure 8.2 Construction of the split graph. The digraph on the left with the laminar family indicated by the two ellipses and the backbone shown in blue results in the split graph on the right (with the lower level in green and the upper level in red). Forward edges (green on the left) are mapped to the lower level, backward edges (red on the left) are mapped to the upper level. Neutral edges (gray on the left) are mapped to both levels. Additional edges (black) connect the two layers.

- For every backward edge $e = (v, w) \in E$, it contains an edge $e^1 = (v^1, w^1)$ with $c^y(e^1) = c^y(e)$.
- For every neutral edge $e = (v, w) \in E$, it contains edges $e^0 = (v^0, w^0)$ and $e^1 = (v^1, w^1)$ with $c^y(e^0) = c^y(e^1) = c^y(e)$.

We write $V^0 := \{v^0 : v \in V\}$ and call $G^{01}[V^0]$ the *lower level* of the split graph G^{01}. Similarly, we write $V^1 := \{v^1 : v \in V\}$ and call $G^{01}[V^1]$ the *upper level* of G^{01}. For a set $W \subseteq V$ we denote by $W^{01} := \{v^j : v \in W,\ j \in \{0, 1\}\}$ the vertex set of G^{01} that corresponds to W.

For any subgraph of G^{01}, we obtain a subgraph of G (its *image*) by replacing both v^0 and v^1 by v and removing loops. Then, obviously, the image of a circuit is an Eulerian graph. The next lemma shows how we can use the split graph to achieve property (ii) of a subtour cover (cf. Definition 7.14).

Lemma 8.2. *If the image of a connected Eulerian multi-subgraph of G^{01} is not local, it contains a vertex of the backbone B.*

Proof. Let D^{01} be a connected Eulerian multi-subgraph of G^{01} such that its image D (a connected Eulerian multi-subgraph of G) is not local. Then D contains a cycle C that is not local. By Lemma 8.1, C (and hence also D) contains a forward edge and a backward edge. Therefore, D^{01} visits both levels of G^{01} and thus contains an edge $e_v^{\updownarrow}$ for some $v \in V(B)$. □

The general idea of the subtour cover algorithm is as follows. First, we map the circulation x (stemming from the given strongly laminar Asymmetric TSP instance $(G, \mathcal{L}, x, y)$) to a circulation z in the split graph. Then, we round the circulation z in the split graph using a procedure similar to the algorithm for the graph subtour cover problem from Section 6.2. Finally, we consider the image of the edge set that results from the rounding in the split graph. Lemma 8.2 will guarantee that every connected component of this image is local or connected to the backbone.

8.2 Witness Flows

Similarly as we can map subgraphs of G^{01} to G, we can also map every flow $z' : E(G^{01}) \to \mathbb{R}_{\geq 0}$ to a flow $x' : E \to \mathbb{R}_{\geq 0}$ in G by setting $x'(e) := z'(e^0) + z'(e^1)$, where we set $z'(e^1) := 0$ for forward edges e and $z'(e^0) := 0$ for backward edges e. If z' is a circulation, then its *image* x' is also a circulation. However, while we can map every circulation in the split graph G^{01} to a circulation in G, not every circulation in G is the image of a circulation in G^{01} (cf. Exercise 8.1). Nevertheless, it turns out that any solution x to the linear program (3.2) is the image of a circulation z in G^{01}. Proving this and showing how we can compute such a circulation z is the topic of this section.

First, we define a flow $f \leq x$, which we will call a *witness flow*. In the construction of z, we will map the witness flow f to the lower level of G^{01} and map the remaining flow $x - f$ to the upper level of G^{01}. See Figure 8.3.

Definition 8.3 (witness flow). Let x' be a circulation in G. Then we call a flow f' in G a *witness flow (for x')* if

(i) $f'(e) = 0$ for every backward edge e;
(ii) $f'(e) = x'(e)$ for every forward edge e;
(iii) $0 \leq f'(e) \leq x'(e)$ for every neutral edge e; and
(iv) $f'(\delta^+(v)) \geq f'(\delta^-(v))$ for all $v \in V \setminus V(B)$.

The concept of a witness flow was introduced by Svensson, Tarnawski, and Végh [2020]. We now show that the pairs (x', f') where f' is a witness flow for the circulation x' in G correspond to circulations in the split graph G^{01}, and vice versa.

For a circulation z' in G^{01}, define $\pi(z') := (x', f')$, where x' is the image of z' in G and f' is the image of the restriction of z' to the lower level $G^{01}[V^0]$ of the split graph.

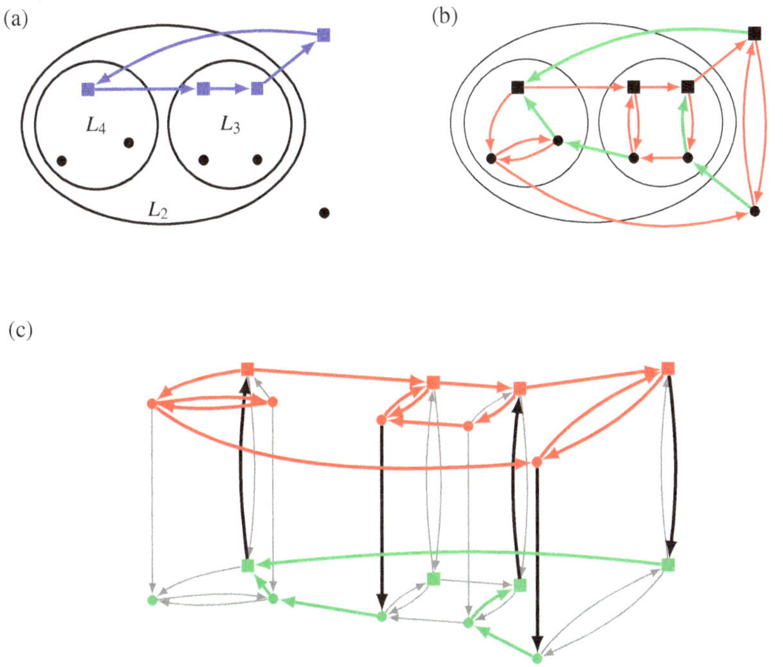

Figure 8.3 An example of the construction of the circulation z in G^{01}. Picture (a) shows the laminar family $\mathcal{L}_{\geq 2} = \{L_2, L_3, L_4\}$ and, in blue, the backbone B. Picture (b) shows a solution x to (3.2) where we have $x_e = \frac{1}{2}$ for all edges; a witness flow f is shown in green. The vertices in $V(B)$ are shown as squares. Every cycle crossing the boundary of a set $L \in \mathcal{L}_{\geq 2}$ contains both a green and a red edge. Picture (c) shows the resulting circulation z in G^{01}, where we have $z_e > 0$ for every thick edge e and $z_e = 0$ for all thin edges. The flow $x - f$ is mapped to the upper level of the split graph (red), and the flow f is mapped to the lower level (green). This picture is taken from Traub and Vygen [2022].

Lemma 8.4. *Let z' be a circulation in G^{01} and $(x', f') := \pi(z')$. Then x' is a circulation in G with $c^y(x') = c^y(z')$, and f' is a witness flow for x'.*

Proof. The first claim being obvious, we show that f' is a witness flow for x'. Property (i) holds because for a backward edge e, the graph G^{01} does not contain an edge e^0. Similarly, (ii) holds because for a forward edge e, the graph G^{01} does not contain an edge e^1. Property (iii) is obvious by construction, and (iv) holds because z' is a circulation and because for $v \in V \setminus V(B)$, the split graph does not contain an edge $e_v^{\uparrow}$. □

Conversely, for a circulation x' in G and a witness flow f' for x', there is a circulation z' in G^{01} with $\pi(z') = (x', f')$.

Lemma 8.5. *Given a circulation x' in G and a witness flow f' for x', we can compute a circulation z' in G^{01} with $\pi(z') = (x', f')$ in linear time.*

Proof. We define z' as follows.

- For every edge e^0 of the lower level of G^{01}, let $z'(e^0) = f'(e)$.
- For every edge e^1 of the upper level of G^{01}, let $z'(e^1) = x'(e) - f'(e)$.
- For every edge $e_v^{\uparrow}$ (for $v \in V(B)$), let $z'(e_v^{\uparrow}) = \max\{0, f'(\delta^-(v)) - f'(\delta^+(v))\}$.
- For every edge $e_v^{\downarrow}$ (for $v \in V$), let $z'(e_v^{\downarrow}) = \max\{0, f'(\delta^+(v)) - f'(\delta^-(v))\}$.

Notice that $x'(e) = z'(e^0)$ for every forward edge e and $x'(e) = z'(e^1)$ for every backward edge e. Moreover, $x'(e) = z'(e^0) + z'(e^1)$ for every neutral edge e. Furthermore, z' indeed defines a circulation in G^{01} because $f'(\delta^+(v)) \geq f'(\delta^-(v))$ for all $v \in V \setminus V(B)$. □

We have seen that pairs (x', f') where f' is a witness flow for x' correspond to circulations in the split graph. The following lemma states a key property of witness flows and can be viewed as the analogue of Lemma 8.2. Recall that χ^F denotes the incidence vector of F.

Lemma 8.6. *Let F be an Eulerian multi-subset of E such that there exists a witness flow f for χ^F. Then every connected component D of (V, F) is local or contains a vertex of the backbone B.*

Proof. Let D be a connected component of (V, F) with $V(D) \cap V(B) = \emptyset$. Since $f \leq \chi^F$, we have $\sum_{v \in V(D)} f(\delta^-(v)) = \sum_{v \in V(D)} f(\delta^+(v))$, and hence property (iv) of a witness flow (Definition 8.3) implies $f(\delta^+(v)) = f(\delta^-(v))$ for all $v \in V(D)$. In other words, f restricted to $E(D)$ is a circulation in D.

Suppose that D is not local. Then by Lemma 8.1, $E(D)$ contains a forward edge $e = (v, w)$. By property (ii), we have $f(e) = \chi_e^F \geq 1$. Let $R_v := \{u \in V : r(u) \leq r(v)\}$. Then $e \in \delta^+(R_v)$. Since f restricted to $E(D)$ is a circulation in D, there is an edge $e' \in \delta^-(R_v)$ with $f(e') > 0$. However, by the definition of R_v, all edges entering R_v are backward edges, and hence property (i) requires $f(e') = 0$. This is a contradiction, implying that D is local. □

Now we prove that we can indeed map any solution x to (3.2) to a circulation in the split graph:

Lemma 8.7 (Svensson, Tarnawski, and Végh [2020]). *Let $(\mathcal{I}, B)$ be a vertebrate pair, with $\mathcal{I} = (G, \mathcal{L}, x, y)$. Then there exists a witness flow f for x, and we can compute one in polynomial time.*

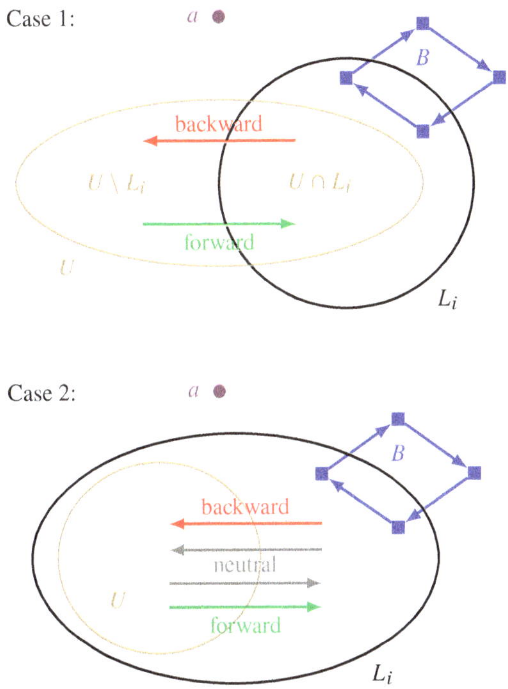

Figure 8.4 Proof of Lemma 8.7 (Case 1 and Case 2). This picture is taken from Traub and Vygen [2022].

Proof. Consider the graph G' that arises from G by adding a new vertex a and edges (a, v) for all $v \in V$ and edges (v, a) for all $v \in V(B)$. Set $l(e') = 0$ and $u(e') = \infty$ for the new edges. Moreover, for $e \in E$ set the lower bound $l(e)$ and the upper bound $u(e)$ according to the requirements from Definition 8.3 – that is, set $u(e) = x(e)$ if e is a forward or neutral edge and $u(e) = 0$ otherwise, and set $l(e) = x(e)$ if e is a forward edge and $l(e) = 0$ otherwise.

Then we are looking for a circulation f' in G' with $l \le f' \le u$. By Hoffman's circulation theorem (Theorem 3.9), this exists if

$$l(\delta^-(U)) \le u(\delta^+(U)) \qquad (8.2)$$

for all $U \subseteq V \cup \{a\}$. We show that this is indeed true. Suppose not, and let U be a minimal set violating (8.2). Since (8.2) obviously holds whenever $a \in U$ or $V(B) \cap U \neq \emptyset$, we have $U \subseteq V \setminus V(B)$. Let i be the largest index so that $U \cap L_i \neq \emptyset$. Recall that $L_i \in \mathcal{L}_{\ge 2}$ and thus $L_i \cap V(B) \neq \emptyset$. We distinguish two cases (see Figure 8.4).

Case 1: $U \setminus L_i \neq \emptyset$.
Then (by the minimality of U) we have $l(\delta^-(U \cap L_i)) \leq u(\delta^+(U \cap L_i))$ and $l(\delta^-(U \setminus L_i)) \leq u(\delta^+(U \setminus L_i))$. Since all edges from $U \setminus L_i$ to $U \cap L_i$ are forward edges and all edges from $U \cap L_i$ to $U \setminus L_i$ are backward edges, we get

$$\begin{aligned} l(\delta^-(U)) + x(\delta^+(U \setminus L_i) \cap \delta^-(U \cap L_i)) \\ &= l(\delta^-(U \cap L_i)) + l(\delta^-(U \setminus L_i)) \\ &\leq u(\delta^+(U \cap L_i)) + u(\delta^+(U \setminus L_i)) \\ &= u(\delta^+(U)) + x(\delta^+(U \setminus L_i) \cap \delta^-(U \cap L_i)) \end{aligned}$$

and hence (8.2), which is a contradiction to the choice of U.

Case 2: $U \subseteq L_i$.
Then $r(u) = i$ for all $u \in U$ and $r(w) \geq i$ for all $w \in L_i$. Hence, $l(\delta^-(U)) \leq x(\delta^-(L_i) \cap \delta^-(U))$ because we have $l(e) > 0$ only for forward edges and because all edges in $\delta^-(U) \setminus \delta^-(L_i)$ are neutral or backward edges. Moreover, edges in $\delta^+(U) \setminus \delta^+(L_i)$ are not backward edges, implying $x(\delta^+(U) \setminus \delta^+(L_i)) = u(\delta^+(U) \setminus \delta^+(L_i)) \leq u(\delta^+(U))$.

Therefore,

$$\begin{aligned} l(\delta^-(U)) &\leq x(\delta^-(L_i) \cap \delta^-(U)) \\ &= x(\delta^-(L_i)) + x(\delta^+(U) \setminus \delta^+(L_i)) - x(\delta^-(L_i \setminus U)) \\ &\leq x(\delta^-(L_i)) + u(\delta^+(U)) - x(\delta^-(L_i \setminus U)). \end{aligned}$$

Since $L_i \setminus U \neq \emptyset$ (because $L_i \cap V(B) \neq \emptyset = U \cap V(B)$), we have $x(\delta^-(L_i \setminus U)) \geq 1$. Moreover, $L_i \in \mathcal{L} \cup \{V\}$ implies $x(\delta(L_i)) \in \{0, 2\}$ and hence $x(\delta^-(L_i)) \leq 1$. Hence, (8.2) follows, which is again a contradiction.

By Theorem 3.11 we can compute f in polynomial time. □

8.3 Rerouting

The first step of our subtour cover algorithm is to apply Lemma 8.7 and Lemma 8.5 in order to obtain a circulation z in the split graph G^{01}. Similar to the proof of Theorem 6.6, before we round to obtain an integral circulation, we first modify the split graph G^{01} to an auxiliary graph $\bar{G}^{01}$ and reroute some flow of z through auxiliary vertices. For the rerouting step, we need the following lemma:

Lemma 8.8. *Let G' be a directed graph and z' a circulation in G'. Let $U \subseteq V(G')$ with $z'(\delta(U)) \geq 2$. Then we can compute in polynomial time a* *multi-set $\mathcal{P}$ of paths in $G'[U]$ and for every $P \in \mathcal{P}$ starting in $s \in U$ and ending in $t \in U$*

- *a weight $\lambda_P > 0$,*
- *an edge $e_P^{\text{in}} = (s', s) \in \delta^-(U)$, and*
- *an edge $e_P^{\text{out}} = (t, t') \in \delta^+(U)$,*

such that $\sum_{P \in \mathcal{P}} \lambda_P = 1$ and

$$\sum_{P \in \mathcal{P}} \lambda_P \cdot \left(\chi^{e_P^{\text{in}}} + \chi^{E(P)} + \chi^{e_P^{\text{out}}}\right) \leq z'.$$

Proof. Contract $V(G') \setminus U$ to a vertex v_{outside}. Then we have $z'(\delta(v_{\text{outside}})) = z'(\delta(U)) \geq 2$. Because z' remains a circulation, by Proposition 3.7 we can compute in polynomial time a set $\mathcal{C}$ of cycles containing v_{outside} and weights $\lambda_C > 0$ for $C \in \mathcal{C}$ with $\sum_{C \in \mathcal{C}} \lambda_C = 1$ such that

$$\sum_{C \in \mathcal{C}} \lambda_C \cdot \chi^{E(C)} \leq z'.$$

After undoing the contraction, each of these circuits results in an edge $e^{\text{in}} = (s', s) \in \delta^-(U)$, an edge $e^{\text{out}} = (t, t') \in \delta^+(U)$, and an s-t-path P in $G'[U]$. □

Recall that a subtour cover instance consists of a vertebrate pair $(\mathcal{I}, B)$ (with backbone B) and a local Eulerian multi-subset H of $E[V \setminus V(B)]$ (cf. Definition 7.14). Let $W_1, \ldots, W_q$ be the vertex sets of the connected components of $(V \setminus V(B), H)$. Thus, $W_1, \ldots, W_q$ are pairwise disjoint subsets of $V \setminus V(B)$. Recall that $W_1^{01}, \ldots, W_q^{01}$ are the corresponding vertex sets in the split graph G^{01}. First, we make the support of z inside each set W_i^{01} acyclic: While there is a circuit C in $G^{01}[W_i^{01}]$ with $\gamma := \min\{z_e : e \in E(C)\} > 0$, reduce z_e by γ for all $e \in E(C)$. Let $\tilde{z}$ be the resulting circulation. We have $\tilde{z}(\delta^-(W_i^{01})) = z(\delta^-(W_i^{01})) = x(\delta^-(W_i)) \geq 1$ for $i = 1, \ldots, q$.

Later, we will round $2\tilde{z}$ to an integer circulation while guaranteeing that we use at least one edge of $\delta^-(W_i^{01})$ for each $i = 1, \ldots, q$. We call $v \in V^{01}$ a *high-throughput vertex* if $\tilde{z}(\delta^-(v)) \geq \frac{1}{2}$ and a *low-throughput vertex* otherwise. If W_i^{01} contains a high-throughput vertex w, we can enforce routing at least one unit of flow through w and hence through W_i^{01}; here, we will use that the support of $\tilde{z}$ inside W_i^{01} is acyclic. If W_i^{01} contains only low-throughput vertices, we will enforce a unit of flow through W_i^{01} by introducing an auxiliary vertex, similar to the proof of Theorem 6.6. See Figure 8.5 for our roadmap.

We construct an auxiliary digraph $\bar{G}$ from G and transform $\tilde{z}$ to a circulation $\bar{z}$ in $\bar{G}^{01}$. Let $i \in \{1, \ldots, q\}$ such that all vertices in W_i^{01} are low-throughput vertices. We add an auxiliary vertex a_i to G and set $r(a_i) := r(v)$ for $v \in W_i$; this is well-defined by the following lemma:

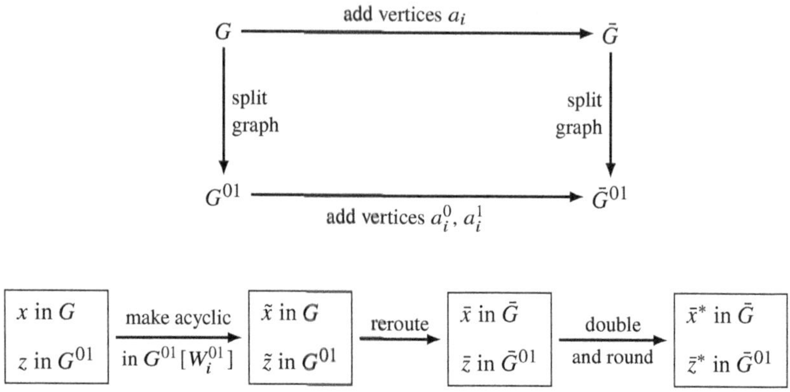

Figure 8.5 Overview of the different graphs and circulations occurring in the subtour cover algorithm. We start with the circulation x in G stemming from the strongly laminar Asymmetric TSP instance $(G, \mathcal{L}, x, y)$. We map x to a circulation z in the split graph G^{01} and perform three steps, ending with an integral circulation $\bar{z}^*$ in $\bar{G}^{01}$. This corresponds to an integral circulation $\bar{x}^*$ in $\bar{G}$, which we will map back to G in the final step of the subtour cover algorithm. A similar picture appeared in Traub and Vygen [2022].

Lemma 8.9. *Let $i \in \{1, \ldots, q\}$ and $v, w \in W_i$. Then $r(v) = r(w)$. Moreover, $G[W_i]$ is strongly connected, and $E(G[W_i])$ is local.*

Proof. Because W_i is a connected component of $(V \setminus V(B), H)$ and H is Eulerian, $G[W_i]$ is strongly connected. Let $L \in \mathcal{L}_{\geq 2}$. Since H is local, we have $H \cap \delta(L) = \emptyset$ and therefore $W_i \subseteq L$ or $W_i \cap L = \emptyset$. This implies $r(v) = \max\{j : v \in L_j\} = \max\{j : w \in L_j\} = r(w)$ and that $E(G[W_i])$ is local. □

For every edge $(v, w) \in \delta^-(W_i)$, we add an edge (v, a_i) of the same cost. Similarly, for every edge $(v, w) \in \delta^+(W_i)$, we add an edge (a_i, w) of the same cost. Note that a new edge is a forward/backward/neutral edge if and only if its corresponding edge in G is forward/backward/neutral. Then the split graph contains new vertices a_i^0 and a_i^1, connected by an edge $e_{a_i}^{\downarrow} = (a_i^1, a_i^0)$ of cost zero.

Moreover, in the split graph, we now reroute $\frac{1}{2}$ unit of flow of the circulation $\tilde{z}$ through one of the auxiliary vertices a_i^0, a_i^1. More precisely, we first apply Lemma 8.8 to the vertex set $U = W_i^{01}$ and the circulation $\tilde{z}$. Then we partition the resulting set $\mathcal{P}$ into sets $\mathcal{P}^0$ and $\mathcal{P}^1$ such that $\mathcal{P}^0$ contains the paths $P \in \mathcal{P}$ for which e_P^{in} is contained in the lower level of the split graph, and $\mathcal{P}^1$ contains the paths $P \in \mathcal{P}$ for which e_P^{in} is contained in the upper level of the split graph.

Since $\sum_{P\in\mathcal{P}} \lambda_P = 1$, we have $\sum_{P\in\mathcal{P}^k} \lambda_P \geq \frac{1}{2}$ for some $k \in \{0, 1\}$. We can choose values $0 \leq \lambda'_P \leq \lambda_P$ such that $\sum_{P\in\mathcal{P}^k} \lambda'_P = \frac{1}{2}$. For every $P \in \mathcal{P}^k$, we do the following (see Figure 8.6 (b)–(c) for an example):

- We decrease the flow on e_P^{in} and increase the flow on its corresponding edge in $\delta^-(a_i^k)$ by λ'_P.
- We decrease the flow on every edge $e \in E(P)$ by λ'_P.
- Let $h = 0$ if e_P^{out} is contained in the lower level of the split graph and $h = 1$ otherwise. We decrease the flow on e_P^{out} and increase the flow on its corresponding edge in $\delta^+(a_i^h)$ by λ'_P.
- Because $W_i \cap V(B) = \emptyset$, the path P contains no edge from the lower to the upper level; hence $h \leq k$. If $h < k$ (i.e., $k = 1$ and $h = 0$), we increase the flow on $e_{a_i}^{\downarrow}$ by λ'_P.

This maintains a circulation in the split graph. We do this transformation successively for all $i \in \{1, \ldots, q\}$ for which all vertices in W_i^{01} are low-throughput vertices. Let $\bar{G}$ be the final graph (resulting from G by the modifications described above), and let $\bar{G}^{01}$ be its split graph. Let $\bar{z}$ be the final circulation in the split graph $\bar{G}^{01}$, and let $\bar{x}$ be its image in $\bar{G}$. See Figure 8.5. We conclude:

Lemma 8.10. *We can compute in polynomial time a circulation $\bar{z}$ in $\bar{G}^{01}$ that satisfies $c^y(\bar{z}) \leq c^y(z)$ and the following property for each $i \in \{1, \ldots, q\}$: either W_i^{01} contains a high-throughput vertex v with $\bar{z}(\delta^-(v)) \geq \frac{1}{2}$, or $\bar{G}$ contains a vertex a_i and*

- *either $\bar{z}(\delta^-(a_i^1)) = \frac{1}{2}$ and $\bar{z}(\delta^-(a_i^0) \setminus \{e_{a_i}^{\downarrow}\}) = 0$*
- *or $\bar{z}(\delta^-(a_i^1)) = 0$ and $\bar{z}(\delta^-(a_i^0) \setminus \{e_{a_i}^{\downarrow}\}) = \frac{1}{2}$.* □

8.4 Rounding

Because we could only reroute $\frac{1}{2}$ unit of flow through a_i^0 or a_i^1, we consider the circulation $2\bar{z}$. We round $2\bar{z}$ to an integral circulation $\bar{z}^*$ as follows:

Lemma 8.11. *We can compute an integral circulation $\bar{z}^*$ in $\bar{G}^{01}$ with*

(i) $0 \leq \bar{z}^*(e) \leq \lceil 2\bar{z}(e) \rceil$ *for all* $e \in E(\bar{G}^{01})$,
(ii) $c^y(\bar{z}^*) \leq c^y(2\bar{z})$,
(iii) $\lfloor 2\bar{z}(\delta^-(v)) \rfloor \leq \bar{z}^*(\delta^-(v)) \leq \lceil 2\bar{z}(\delta^-(v)) \rceil$ *for all* $v \in V(\bar{G}^{01})$, *and*
(iv) $\lfloor 2\bar{z}(\delta^-(v^0) \setminus \{e_v^{\downarrow}\}) \rfloor \leq \bar{z}^*(\delta^-(v^0) \setminus \{e_v^{\downarrow}\}) \leq \lceil 2\bar{z}(\delta^-(v^0) \setminus \{e_v^{\downarrow}\}) \rceil$ *for all* $v \in V(\bar{G})$,

in polynomial time.

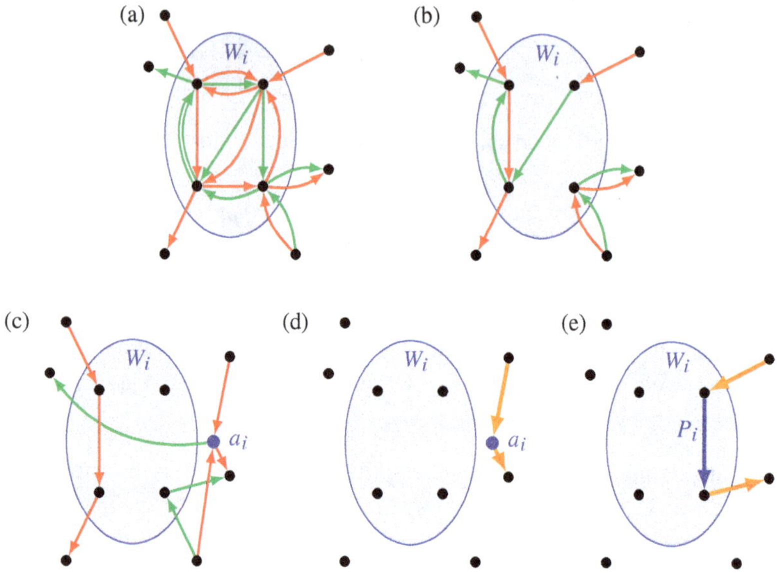

Figure 8.6 Example of the construction of the subtour cover F from the witness flow f. On all pictures, a set W_i (blue and filled) is shown. The pictures show only edges with at least one endpoint in W_i. Picture (a) shows (parts of) a possible solution x to (3.2) (green and red) and a witness flow f (green). The edges drawn with a single line have value $\frac{1}{4}$; the edges drawn with a double line have value $\frac{1}{2}$. Recall that red edges correspond to flow on the upper level of the split graph, and green edges correspond to flow on the lower level. Picture (b) shows possible flows $(\tilde{x}, \tilde{f}) = \pi(\tilde{z})$, where the circulation $\tilde{z}$ in the split graph has acyclic support in $G^{01}[W_i^{01}]$. Picture (c) shows a possible circulation $\bar{x}$ in $\bar{G}$ resulting from the rerouting of flow through a_i (blue); the witness flow $\bar{f}$ is again shown in green. Picture (d) shows in orange a possible integral circulation $\bar{x}^*$ in $\bar{G}$; the orange edges are elements of the edge set $\bar{F}$ with $\chi^{\bar{F}} = \bar{x}^*$. Picture (e) shows the result of mapping $\bar{F}$ back to G. In blue, the path P_i in $G[W_i]$ is shown; it completes the orange edges to a circulation.

Proof. In order to find an integral circulation with (i), (ii), and (iii), we could apply Corollary 3.12 to the fractional flow $2\bar{z}$. In order to achieve property (iv), we modify the flow network as follows.

For every vertex v^0 on the lower level of $\bar{G}^{01}$ (including the auxiliary vertices on the lower level), we add a vertex v^0_- and an edge (v^0_-, v^0). Then we replace every edge (w, v^0) with $w \neq v^1$ by an edge (w, v^0_-). Setting $\bar{z}(v^0_-, v^0) := \bar{z}(\delta^-(v^0) \setminus \{e_v^{\downarrow}\})$ maintains a circulation. When rounding $2\bar{z}$ to a circulation, we can impose a lower bound $\lfloor 2\bar{z}(\delta^-(v^0) \setminus \{e_v^{\downarrow}\}) \rfloor$ and an upper bound $\lceil 2\bar{z}(\delta^-(v^0) \setminus \{e_v^{\downarrow}\}) \rceil$ on the flow along (v^0_-, v^0). Applying Corollary 3.12

to this modified flow network and then contracting the auxiliary edges (v^0_-, v^0) leads to a circulation fulfilling also (iv). □

Let $(\bar{x}, \bar{f}) := \pi(\bar{z})$ and $(\bar{x}^*, \bar{f}^*) := \pi(\bar{z}^*)$; see Figure 8.5. Then by Lemma 8.4, the flows $\bar{f}$ and $\bar{f}^*$ are witness flows for the circulations $\bar{x}$ and $\bar{x}^*$, respectively, in $\bar{G}$. Let $\bar{F}$ be the multi-set of edges in $\bar{G}$ with $\chi^{\bar{F}} = \bar{x}^*$ (see Figure 8.6 (d)). Then $\bar{F}$ is Eulerian because $\bar{x}^*$ is a circulation.

Next, we observe some important properties of $\bar{F}$. First,

$$\begin{aligned} c^y(\bar{F}) &= c^y(\bar{x}^*) = c^y(\bar{z}^*) \le 2 \cdot c^y(\bar{z}) \le 2 \cdot c^y(\tilde{z}) \\ &\le 2 \cdot c^y(z) = 2 \cdot c^y(x) = 2 \cdot \mathrm{LP}(\mathcal{I}). \end{aligned} \tag{8.3}$$

Lemma 8.12. *Let $i \in \{1, \ldots, q\}$ such that W_i^{01} contains only low-throughput vertices. Then $\bar{G}$ contains an auxiliary vertex a_i, and we have $|\delta^-_{\bar{F}}(a_i)| = 1$.*

Proof. We have

$$|\delta^-_{\bar{F}}(a_i)| = \bar{x}^*(\delta^-(a_i)) = \bar{z}^*(\delta^-(a_i^1)) + \bar{z}^*(\delta^-(a_i^0) \setminus \{e^{\downarrow}_{a_i}\}).$$

By Lemma 8.10 and properties (iii) and (iv) of Lemma 8.11, we have either $\bar{z}^*(\delta^-(a_i^1)) = 1$ or $\bar{z}^*(\delta^-(a_i^0) \setminus \{e^{\downarrow}_{a_i}\}) = 1$, and the other term is 0. □

Lemma 8.13. *Let $i \in \{1, \ldots, q\}$ such that W_i^{01} contains a high-throughput vertex. Then $|\delta^-_{\bar{F}}(W_i)| \ge 1$.*

Proof. Let $w_i \in W_i^{01}$ be a high-throughput vertex. Due to property (iii) of Lemma 8.11, we have $\bar{z}^*(\delta^-(w_i)) \ge \lfloor 2\bar{z}(\delta^-(w_i)) \rfloor \ge 1$. Moreover, because the support of $\bar{z}$ in $G^{01}[W_i^{01}]$ is acyclic by construction, property (i) of Lemma 8.11 implies that the support of $\bar{z}^*$ in $G^{01}[W_i^{01}]$ is acyclic as well. Using that $\bar{z}^*$ is a circulation, this implies $|\delta^-_{\bar{F}}(W_i)| = \bar{z}^*(\delta^-(W_i^{01})) \ge \bar{z}^*(\delta^-(w_i)) \ge 1$. □

Lemma 8.14. *Let $v \in V \setminus V(B)$ with $y_v > 0$. Then $|\delta^-_{\bar{F}}(v)| \le 3$.*

Proof. By properties (iii) and (iv) of Lemma 8.11, we have

$$\begin{aligned} |\delta^-_{\bar{F}}(v)| = \bar{x}^*(\delta^-(v)) &= \bar{z}^*(\delta^-(v^1)) + \bar{z}^*(\delta^-(v^0) \setminus \{e^{\downarrow}_v\}) \\ &\le \lceil 2\bar{z}(\delta^-(v^1)) \rceil + \lceil 2\bar{z}(\delta^-(v^0) \setminus \{e^{\downarrow}_v\}) \rceil \\ &\le \lceil 2x(\delta^-(v)) \rceil + 1, \end{aligned}$$

where we used that $\bar{z}(\delta^-(v^1)) + \bar{z}(\delta^-(v^0) \setminus \{e^{\downarrow}_v\}) \le x(\delta^-(v))$. Because $y_v > 0$, we have $\{v\} \in \mathcal{L}$ and thus $x(\delta^-(v)) = 1$. □

In certain cases, we can strengthen the bound from Lemma 8.14.

Lemma 8.15. *Let $v \in V \setminus V(B)$ such that v^0 and v^1 are low-throughput vertices. Then $|\delta^-_{\bar{F}}(v)| \le 2$.*

Proof. By properties (iii) and (iv) of Lemma 8.11, we have

$$\begin{aligned}|\delta^-_{\bar{F}}(v)| &= \bar{x}^*(\delta^-(v)) = \bar{z}^*(\delta^-(v^1)) + \bar{z}^*(\delta^-(v^0) \setminus \{e_v^{\downarrow}\}) \\ &\le \lceil 2\bar{z}(\delta^-(v^1))\rceil + \lceil 2\bar{z}(\delta^-(v^0)\rceil \\ &\le \lceil 2\bar{z}(\delta^-(v^1))\rceil + \lceil 2\bar{z}(\delta^-(v^0))\rceil \\ &\le 2,\end{aligned}$$

where the last inequality holds because v^0 and v^1 are low-throughput vertices. □

8.5 Mapping Back to G

Finally, we show how we can turn the multi-subset $\bar{F}$ of $E(\bar{G})$ into a subtour cover in G. See Figure 8.6 (d)–(e). For every $i \in \{1, \dots, q\}$ for which W_i^{01} contains no high-throughput vertex and hence $\bar{G}$ contains an auxiliary vertex a_i, we do the following. By Lemma 8.12, the auxiliary vertex a_i has exactly one incoming edge in $\bar{F}$, and because $\bar{F}$ is Eulerian, a_i also has exactly one outgoing edge. We replace all the edges in $\delta_{\bar{F}}(a_i)$ by their corresponding edges in G. This replacement removes the two edges in $\delta_{\bar{F}}(a_i)$ and adds one edge $(v, s) \in \delta^-(W_i)$ and one edge $(t, w) \in \delta^+(W_i)$; to obtain an Eulerian edge set, we add an arbitrary s-t-path P_i in $G[W_i]$. Since $G[W_i]$ is strongly connected, such a path exists, and $E(P_i)$ is local (cf. Lemma 8.9).

Let F be the resulting Eulerian multi-set of edges in G after applying this procedure for all sets W_i (with $i \in \{1, \dots, q\}$) that do not contain a high-throughput vertex. In the following, we show that returning this set F yields a $(3, 2, 1)$-algorithm for subtour cover.

Lemma 8.16. *The Eulerian multi-edge set F is a subtour cover for H in G.*

Proof. First, we note that $F \cap \delta(W) \neq \emptyset$ for every vertex set W of a connected component of $(V \setminus V(B), H)$ follows from Lemma 8.12 and Lemma 8.13.

Because $\bar{f}^*$ is a witness flow for $\bar{x}^* = \chi^{\bar{F}}$, Lemma 8.6 implies that every connected component of $(V(\bar{G}), \bar{F})$ is local or connected to the backbone. Here, we consider the auxiliary vertices a_i as elements of the same sets of $\mathcal{L}_{\ge 2}$ as all vertices in W_i (cf. Lemma 8.9).

It remains to show that also every connected component of (V, F) is local or connected to the backbone. By Lemma 8.9, the edge sets of the paths P_i within W_i (that we used to obtain F from $\bar{F}$) are local. Because the replacement of the two edges incident to an auxiliary vertex a_i by their original edges and the path P_i never disconnects vertices that were in the same connected component before, we conclude that also every connected component of $(V, \bar{F})$ is local or connected to the backbone. □

Theorem 8.17. *There is a* $(3, 2, 1)$*-algorithm for subtour cover.*

Proof. We return the subtour cover F we constructed above. By Lemma 8.16, it remains to show that

$$c^y(F) \le 2 \cdot \mathrm{LP}(\mathcal{I}) + \sum_{v \in V \setminus V(B)} 2y_v, \tag{8.4}$$

and for every connected component D of (V, F) with $V(D) \cap V(B) = \emptyset$, we have

$$c^y(E(D)) \le 3 \cdot \sum_{v \in V(D)} 2y_v. \tag{8.5}$$

Since the edge set of each path P_i is within W_i and thus local (cf. Lemma 8.9),

$$c^y(E(P_i)) = \sum_{v \in V(P_i)} |E(P_i) \cap \delta(v)| \cdot y_v \le \sum_{v \in W_i} 2y_v.$$

Moreover, the sets W_i $(i = 1, \ldots, q)$ are pairwise disjoint, and $V(B) \cap W_i = \emptyset$ for $i = 1, \ldots, q$. Therefore, we obtain $\sum_{i=1}^{q} c^y(E(P_i)) \le \sum_{v \in V \setminus V(B)} 2y_v$. Together with (8.3), this implies (8.4).

Finally, we prove (8.5). Let D be a connected component of (V, F) with $V(D) \cap V(B) = \emptyset$. Because F is a subtour cover, $E(D)$ is local and hence $c^y(E(D)) = \sum_{v \in V(D)} |\delta_F^-(v)| \cdot 2y_v$. Moreover, because the sets $W_1, \ldots, W_q$ are pairwise disjoint, we have $|\delta_F^-(v)| \le |\delta_{\bar{F}}^-(v)| + 1$ for every vertex $v \in V(D)$.

Let $v \in V(D)$ with $y_v > 0$, and let $i \in \{1, \ldots, q\}$ such that $v \in W_i$. If v^0 and v^1 are low-throughput vertices, we have $|\delta_F^-(v)| \le |\delta_{\bar{F}}^-(v)| + 1 \le 3$ by Lemma 8.15. Otherwise, we have not added an auxiliary vertex a_i and thus have not added a path P_i when transforming $\bar{F}$ to F. Thus, in this case $|\delta_F^-(v)| = |\delta_{\bar{F}}^-(v)| \le 3$ by Lemma 8.14.

We have shown $|\delta_F^-(v)| \le 3$ for every vertex $v \in V(D)$ with $y_v > 0$, and this implies $c^y(E(D)) = \sum_{v \in V(D)} |\delta_F^-(v)| \cdot 2y_v \le \sum_{v \in V(D)} 3 \cdot 2y_v$. □

This yields a constant-factor approximation algorithm for the Asymmetric TSP as follows. Theorem 8.17 yields a $(3, 2, 1)$-algorithm for subtour cover, and by Theorem 7.18, this implies that there is a $(2, 13 + \varepsilon)$-algorithm for vertebrate pairs. Using Theorem 7.12, we then obtain a $(21 + \varepsilon)$-approximation algorithm for the Asymmetric TSP for any fixed $\varepsilon > 0$. In the next section, we strengthen this to a $(17 + \varepsilon)$-approximation algorithm.

Let us summarize our subtour cover algorithm: We started with the LP solution x, mapped it to a circulation z in the split graph (using a witness flow), and transformed z in several steps (cf. Figure 8.5) to an integral circulation $\bar{z}^*$ in the extended split graph $\bar{G}^{01}$. The image $\bar{x}^*$ of $\bar{z}^*$ corresponds to an Eulerian multi-edge set $\bar{F}$, which we mapped back to a subtour cover F in G. In the final version of their paper, Svensson, Tarnawski, and Végh [2020] described an alternative way to transform x to $\bar{x}^*$. Instead of working in the split graph directly, they guarantee that at any stage there exists a witness flow (which is equivalent); see Exercise 8.3.

8.6 Better Subtour Covers by Acyclic Witness Flows

In this section, we strengthen Theorem 8.17 and present a $(2, 2, 1)$-algorithm for subtour cover. The key difference to the subtour algorithm discussed in the previous sections of this chapter is that we will only work with witness flows with acyclic support, an idea introduced by Traub and Vygen [2022]. More precisely, we choose the witness flow f such that it has acyclic support, and we will make sure that the witness flows $\bar{f}$ and $\bar{f}^*$ that we obtain after rerouting and rounding (see Figure 8.5) have acyclic support as well. Then every vertex $v \in V$ for which the integral circulation $\bar{z}^*$ in the split graph fulfills $\bar{z}^*(\delta^-(v^0)) > 0$ will be connected to the backbone by the subtour cover F. This allows us to strengthen Lemmas 8.14 and 8.15 and thus leads to an improved bound on $|\delta^-_{\bar{F}}(v)|$.

We first show that we can choose the witness flow f with acyclic support. The second property of f in Lemma 8.18 will be needed to maintain an acyclic support of the witness flow during later steps of the subtour cover algorithm – in particular, the rerouting. Recall that $W_1, \dots, W_q$ are the vertex sets of the connected components of $(V \setminus V(B), H)$.

Lemma 8.18. *Let $(\mathcal{I}, B)$ be a vertebrate pair, with $\mathcal{I} = (G, \mathcal{L}, x, y)$. Then we can compute in polynomial time a witness flow f for x such that*

(v) *the support of f is acyclic, and*
(vi) $\sum_{i=1}^q f(\delta(W_i)) \le \sum_{i=1}^q f'(\delta(W_i))$ *for every witness flow f' for x.*

Proof. We first compute a witness flow $\tilde{f}$ by minimizing $\sum_{i=1}^q f(\delta(W_i))$ subject to the constraints (i) – (iv) from Definition 8.3. This linear program is feasible by Lemma 8.7. Then the flow $\tilde{f}$ fulfills property (vi).

To compute the flow f, we minimize $\sum_{e\in E} f(e)$ subject to the constraints (i) – (iv) and $f(e) \le \tilde{f}(e)$ for all $e \in E$. This linear program is feasible because $\tilde{f}$ is a feasible solution. Then f is a witness flow for x with $\sum_{i=1}^q f(\delta(W_i)) \le \sum_{i=1}^q \tilde{f}(\delta(W_i))$. Since the flow $\tilde{f}$ fulfills property (vi), the same holds for the flow f.

Suppose f does not fulfill (v) – that is, the support of f is not acyclic. Then there is a cycle $C \subseteq E$ with $f(e) > 0$ for all $e \in C$. As f fulfills (i), the set C does not contain any backward edge. This implies that C also contains no forward edge because C is a cycle. Let $\varepsilon := \min_{e \in C} f(e)$. For $e \in E$, we set $f'(e) := f(e) - \varepsilon \leq \tilde{f}(e)$ if $e \in C$ and set $f'(e) := f(e) \leq \tilde{f}(e)$ otherwise. Because C contains neither forward nor backward edges, f' fulfills (i) and (ii). By the choice of ε, we have $f'(e) \geq 0$ for all $e \in E$, implying (iii). Finally, $f'(\delta^+(v)) - f'(\delta^-(v)) = f(\delta^+(v)) - f(\delta^-(v)) \geq 0$ for all $v \in V \setminus V(B)$, where we used that C is a cycle and f fulfills (iv). This shows that f' is a witness flow and $f'(e) \leq \tilde{f}(e)$ for all $e \in E$, but $\sum_{e \in E} f'_e < \sum_{e \in E} f_e$, a contradiction to the choice of f. □

We now work with a flow f as in Lemma 8.18. As before, we apply Lemma 8.5 to construct z from x and f. Again, we first make the support of z inside each set W_i^{01} acyclic: While there is a circuit C in $G^{01}[W_i^{01}]$ with $\gamma := \min\{z_e : e \in E(C)\} > 0$, reduce z_e by γ for all $e \in E(C)$. Let the resulting circulation be $\tilde{z}$. Then we again call $v \in V^{01}$ a *high-throughput vertex* if $\tilde{z}(\delta^-(v)) \geq \frac{1}{2}$ and a *low-throughput vertex* otherwise.

If for some $i \in \{1, \ldots, q\}$, the set W_i^{01} contains only low-throughput vertices, we proceed similar to the earlier subtour cover algorithm – that is, we introduce an auxiliary vertex a_i and reroute some flow through a_i. Applying the same rerouting procedure as earlier could however lead to cycles in the support of the witness flow (this will become clear in the proof of Lemma 8.20). Thus, in order to maintain an acyclic witness flow, we use a slightly different rerouting procedure.

To this end, we denote by G_f the residual graph of the witness flow f in the graph G with edge capacities $u = x$ (see Definition 2.4). For $i \in \{1, \ldots, q\}$, let $\hat{W}_i$ be the vertex set of the first strongly connected component of $G_f[W_i]$ in some topological order – that is, $G_f[\hat{W}_i]$ is a strongly connected component of $G_f[W_i]$ such that no edge of $G_f[W_i]$ enters $\hat{W}_i$ (cf. Proposition 6.4). Instead of rerouting some flow going through W_i as we did earlier, we will now reroute flow going through $\hat{W}_i$. To show that $\tilde{z}$ has at least one unit of flow entering (and leaving) $\hat{W}_i^{01}$, we observe that reducing z to $\tilde{z}$ did not reduce the flow on any edge in $\delta(\hat{W}_i^{01})$.

Lemma 8.19. *Let C be any cycle in $G^{01}[W_i^{01}]$ in the support of z. Then C does not contain any edge of $\delta(\hat{W}_i^{01})$.*

Proof. By the definition of $\hat{W}_i \subseteq V$ and the definition of the residual graph (Definition 2.4), we have $f(e) = x(e)$ for every edge $e \in \delta^-(\hat{W}_i) \cap E[W_i]$, which implies $z(e^1) = x(e) - f(e) = 0$. Similarly, we have $f(e) = 0$ for every edge $e \in \delta^+(\hat{W}_i) \cap E[W_i]$, which implies $z(e^0) = f(e) = 0$. Hence, every edge in the support of z that enters $\hat{W}_i^{01}$ in $G^{01}[W_i^{01}]$ is contained in the lower level of the split graph, and every edge in the support of z that leaves $\hat{W}_i^{01}$ in $G^{01}[W_i^{01}]$ is contained in the upper level of the split graph. This shows that any cycle in $G^{01}[W_i^{01}]$ in the support of z that contains an edge of $\delta(\hat{W}_i^{01})$ must visit both levels. However, such a cycle cannot exist because W_i contains no vertex of the backbone and hence $G^{01}[W_i^{01}]$ contains no edge from the lower to the upper level. □

Now we construct an auxiliary digraph $\bar{G}$ from G and transform $\tilde{z}$ to a circulation $\bar{z}$ in $\bar{G}^{01}$ as follows. Let $i \in \{1, \ldots, q\}$ such that all vertices in W_i^{01} are low-throughput vertices. We add an auxiliary vertex a_i to G and set $r(a_i) := r(v)$ for $v \in W_i$; this is well-defined by Lemma 8.9. For every edge $(v, w) \in \delta^-(\hat{W}_i)$, we add an edge (v, a_i) of the same cost. Similarly, for every edge $(v, w) \in \delta^+(\hat{W}_i)$, we add an edge (a_i, w) of the same cost. Then we reroute $\frac{1}{2}$ unit of flow of the circulation $\tilde{z}$ through one of the auxiliary vertices a_i^0, a_i^1 in the split graph. By Lemma 8.19, we have $\tilde{z}(\delta(\hat{W}_i^{01})) = z(\delta(\hat{W}_i^{01})) = x(\delta(\hat{W}_i)) \geq 2$. Hence, we can do this in the same way as described in Section 8.3 except that we apply Lemma 8.8 to the vertex set $U = \hat{W}_i^{01}$ instead of W_i^{01}.

We do this transformation successively for all $i \in \{1, \ldots, q\}$ for which all vertices in W_i^{01} are low-throughput vertices. Let $\bar{G}$ be the final graph and $\bar{G}^{01}$ its split graph. Let $\bar{z}$ be the final circulation in the split graph $\bar{G}^{01}$ and let $(\bar{x}, \bar{f}) := \pi(\bar{z})$. See Figures 8.5 and 8.7.

Then, using Lemma 8.11, we round $2\bar{z}$ to an integral circulation $\bar{z}^*$ with the properties (i) – (iv) of that lemma. Let $(\bar{x}^*, \bar{f}^*) := \pi(\bar{z}^*)$ and let $\bar{F}$ be the multi-set of edges in $\bar{G}$ with $\chi^{\bar{F}} = \bar{x}^*$ (see Figure 8.7 (c)). Then $\bar{F}$ is Eulerian because $\bar{x}^*$ is a circulation. As before, we have $c^y(\bar{F}) \leq 2 \cdot \mathrm{LP}(\mathcal{I})$ (cf. (8.3)), $|\delta^-_{\bar{F}}(a_i)| = 1$ for every $i \in \{1, \ldots, q\}$ for which W_i^{01} contains only low-throughput vertices (Lemma 8.12), and $|\delta^-_{\bar{F}}(W_i)| \geq 1$ for every other $i \in \{1, \ldots, q\}$ (Lemma 8.13).

Next we show that we maintain acyclic witness flows during the rerouting and rounding. Here, we crucially use that the witness flow f has property (vi) from Lemma 8.18. See Figure 8.8 (a)–(b) for an illustration.

Lemma 8.20. *The flows $\bar{f}$ and $\bar{f}^*$ have acyclic support.*

Proof. Since the support of $\bar{f}^*$ is contained in the support of $\bar{f}$ by (i) of Lemma 8.11, it suffices to show that $\bar{f}$ has acyclic support. Suppose the support of $\bar{f}$ contains a cycle $\bar{C}$. Then there exists $i \in \{1, \ldots, q\}$ such that $a_i \in V(\bar{C})$ because otherwise $\bar{C}$ is contained in the support of f (which is acyclic). Let $\bar{e} = (a_i, v) \in E(\bar{C})$, and let $e \in \delta^+(\hat{W}_i)$ be the edge of G corresponding to $\bar{e}$. Then $f(e) > 0$, and hence the residual graph G_f contains an

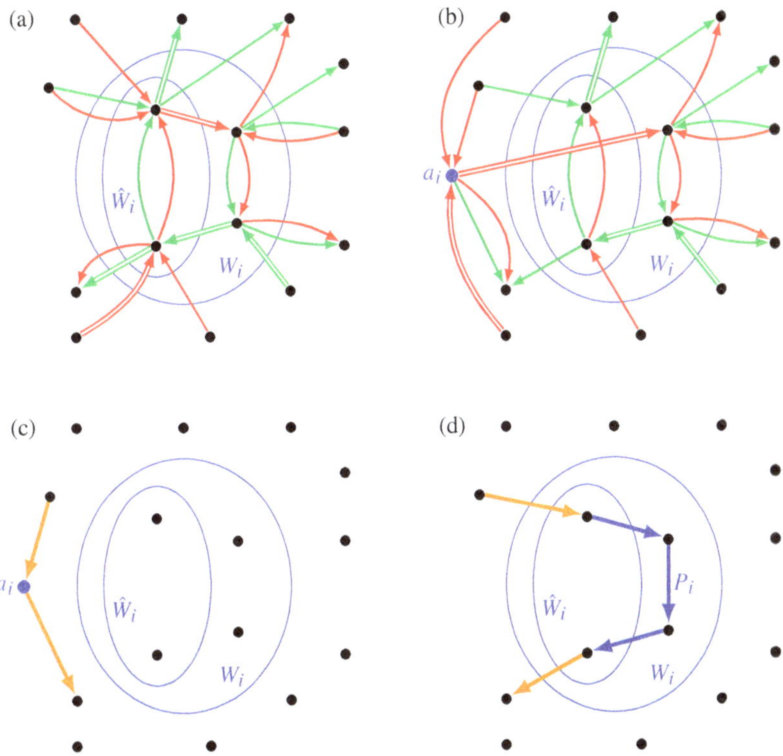

Figure 8.7 Example of the construction of the solution F from the circulation $\tilde{z}$ in the split graph. In all pictures, a set W_i (blue with white interior) and the subset $\hat{W}_i$ (blue and filled) is shown. The pictures show only edges with at least one endpoint in W_i. Picture (a) shows (parts of) $(\tilde{x}, \tilde{f}) = \pi(\tilde{z})$ where $\tilde{f}$ (green) is a witness flow for $\tilde{x}$ (green and red). The edges drawn with a single line have value $\frac{1}{8}$; the edges drawn with a double line have value $\frac{1}{4}$. Note that W_i^{01} contains no high-throughput vertex. Picture (b) shows a possible flow $\bar{x}$ in $\bar{G}$ resulting from rerouting flow through a_i (blue); the witness flow $\bar{f}$ is shown in green. Picture (c) shows in orange a possible integral flow $\bar{x}^*$ in $\bar{G}$. The orange edges are elements of the edge set $\bar{F}$ with $\chi^{\bar{F}} = \bar{x}^*$. Picture (d) shows the result of mapping $\bar{F}$ back to G. In blue, the path P_i in $G[W_i]$ is shown; it completes the orange edges to a circulation. A similar picture appeared in Traub and Vygen [2022].

edge $e^{\leftarrow} \in \delta^-_{G_f}(\hat{W}_i)$. We have $e^{\leftarrow} \in \delta^-_{G_f}(W_i)$ since $\hat{W}_i$ is the vertex set of the first strongly connected component of $G_f[W_i]$. This shows $E(\bar{C}) \cap \delta(W_i \cup \{a_i\}) \neq \emptyset$ for some $i \in \{1, \dots, q\}$.

We claim that we can map $\bar{C}$ to a closed walk C in the residual graph G_f. See Figure 8.8 (b)–(c). We first map every edge of the cycle $\bar{C}$ to its corresponding

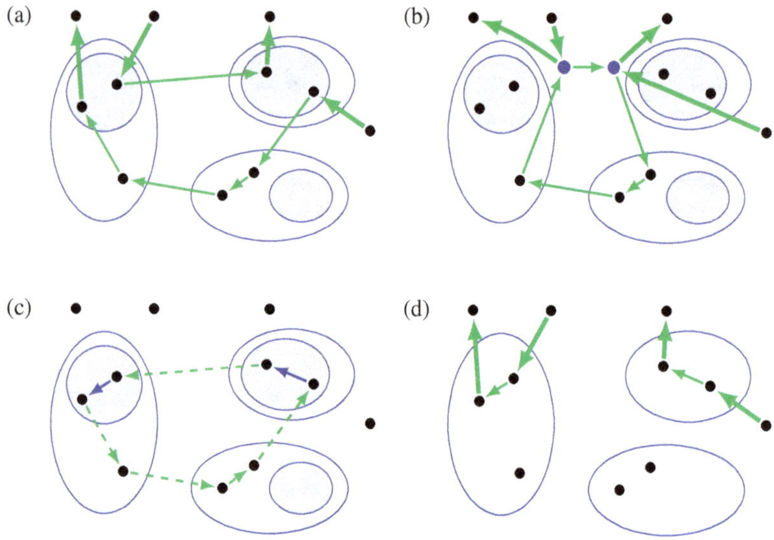

Figure 8.8 Illustration of the proof of Lemma 8.20. Three sets W_i are shown in blue with white interior; pictures (a)–(c) also show their subsets $\hat{W}_i$ (blue and filled). Picture (a) shows (parts of) an arbitrary witness flow f (green). The thin edges are neutral; the thick edges can be neutral or forward edges. This witness flow f will not be chosen by our algorithm; it does not minimize $\sum_{i=1}^{q} f(\delta(W_i))$. Picture (b) shows what would happen if we chose this flow anyway. We see a possible result of rerouting this flow through the vertices $a_i \in V(\bar{G})$ (shown in blue). In this example, the support of $\bar{f}$ contains a cycle $\bar{C}$. Picture (c) shows a corresponding closed walk C in the residual graph G_f. The blue edges show paths inside the sets $\hat{W}_i$; these exist because $G_f[\hat{W}_i]$ is strongly connected. Picture (d) shows the flow resulting from f by augmenting along C. The augmentation decreased $\sum_{i=1}^{q} f(\delta(W_i))$ but did not change the flow on forward edges. This picture is taken from Traub and Vygen [2022].

edge in G. Notice that the resulting edge set D is not necessarily a cycle: If $a_i \in V(\bar{C})$ for some $i \in \{1, \ldots, q\}$, then D contains an edge entering $\hat{W}_i$ and an edge leaving $\hat{W}_i$, but it might be disconnected in between.

We have $f(e) > 0$ for every edge $e \in D$. Thus, by reversing all edges in D, we obtain edges in G_f (with positive residual capacity u_f). Moreover, we can complete this edge set to a closed walk C in G_f (with positive residual capacity u_f) by adding only edges of $G_f[\hat{W}_i]$ for $i \in \{1, \ldots, q\}$; this is possible because for every $i \in \{1, \ldots, q\}$, the subgraph $G_f[\hat{W}_i]$ is strongly connected by the choice of $\hat{W}_i$. We found a closed walk C in G_f. Let $i \in \{1, \ldots, q\}$ such that $E(\bar{C}) \cap \delta(W_i \cup \{a_i\}) \neq \emptyset$. Then $E(C) \cap \delta(W_i) \neq \emptyset$, where $E(C)$ denotes the footprint of C.

Also note that $r(v) \geq r(w)$ for all edges (v, w) of C: Every edge (v, w) of C has a corresponding edge $(w, v) \in E(G)$ with $f(e) > 0$, or it has both endpoints in the same set $\hat{W}_i \subseteq W_i$. In the first case, we can conclude that (w, v) is not a backward edge and hence $r(w) \leq r(v)$. In the latter case, $r(v) = r(w)$ by Lemma 8.9. Since C is a closed walk, we conclude that $r(v) = r(w)$ for all vertices v and w in C.

We augment f along the closed walk C by some sufficiently small but positive amount. Augmenting f along C by some $\varepsilon > 0$ means that for every edge $e \in E$ with $e \in C$, we increase $f(e)$ by ε, and for every edge $e \in E$ with $e^{\leftarrow} \in C$, we decrease $f(e)$ by ε. Because $r(v) = r(w)$ for all $v, w \in V(C)$, this augmentation changes the flow f only on neutral edges. Hence, we can augment f by $\varepsilon := \min_{e \in C} u_f(e) > 0$ and maintain a witness flow.

We claim that the augmentation decreases $\sum_{i=1}^{q} f(\delta(W_i))$, which is a contradiction to our choice of f. See Figure 8.8 (d). All edges of C that are contained in a cut $\delta(W_i)$ for some $i \in \{1, \ldots, q\}$ result from mapping the edges of the cycle $\bar{C}$ in $\bar{G}$ to G and reversing them; for these edges, the augmentation decreases the flow value. Therefore, augmenting f along C decreases the flow value on all edges in $E(C) \cap (\delta(W_1) \cup \cdots \cup \delta(W_q))$, and we have already shown that this set is nonempty. □

Next we show how the fact that the witness flow $\bar{f}^*$ is acyclic leads to an improved bound on $|\delta^-_{\bar{F}}(v)|$. We need the following observation:

Lemma 8.21. *Let $\bar{D}$ be a connected component of the graph $(V(\bar{G}), \bar{F})$ with $V(\bar{D}) \cap V(B) = \emptyset$. Then $\bar{f}^*(E(\bar{D})) = 0$.*

Proof. Because $\bar{f}^*$ is a witness flow and $V(\bar{D}) \cap V(B) = \emptyset$, we have $\bar{f}^*(\delta^-(v)) \leq \bar{f}^*(\delta^+(v))$ for every $v \in V(\bar{D})$. Since

$$\bar{f}^*(E(\bar{D})) = \sum_{v \in V(\bar{D})} \bar{f}^*(\delta^-(v)) \leq \sum_{v \in V(\bar{D})} \bar{f}^*(\delta^+(v)) = \bar{f}^*(E(\bar{D})),$$

we have $\bar{f}^*(\delta^-(v)) = \bar{f}^*(\delta^+(v))$ for every $v \in V(\bar{D})$. In other words, $\bar{f}^*$ restricted to $E(\bar{D})$ is a circulation. Because the support of $\bar{f}^*$ is acyclic by Lemma 8.20, this implies $\bar{f}^*(E(\bar{D})) = 0$. □

This allows us to strengthen the bounds from Lemmas 8.14 and 8.15 for vertices that $\bar{F}$ does not connect to the backbone. Notice that these are the only vertices for which we applied Lemmas 8.14 and 8.15.

Lemma 8.22. *Let $\bar{D}$ be a connected component of the graph $(V(\bar{G}), \bar{F})$ with $V(\bar{D}) \cap V(B) = \emptyset$, and let $v \in V(\bar{D})$ with $y_v > 0$. Then $|\delta^-_{\bar{F}}(v)| \leq 2$. If v^1 is a low-throughput vertex, $|\delta^-_{\bar{F}}(v)| \leq 1$.*

Proof. We have $|\delta^-_{\bar{F}}(v)| = \bar{x}^*(\delta^-(v)) = \bar{z}^*(\delta^-(v^1)) + \bar{z}^*(\delta^-(v^0) \setminus \{e^{\downarrow}_v\}) = \bar{z}^*(\delta^-(v^1)) + \bar{f}^*(\delta^-(v)) = \bar{z}^*(\delta^-(v^1))$, where we used Lemma 8.21. Therefore, by (iii) of Lemma 8.11, we get

$$|\delta^-_{\bar{F}}(v)| = \bar{z}^*(\delta^-(v^1)) \leq \lceil 2\bar{z}(\delta^-(v^1))\rceil.$$

If v^1 is a low-throughput vertex, this implies $|\delta^-_{\bar{F}}(v)| \leq \lceil 2\bar{z}(\delta^-(v^1))\rceil \leq 1$. Otherwise, $y_v > 0$ implies $\{v\} \in \mathcal{L}$ and hence $x(\delta^-(v)) = 1$. Therefore, $|\delta^-_{\bar{F}}(v)| \leq \lceil 2\bar{z}(\delta^-(v^1))\rceil \leq \lceil 2x(\delta^-(v))\rceil = 2$. □

Now we show that despite the modification of the rerouting procedure, $\bar{F}$ enters and leaves every set $W_i \cup \{a_i\}$ for which W^{01} contains no high-throughput vertex.

Lemma 8.23. *Let $i \in \{1, \ldots, q\}$ such that W_i^{01} contains no high-throughput vertex. Then $\bar{F} \cap \delta(W_i \cup \{a_i\}) \neq \emptyset$.*

Proof. As in Lemma 8.12, there exists an edge $\bar{e} = (v, a_i) \in \bar{F}$. If $v \notin W_i$, we have $\bar{e} \in \bar{F} \cap \delta(W_i \cup \{a_i\})$. Otherwise, the edge e of G that corresponds to $\bar{e}$ belongs to $\delta^-(\hat{W}_i) \setminus \delta^-(W_i)$. Therefore, we have $f(e) = x(e)$ as otherwise the residual graph G_f contained e as well, contradicting the choice of $\hat{W}_i$. This implies

$$\bar{z}^*(\bar{e}^1) \leq \lceil 2\bar{z}(\bar{e}^1)\rceil \leq \lceil 2\bar{z}(e^1)\rceil \leq \lceil 2z(e^1)\rceil = \lceil 2(x(e) - f(e))\rceil = 0.$$

But then

$$\bar{f}^*(\bar{e}) = \bar{z}^*(\bar{e}^0) = \bar{z}^*(\bar{e}^0) + \bar{z}^*(\bar{e}^1) = \bar{x}^*(\bar{e}) \geq 1$$

because $\bar{e} \in \bar{F}$. By Lemma 8.21, this implies that the connected component $\bar{D}$ of $(V(\bar{G}), \bar{F})$ that contains a_i also contains a vertex $w \in V(B)$. Since $W_i \cap V(B) = \emptyset$, this completes the proof. □

Due to the improved bound from Lemma 8.22 (replacing the bounds from Lemma 8.15 and Lemma 8.14), we obtain a $(2, 2, 1)$-algorithm for subtour cover instead of a $(3, 2, 1)$-algorithm:

Theorem 8.24. *There is a $(2, 2, 1)$-algorithm for subtour cover.*

Proof. As before (cf. Section 8.5), we can map $\bar{F}$ as constructed earlier to a subtour cover solution F. Note that the path P_i that we add in $G[W_i]$ for each i for which we rerouted flow through the auxiliary vertex a_i is not necessarily in $\hat{W}_i$. Lemma 8.23 guarantees that after replacing the edges incident to a_i (and adding P_i), we have $\delta_F(W_i) \neq \emptyset$. For $i \in \{1, \ldots, q\}$ for which W^{01} contains a high-throughput vertex, we have $\bar{F} \cap \delta(W_i) \neq \emptyset$ as before (cf. Lemma 8.13).

Again, every connected component of $(V, \bar{F})$ is local or connects to the backbone (due to the witness flow $\bar{f}^*$, cf. Lemma 8.6), and the same holds for every connected component of (V, F) because the edge sets of the paths P_i are local. We conclude that F is a subtour cover.

As before, we have the bound (8.3) on the cost of $\bar{F}$, and the total cost of the paths P_i is at most $\sum_{v \in V \setminus V(B)} 2y_v$. Hence, again,

$$c^y(F) \leq 2 \cdot \mathrm{LP}(\mathcal{I}) + \sum_{v \in V \setminus V(B)} 2y_v.$$

Finally, let D be a connected component of (V, F) with $V(D) \cap V(B) = \emptyset$. Then $E(D)$ is local and hence $c^y(E(D)) = \sum_{v \in V(D)} |\delta^-_F(v)| \cdot 2y_v$. If v^1 is a high-throughput vertex, Lemma 8.22 implies $|\delta^-_F(v)| = |\delta^-_{\bar{F}}(v)| \leq 2$. If v^1 is a low-throughput vertex, Lemma 8.22 implies $|\delta^-_F(v)| \leq |\delta^-_{\bar{F}}(v)| + 1 \leq 2$. We conclude

$$c^y(E(D)) \leq 2 \cdot \sum_{v \in V(D)} 2y_v$$

as required. □

From this, we obtain a $(17+\varepsilon)$-approximation algorithm for the Asymmetric TSP.

Theorem 8.25. *For every $\varepsilon > 0$, there is a polynomial-time algorithm that computes for any given instance (G, c) of the* Asymmetric *TSP a solution of cost at most $17 + \varepsilon$ times the cost of an optimum solution to* (3.2).

Proof. Theorem 8.24 yields a $(2, 2, 1)$-algorithm for subtour cover, and by Theorem 7.18, this implies that there is a $(2, 9 + \varepsilon)$-algorithm for vertebrate pairs. Using Theorem 7.12, we then obtain a polynomial-time algorithm that finds a solution of cost at most $17 + \varepsilon$ times the value of (3.2) for any given instance (G, c) of the Asymmetric TSP. □

An immediate consequence of Theorem 8.25 is an upper bound of 17 on the integrality ratio of (3.2).

Corollary 8.26. *The integrality ratio of* (3.2) *is at most* 17. □

Using the observation from Exercise 7.7, one could slightly improve Theorem 8.25 and Corollary 8.26, but the improvement would be less than 0.2. Can we at least obtain the approximation ratio that we know for Asymmetric Graph TSP?

Open Problem 8.27. Obtain an approximation ratio of $(8 + \varepsilon)$ or better for the Asymmetric TSP.

Exercises

8.1 Show that in general not every circulation in G is the image of a circulation in G^{01}.

8.2 In the algorithm(s) described in this chapter, we first map x to a circulation in the split graph and then make the support of z within each W_i^{01} acyclic, resulting in $\tilde{z}$. Suppose we instead first make the support of x within each W_i acyclic and obtain $\tilde{x}$. Show that then it might be impossible to map $\tilde{x}$ to a circulation $\tilde{z}$ in the split graph.

8.3 Let $G = (V, E)$ be a digraph, $c : E \to \mathbb{R}$, and let $f, g : E \to \mathbb{R}_{\geq 0}$ such that $f + g$ is a circulation in G. Prove that there are $f^*, g^* : E \to \mathbb{Z}_{\geq 0}$ such that the following properties hold:

- $f^* + g^*$ is a circulation in G;
- $c(f^* + g^*) \leq c(f + g)$;
- $f^*(\delta^+(v)) \geq f^*(\delta^-(v))$ for all $v \in V$ with $f(\delta^+(v)) \geq f(\delta^-(v))$;
- $\lfloor f(\delta^-(v)) \rfloor \leq f^*(\delta^-(v)) \leq \lceil f(\delta^-(v)) \rceil$ for all $v \in V$;
- $\lfloor g(\delta^-(v)) \rfloor \leq g^*(\delta^-(v)) \leq \lceil g(\delta^-(v)) \rceil$ for all $v \in V$;
- $f^*(e) = 0$ for all $e \in E$ with $f(e) = 0$;
- $g^*(e) = 0$ for all $e \in E$ with $g(e) = 0$.

(Svensson, Tarnawski, and Végh [2020])

8.4 Devise a combinatorial polynomial-time algorithm to transform a given witness flow into a witness flow with acyclic support. Do not use linear programming or Theorem 3.11.

8.5 Let (G, l, u, c) be an instance of the minimum-cost circulation problem (cf. Theorem 3.11) with $l(e) = 0$ for all $e \in E$. Let f be a circulation in G with $0 \leq f \leq u$. Show that f is a minimum-cost circulation if and only if the residual graph contains no circuit with negative total weight, where the weight of $e^{\leftarrow}$ is $-c(e)$.

8.6 Consider the $(2, 2, 1)$-algorithm for subtour cover described in this chapter. Show that for instances where $\tilde{z} = z$, this is actually a $(2, 2, 0)$-algorithm for subtour cover.

9

Asymmetric Path TSP

A natural generalization of the (asymmetric) traveling salesman problem arises when we are given a start vertex s and an end vertex t and ask for a tour that begins in s and ends in t, rather than a round trip. As before, there is the version with triangle inequality where we ask for a path from s to t that contains every vertex exactly once, or the equivalent version in which we ask for a walk from s to t that visits all vertices at least once.

While this problem seems to be harder, we will see in this chapter that it can be tackled by similar techniques. In particular, we show black-box reductions to Asymmetric TSP and prove, as new results, the best-known approximation ratios and bounds on the integrality ratio.

9.1 Overview

Let us first introduce the two equivalent versions of the problem. If the triangle inequality holds, we can ask for a path from s to t that visits every vertex exactly once:

Problem 9.1 (Asymmetric Path TSP with Triangle Inequality).

Instance: A finite set V (of cities), a distance function $c : V \times V \to \mathbb{R}_{\geq 0} \cup \{\infty\}$ such that $c(u,w) \leq c(u,v) + c(v,w)$ for all $u, v, w \in V$, and two elements s and t of V.

Task: Compute a list $v_1, v_2, \ldots, v_n$ with $v_1 = s$ and $v_n = t$ that contains every element of V exactly once and minimizes $\sum_{i=2}^{n} c(v_{i-1}, v_i)$.

Note that infinite distances are allowed. For example, it might not be possible to reach s from t at finite cost.

Without the triangle inequality, we ask for a walk from s to t that visits every vertex at least once. The footprint of such a walk is called an s-t-tour:

Definition 9.2 (s-t-tour)**.** An s-t-*tour* in a digraph $G = (V, E)$ is a multi-subset F of E such that $(V, F \mathbin{\dot\cup} \{(t, s)\})$ is connected and Eulerian.

By Euler's Theorem 1.6, the s-t-tours are precisely the footprints of the walks from s to t that visit every vertex at least once (cf. Lemma 1.7). Hence, we can also define the problem as follows:

Problem 9.3 (Asymmetric Path TSP)**.**

Instance: A directed graph $G = (V, E)$, a cost function $c : E \to \mathbb{R}_{\geq 0}$, and two vertices $s, t \in V$.

Task: Compute an s-t-tour in G with minimum cost (or decide that there is no s-t-tour in G).

It is easy to decide whether a given digraph contains an s-t-tour (see Exercise 9.2). As in Section 1.2, we note:

Proposition 9.4. *Asymmetric Path TSP with Triangle Inequality (Problem 9.1) and Asymmetric Path TSP (Problem 9.3) are equivalent.*

Proof. As in Proposition 1.12, except that we have s-t-tours instead of tours and Hamiltonian s-t-paths instead of Hamiltonian circuits. When constructing an instance of Asymmetric Path TSP, we only include edges (v, w) with $c(v, w) < \infty$. □

Let us first describe a very easy n-approximation algorithm:

Proposition 9.5. *There is an n-approximation algorithm for Asymmetric Path TSP with Triangle Inequality.*

Proof. Starting with the sequence s, t, successively insert all other cities in an arbitrary order: When inserting city v, choose a position (but not before s or after t) to minimize the cost of the resulting sequence (where the cost of sequence $v_1, \ldots, v_k$ is $\sum_{i=2}^{k} c(v_{i-1}, v_i)$). Since v is between s and t in an optimum solution, it is between v_{i-1} and v_i in an optimum solution for some $i \in \{2, \ldots, k\}$ for the current sequence $v_1, \ldots, v_k$. Inserting v there increases the cost by at most $c(v_{i-1}, v) + c(v, v_i) \leq \text{OPT}$. □

This bound for the cheapest insertion heuristic is essentially tight (see Exercises 9.3 and 9.4 for details).

In this chapter, we describe relations between the approximation ratios of Asymmetric Path TSP and Asymmetric TSP as well as their integrality ratios.

Table 9.1 *Approximation ratios and upper bounds on the integrality ratio of* (9.1) *for* Asymmetric Path *TSP in the order of their discovery. The approximation ratios in line 3, 6, 8, and 10 follow from the black-box reduction by Feige and Singh [2007]; see Section 9.2. (R) means randomized; this algorithm computes a random s-t-tour, and the approximation ratio compares its expected cost to* OPT. *Moreover, ε stands for an arbitrarily small positive constant.*

Approximation Ratio	Integrality Ratio	Year	Reference	Chapter
$2\sqrt{n}$	–	2005	Lam and Newman [2008]	Exercise 9.5
$2.78 \log_2 n$	–	2006	Chekuri and Pál [2007]	–
$1.34 \log_2 n$	–	2006	Feige and Singh [2007]	9.2
–	$O(\sqrt{n})$	2008	Nagarajan and Ravi [2008]	–
–	$1 + 2\log_2 n$	2009	Friggstad, Salavatipour, and Svitkina [2013]	–
$O(\frac{\log n}{\log \log n})$ (R)	–	2009	Asadpour et al. [2017]	5, 9.2
–	$O(\frac{\log n}{\log \log n})$	2012	Friggstad, Gupta, and Singh [2016]	–
1012	–	2018	Svensson, Tarnawski, and Végh [2020]	6–8, 9.2
–	1273	2018	Köhne, Traub, and Vygen [2020]	9.5, 9.6
$44 + \varepsilon$	85	2019	Traub and Vygen [2022]	6–8, 9.2, 9.6
$43 + \varepsilon$	43	2019	Traub [2020a]	9.5, 9.7
$17 + \varepsilon$	17	2021	this book	9.5, 9.7

As far as we know today, Asymmetric Path TSP might be more difficult, but at most by a small constant factor. We will also derive a direct approximation algorithm, using results from the previous chapters.

Table 9.1 summarizes the state of the art, also with respect to the integrality ratio of the natural linear programming relaxation (see Section 9.3). The first nontrivial approximation algorithms by Lam and Newman [2008] and Chekuri and Pál [2007] obtained approximation ratios of $O(\sqrt{n})$ and $O(\log n)$, respectively. Feige and Singh [2007] showed that any α-approximation algorithm for Asymmetric TSP implies an $\alpha(2 + \varepsilon)$-approximation algorithm for Asymmetric Path TSP; we will prove their result in Section 9.2. Thus, the better approximation algorithms for Asymmetric TSP by Feige and Singh [2007], Asadpour et al. [2017], and Svensson, Tarnawski, and Végh [2020] implied better approximation algorithms for Asymmetric Path TSP, too.

However, none of these results implied an upper bound on the integrality ratio of the natural LP relaxation, which we will discuss in Section 9.3. Such bounds were obtained by Nagarajan and Ravi [2008], Friggstad, Salavatipour, and Svitkina [2013], and Friggstad, Gupta, and Singh [2016]. Köhne, Traub, and Vygen [2020] showed that the integrality ratio is at most $4\rho - 3$, where ρ is the integrality ratio of the Asymmetric TSP LP. We will present this in Sections 9.5 and 9.6. Together with the results of the previous chapters, this yields a constant upper bound. Traub [2020a] found a more elegant reduction that also yields a better upper bound. At the end of this chapter, in Section 9.7, we improve the bounds further.

9.2 Reduction to Asymmetric TSP

Any α-approximation algorithm for Asymmetric Path TSP implies an α-approximation algorithm for Asymmetric TSP: Simply take an arbitrary city v of an Asymmetric TSP instance, split it into two copies v_s and v_t, and find an α-approximate v_s-v_t-tour. In other words, requiring that s and t are two distinct vertices is not important in Definition 9.3, but it is sometimes convenient to simplify notation.

Feige and Singh [2007] showed that, conversely, any α-approximation algorithm for Asymmetric TSP implies an $\alpha(2+\varepsilon)$-approximation algorithm for Asymmetric Path TSP, for any $\varepsilon > 0$. The core of their proof is the following lemma. Figure 9.1 shows an example.

Lemma 9.6 (Feige and Singh [2007]). *Let (V, c, s, t) be an instance of Asymmetric Path TSP with Triangle Inequality whose optimum solution has cost* OPT. *Let $P^1, \ldots, P^k$ be s-t-paths of total cost L such that no vertex except s and t belongs to more than one of them. Then there is an s-t-path P that contains all vertices of P^i in the same order as P^i (for all $i = 1, \ldots, k$) and that has cost at most $L + k$* OPT.

Proof. Let R be an s-t-path that contains $V(P^1) \cup \cdots \cup V(P^k)$ and has cost at most OPT. Let $R^1, \ldots, R^k$ be k copies of R. We construct a path P as desired as follows. Initially, P contains only s, and all other vertices (that do not yet belong to P) are called future vertices. The endpoint of P is called the current vertex. It will always belong to (at least) one of the given paths. The first vertex of P^j that does not yet belong to P is called the candidate of P^j.

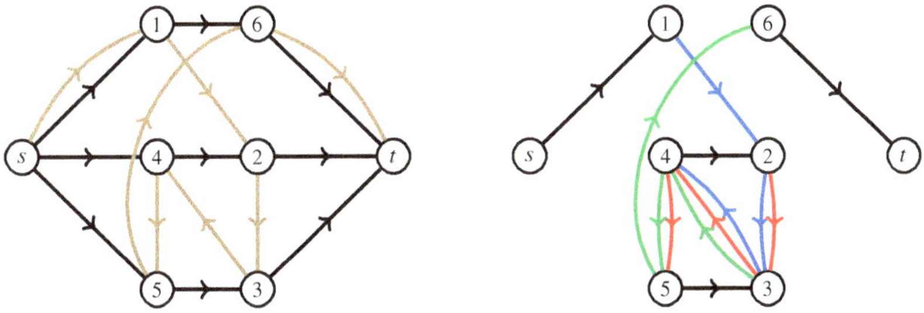

Figure 9.1 An example for the proof of Lemma 9.6. Here $k = 3$. Left-hand side: The initial s-t-paths P^1, P^2, P^3 are shown in black; they go from left to right. The Hamiltonian s-t-path R (brown) visits the vertices in the order of the numbers. Right-hand side: In the proof, we combine these to a path that visits s, 1, 4, 2, 5, 3, 6, t in this order. Besides some edges of P^1, P^2, P^3, it uses the segment from 1 to 4 from the first (blue) copy of R, the segment from 2 to 5 from the second (red) copy, and the segment from 3 to 6 from the third (green) copy.

Let v be the current vertex, and let $j \in \{1, \ldots, k\}$ such that $v \in P^j$. Let $i \in \{1, \ldots, k\} \setminus \{j\}$ be the index of the path P^i whose candidate x comes earliest on $R_{[v,t]}$ (the subpath of R from v to t; this will always contain all candidates). We show how to extend P. Let w be the last vertex of P^j that belongs to $R_{[s,x]}$. Take $P^j_{[v,w]}$, append it to P, and remove it from P^j (so that P^j now starts with w; the vertices up to w are no longer future vertices). Append the edge (w, x) to P (then x is no longer a future vertex) and remove $R_{[w,x]}$ from R^j. Now x is the new current vertex.

This procedure maintains the following invariants:

(i) P contains exactly the vertices from the original P^h that precede its candidate, in the same order, for all $h \in \{1, \ldots, k\}$.

(ii) For the current vertex $v \in V(P^j)$, the subpath $R_{[v,t]}$ contains the candidate of P^h for all $h \in \{1, \ldots, k\} \setminus \{j\}$.

(iii) R^h contains a path from any future or current vertex of P^h to t, for all $h \in \{1, \ldots, k\}$.

(iv) The total cost of P, $P^1, \ldots, P^k$, and $R^1, \ldots, R^k$ is at most $L + k\,\mathrm{OPT}$.

Here, (i) is obvious, (ii) follows from the choice of i and x, and (iii) follows from the choice of w. For (iv), we use the triangle inequality: Whenever we add an edge to P, we remove a corresponding subpath from P^j or R^j. Invariants (i) and (iii) imply that these subpaths still exist at that time.

When the procedure terminates, the claim follows from (i) and (iv). □

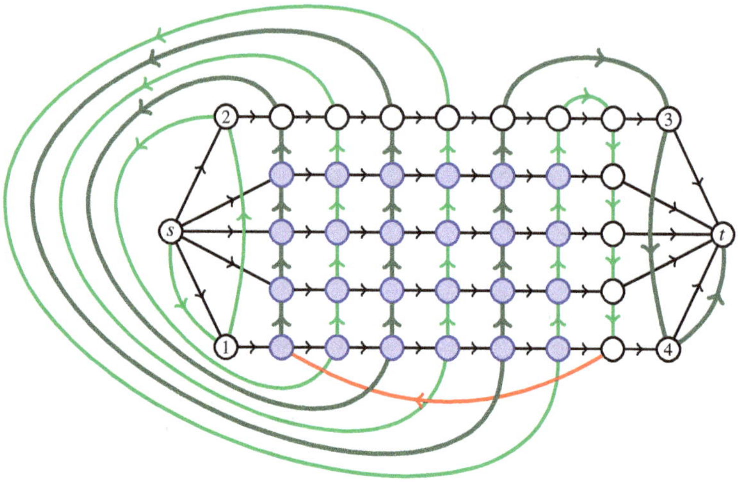

Figure 9.2 A tight example for Lemma 9.6, based on an almost tight example by Feige and Singh [2007]. The initial s-t-paths $P^1, \ldots, P^k$ are shown in black (going from left to right). A Hamiltonian s-t-path R traverses the light green edges, then the red edge, then the dark green edges. Note that the red edge is the only one that goes from right to left. The middle grid (without the vertices s, 1, 2, 3, 4, t) has kl vertices; here, $k = 5$ and $l = 7$. Any s-t-tour in this graph must visit s, 1, 2, 3, 4, t in this order. In between 2 and 3, it will visit the vertices in the grid in some order but must respect the order of the P^i. Our s-t-tour must use the red edge before visiting any of the $(k-1)(l-1)$ blue vertices, and then it can visit at most $k+l-3$ of them before using the red edge again. Finally, after visiting the last of the blue vertices, it will visit the successor on the corresponding path P^i (in the rightmost column of the grid), and then it needs to use the red edge again in order to reach vertex 3. This makes us use the red edge at least $1 + \lceil \frac{(k-1)(l-1)}{k+l-3} \rceil$ times. For $l \geq k^2$, this is k times. One can also see that an s-t-tour uses the red edge more than k times unless it uses almost all black edges. Therefore, if the cost of the red edge equals the total cost of the black edges, and the green edges have cost 0, the bound in Lemma 9.6 is tight (when considering the metric closure).

The bound in Lemma 9.6 is tight for all $k \geq 2$: see Figure 9.2. A path P as in Lemma 9.6 can be found by dynamic programming in polynomial time if k is a constant. In general, we get:

Lemma 9.7. *Let $\varepsilon > 0$ be a constant. Let (V, c, s, t) be an instance of* Asymmetric Path TSP with Triangle Inequality *whose optimum solution has cost* OPT. *Let $P^1, \ldots, P^r$ be s-t-paths of total cost L. Then one can find an s-t-path P with vertex set $V(P^1) \cup \cdots \cup V(P^r)$ and cost at most $L + (1+\varepsilon)r$ OPT in polynomial time.*

Proof. Without loss of generality, let $\varepsilon \le 1$. Moreover, by taking shortcuts, we can assume that every vertex except s and t appears in at most one of the paths.

If $r = 1$, return $P = P^1$. Otherwise, let $k := \min\left\{r, 1 + \lceil\frac{1}{\varepsilon}\rceil\right\}$. Replace k of the paths, say $P^1, \ldots, P^k$, by a cheapest s-t-path that contains all vertices of P^i in the same order as P^i (for all $i = 1, \ldots, k$).

This can be found by dynamic programming in $O(k^2 n^k)$ time as follows. For all $n_1, \ldots, n_k$ and all $i \in \{1, \ldots, k\}$, we store a shortest path from s to the n_i-th vertex of P^i that contains, for all $j \in \{1, \ldots, k\}$, the first n_j vertices of P^j in the same order as P^j. We compute these paths in an order of increasing $n_1 + \cdots + n_k$, each in $O(k)$ time using the previously computed information.

By Lemma 9.6, the total cost increases by at most k OPT, and the number of paths decreases by $k - 1$. We iterate this until only one path remains.

If this procedure stops after l iterations, the total cost increases by at most $(l + r - 1)$ OPT, and the total running time is $O(lk^2 n^k) = O(n^{3+\frac{1}{\varepsilon}})$. Now $l = \lceil\frac{r-1}{k-1}\rceil < 1 + \varepsilon r$. □

Now we can prove the main result of Feige and Singh [2007]:

Theorem 9.8 (Feige and Singh [2007]). *If there is an α-approximation algorithm for* Asymmetric *TSP, then there is an $\alpha(2 + \varepsilon)$-approximation algorithm for* Asymmetric Path *TSP, for any $\varepsilon > 0$.*

Proof. By Proposition 9.4, it suffices to consider instances (V, c, s, t) of Asymmetric Path TSP with Triangle Inequality. We may assume that there is a solution of finite cost OPT.

Let γ_0 be the cost of an s-t-tour obtained by an n-approximation algorithm (such as cheapest insertion; cf. Proposition 9.5). Then $\text{OPT} \le \gamma_0 \le n\,\text{OPT}$. Let $\varepsilon' := \frac{\varepsilon}{2}$. Starting with $\gamma = \gamma_0$, we will compute $\lceil\log_{1+\varepsilon'} n\rceil$ s-t-tours, each for a different value of γ. Each time, we decrease γ by a factor $1 + \varepsilon'$, so in the end, $\gamma \le n\,\text{OPT}/(1 + \varepsilon')^{\log_{1+\varepsilon'} n} = \text{OPT}$. Therefore, for one of the γ-values, we have $\text{OPT} \le \gamma \le (1 + \varepsilon')\,\text{OPT}$. We output the best of all s-t-tours and show that the one computed when $\text{OPT} \le \gamma \le (1 + \varepsilon')\,\text{OPT}$ is cheap enough.

Let (G, c) be the following Asymmetric TSP instance. The digraph G has vertex set V and edge set $E \cup \{e_{\text{back}}\}$, where $E = \{(v, w) \in V \times V : t \ne v \ne w \ne s, c(v, w) < \infty\}$ and $e_{\text{back}} = (t, s)$. We set $c(e) = c(v, w)$ for $e = (v, w) \in E$ and $c(e_{\text{back}}) = \gamma$. Note that (G, c) contains a tour of cost $\text{OPT} + \gamma$.

Use an α-approximation algorithm for Asymmetric TSP to find a tour of cost at most $\alpha(\text{OPT} + \gamma)$. This tour uses $r \ge 1$ copies of the edge e_{back} (at least one because s has no other entering edges); removing them and taking shortcuts (Lemma 1.7) yields s-t-paths $P^1, \ldots, P^r$ such that every vertex belongs to at least one of them. Assuming $\mathrm{OPT} \le \gamma \le (1+\varepsilon')\,\mathrm{OPT}$, we have $r = \frac{1}{\gamma} r\gamma \le \frac{1}{\gamma}\alpha(\mathrm{OPT}+\gamma) = \alpha(1+\frac{\mathrm{OPT}}{\gamma}) \le 2\alpha$.

Then apply Lemma 9.7 with error parameter ε'. The resulting s-t-path contains all vertices and has cost at most

$$\begin{aligned}
&\alpha(\mathrm{OPT}+\gamma) - r\gamma + (1+\varepsilon')r\,\mathrm{OPT}\\
=\ &\alpha(\mathrm{OPT}+\gamma) + ((1+\varepsilon')\mathrm{OPT}-\gamma)r\\
\le\ &\alpha(\mathrm{OPT}+\gamma) + ((1+\varepsilon')\mathrm{OPT}-\gamma)2\alpha\\
=\ &\alpha(\mathrm{OPT}-\gamma) + (1+\varepsilon')2\alpha\,\mathrm{OPT}\\
\le\ &(1+\varepsilon')2\alpha\,\mathrm{OPT}\\
=\ &\alpha(2+\varepsilon)\,\mathrm{OPT},
\end{aligned}$$

where we used $\gamma \le (1+\varepsilon')\mathrm{OPT}$ and $r \le 2\alpha$ in the first inequality and $\gamma \ge \mathrm{OPT}$ in the second inequality. □

For integer weight functions c (and constant α), one can get rid of the ε in Theorem 9.8 by a kind of binary search (in weakly polynomial time); see Exercise 9.7.

Feige and Singh [2007] described a $(\frac{2}{3}\log_2 n)$-approximation algorithm for Asymmetric TSP and hence got a $((\frac{4}{3}+\varepsilon)\log_2 n)$-approximation algorithm for Asymmetric Path TSP. Moreover, the better approximation algorithms for Asymmetric TSP discovered later led to better approximation algorithms for Asymmetric Path TSP immediately (cf. Table 9.1).

9.3 Linear Programming Relaxation

In the rest of this chapter, we will work with the Asymmetric Path TSP and will not require the triangle inequality. First, it is easy to adapt the linear programming relaxation (3.2):

$$\begin{aligned}
\min\ & c(x)\\
\text{subject to}\quad & x(\delta^-(v)) - x(\delta^+(v)) = \begin{cases} -1, & \text{if } v = s\\ 1, & \text{if } v = t\\ 0, & \text{if } v \in V\setminus\{s,t\}\end{cases}\\
& x(\delta(U)) \ge 2 \quad \text{for } \emptyset \ne U \subseteq V\setminus\{s,t\}\\
& x_e \ge 0 \quad \text{for } e \in E.
\end{aligned} \tag{9.1}$$

The integral solutions to (9.1) are the incidence vectors of s-t-tours. It is obvious that the integrality ratio of (9.1) is no smaller than the integrality ratio of (3.2). Hence, by Theorem 3.18, it is at least 2. This is also shown by the simpler example in Figure 9.3, due to Friggstad, Gupta, and Singh [2016].

The constraints $x(\delta(U)) \geq 2$ for $\emptyset \neq U \subseteq V \setminus \{s,t\}$ can be replaced equivalently by $x(\delta^-(U)) \geq 1$ for $\emptyset \neq U \subseteq V \setminus \{s\}$, or by $x(\delta^+(U)) \geq 1$ for $\emptyset \neq U \subseteq V \setminus \{t\}$. This does not change the set of feasible solutions. If G is a complete graph and c satisfies the triangle inequality, one could also add degree constraints without changing the value of the LP (see Exercise 9.8).

The following lemma describes the structure of the support graph G of feasible solutions to (9.1); this will be useful later.

Lemma 9.9 (Svensson, Tarnawski, and Végh [2020]). *Let (G, c, s, t) be an* Asymmetric Path TSP *instance. Let $V_1, \dots, V_l$ be the strongly connected components of G numbered in topological order. Let x be a feasible solution to* (9.1). *Then we have $s \in V_1$, $t \in V_l$, and*

$$x(\delta^+(V_i) \cap \delta^-(V_j)) = \begin{cases} 1 & \text{if } j = i+1 \\ 0 & \text{otherwise.} \end{cases} \tag{9.2}$$

Proof. Let $V_1, \dots, V_l$ be a topological order of the strongly connected components of G. Then $\emptyset = \delta^-(V_1) = \delta^+(V_l)$. Moreover, we have $x(\delta^-(U)) \geq 1$ for all $\emptyset \neq U \subseteq V \setminus \{s\}$; hence, every vertex is reachable from s. Similarly, t is reachable from every vertex. Therefore, we have $s \in V_1$ and $t \in V_l$.

We now show (9.2) by induction on i. For $i = 1 < l$, we have $x(\delta^+(V_i)) = 1$ because $\delta^-(V_1) = \emptyset$, $s \in V_1$, and $t \notin V_1$. This implies

$$1 \leq x(\delta^-(V_2)) = x(\delta^+(V_1) \cap \delta^-(V_2)) \leq x(\delta^+(V_1)) = 1,$$

which settles the base case of the induction.

Now let $i \in \{2, \dots, l-1\}$. We have $x(\delta^-(V_i)) = x(\delta^+(V_i))$ because V_i contains neither s nor t. Moreover, $\delta^-(V_{i+1}) \subseteq \delta^+(V_1) \cup \dots \cup \delta^+(V_i)$ because $V_1, \dots, V_l$ is a topological order. Using the induction hypothesis in the first and last equation, we get

$$1 \leq x(\delta^-(V_{i+1})) = x(\delta^+(V_i) \cap \delta^-(V_{i+1})) \leq x(\delta^+(V_i)) = x(\delta^-(V_i)) = 1.$$

We have equality throughout and conclude

$$x(\delta^+(V_i) \cap \delta^-(V_{i+1})) = x(\delta^+(V_i)) = 1. \qquad \square$$

The dual LP of (9.1) is the same as (4.5) except for the term $a_t - a_s$ in the objective function and the omission of variables y_U for sets U that contain exactly one of s and t:

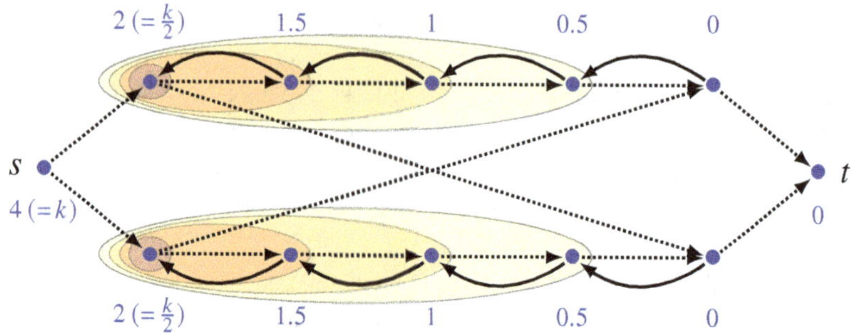

Figure 9.3 Example by Friggstad, Gupta, and Singh [2016] with integrality ratio approaching 2 as the number of vertices increases. Setting $x_e := \frac{1}{2}$ for all shown edges defines a feasible solution to (9.1). If the $2k$ curved edges have cost 1 and the dotted edges have cost 0, we have LP $= c(x) = k$, but any s-t-tour costs at least $2k-1$. (In the figure, $k=4$.) Setting $y_U = \frac{1}{2}$ for the vertex sets U indicated by the ellipses and a_v as shown in blue defines an optimum solution to the dual LP (9.3). This dual solution has strongly laminar support. This picture is taken from Köhne, Traub, and Vygen [2020] (with permission from Springer Nature).

$$
\begin{array}{llll}
\max \ a_t - a_s + \displaystyle\sum_{\emptyset \neq U \subseteq V \setminus \{s,t\}} 2\,y_U & & & \\
\text{subject to} \quad a_w - a_v + \displaystyle\sum_{\substack{\emptyset \neq U \subseteq V \setminus \{s,t\}:\\ e \in \delta(U)}} y_U & \le & c(e) & (e = (v,w) \in E) \\
\qquad\qquad\qquad\qquad\qquad\qquad y_U & \ge & 0 & (\emptyset \neq U \subseteq V \setminus \{s,t\}).
\end{array}
\tag{9.3}
$$

The primal and the dual LP can be solved in polynomial time, as in Proposition 4.3.

Almost exactly like in Lemma 7.2, we always have an optimum dual solution with *strongly laminar support* – that is, $\mathcal{L} = \{U : y_U > 0\}$ is a laminar family and $G[U]$ is strongly connected for every $U \in \mathcal{L}$. Figure 9.3 also shows an example of an optimum dual solution that has strongly laminar support.

Lemma 9.10. *Let (G, c, s, t) be an instance of the* Asymmetric Path TSP*. Moreover, let (a, y) be an optimum solution to* (9.3)*. Then we can compute in polynomial time (a', y') such that*

- *(a', y') is an optimum solution to* (9.3),
- *(a', y') has strongly laminar support, and*
- *$a'_s = a_s$ and $a'_t = a_t$.*

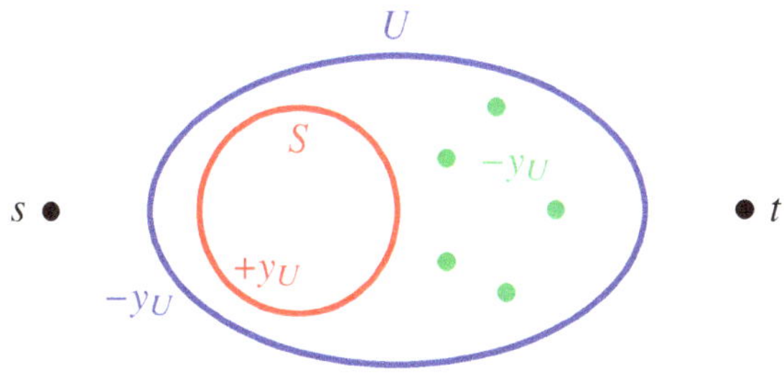

Figure 9.4 Illustration of the proof of Lemma 9.10. A set $U \subsetneq V \setminus \{s,t\}$ is shown in blue. The vertex set S of a strongly connected component of $G[U]$ with $\delta^-(S) \subseteq \delta^-(U)$ is shown in red. We modify our dual solution by increasing the dual variable corresponding to S, decreasing the dual variable corresponding to U (blue), and decreasing the variables corresponding to the vertices in $U \setminus S$ (green).

Proof. First, we apply uncrossing to y exactly as in the proof of Lemma 4.13. Henceforth, we may assume that the suppport of y is a laminar family of nonempty subsets of $V \setminus \{s,t\}$.

As long as there is a set U with $y_U > 0$, but $G[U]$ is not strongly connected, we do the following. Let U be a minimal set with $y_U > 0$ and such that $G[U]$ is not strongly connected. Moreover, let $G[S]$ be a strongly connected component of $G[U]$ with $\delta^-(S) \subseteq \delta^-(U)$; note that S exists and can be computed by Proposition 6.4.

Define a dual solution (a', y') as follows (see Figure 9.4). We set $y'_U := 0$, $y'_S := y_S + y_U$, and $y'_W := y_W$ for other sets W. Moreover, $a'_v := a_v - y_U$ for $v \in U \setminus S$ and $a'_v := a_v$ for all other vertices v. The only edges $e = (v,w)$ for which $a'_w - a'_v + \sum_{U:e\in\delta(U)} y'_U > a_w - a_v + \sum_{U:e\in\delta(U)} y_U$ are edges from $U \setminus S$ to S. However, such edges do not exist by choice of S. Hence, (a', y') is a feasible dual solution. Since $a'_s = a_s$ and $a'_t = a_t$ and $\sum_{\emptyset\neq U\subsetneq V} 2y'_U = \sum_{\emptyset\neq U\subsetneq V} 2y_U$, it is also optimum.

We now show that the support of y' remains laminar. Suppose there is a set W in the support of y' that crosses S. Then W must be in the support of y and hence a subset of U because the support of y is laminar. By the minimal choice of U, $G[W]$ is strongly connected. But this implies that G contains an edge from $W \setminus S$ to $W \cap S$, contradicting $\delta^-(S) \subseteq \delta^-(U)$.

We have now decreased the number of sets U in the support for which $G[U]$ is not strongly connected. Because the support of y is laminar, it has at most $2|V|$ elements by Proposition 4.8. Therefore, after iterating the procedure at most $2|V|$ times, the dual solution has the desired properties. □

9.4 Asymmetric Graph Path TSP

Asymmetric Graph Path TSP is the special case of Asymmetric Path TSP where $c(e) = 1$ for all $e \in E$. In this section, we will prove a constant upper bound on the integrality ratio for such unit-weight instances. Our proof will also yield a slightly better approximation ratio than what is implied by the Feige–Singh reduction (Theorem 9.8) combined with Theorem 6.12. We start with a better black-box reduction:

Theorem 9.11 (Köhne, Traub, and Vygen [2020], Traub [2020a]). *Let ρ be the integrality ratio of the Asymmetric Graph TSP LP* (6.1). *Then the integrality ratio of* (9.1) *for the Asymmetric Graph Path TSP is at most* $2\rho - 1$.

Suppose there exists an α-approximation algorithm for Asymmetric Graph TSP. Then there is a $(2\alpha - 1)$-approximation algorithm for Asymmetric Graph Path TSP.

Proof. For an instance $\mathcal{I} = (G, s, t)$ of Asymmetric Graph Path TSP, we first compute a topological order $V_1, \ldots, V_l$ of the strongly connected components of G. By Lemma 9.9, no s-t-tour can use an edge from V_i to V_j for any $j \notin \{i, i+1\}$; moreover, such edges cannot belong to the support of any LP solution. Hence, we delete such edges, which does not change $\mathrm{OPT}(\mathcal{I})$ (the minimum cardinality of an s-t-tour) or $\mathrm{LP}(\mathcal{I})$ (the value of (9.1)).

Let the instance $\mathcal{I}' = G'$ of Asymmetric Graph TSP result from this graph by adding a t-s-path P_{back} of length $n - 1$, where all inner vertices of this path are new vertices not contained in G; here, n is the number of vertices of G. Then $\mathrm{LP}(\mathcal{I}') \leq \mathrm{LP}(\mathcal{I}) + (n - 1)$ and $\mathrm{OPT}(\mathcal{I}') \leq \mathrm{OPT}(\mathcal{I}) + (n - 1)$, where $\mathrm{LP}(\mathcal{I}')$ and $\mathrm{OPT}(\mathcal{I}')$ denote the LP value of (3.2) and the minimum cardinality of a tour in instance $\mathcal{I}'$.

If we have a tour R in G' with at most $\alpha \cdot \mathrm{OPT}(\mathcal{I}')$ or $\rho \cdot \mathrm{LP}(\mathcal{I}')$ edges, we remove the copies of P_{back} and obtain r (at least one) s-t-walks of total cost at most $\alpha \cdot \mathrm{OPT}(\mathcal{I}') - r(n - 1)$ or $\rho \cdot \mathrm{LP}(\mathcal{I}') - r(n - 1)$. Note that each of them visits all the sets $V_1, \ldots, V_l$ in this order (we have removed edges that skip sets V_i).

Next, we iteratively replace two of the walks to a single s-t-walk until only one remains. We merge two s-t-walks P_1 and P_2 to a single s-t-walk P that contains all vertices of P_1 and P_2 as follows (see Figure 9.5). For odd i, we follow P_1 from the first vertex that P_1 visits in V_i until the last vertex that P_1 visits in V_i, then add a path to the first vertex that P_2 visits in V_i, then follow P_2 until the last vertex in V_i. If $i < l$, we add the edge of P_2 that goes from V_i to V_{i+1}. For even i, we exchange the roles of P_1 and P_2.

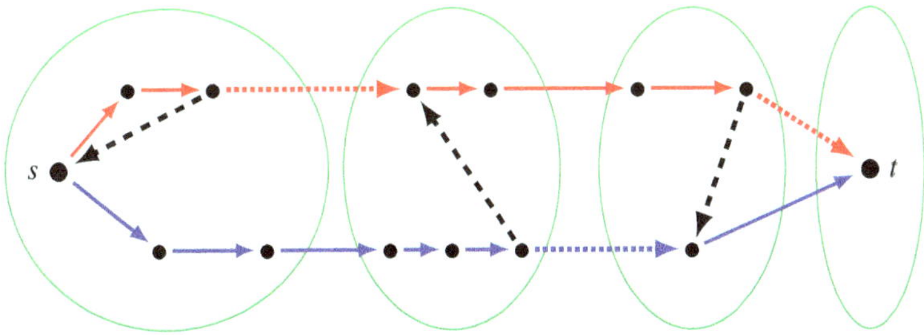

Figure 9.5 Construction of the s-t-walk P in the proof of Theorem 9.11. The s-t-walks P_1 and P_2 are shown with solid and dotted lines. (Here, P_1 is the red walk at the top, and P_2 is shown in blue at the bottom.) The vertex sets $V_1, \ldots, V_l$ of the strongly connected components are indicated by the green ellipses. The red and blue solid edges of the walks P_1 and P_2 are those that are used in the walk P. The dashed black arrows indicate the paths within the sets V_i that we add. This picture is taken from Köhne, Traub, and Vygen [2020] (with permission from Springer Nature).

The walk P that we constructed consists of edges of P_1, edges of P_2, and vertex-disjoint paths, one in each strongly connected component $G[V_i]$. Hence, $|E(P)| \leq |E(P_1)| + |E(P_2)| + (n-1)$.

Iterating until only one walk remains yields an s-t-tour of total cost at most

$$\alpha \cdot \mathrm{OPT}(\mathcal{I}') - (n-1) \;\leq\; \alpha \cdot \mathrm{OPT}(\mathcal{I}) + (\alpha - 1)(n-1)$$

or at most

$$\rho \cdot \mathrm{LP}(\mathcal{I}') - (n-1) \;\leq\; \rho \cdot \mathrm{LP}(\mathcal{I}) + (\rho - 1)(n-1).$$

This completes the proof because $n - 1 \leq \mathrm{LP}(\mathcal{I}) \leq \mathrm{OPT}(\mathcal{I})$. □

We conclude:

Corollary 9.12. *The integrality ratio of* (9.1) *for* Asymmetric Graph Path TSP *is at most* 15. *For every* $\varepsilon > 0$, *there is a* $(15 + \varepsilon)$*-approximation algorithm for* Asymmetric Graph Path TSP.

Proof. By Theorem 6.12, there is an $(8 + \frac{\varepsilon}{2})$-approximation algorithm for Asymmetric Graph TSP, and the integrality ratio of (6.1) is at most 8. Applying Theorem 9.11 completes the proof. □

Table 9.2 summarizes the history and the state of the art.

Table 9.2 *Approximation ratios and upper bounds on the integrality ratio of* (9.1) *for* Asymmetric Graph Path *TSP in the order of their discovery. Only results that are better than for general* Asymmetric Path *TSP are shown. Again,* ε *stands for an arbitrarily small positive constant.*

Approximation Ratio	Integrality Ratio	Year	Reference	Chapter
$54 + \varepsilon$	–	2015	Svensson [2015]	6, 9.2
–	25	2018	Köhne, Traub, and Vygen [2020]	6, 9.6
$25 + \varepsilon$	25	2019	Traub [2020a]	6, 9.4
$15 + \varepsilon$	15	2021	this book	6, 9.4

9.5 Reducing to Strongly Laminar Instances

Similar to Asymmetric TSP, we now also define strongly laminar instances of Asymmetric Path TSP. We will reduce general instances to strongly laminar instances, albeit this will here lose a factor of 2. This section is based on Traub [2020a] and Köhne, Traub, and Vygen [2020].

Definition 9.13 (strongly laminar Asymmetric Path TSP instance)**.** A *strongly laminar* Asymmetric Path TSP *instance* is a tuple $(G, s, t, \mathcal{L}, x, y)$, where

- $G = (V, E)$ is a directed graph,
- $s, t \in V$ with $s \neq t$,
- $\mathcal{L}$ is a laminar family of subsets of $V \setminus \{s, t\}$ such that $G[U]$ is strongly connected for all $U \in \mathcal{L}$,
- x is a feasible solution to (9.1) such that $x(\delta(U)) = 2$ for all $U \in \mathcal{L}$ and $x_e > 0$ for all $e \in E$, and
- $y : \mathcal{L} \to \mathbb{R}_{\geq 0}$.

This induces the Asymmetric Path TSP instance (G, c^y, s, t), where c^y is the *induced cost function* defined by $c^y(e) := \sum_{U \in \mathcal{L}: e \in \delta(U)} y_U$ for $e \in E$.

A solution to a strongly laminar Asymmetric Path TSP instance $\mathcal{I}$ is a solution to its induced Asymmetric Path TSP instance $\mathcal{I}' = (G, c^y, s, t)$. Note that for the induced instance, by complementary slackness (Corollary 4.2), x is an optimum solution to (9.1), and $(0, y)$ is an optimum solution to (9.3). We define $\text{LP}(\mathcal{I}) := c^y(x) = \sum_{L \in \mathcal{L}} 2y_L$ (the equality holds by the LP Duality Theorem 4.1).

The goal of this section is to give a reduction of general instances of ASYMMETRIC PATH TSP to strongly laminar ones. However, in contrast to ASYMMETRIC TSP, this reduction will lose a factor 2 in the approximation guarantees and integrality ratios. More precisely, if the integrality ratio for strongly laminar instances is at most ρ, then the integrality ratio for general instances is at most $2\rho - 1$. The same holds for approximation ratios (see Theorem 9.18).

This works as follows. For an instance $\mathcal{I}$ of ASYMMETRIC PATH TSP, let x and y be optimum primal and dual LP solutions (to (9.1) and (9.3), respectively). We may assume that the support of y is strongly laminar (cf. Lemma 9.10). Delete all edges outside the support of x so that $c(e) = c^y(e) + a_w - a_v$ for every edge $e = (v, w)$ by complementary slackness. Then for every feasible solution x' to (9.1), we have

$$\begin{aligned} c(x') &= \sum_{e=(v,w)\in E} \left(c^y(e) + a_w - a_v\right) x'_e \\ &= \sum_{e\in E} c^y(e)x'_e + \sum_{v\in V} a_v \left(x'(\delta^-(v)) - x'(\delta^+(v))\right) \\ &= c^y(x') + a_t - a_s. \end{aligned}$$

Now if F is an s-t-tour with $c^y(F) \le \gamma \cdot c^y(x)$, then

$$\begin{aligned} c(F) &= c^y(F) + a_t - a_s \\ &\le \gamma \cdot c^y(x) + a_t - a_s \\ &= \gamma \cdot (c(x) + a_s - a_t) + a_t - a_s \\ &= \gamma \cdot c(x) + (\gamma - 1)(a_s - a_t). \end{aligned}$$

Hence, the key step will be to bound $a_s - a_t$.

In the rest of this section, we show that there always exists an optimum dual solution with $a_s - a_t \le \text{LP}$, where we again denote by LP the value of the primal and dual LPs. This important lemma will be exploited again in Section 9.6. However, Figure 9.6 shows that we cannot bound $a_s - a_t$ by LP for an arbitrary optimum dual solution (a, y). Thus, we will work with an optimum dual solution (a, y) with $a_s - a_t$ minimum. Note that this minimum is attained because for every feasible dual solution (a, y), we have $\text{LP} \ge a_t - a_s + \sum_{\emptyset \ne U \subseteq V\setminus\{s,t\}} 2y_U \ge a_t - a_s$ and hence $a_s - a_t \ge -\text{LP}$.

The following lemma describes an important property of optimum dual LP solutions with minimum $a_s - a_t$. We will later show that this property implies $a_s - a_t \le \text{LP}$.

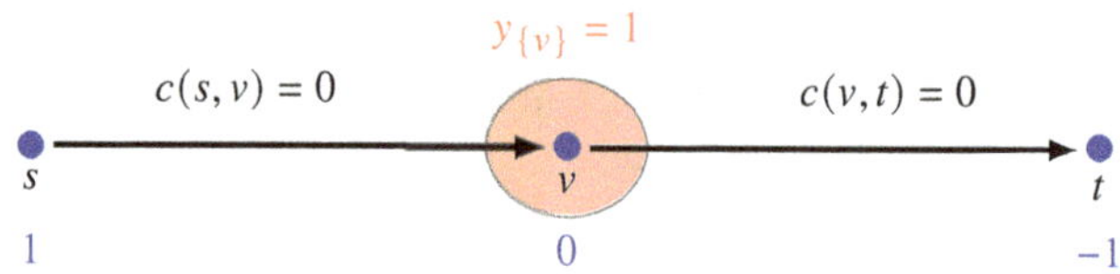

Figure 9.6 Example of an instance with LP = 0 and an optimum dual solution with $a_s - a_t = 2$. The blue numbers below the vertices show the dual variables $a_s = 1$, $a_v = 0$, and $a_t = -1$. Of course, this instance has another optimum dual solution in which all variables are zero. This picture is taken from Köhne, Traub, and Vygen [2020] (with permission from Springer Nature).

Lemma 9.14. *Let (G, c, s, t) be an* Asymmetric Path TSP *instance, where G is the support graph of an optimum solution to* (9.1)*. Moreover, let (a, y) be an optimum solution to* (9.3) *such that $a_s - a_t$ is minimum. Let $\bar{U} \subseteq V \setminus \{s,t\}$ such that every s-t-path in G enters (and leaves) $\bar{U}$ at least once. Then $y_{\bar{U}} = 0$.*

Proof. Suppose $y_{\bar{U}} > 0$ and let $\varepsilon := y_{\bar{U}}$. Let R be the set of vertices reachable from s in $G - \bar{U}$ (the subgraph of G induced by $V \setminus \bar{U}$). We define a dual solution $(\bar{a}, \bar{y})$ as follows:

$$\bar{y}_U := \begin{cases} y_U - \varepsilon & \text{if } U = \bar{U} \\ y_U & \text{else} \end{cases}$$

$$\bar{a}_v := \begin{cases} a_v - \varepsilon & \text{if } v \in R \\ a_v & \text{if } v \in \bar{U} \\ a_v + \varepsilon & \text{else.} \end{cases}$$

See Figure 9.7 for an illustration.

We claim that $(\bar{a}, \bar{y})$ is an optimum (and feasible) solution to (9.3). Note that $t \in V \setminus (R \cup \bar{U})$ and thus $\bar{a}_t = a_t + \varepsilon$. Since $s \in R$, we have $\bar{a}_s - \bar{a}_t < a_s - a_t$. Thus, if $(\bar{a}, \bar{y})$ is indeed optimum (and feasible), we obtain a contradiction to our choice of the dual solution (a, y).

First, we observe that $(\bar{a}, \bar{y})$ and (a, y) have the same objective value since

$$\bar{a}_t - \bar{a}_s + \sum_{\emptyset \neq U \subseteq V \setminus \{s,t\}} 2\bar{y}_U = (a_t + \varepsilon) - (a_s - \varepsilon) + \sum_{\emptyset \neq U \subseteq V \setminus \{s,t\}} 2y_U - 2\varepsilon.$$

By our choice of ε, the vector $\bar{y}$ will be nonnegative.

Now consider an edge $e = (v, w) \in E(G)$. We need to show that

$$\bar{a}_w - \bar{a}_v + \sum_{U: e \in \delta(U)} \bar{y}_U \le c(e).$$

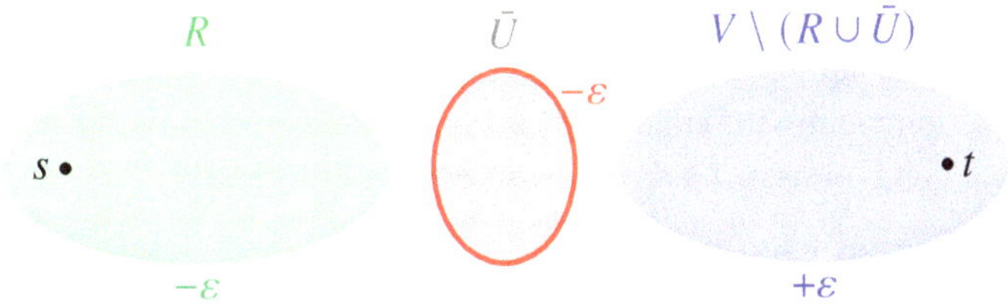

Figure 9.7 Modifying the dual solution in the proof of Lemma 9.14. The green and blue numbers in the bottom indicate the change of the dual variables corresponding to the vertices. In red, the decrease of the variable $y_{\bar{U}}$ is indicated. There is no edge from R to $V \setminus (R \cup \bar{U})$. This picture is taken from Köhne, Traub, and Vygen [2020] (with permission from Springer Nature).

The only edges for which $\bar{a}_w - \bar{a}_v + \sum_{U:e\in\delta(U)} \bar{y}_U$ is greater than $a_w - a_v + \sum_{U:e\in\delta(U)} y_U$ (which is at most $c(e)$) are those from R to $V \setminus (R \cup \bar{U})$. However, such edges do not exist by definition of R.

This shows that $(\bar{a}, \bar{y})$ is an optimum dual solution and $\bar{a}_s - \bar{a}_t < a_s - a_t$, a contradiction. Hence, $y_{\bar{U}} = 0$. □

We will need the following variant of Menger's theorem (Menger [1927], cf. Exercise 2.2).

Lemma 9.15. *Let $G = (V, E)$ be a directed graph and $s, t \in V$ such that t is reachable from s in G. Let $U \subseteq V \setminus \{s, t\}$ such that for every vertex $u \in U$, there exists an s-t-path in $G - u$. Then there exist two s-t-paths P_1 and P_2 in G such that no vertex $u \in U$ is contained in both P_1 and P_2.*

Proof. We construct a graph G' that arises from G by processing the vertices $u \in U$ one by one as follows: We split u into two vertices u^- and u^+ that are connected by an edge $e_u := (u^-, u^+)$; moreover, entering edges (v, u) are replaced by (v, u^-), and leaving edges (u, v) are replaced by (u^+, v).

In the graph G', we now define integral edge capacities. Every edge e_u for $u \in U$ has capacity one. All other edges (i.e., all edges corresponding to edges of G) have infinite capacity.

Since for every vertex $u \in U$, there exists an s-t-path in $G - u$, we conclude that for every $u \in U$, there exists an s-t-path in $G' - e_u$. Thus, the minimum capacity of an s-t-cut in G' is at least two. Hence, by Theorem 2.5 and Corollary 2.6, there exists an integral s-t-flow of value 2 in G' with the defined edge capacities. This flow can be decomposed into two s-t-paths, P'_1 and P'_2, and possibly some cycles (like in Proposition 3.7). By the choice of the edge capacities, no edge e_u for $u \in U$ occurs in both paths. Since the edge e_u is the only outgoing edge of u^- and the only incoming edge of u^+, an s-t-path containing u^- or u^+ must contain e_u, and at most one of P'_1 and P'_2 can do so.

Hence, contracting the edges e_u (for $u \in U$) yields two s-t-paths P_1 and P_2 in G such that no vertex $u \in U$ is contained in both P_1 and P_2. □

We will now continue to work with a dual solution (a, y) that minimizes $a_s - a_t$. By Lemma 9.10, we can assume in addition that (a, y) has strongly laminar support. For an illustration of the following lemma, see Figure 9.8.

Lemma 9.16. *Let (G, c, s, t) be an* Asymmetric Path TSP *instance, where G is the support graph of an optimum solution to* (9.1). *Moreover, let (a, y) be an optimum solution to* (9.3) *such that $a_s - a_t$ is minimum and y has strongly laminar support.*

Then G contains two s-t-paths P_1 and P_2 such that for every set U in the support of y, we have $|E(P_1) \cap \delta(U)| + |E(P_2) \cap \delta(U)| \le 2$.

Proof. Let $\mathcal{L}$ denote the support of y. By Lemma 9.14, for every set $U \in \mathcal{L}$, there is an s-t-path in G that visits no vertex in U. We contract each maximal set of $\mathcal{L}$. Using Lemma 9.15, we can find two s-t-paths such that no vertex arising from the contraction of a set is visited by both paths.

Now we revert the contraction operations. We complete the edge sets of the two s-t-paths we found before (which are not necessarily connected anymore after undoing the contraction) to paths P_1 and P_2 with the desired properties. To see that this is possible, let v be the end vertex of the edge e^{in} entering a contracted set U, and let w be the start vertex of the edge e^{out} leaving U. By Lemma 7.5, there is a nice v-w-path $P_{v,w}$. This path is completely contained in $G[U]$ and enters and leaves every set $U' \in \mathcal{L}$ with $U' \subsetneq U$ at most once. Moreover, if $e^{\text{in}} \in \delta^-(U')$, then $v \in U'$ and the nice v-w-path $P_{v,w}$ does not enter U'. Similarly, if $e^{\text{out}} \in \delta^+(U')$, then $w \in U'$ and $P_{v,w}$ never leaves U'. □

We finally show the key lemma of this section.

Lemma 9.17 (Köhne, Traub, and Vygen [2020]). *Let $I = (G, c, s, t)$ be an* Asymmetric Path TSP *instance, where G is the support graph of an optimum solution to* (9.1). *Let* LP *denote the value of* (9.1). *Then there is an optimum solution (a, y) to* (9.3) *with strongly laminar support and $a_s - a_t \le$* LP.

Proof. Let (a, y) be an optimum solution to (9.3) that has strongly laminar support and minimum $a_s - a_t$. Note that such an optimum dual solution exists by Lemma 9.10.

As before, we define the c^y cost of an edge $e = (v, w)$ to be

$$c^y(e) \;=\; \sum_{U: e \in \delta(U)} y_U \;=\; c(e) + a_v - a_w. \tag{9.4}$$

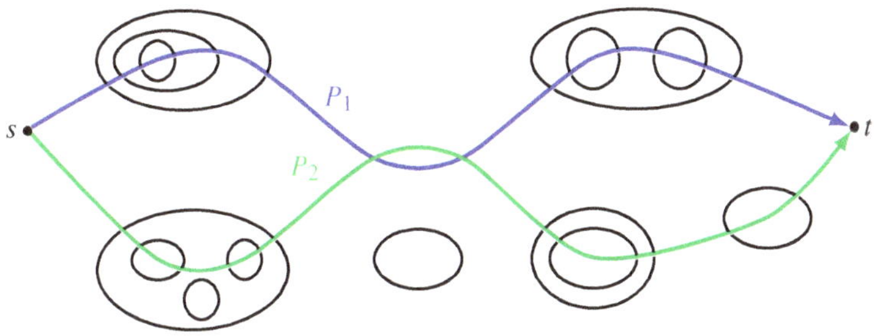

Figure 9.8 The paths P_1 and P_2 as in Lemma 9.16. In black, the vertex sets in the support of y are shown. The s-t-paths P_1 and P_2 are not necessarily internally disjoint, but they never both visit the same set U with $y_U > 0$. This picture is taken from Köhne, Traub, and Vygen [2020] (with permission from Springer Nature).

By Lemma 9.16, G contains two s-t-paths P_1 and P_2 such that

$$c^y(E(P_1)) + c^y(E(P_2)) \leq \sum_{\emptyset \neq U \subseteq V \setminus \{s,t\}} 2y_U.$$

Then, using (9.4),

$$\begin{aligned} 0 &\leq c(E(P_1)) + c(E(P_2)) \\ &= \sum_{e=(v,w)\in E(P_1)} (c^y(e) + a_w - a_v) + \sum_{e=(v,w)\in E(P_2)} (c^y(e) + a_w - a_v) \\ &= c^y(E(P_1)) - (a_s - a_t) + c^y(E(P_2)) - (a_s - a_t) \\ &\leq \sum_{\emptyset \neq U \subseteq V \setminus \{s,t\}} 2y_U - 2(a_s - a_t), \end{aligned}$$

implying

$$a_s - a_t \leq \sum_{\emptyset \neq U \subseteq V \setminus \{s,t\}} 2y_U - (a_s - a_t) = \text{LP}.$$

□

This bound is tight – for example, for the instance in Figure 9.3 (Exercise 9.12), in which the integrality ratio is arbitrarily close to the best-known lower bound of 2.

We remark (although we will not need it) that the above proof can be adapted to show that Lemma 9.17 also holds for general instances of Asymmetric Path TSP; this is Exercise 9.10. Moreover, Exercise 9.11 gives a characterization of the minimum value of $a_s - a_t$.

Now we obtain the main result of this section:

Theorem 9.18 (Traub [2020a]). *Let $\alpha \geq 1$. Suppose there is a polynomial-time algorithm that computes a solution of cost at most $\alpha \cdot \mathrm{LP}(\mathcal{I}')$ for any given strongly laminar* Asymmetric Path TSP *instance $\mathcal{I}'$. Then there is a polynomial-time algorithm that computes a solution of cost at most $(2\alpha - 1)$ times the optimum value of* (9.1) *for any given instance $\mathcal{I}$ of* Asymmetric Path TSP.

If ρ is the integrality ratio of (9.1) *restricted to strongly laminar instances, then the integrality ratio of* (9.1) *(for general instances) is at most $2\rho - 1$.*

Proof. Given an Asymmetric Path TSP instance $\mathcal{I} = (G, c, s, t)$, first compute an optimum solution x^* to (9.1). We may assume that the given graph $G = (V, E)$ is the support graph of x^*, so $x^*_e > 0$ for all $e \in E$. (This is because omitting edges e with $x^*_e = 0$ does not change the LP value and can only increase the cost of an optimum integral solution.) Moreover, let (a, y) be an optimum solution to (9.3) with strongly laminar support $\mathcal{L}$ and $a_s - a_t \leq c(x^*)$; this is guaranteed to exist by Lemma 9.17.

Since $x^*_e > 0$ for all $e \in E$, we have $c(e) = a_w - a_v + \sum_{U: e \in \delta(U)} y_U$ for all $e = (v, w) \in E$ by complementary slackness (Corollary 4.2). Then $\mathcal{I}' = (G, s, t, \mathcal{L}, x^*, y)$ is a strongly laminar Asymmetric Path TSP instance with the induced cost function c^y, where for $e = (v, w) \in E$, we have $c^y(e) = \sum_{U: e \in \delta(U)} y_U = c(e) + a_v - a_w$.

Now, for every feasible solution x to (9.1), we have

$$c(x) \;=\; \sum_{e \in E} x_e \cdot c(e) \;=\; \sum_{e=(v,w) \in E} x_e \cdot (a_w - a_v + c^y(e)) \;=\; c^y(x) + a_t - a_s,$$

where we used that x is an s-t-flow of value 1. Now let $\gamma \geq 1$ and let x' be a feasible solution to (9.1) such that $c^y(x') \leq \gamma \cdot c^y(x^*)$. Then

$$\begin{aligned} c(x') &= c^y(x') + a_t - a_s \\ &\leq \gamma \cdot c^y(x^*) + a_t - a_s \\ &= \gamma \cdot (c(x^*) + a_s - a_t) + a_t - a_s \\ &= \gamma \cdot c(x^*) + (\gamma - 1) \cdot (a_s - a_t) \\ &\leq \gamma \cdot c(x^*) + (\gamma - 1) \cdot c(x^*). \end{aligned}$$

This now implies the theorem as follows. For $\gamma = \alpha$ and x' the incidence vector of a solution to the strongly laminar instance $\mathcal{I}'$ with $c^y(x') \leq \alpha \cdot c^y(x^*)$, we obtain the claimed bound on the approximation ratio with respect to (9.1).

Moreover, for $\gamma = \rho$ and x' the incidence vector of an optimum solution to the strongly laminar instance $\mathcal{I}'$, we obtain the claimed bound on the integrality ratio. ◻

9.6 Another Black-Box Reduction to Asymmetric TSP

In this section, we give another black-box reduction. In contrast to the one presented in Section 9.2, the one presented here will show that the integrality ratio for Asymmetric Path TSP is less than four times the integrality ratio for Asymmetric TSP. This result, due to Köhne, Traub, and Vygen [2020], was the first proof that the integrality ratio of the LP relaxation (9.1) of Asymmetric Path TSP is constant. With the preparations from Section 9.5, the proof is now easy.

The following procedure is similar to one step ("inducing on a tight set") of the approximation algorithm for Asymmetric TSP by Svensson, Tarnawski, and Végh [2020]. For strongly laminar instances, it yields a better bound than Lemma 9.6.

Lemma 9.19. *Let $\mathcal{I} = (G, s, t, \mathcal{L}, x, y)$ be a strongly laminar Asymmetric Path TSP instance. Let P_1 and P_2 be s-t-walks in G with total cost L. Then there is a single s-t-walk P in G that contains every vertex of P_1 and P_2 and has cost at most $L + \mathrm{LP}(\mathcal{I})$.*

Proof. We proceed similarly to the proof of Theorem 9.11. Let $V_1, \ldots, V_l$ be the vertex sets of the strongly connected components of G in topological order. Let P_i^j be the section of P_i that visits vertices in V_j (for $i = 1, 2$ and $j = 1, \ldots, l$). By Lemma 9.9 (and since G is the support graph of x by Definition 9.13), none of these sections of P_i is empty. (Such a section might consist of a single vertex and no edges, but it has to contain at least one vertex.)

Because $G[L]$ is strongly connected for every $L \in \mathcal{L}$ and the sets V_i for $i \in \{1, \ldots, l\}$ are the vertex sets of the strongly connected components of G, we have that $\mathcal{L}' := \mathcal{L} \cup \{V_1, \ldots, V_l\}$ is a laminar family (by Proposition 6.3). Moreover, $G[L]$ is strongly connected for every set $L \in \mathcal{L}'$. Therefore, we can apply Lemma 7.5 to the laminar family $\mathcal{L}'$.

We consider nice paths R^j in G for $j = 1, \ldots, l$ that we will use to connect the walks P_1^j and P_2^j to a single walk visiting all vertices in V_j. See Figure 9.5. If j is odd, let R^j be a nice path from the last vertex of P_1^j to the first vertex of P_2^j. If j is even, let R^j be a nice path from the last vertex of P_2^j to the first vertex of P_1^j. (Such paths exist by Lemma 7.5.)

We now construct our s-t-walk P that will visit every vertex of P_1 and P_2. We start by setting $P = s$ and then add for $j = 1, \ldots, l$ all the vertices in V_j to P as follows. If j is odd, we append P_1^j and R^j and then P_2^j. If j is even, we append P_2^j and R^j and then P_1^j. Note that when moving from one connected component V_j to the next component V_{j+1}, we use an edge from either P_1 (if j

is even) or P_2 (if j is odd). Then P is, indeed, an s-t-walk in G and contains every vertex of P_1 and P_2.

We now bound the cost of the walk P. It is constructed from pieces of P_1 and P_2 and the paths R^j. Each of the paths R^j can only contain vertices of V_j. Two paths R^j and $R^{j'}$, such that $j \neq j'$, can never both enter or both leave the same element of the laminar family $\mathcal{L}$ because otherwise, they would contain vertices of the same strongly connected component of G. Thus every element of $\mathcal{L}$ is entered at most once and left at most once on all the paths R^j used in the construction of P, and the total cost of these paths is at most $\sum_U 2y_U = \mathrm{LP}(\mathcal{I})$. Consequently, we have $c^y(E(P)) \leq L + \mathrm{LP}(\mathcal{I})$ as claimed. □

Our reduction first transforms an instance and a solution to (9.1) to an instance and a solution to (3.2) and then works with an integral solution to this Asymmetric TSP instance. The proof of the following lemma is similar to the proofs of Theorems 9.8 and 9.11.

Lemma 9.20 (Traub [2020a]). *Let ρ be the integrality ratio of* (3.2). *Then the integrality ratio of* (9.1) *restricted to strongly laminar instances is at most* $2\rho - 1$.

Proof. Let $\mathcal{I} = (G, s, t, \mathcal{L}, x, y)$ be a strongly laminar Asymmetric Path TSP instance. Consider the strongly laminar Asymmetric TSP instance $\mathcal{I}' = (G', \mathcal{L}', x, y)$ that arises from $\mathcal{I}$ as follows. We add a new vertex v to G and two edges (t, v) and (v, s); we set $x_{(t,v)} = x_{(v,s)} = 1$. Moreover, we add the set $\{v\}$ to $\mathcal{L}$ and set $y_{\{v\}} = \frac{1}{2} \cdot \mathrm{LP}(\mathcal{I})$. Then $\mathrm{LP}(\mathcal{I}') = 2 \cdot \mathrm{LP}(\mathcal{I})$. Hence, there is a tour (a solution to $\mathcal{I}'$) with cost at most $2\rho \cdot \mathrm{LP}(\mathcal{I})$.

Let R be such a solution. Then R has to use (t, v) and (v, s) at least once, since it has to visit v. By deleting all copies of (t, v) and (v, s) from R, we get $r \geq 1$ s-t-walks in G with total cost at most $2\rho \cdot \mathrm{LP}(\mathcal{I}) - r \cdot \mathrm{LP}(\mathcal{I})$ such that every vertex of G is visited by at least one of them. As long as $r > 1$, by Lemma 9.19, we can replace two of the s-t-walks by a single one, increasing the cost by at most $\mathrm{LP}(\mathcal{I})$ and decreasing r by one. We end up with a single s-t-walk P in G that contains every vertex of G and has cost $c^y(E(P)) \leq 2\rho \cdot \mathrm{LP}(\mathcal{I}) - \mathrm{LP}(\mathcal{I})$. This walk is an s-t-tour (a solution to $\mathcal{I}$), and thus the integrality ratio of (9.1) restricted to strongly laminar instances is at most $2\rho - 1$. □

We will now prove the main result of this section.

Theorem 9.21 (Köhne, Traub, and Vygen [2020]). *Let ρ be the integrality ratio of* (3.2). *Then the integrality ratio of* (9.1) *is at most* $4\rho - 3$.

Proof. By Lemma 9.20, the integrality ratio of (9.1) restricted to strongly laminar ASYMMETRIC PATH TSP instances is at most $2\rho - 1$. By Theorem 9.18, this implies that the integrality ratio of (9.1) is at most $4\rho - 3$. □

9.7 Algorithms for Asymmetric Path TSP

In this section, largely based on Traub [2020a], we show that the integrality ratio of (9.1) is at most 17. Here we do not use a black-box reduction to ASYMMETRIC TSP anymore, but instead we use the algorithm for vertebrate pairs from Section 7.5.

First we show an approximation algorithm for strongly laminar instances with the same ratio as for ASYMMETRIC TSP. By Theorem 9.18, this yields a slightly better approximation ratio for ASYMMETRIC PATH TSP than applying the Feige–Singh reduction (Theorem 9.8): For $\kappa = 2$ and $\eta = 9 + \varepsilon$ (which we have by Theorems 8.24 and 7.18), we get an approximation ratio $17 + \varepsilon$ for strongly laminar ASYMMETRIC PATH TSP instances and $33 + \varepsilon$ for general instances (instead of $34 + \varepsilon$ by the Feige–Singh reduction). At the end of this section, we will improve on this further by dealing directly with general instances (see Theorem 9.23).

Theorem 9.22 (Traub [2020a]). *Let $\kappa, \eta \geq 0$ with $\kappa + \eta \geq 1$. Suppose we have a (κ, η)-algorithm for vertebrate pairs. Then there is a polynomial-time algorithm that computes a solution of cost at most*

$$(3\kappa + \eta + 2) \cdot \mathrm{LP}(\mathcal{I})$$

for any given strongly laminar ASYMMETRIC PATH TSP instance.

Proof. Given an instance $\mathcal{I} = (G, s, t, \mathcal{L}, x, y)$, let $V_1, \ldots, V_l$ be the vertex sets of the strongly connected components of G numbered in topological order. We construct a strongly laminar ASYMMETRIC TSP instance $\mathcal{I}' = (G', \mathcal{L}', x', y')$. Let G' arise from $G = (V, E)$ by adding a new vertex v and two edges (t, v) and (v, s). We set $x'_{(t,v)} = x'_{(v,s)} = 1$ and $x'_e = x_e$ for all $e \in E$. Then x' is a feasible solution to (3.2). We define $\mathcal{L}' = \mathcal{L} \cup \{V_1, \ldots, V_l\}$. We set $y'_{V_i} = 0$ for all $i \in \{1, \ldots, l\}$ with $V_i \notin \mathcal{L}$, and $y'_L = y_L$ for $L \in \mathcal{L}$. The reason that we add the sets V_i to $\mathcal{L}'$ is that we will soon call Algorithm 7.8 for $W = V_i$ $(i = 1, \ldots, l)$.

We claim that $\mathcal{I}'$ is a strongly laminar ASYMMETRIC TSP instance. First, we observe that $\emptyset \neq L \subsetneq V$ and $G'[L]$ is strongly connected for all $L \in \mathcal{L}'$ by construction and that G' is strongly connected by Lemma 9.9. Moreover, by Lemma 9.9, we have $x'(\delta(V_i)) = 2$ for all $i \in \{1, \ldots, l\}$. Furthermore, for

$L \in \mathcal{L}$, we have $s, t \notin L$ and hence $x'(\delta(L)) = x(\delta(L)) = 2$. Moreover, $\mathcal{L}'$ is a laminar family by Proposition 6.3 (applied to G). This shows that $\mathcal{I}'$ is a strongly laminar Asymmetric TSP instance.

By Lemmas 7.9 and 7.11, we can compute in polynomial time a tour F_i in $G[V_i]$ of cost at most

$$c^y(F_i) = c^{y'}(F_i) \leq (2\kappa + 2) \cdot \text{value}(V_i) + (\kappa + \eta) \cdot (\text{value}(V_i) - D_{V_i}) \quad (9.5)$$

for every $i \in \{1, \ldots, l\}$.

By Definition 9.13, we have $x_e > 0$ for all $e \in E$. Therefore, by Lemma 9.9, any s-t-path in G visits every strongly connected component $G[V_i]$ of G exactly once. Hence, the union of the edge set of any s-t-path (for example, a shortest s-t-path P) and $F_1 \,\dot\cup \ldots, \dot\cup\, F_l$ is an s-t-tour F. We now bound the cost of P. To this end, we consider the s-t-path P' that results from P by replacing, for each $i = 1, \ldots, l$, the maximal subpath of P within V_i, say from u_i to v_i, by the nice u_i-v_i-path P_{u_i,v_i}, which satisfies

$$\sum_{L \in \mathcal{L}: u_i \in L \subsetneq V_i} y_L + \sum_{L \in \mathcal{L}: v_i \in L \subsetneq V_i} y_L + c^y(E(P_{u_i,v_i})) \leq D_{V_i}$$

(see Section 7.3, in particular (7.1)). We get

$$c^y(E(P)) \leq c^y(E(P')) = c^{y'}(E(P')) \leq \sum_{i=1}^{l} \left(D_{V_i} + y'_{V_i}\right).$$

Therefore, using (9.5), we obtain

$$\begin{aligned} c^y(F) &\leq \sum_{i=1}^{l} \left(D_{V_i} + y'_{V_i} + (2\kappa + 2) \cdot \text{value}(V_i) + (\kappa + \eta) \cdot (\text{value}(V_i) - D_{V_i})\right) \\ &\leq \sum_{i=1}^{l} \left(y'_{V_i} + (3\kappa + \eta + 2) \cdot \text{value}(V_i)\right) \\ &\leq (3\kappa + \eta + 2) \cdot \text{LP}(\mathcal{I}). \end{aligned}$$

Here, we used that $\text{LP}(\mathcal{I}) = \sum_{L \in \mathcal{L}} 2y_L = \sum_{i=1}^{l} (y'_{V_i} + \text{value}(V_i))$. □

By the reduction from general to strongly laminar instances (Theorem 9.18), this yields the ratio $6\kappa + 2\eta + 3$ for general Asymmetric Path TSP instances. In this reduction, we lose essentially a factor of 2, but this is tight only if $a_s - a_t$ equals the LP value, where (a, y) is a solution to the dual LP (9.3) with $a_s - a_t$ minimum and strongly laminar support (see the proof of Theorem 9.18). By the proof of Lemma 9.17, this happens only if there are two s-t-paths that cost nothing and visit every set in the support of y. In this case, however, we will be able to construct an excellent backbone from these two paths.

Exploiting this idea, we obtain the same approximation ratio for general ASYMMETRIC PATH TSP instances as for ASYMMETRIC TSP:

Theorem 9.23. *Let $\kappa, \eta \geq 0$ with $\kappa + \eta \geq 1$. Suppose we have a (κ, η)-algorithm for vertebrate pairs. Then there is a polynomial-time algorithm that computes a solution of cost at most*

$$\max\{3\kappa + \eta + 2,\ 4\kappa + 7\}$$

times the value of (9.1) *for any given ASYMMETRIC PATH TSP instance.*

Proof. Let (G, c, s, t) be an ASYMMETRIC PATH TSP instance with $G = (V, E)$, let x be an optimum solution to (9.1), and let (a, y) be an optimum solution to (9.3) that has strongly laminar support and minimizes $a_s - a_t$. Note that such a dual solution exists by Lemma 9.10; furthermore, x and y can be computed in polynomial time. Let $\mathrm{LP} = c(x)$ denote the value of those LPs. We may assume that $x_e > 0$ for all $e \in E$ because removing edges e with $x_e = 0$ does not change the LP value. Let $V_1, \ldots, V_l$ be the vertex sets of the strongly connected components of G numbered in topological order.

We define a strongly laminar ASYMMETRIC TSP instance $\mathcal{I}' = (G', \mathcal{L}', x', y')$. As in the proof of Theorem 9.22, let G' arise from G by adding a new vertex v and edges (t, v) and (v, s). We set $x'_{(t,v)} = x'_{(v,s)} = 1$ and $x'_e = x_e$ for all $e \in E$. Then x' is a feasible solution to (3.2) for (G', c). We define $\mathcal{L}' = \mathcal{L} \cup \{V_1, \ldots, V_l\}$. We set $y'_{V_i} = 0$ for all $i \in \{1, \ldots, l\}$ with $V_i \notin \mathcal{L}$ and $y'_L = y_L$ for $L \in \mathcal{L}$. Again, $\mathcal{I}'$ is a strongly laminar ASYMMETRIC TSP instance. We again denote by $c^y = c^{y'}$ the induced cost function.

We first construct an s-t-walk Q that is not too expensive but visits many sets $L \in \mathcal{L}$. This walk Q will be part of our final s-t-tour, and it will also be the essential part of the backbone for the vertebrate pairs algorithm.

To construct Q, let P_1 and P_2 be (edge sets of) two s-t-paths as in Lemma 9.16. We merge them to a single walk as follows (cf. Figure 9.9). Let P_i^j be the section of P_i that visits vertices in V_j (for $i = 1, 2$ and $j = 1, \ldots, l$). Since $x_e > 0$ for all $e \in E$, Lemma 9.9 implies that none of these sections of P_i is empty. (Such a section might consist of a single vertex and no edges, but it has to contain at least one vertex.) Let $u_i^*, v_i^* \in V_i$ such that $D_{V_i}(u_i^*, v_i^*) = D_{V_i}$ (see Section 7.3). Now we can apply Lemma 7.5 to the laminar family $\mathcal{L}'$ and construct Q as follows, starting with the single vertex s.

If j is odd, let R^j consist of P_1^j, a nice path from the last vertex of P_1^j to u_i^*, the nice path $P_{u_i^*, v_i^*}$ from u_i^* to v_i^*, a nice path from v_i^* to the first vertex of P_2^j, and P_2^j plus (unless $j = l$) the following edge of P_2 (that enters V_{j+1}). If j is

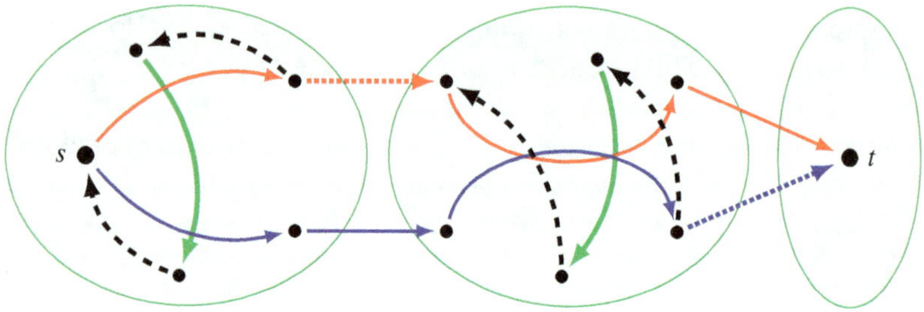

Figure 9.9 Construction of the walk Q. The s-t-paths P_1 (red) and P_2 (blue) are shown with solid and dotted lines. The vertex sets $V_1, \ldots, V_l$ of the strongly connected components are indicated by the green ellipses (here $l = 3$, and V_3 contains only t). The thick green curves illustrate the nice paths $P_{u_i^*, v_i^*}$ that we include in the backbone. The red and blue solid parts of the paths P_i are those that are used for the backbone; the dotted edges are omitted. The dashed black arrows indicate the paths that we add.

even, we swap the roles of P_1 and P_2, starting with P_2^j and ending with P_1^j plus (unless $j = l$) the following edge of P_1 (that enters V_{j+1}).

We constructed an s-t-walk Q with $c^y(Q) \le c^y(P_1) + c^y(P_2) + \sum_{i=1}^{l} 3D_{V_i}$. Our backbone B is the multi-graph that corresponds to this walk Q plus the two auxiliary edges (t, v) and (v, s).

Now consider the variant of Algorithm 7.8 that takes B as the backbone instead of recomputing it in Step (2) (see Lemma 7.13). We apply this to $W = V_i$ for $i = 1, \ldots, l$ and the cost function c^y. We get an Eulerian multi-edge set F_i such that $(V(B), E(B) \cup F_i)[V_i]$ is connected. By Lemma 7.13,

$$c^y(F_i) \;\le\; \kappa \cdot D_{V_i} + \kappa \cdot \mathrm{value}(V_i) + \sum_{L \in \mathcal{L}: L \subsetneq V_i,\, V(B) \cap L = \emptyset} (2\kappa + \eta + 2) \cdot 2y_L$$

for every $i \in \{1, \ldots, l\}$.

Now $\sum_{L \in \mathcal{L}: L \subsetneq V_i,\, V(B) \cap L = \emptyset} 2y_L \le \mathrm{value}(V_i) - D_{V_i}$ because B contains the nice u_i^*-v_i^*-path $P_{u_i^*, v_i^*}$ and $D_{V_i}(u_i^*, v_i^*) = D_{V_i}$. Thus,

$$c^y(F_i) \;\le\; (2\kappa + 3) \cdot \mathrm{value}(V_i) - 3 \cdot D_{V_i} + \sum_{L \in \mathcal{L}: L \subsetneq V_i,\, V(B) \cap L = \emptyset} (\kappa + \eta - 1) \cdot 2y_L$$

for $i = 1, \ldots, l$. Recall that $c^y(P_1) + c^y(P_2) \le \sum_{L \in \mathcal{L}} 2y_L = \mathrm{value}(V)$ by Lemma 9.16. Let $\tau \in [0, 1]$ such that $c^y(P_1) + c^y(P_2) = \tau \cdot \mathrm{value}(V)$. Then $c^y(Q) \le \tau \cdot \mathrm{value}(V) + \sum_{i=1}^{l} 3D_{V_i}$ and

$$\sum_{i=1}^{l} \sum_{\substack{L \in \mathcal{L}: L \subsetneq V_i, \\ V(B) \cap L = \emptyset}} 2y_L \;\le\; \sum_{L \in \mathcal{L}: V(B) \cap L = \emptyset} 2y_L \;=\; \mathrm{value}(V) - \tau \cdot \mathrm{value}(V)$$

because the union of two paths P_1 and P_2 does not enter or leave any set in the support of y more than once.

Hence, the disjoint union of Q and $F_1, \dots, F_l$ is an s-t-tour F of total cost at most

$$\begin{aligned} c(F) &= c^y(F) - (a_s - a_t) \\ &= c^y(Q) + \sum_{i=1}^{l} c^y(F_i) - (a_s - a_t) \\ &\le \tau \cdot \text{value}(V) + (2\kappa + 3) \cdot \text{value}(V) + (\kappa + \eta - 1)(1 - \tau) \cdot \text{value}(V) \\ &\quad - (a_s - a_t) \\ &= \text{LP} + (2\kappa + 3) \cdot \text{value}(V) + (\kappa + \eta - 2)(1 - \tau) \cdot \text{value}(V). \end{aligned}$$

Here we used $\kappa + \eta \ge 1$ in the inequality and $\text{LP} = \text{value}(V) - (a_s - a_t)$ in the last equation. Moreover, $0 \le c(P_1) + c(P_2) = c^y(P_1) + c^y(P_2) - 2(a_s - a_t) = \tau \cdot \text{value}(V) - 2(a_s - a_t) = (\tau - 2)\text{value}(V) + 2\,\text{LP}$ yields $\text{value}(V) \le \frac{2}{2-\tau}\,\text{LP}$. We conclude

$$\begin{aligned} c(F) &\le \text{LP} + \big(2\kappa + 3 + (\kappa + \eta - 2)(1 - \tau)\big) \cdot \text{value}(V) \\ &\le \Big(1 + \tfrac{2}{2-\tau}\big(2\kappa + 3 + (\kappa + \eta - 2)(1 - \tau)\big)\Big) \cdot \text{LP} \\ &= \Big((2\kappa + 2\eta - 3) + \tfrac{2}{2-\tau}(\kappa + 5 - \eta)\Big) \cdot \text{LP}. \end{aligned}$$

For $\eta \ge \kappa + 5$, the right-hand side is maximized for $\tau = 0$ and hence at most $(3\kappa + \eta + 2) \cdot \text{LP}$. For $\eta \le \kappa + 5$, the right-hand side is maximized for $\tau = 1$ and hence at most $(4\kappa + 7) \cdot \text{LP}$. □

Corollary 9.24. *For every $\varepsilon > 0$, there is a $(17 + \varepsilon)$-approximation algorithm for* Asymmetric Path TSP. *The integrality ratio of* (9.1) *is at most* 17.

Proof. By Theorems 8.24 and 7.18, there is a $(2, 9+\varepsilon)$-algorithm for vertebrate pairs. Hence, Theorem 9.23 implies a $(17 + \varepsilon)$-approximation algorithm for Asymmetric Path TSP.

Suppose there is an instance $\mathcal{I}$ of Asymmetric Path TSP where $\frac{\text{OPT}(\mathcal{I})}{\text{LP}(\mathcal{I})} > 17$. Then there exists an $\varepsilon > 0$ such that $\frac{\text{OPT}(\mathcal{I})}{\text{LP}(\mathcal{I})} > 17+\varepsilon$. However, by Theorem 9.23, we can compute an integral solution for $\mathcal{I}$ with cost at most $(17 + \varepsilon) \cdot \text{LP}(\mathcal{I}) < \text{OPT}(\mathcal{I})$, a contradiction. □

We currently have the same upper bound on the integrality ratio for (9.1) as for (3.2). However, we do not know whether the two ratios are the same:

Open Problem 9.25. Does the LP relaxation (9.1) of the ASYMMETRIC PATH TSP have the same integrality ratio as the LP relaxation (3.2) of the ASYMMETRIC TSP?

Exercises

9.1 Consider the variants of ASYMMETRIC PATH TSP where s and/or t are not part of the input but can be chosen among all vertices. Show that all of them are equivalent to ASYMMETRIC PATH TSP.

9.2 Characterize the digraphs G that contain an s-t-tour and describe an $O(n^3)$-time algorithm to decide this.
Note: Even a linear-time algorithm exists.

9.3 Show that the cheapest insertion heuristic from Proposition 9.5 yields a solution of cost at most $\max\{1, (n-3)\} \cdot \mathrm{OPT}$ for any given instance of the ASYMMETRIC PATH TSP WITH TRIANGLE INEQUALITY.

9.4 Show that the approximation ratio of the cheapest insertion heuristic from Proposition 9.5 cannot be bounded by $o(n)$.
(Frieze, Galbiati, and Maffioli [1982])

9.5 Devise a combinatorial $(2\sqrt{n})$-approximation algorithm with running time $O(n^{3.5})$ for the ASYMMETRIC PATH TSP WITH TRIANGLE INEQUALITY.
Hint: First compute a cheapest edge set F such that $F \dot{\cup} \{(t, s)\}$ is a cycle cover, similarly to the proof of Lemma 1.34. If (V, F) has more than $\sqrt{n}$ connected components, buy all edges in F except those on the s-t-path in (V, F), delete all but one vertex from every circuit in (V, F), and iterate (like in Algorithm 1.35). Otherwise, successively pick an arbitrary vertex v from each circuit in (V, F) and insert it into the s-t-path in (V, F) in a cheapest possible way.
(Lam and Newman [2008])

9.6 Let α be a constant. Prove that if there is an α-approximation algorithm for ASYMMETRIC GRAPH TSP, then there is a (2α)-approximation algorithm for ASYMMETRIC GRAPH PATH TSP.
Hint: Guess OPT and note that r in the proof of Theorem 9.8 is now bounded by a constant.

9.7 Extend Exercise 9.6 to show that there is a weakly polynomial-time algorithm for integer weights rather than unit weights. To this end, modify the proof of Theorem 9.8 by doing a kind of binary search as follows. Starting with a lower bound $L = 0$ and an upper bound $U \geq \mathrm{OPT}$, trying $\gamma = L$ and $\gamma = U$ does the job unless we get $r_L > \alpha$ paths for $\gamma = L$ and $r_U < \alpha$ paths for $\gamma = U$. Next, try $\gamma = \lfloor \frac{L+U}{2} \rfloor$. If $r_\gamma > \alpha$, update the lower bound $L := \gamma$; if $r_\gamma < \alpha$, update the upper bound $U := \gamma$. Stop when $r_\gamma = \alpha$ or $U = L + 1$ and output the best solution found.

9.8 Consider the LP (9.1) for instances of the ASYMMETRIC PATH TSP WITH TRIANGLE INEQUALITY, where $E = \{(v, w) \in V \times V : c(v, w) < \infty\}$. Show that adding degree constraints $x(\delta^-(v)) = 1$ for $v \in V \setminus \{s\}$ and $x(\delta^-(s)) = 0$ would not change the value of the LP.

9.9 Let $G = (V, E)$ be an undirected graph and $1 \leq k < |V|$. Show that $G - U$ is connected for all $U \subsetneq V$ with $|U| < k$ if and only if for all $v, w \in V$, there exist k paths from v to w in G such that no pair of these paths shares any vertex except v and w.
Hint: Proceed similarly as in the proof of Lemma 9.15.
(Whitney [1932b])

9.10 Show that Lemma 9.17 also holds for general instances of ASYMMETRIC PATH TSP.
Hint: To adapt the proof, work with the subgraph of G that contains all edges of G for which the dual constraint is tight. This requires adapting the proof of Lemma 9.14.
(Köhne, Traub, and Vygen [2020])

9.11 Let $\mathcal{I} = (G, c, s, t)$ be an ASYMMETRIC PATH TSP instance and let $\gamma \geq 0$. Now consider the instance $\mathcal{I}' = (G + e_{\text{back}}, c, s, t)$ where we add an edge $e_{\text{back}} = (t, s)$ with $c(e_{\text{back}}) := \gamma$. Show that $\text{LP}(\mathcal{I}) = \text{LP}(\mathcal{I}')$ if and only if there exists an optimum solution (a, y) to (9.3) for the instance $\mathcal{I}$ with $a_s - a_t \leq \gamma$.
(Köhne, Traub, and Vygen [2020])

9.12 Show that the instance in Figure 9.3 has no optimum dual LP solution (a, y) with $a_s - a_t < \text{LP}$.

10

Parity Correction of Random Trees

The random sampling approach described in Chapter 5 that yielded the first $o(\log n)$-approximation algorithm for the ASYMMETRIC TSP has also been used successfully for the SYMMETRIC TSP. First, Oveis Gharan, Saberi, and Singh [2011] obtained the first algorithm with approximation ratio less than $\frac{3}{2}$ for GRAPH TSP. More recently, Karlin, Klein, and Oveis Gharan [2021] proved that essentially the same algorithm has approximation ratio less than $\frac{3}{2}$ for the general SYMMETRIC TSP. The algorithm is simple, but its analysis is very complicated. While for GRAPH TSP we know simpler and better algorithms today (see Chapters 12 and 13), the random sampling algorithm is still the best-known approximation algorithm for SYMMETRIC TSP.

The algorithm samples a spanning tree from an (approximately) marginal-preserving λ-uniform distribution and then proceeds with parity correction like Christofides' algorithm. In Section 10.1, we discuss the analysis by Oveis Gharan, Saberi, and Singh [2011] for GRAPH TSP. Then, in the remaining part of this chapter, we present the first part of the analysis by Karlin, Klein, and Oveis Gharan [2021], with some simplifications suggested by Drees [2022]. The main point is to reduce the set of relevant cuts that need to be considered to bound the cost of parity correction and obtain a nice structure that will be exploited in Chapter 11.

10.1 Random Sampling for Graph TSP

In Chapter 5, we studied an algorithm for ASYMMETRIC TSP that samples a spanning tree S from the maximum entropy distribution and extends it to a tour. Essentially the same can be done for SYMMETRIC TSP. This was first analyzed by Oveis Gharan, Saberi, and Singh [2011] for GRAPH TSP, for which they obtained the first improvement upon Christofides' algorithm.

Their analysis is based on the following observation:

Lemma 10.1. *Let (V, c) be an instance of the* Symmetric TSP with Triangle Inequality*, and let $0 < \eta \le 1$. Let S be a random spanning tree, and let x be a feasible solution to the subtour LP (2.2). Call an edge $e \in E = \binom{V}{2}$* good *if e does not belong to any* odd(S)*-cut C with $x(C) \le 2 + \eta$. Then the expected cost of a minimum-cost* odd(S)*-join is at most $\sum_{e \in E}(\frac{1}{2} - \frac{\eta}{6} \cdot \mathbb{P}[e \text{ is good}]) \cdot x_e \cdot c(e)$.*

Proof. We define a vector $y^S \in \mathbb{R}^E_{\ge 0}$ by setting $y^S_e := (\frac{1}{2} - \frac{\eta}{6})x_e$ if e is good and setting $y^S_e := \frac{1}{2}x_e$ otherwise. Then y^S is a feasible solution to the odd(S)-join polyhedron (2.9) because for every cut C with $x(C) \ge 2 + \eta$, we have $y^S(C) \ge (\frac{1}{2} - \frac{\eta}{6}) \cdot x(C) \ge 1$, and for every odd$(S)$-cut C with $x(C) \le 2 + \eta$, we have $y^S(C) = \frac{1}{2}x(C) \ge 1$. Therefore, by Theorem 2.19, a minimum-cost odd(S)-join costs at most $c(y^S)$. The expected cost of y^S is $\sum_{e \in E}(\frac{1}{2} - \frac{\eta}{6} \cdot \mathbb{P}[e \text{ is good}]) \cdot x_e \cdot c(e)$. □

If one could show that there is a sufficiently large set of edges that are good with a constant probability, Lemma 10.1 would imply that for Graph TSP, the expected cost of completing the spanning tree S to a tour is at most $(\frac{1}{2} - \delta)c(x)$ for some constant $\delta > 0$, leading to an improvement over Christofides' algorithm. Oveis Gharan, Saberi, and Singh [2011] showed that for the maximum entropy distribution (see Section 5.4), this is the case unless there are many edges e with x_e very close to 1. (The constants in Theorem 10.2 are not best possible, but the improvement is tiny anyway.)

Theorem 10.2 (Oveis Gharan, Saberi, and Singh [2011]). *Let (V, c) be an instance of the* Symmetric TSP with Triangle Inequality *and $n = |V|$. Let x be a feasible solution to the subtour LP (2.2), and let (V, S) be a spanning tree picked at random according to the maximum entropy distribution μ on $\mathcal{S}$ with $\sum_{S \in \mathcal{S}: e \in S} \mu(S) = \frac{n-1}{n}x_e$ for all $e \in E = \binom{V}{2}$. Call an edge $e \in E$* good *if e does not belong to any* odd(S)*-cut $\delta(U)$ with $x(\delta(U)) \le 2 + 10^{-15}$. Then at least one of the following holds:*

(i) *There is a subset E^* of edges with $x(E^*) \ge 10^{-12}n$ such that for each $e \in E^*$, the probability that e is good is at least 10^{-24}.*
(ii) *There are at least $\frac{19}{20}n$ edges e with $x_e \ge 1 - 10^{-7}$.*

The proof of this theorem is very long and not presented here. It uses deep results about random spanning trees and the structure of near-minimum cuts. These were the first steps towards the improvement for the general Symmetric TSP, which we will present in much more detail shortly.

Following Oveis Gharan, Saberi, and Singh [2011], we now show that Theorem 10.2 implies a better approximation ratio for the Graph TSP:

Corollary 10.3 (Oveis Gharan, Saberi, and Singh [2011]). *There is a randomized* $(\frac{3}{2} - 10^{-53})$*-approximation algorithm for* Graph TSP.

Proof. Let G be a Graph TSP instance. We work in the metric closure (cf. Proposition 1.12) and let x be an optimum solution to (2.2), where $c(v, w) = \mathrm{dist}_G(v, w)$ for all $v, w \in V$.

We first check whether we are in case (ii) of Theorem 10.2. If so, let I be the set of edges e with $x_e \geq 1 - 10^{-7}$. The edges in I form vertex-disjoint paths and circuits. Moreover, each circuit in I has length at least 10^7 (or contains all vertices) because for the vertex set U of such a circuit, we either have $U = V$ or

$$2 \leq x(\delta(U)) = 2|U| - 2 \cdot x(E[U]) \leq 2|U| - 2|U| \cdot (1 - 10^{-7}) = 2 \cdot 10^{-7}|U|.$$

Remove one edge from each circuit in I and add edges of G (each of cost 1) to obtain a spanning tree (V, S). Note that $c(S \setminus I) = |S \setminus I| < (\frac{1}{20} + 10^{-7})n \leq (\frac{1}{20} + 10^{-7})c(x)$. Hence, we can bound the cost of S by

$$c(S) = c(S \cap I) + c(S \setminus I) \leq \sum_{e \in S} \frac{x_e \cdot c(e)}{1 - 10^{-7}} + \left(\tfrac{1}{20} + 10^{-7}\right)c(x).$$

Now we add a minimum-cost odd(S)-join J. To bound $c(J)$, let $y_e := \frac{1}{3}$ for $e \in S$ and $y_e := \frac{2}{3}x_e$ for $e \in \binom{V}{2} \setminus S$. We show that y is a feasible solution to the odd(S)-join polyhedron (2.9). For any vertex set U with $|U \cap \mathrm{odd}(S)|$ odd, we have $|\delta(U) \cap S|$ odd by Lemma 2.20. If $|\delta(U) \cap S| = 1$, then $y(\delta(U)) \geq \frac{1}{3} + y(\delta(U) \setminus S) \geq \frac{1}{3} + \frac{2}{3}(x(\delta(U)) - 1) \geq 1$. If $|\delta(U) \cap S| \geq 3$, then $y(\delta(U)) \geq 3 \cdot \frac{1}{3} = 1$. Hence, y is a feasible solution to the odd(S)-join polyhedron (2.9) and thus by Theorem 2.19, we have $c(J) \leq c(y) = \frac{1}{3}c(S) + \frac{2}{3}\sum_{e \in \binom{V}{2} \setminus S} c(e)x_e$. We conclude

$$\begin{aligned} c(S \mathbin{\dot\cup} J) &\leq \tfrac{4}{3}c(S) + \tfrac{2}{3} \sum_{e \in \binom{V}{2} \setminus S} x_e \cdot c(e) \\ &\leq \tfrac{4}{3} \cdot \tfrac{c(x)}{1 - 10^{-7}} + \tfrac{4}{3} \cdot \left(\tfrac{1}{20} + 10^{-7}\right)c(x) \\ &\leq \left(\tfrac{7}{5} + 10^{-6}\right)c(x). \end{aligned}$$

Now assume that we are not in case (ii), so (i) of Theorem 10.2 must hold for x. Recall that $\frac{n-1}{n}x$ is in the relative interior of the spanning tree polytope of the complete graph on V (see Proposition 2.17). First assume that we sample a spanning tree S according to the maximum entropy distribution μ with marginals exactly $\frac{n-1}{n}x$. Note that this is possible only approximately, and we will bound the difference in the end.

We add an odd(S)-join to S like Christofides' algorithm. The expected cost of S is less than $c(x)$, and we now bound the cost of the odd(S)-join. Because of (i), we can apply Lemma 10.1 for $\eta = 10^{-15}$ to bound the expected cost of the odd(S)-join by

$$\sum_{e \in \binom{V}{2}} \left(\tfrac{1}{2} - 10^{-16} \cdot \mathbb{P}[e \text{ is good}]\right) \cdot x_e \cdot c(e)$$
$$\leq \tfrac{1}{2} \cdot c(x) - 10^{-16} \cdot 10^{-24} \cdot \sum_{e \in E^*} x_e \cdot c(e)$$
$$\leq \tfrac{1}{2} \cdot c(x) - 10^{-52} n,$$

where we used $c(e) \geq 1$ for all $e \in E^*$. Moreover, we have $c(x) \leq \mathrm{OPT} \leq 2n$ because for graph metrics, the solution produced by the double tree algorithm (see Proposition 1.22) costs less than $2n$. Then we can bound the expected cost of the odd(S)-join by $\frac{1}{2}(1 - 10^{-52})c(x)$.

In our algorithm, we actually sample a spanning tree S from a λ-uniform distribution μ^λ such that $\mathbb{P}[e \in S] \leq \frac{n-1}{n} x_e + n^{-4} 10^{-159}$ for all e (cf. Corollary 5.23 and Theorem 5.25). Then the expected cost of S is still less than $c(x)$. Moreover, by Theorem 5.24, $\sum_{S \in \mathcal{S}} |\mu^\lambda(S) - \mu(S)| \leq 10^{-53}$. Parity correction never costs more than $n - 1$ in a Graph TSP instance because any spanning tree contains an odd(S)-join by Proposition 1.27. Hence, the expected cost of parity correction with respect to the distribution μ^λ is at most $10^{-53} n$ larger than in the analysis above. □

Note that we used properties of the Graph TSP in both cases, (i) and (ii). Although the improvement over Christofides' algorithm is tiny (in case (i)), this result received a lot of interest. Czeller and Pap [2014] showed that essentially the same proof yields an approximation ratio slightly better than $\frac{3}{2}$ for metric closures of graphs with edge weights between 1 and β, for any fixed constant β.

The result by Oveis Gharan, Saberi, and Singh [2011] has initiated a sequence of works on Graph TSP that yielded, within less than a year, much better approximation ratios with completely different techniques. We will describe these in Chapters 12 and 13.

Karlin, Klein, and Oveis Gharan [2020] gave a 1.49993-approximation algorithm for the half-integral TSP, which is the special case of the Symmetric TSP where an optimum solution x^* to the subtour LP (2.2) fulfills $x^* \in \{0, \frac{1}{2}, 1\}$ for all $e \in \binom{V}{2}$. For this special case, the algorithm from Karlin, Klein, and Oveis Gharan [2020] was later improved to a 1.498305-approximation algorithm by Gupta et al. [2022], using ideas from Haddadan and Newman [2023]. See Exercise 10.1.

Theorem 10.2 and its proof were also the first step towards an improvement for the general Symmetric TSP. This result due to Karlin, Klein, and Oveis Gharan [2021] will be the content of the rest of this chapter and the next chapter.

10.2 The Karlin–Klein–Oveis Gharan Algorithm

We will now describe the algorithm for which Karlin, Klein, and Oveis Gharan [2021] proved an approximation ratio better than $\frac{3}{2}$ for SYMMETRIC TSP. Before we present the main algorithm, we describe a small preprocessing step that allows us to assume that there exists an edge e_0 with $x^*_{e_0} = 1$ and $c(e_0) = 0$ and to decompose the rest of x^* into incidence vectors of spanning trees.

Proposition 10.4. *Let (V, c) be an instance of SYMMETRIC TSP WITH TRIANGLE INEQUALITY, and let x^* be an optimum solution to the subtour LP (2.2). Take an arbitrary vertex $r \in V$ and replace it by two copies u_0 and v_0 with $c(u_0, v_0) = 0$. Set $x^*_{\{u_0,v_0\}} := 1$. For each $w \in V \setminus \{r\}$, set $x^*_{\{u_0,w\}} = x^*_{\{v_0,w\}} = \frac{1}{2}x^*_{\{r,w\}}$. Then x^* is a feasible solution to the subtour LP (2.2) for the new instance.*

Proof. Cuts that do not separate u_0 and v_0 are not affected, $x^*(\delta(u_0)) = x^*(\delta(v_0)) = 2$ due to the even split, and for a vertex set U with $\{u_0\} \subsetneq U \subsetneq V \setminus \{v_0\}$, we have $2x^*(\delta(U)) = x^*(\delta(U \setminus \{u_0\}) + x^*(\delta(U \cup \{v_0\})) - x^*(\delta(u_0)) - x^*(\delta(v_0)) + 4x^*_{\{u_0,v_0\}} \geq 4$. □

The new instance has an optimum solution that visits u_0 and v_0 consecutively (i.e., traverses $e_0 = \{u_0, v_0\}$). Once we have done the preprocessing, we will also call the new instance (V, c). In this and the next chapter, we denote by $\mathcal{S}$ the set of edge sets of trees with vertex set V that do not contain the edge e_0. The algorithm samples a tree from $\mathcal{S}$ and does parity correction as in Christofides' algorithm. See Algorithm 10.5 for a detailed description. Recall that $S \sim \mu$ means that S is sampled from the probability distribution μ.

The results of Chapter 5 imply that the algorithm can actually be implemented:

Proposition 10.6. *Algorithm 10.5 can be implemented to run in polynomial time.*

Proof. Step (1), solving the subtour LP, can be done by Corollary 2.11.

Next, we note that x^* (restricted to E) is in the spanning tree polytope of (V, E). Indeed, checking the constraints of (2.8) (cf. Theorem 2.16) is easy: We have $2x^*(E) = \sum_{v\in V} x^*(\delta(v)) - 2x^*(e_0) = 2n - 2$ and $\sum_{e\in E[U]} x^*_e \leq \frac{1}{2}(\sum_{v\in U} x^*(\delta(v)) - x^*(\delta(U))) = \frac{1}{2}(2|U| - x^*(\delta(U))) \leq |U| - 1$ for any $\emptyset \neq U \subsetneq V$.

Hence, we can apply Corollary 5.23 to implement Step (3) – that is, compute $\lambda_e > 0$ for $e \in E$ such that the λ-uniform distribution on $\mathcal{S}$ preserves the marginals up to a factor of at most $1 + \frac{\varepsilon_\mu^3}{n^4}$. Sampling a tree from the λ-uniform distribution μ^λ (Step (4)) is now easy using Theorem 5.25.

Finally, we apply Theorem 1.29 to compute a minimum-cost odd(S)-join (Step (5)). □

Algorithm 10.5: Max-Entropy Sampling and Parity Correction

Input: an instance (V, c) of the SYMMETRIC TSP WITH TRIANGLE INEQUALITY, a constant $\varepsilon_\mu > 0$

Output: a tour in the complete graph on V

(1) Let x^* be an optimum solution to the subtour LP (2.2).
(2) Perform the preprocessing described in Proposition 10.4 so that there exists an edge $e_0 \in \binom{V}{2}$ with $x^*_{e_0} = 1$ and $c(e_0) = 0$. Let now (V, c) denote the new instance, $n = |V|$, and $E := \left\{e \in \binom{V}{2} \setminus \{e_0\} : x^*_e > 0\right\}$.
(3) Let $\mathcal{S}$ denote the set of edge sets of spanning trees in (V, E). Find $\lambda_e > 0$ $(e \in E)$ so that the λ-uniform distribution μ^λ on $\mathcal{S}$ satisfies $\mathbb{P}_{S \sim \mu^\lambda}[e \in S] \leq \left(1 + \frac{\varepsilon_\mu^3}{n^4}\right) x^*_e$ for all $e \in E$.
(4) Sample a spanning tree $S \in \mathcal{S}$ so that each S is sampled with probability $\mu^\lambda(S)$.
(5) Compute a minimum-cost odd(S)-join J in the complete graph on V and output the tour $S \,\dot{\cup}\, J$.

Since the λ-uniform distribution μ^λ obtained in Step (3) is almost marginal-preserving, it is very close to the maximum entropy distribution μ. Later, we will choose the constant $\varepsilon_\mu > 0$ small enough so that the difference between μ^λ and μ is negligible. We will again bound the tiny difference by Theorem 5.24.

All probabilities and expectations in this and the next chapter (unless explicitly stated otherwise) are with respect to the maximum entropy distribution μ (with marginals x^*). For example, we simply write $\mathbb{P}[e \in S] = x^*_e$ for $e \in E$ and $\mathbb{E}[\chi^S] = x^*$ (in the latter equation, we implicitly restrict x^* to E).

In this chapter, it is actually sufficient to assume that μ is an arbitrary marginal-preserving distribution. Only in the next chapter will we need stronger properties of the maximum entropy distribution.

10.3 An Almost Laminar Family of Near-Minimum Cuts

We now start analyzing the approximation ratio of Algorithm 10.5. Since the expected cost of the spanning tree S is roughly $c(x^*)$, and $c(x^*) \leq \mathrm{OPT}$, our goal is to bound the cost of parity correction in Step (5) of Algorithm 10.5 by less than $\frac{1}{2}\mathrm{OPT}$.

Let $e_0 = \{u_0, v_0\}$ be the special edge introduced by the preprocessing, and let $v_1, \ldots, v_{n-1}, u_0, v_0$ be an optimum solution to our instance; we can also view it

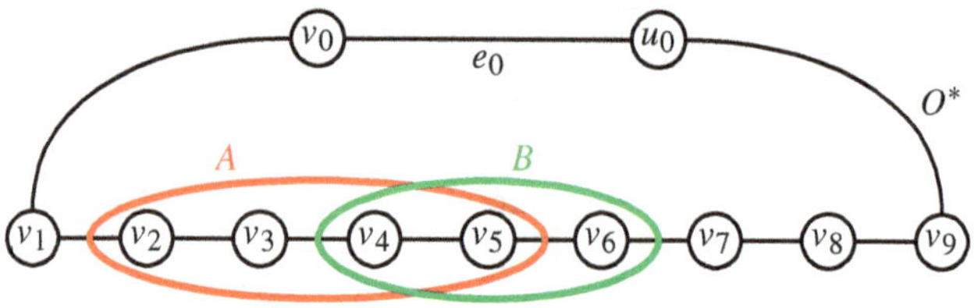

Figure 10.1 The optimum Hamiltonian cycle O^* is shown in black. Two intervals $A = \{v_2, v_3, v_4, v_5\}$ and $B = \{v_4, v_5, v_6\}$ are shown in red and green. A crosses B on the left.

as a Hamiltonian cycle O^* in the complete graph on V. See Figure 10.1. Let o^* be the incidence vector of O^*.

Following Karlin, Klein, and Oveis Gharan [2021], we will define a parity correction vector $y^S \in \mathbb{R}_{\geq 0}^{\binom{V}{2}}$ for each $S \in \mathcal{S}$ and start with

$$y^o \;=\; \tfrac{1}{4}(x^* + o^*) + \chi^{e_0}. \tag{10.1}$$

Note that $y^o \geq 0$ and $y^o(\delta(A)) \geq 1$ for all $\emptyset \neq A \subsetneq V$, and hence $c(y^o)$ bounds the cost of parity correction for any tree by Theorem 2.19. We have $c(y^o) = \frac{1}{4}c(x^*) + \frac{1}{4}\mathrm{OPT} + 0 \leq \frac{1}{2}\mathrm{OPT}$. This is a variant of Wolsey's analysis and re-proves the approximation ratio $\frac{3}{2}$. To obtain y^S from y^o, we will add and subtract three vectors that depend on S, to be defined later. Ideally we would like to subtract ηx^* for some constant $\eta > 0$ (and add only very little), but we must make sure that $y^S(\delta(A)) \geq 1$ whenever $|A \cap \mathrm{odd}(S)|$ is odd, or equivalently whenever $|\delta_S(A)|$ is odd (cf. Lemma 2.20).

Therefore, for a constant $0 < \eta \leq \frac{1}{8}$ to be fixed later, call a nonempty set $A \subseteq V \setminus \{u_0\}$ an *η-mincut* if $y^o(\delta(A)) < 1 + \eta$. Only constraints $y^S(\delta(A)) \geq 1$ corresponding to η-mincuts A can become violated if y^S is similar to y^o. Note that an η-mincut A is a vertex set, while the induced cut $\delta(A)$ is an edge set.

The use of o^* in the parity correction vector implies that we have to consider only cuts with a simple structure. Subsets of the form $\{v_i, v_{i+1}, \ldots, v_j\}$ for some $1 \leq i \leq j \leq n-1$ are called *intervals*. We say that an interval A is *left* of an interval B (and B is *right* of A) if there are $v_i \in A \setminus B$ and $v_j \in B \setminus A$ with $i < j$. (See Figure 10.1.) For an interval $A = \{v_i, v_{i+1}, \ldots, v_j\}$, we write $\delta^{\mathrm{left}}(A) := \delta(A) \cap \delta(\{v_1, \ldots, v_{i-1}\})$ and $\delta^{\mathrm{right}}(A) := \delta(A) \cap \delta(\{v_{j+1}, \ldots, v_{n-1}\})$ for the incident edges to the left and to the right, respectively.

Proposition 10.7. *All 4η-mincuts are intervals.*

Proof. If a nonempty set $A \subsetneq V$ is not an interval, then $e_0 \in \delta(A)$ or $o^*(\delta(A)) \geq 4$; moreover, $x^*(\delta(A)) \geq 2$. Hence, $y^o(\delta(A)) \geq \frac{3}{2} \geq 1 + 4\eta$. □

In particular, every 4η-mincut is a subset of $V \setminus \{u_0, v_0\}$. We say that (for a tree $S \in \mathcal{S}$) a nonempty set $A \subseteq V \setminus \{u_0, v_0\}$ *induces a tree* if $(A, S[A])$ is a tree. This is very likely for η-mincuts:

Proposition 10.8. *If A is an η-mincut, then $x^*(\delta(A)) < 2 + 4\eta$ and*

$$\mathbb{P}[A \text{ induces a tree}] > 1 - 2\eta.$$

Proof. Let A be an η-mincut. By Proposition 10.7, A is an interval and hence $o^*(\delta(A)) = 2$. We have $x^*(\delta(A)) = 4y^o(\delta(A)) - o^*(\delta(A)) < 4(1+\eta) - 2 = 2 + 4\eta$ and hence $x^*(E[A]) = \frac{1}{2}\left(\sum_{v \in A} x^*(\delta(v)) - x^*(\delta(A))\right) = |A| - \frac{1}{2}x^*(\delta(A)) > |A| - 1 - 2\eta$. Since $\mathbb{E}[\chi^S] = x^*$ and $|S[A]| \leq |A| - 1$ for all $S \in \mathcal{S}$, we have $|S[A]| < |A| - 1$ with probability less than 2η. □

As in Definition 4.6, we say that two subsets A and B of $V \setminus \{u_0, v_0\}$ *cross* if the sets $A \setminus B$, $A \cap B$, and $B \setminus A$ are all nonempty. If in addition A is left of B, we say that *A crosses B on the left* and *B crosses A on the right* (see Figure 10.1). A family of subsets of $V \setminus \{u_0, v_0\}$ is called *almost laminar* if for any of its sets A, B, C such that A crosses B and A crosses C, we have $A \cup B = A \cup C$. See the top part of Figure 10.2 for an example.

Proposition 10.9. *If A is an η-mincut and B is a ζ-mincut that crosses A, then $A \setminus B$, $A \cap B$, $B \setminus A$, and $A \cup B$ are all $(\eta + \zeta)$-mincuts.*

Proof. As in the proof of Proposition 4.5, we deduce from Proposition 4.4 and $y^o(\delta(U)) \geq 1$ for all $\emptyset \neq U \subsetneq V$ that $1 + 1 \leq y^o(\delta(A \cap B)) + y^o(\delta(A \cup B)) \leq y^o(\delta(A)) + y^o(\delta(B)) \leq 1 + \eta + 1 + \zeta$, which implies that $A \cup B$ and $A \cap B$ are $(\eta + \zeta)$-mincuts. For $A \setminus B$ and $B \setminus A$, apply the above to A and $V \setminus B$. □

An η-mincut A is called *S-ideal* for a tree $S \in \mathcal{S}$ if $|\delta_S^{\text{left}}(A)| = |\delta_S^{\text{right}}(A)| = 1$ and $\delta_S(A) \cap \delta(\{u_0, v_0\}) = \emptyset$. In particular, this implies that $|\delta_S(A)|$ is even. An η-mincut A is called *irrelevant* if $\mathbb{P}[A \text{ is } S\text{-ideal}] \geq 1 - 16\eta$. Because irrelevant cuts are almost always even, they are easy to deal with as we will see in Theorem 10.12.

Lemma 10.10. *If an η-mincut is crossed on the left and on the right by an η-mincut and a 2η-mincut, respectively, then it is irrelevant.*

Proof. Let A be crossed by L on the left and R on the right and $y^o(\delta(A)) < 1+\eta$, $y^o(\delta(L)) < 1 + \eta$, and $y^o(\delta(R)) < 1 + 2\eta$. By Proposition 10.8, we have $x^*(\delta(A)) < 2 + 4\eta$.

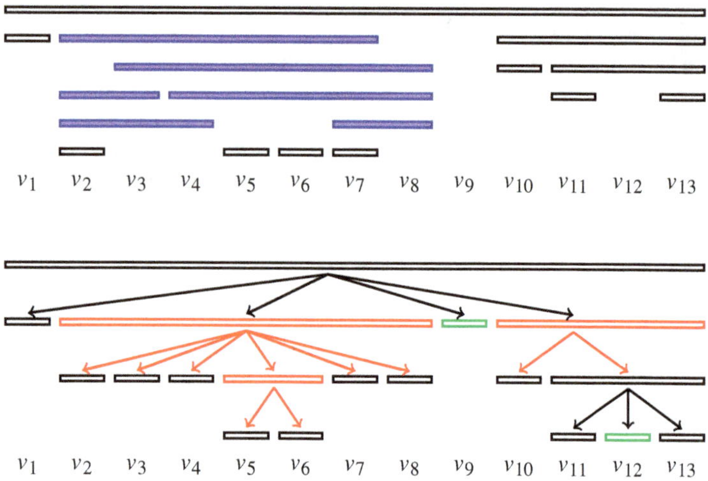

Figure 10.2 Top: The almost laminar family $\mathcal{L}$ of the η-mincuts that are not irrelevant could look like this. Each set is shown as a horizontal bar representing the interval. Here there is only one nontrivial component, formed by subsets of $\{v_2, \ldots, v_8\}$, shown as blue-filled bars. Bottom: The hierarchy $\mathcal{H}$ corresponding to $\mathcal{L}$. The parent–child relation is shown by arrows. This hierarchy contains three polygons, shown in red. The polygon corresponding to the nontrivial component is $\{v_2, \ldots, v_8\}$; its atoms are $\{v_2\}$, $\{v_3\}$, $\{v_4\}$, $\{v_5, v_6\}$, $\{v_7\}$, and $\{v_8\}$. The two green singletons are irrelevant but still part of the hierarchy.

If $|\delta_S(A)| = 1$, then L or R does not induce a tree, which happens with total probability less than 6η by Proposition 10.8. Therefore, the probability that $|\delta_S(A)| > 2$ is at most $x^*(\delta(A)) - 2 + 6\eta < 10\eta$. If L and R induce a tree and $|\delta_S(A)| \leq 2$, then A is S-ideal. By the union bound, this happens with probability at least $1 - 16\eta$. □

Lemma 10.11. *The family of η-mincuts that are not irrelevant is almost laminar.*

Proof. Suppose A, B, C are three sets in this family such that A crosses B and C but $A \cup B \neq A \cup C$. Recall that all these sets are intervals by Proposition 10.7. By Lemma 10.10 and symmetry, we can assume that both B and C cross A at the right, as well as that $B \setminus A \subsetneq C \setminus A$. By Proposition 10.9, $C \setminus A$ is a 2η-mincut. Then the η-mincut A crosses B on the left and the 2η-mincut $C \setminus A$ crosses B on the right. By Lemma 10.10, B is irrelevant, a contradiction. □

We now show that irrelevant sets can be ignored by adding some slack to the parity correction vectors. These slack vectors will be very cheap (for small η).

Theorem 10.12. *There are vectors $\tilde{s}^S \in \mathbb{R}_{\geq 0}^{\binom{V}{2}}$ for $S \in \mathcal{S}$ such that:*

- $\mathbb{E}[\tilde{s}_e^S] \leq 64\eta^2 o_e^*$ *for all* $e \in \binom{V}{2}$*; and*
- $\tilde{s}^S(\delta(A)) \geq \eta$ *for every* $S \in \mathcal{S}$ *and every irrelevant set* A *with* $|\delta_S(A)|$ *odd.*

Proof. For every edge $e = \{v_i, v_{i+1}\}$ $(i = 1, \ldots, n-2)$ of $O^* \setminus (\delta(u_0) \cup \delta(v_0))$, consider the largest irrelevant set A with $e \in \delta^{\text{right}}(A)$, the smallest irrelevant set A with $e \in \delta^{\text{right}}(A)$, the largest irrelevant set A with $e \in \delta^{\text{left}}(A)$, and the smallest irrelevant set A with $e \in \delta^{\text{left}}(A)$. If any of these exists and is not S-ideal, we set $\tilde{s}_e^S := \eta$, otherwise set $\tilde{s}_e^S := 0$. The probability of the former event is at most $4 \cdot 16\eta$, which implies $\mathbb{E}[\tilde{s}^S] \leq 64\eta^2 o^*$.

For any irrelevant set B with $\tilde{s}^S(\delta(B)) = 0$, consider the smallest and largest irrelevant sets A and C with $\delta_{O^*}^{\text{left}}(A) = \delta_{O^*}^{\text{left}}(B) = \delta_{O^*}^{\text{left}}(C)$. By the definition of $\tilde{s}^S$, we have that A and C are S-ideal and hence S contains precisely one edge in $\delta^{\text{left}}(A)$ and in $\delta^{\text{left}}(C)$, and this is then also the only edge of S in $\delta^{\text{left}}(B)$. Similarly, S contains exactly one edge in $\delta^{\text{right}}(B)$. Moreover, S contains no edge in $\delta(C) \cap \delta(\{u_0, v_0\}) \supseteq \delta(B) \cap \delta(\{u_0, v_0\})$. So B is S-ideal, and hence $|\delta_S(B)| = 2$. □

To form the parity correction vector y^S for a tree S, we will first add $\tilde{s}^S$ from Theorem 10.12 to y^o. Since our goal is to subtract (at least in expectation) $\Theta(\eta x_e^*)$ from every edge e in order to make y^S cheaper, the cost of $\tilde{s}$ will be negligible (on average, for sufficiently small η). Since we never subtract much, only η-mincuts are dangerous, and irrelevant cuts that are odd have enough slack after having added $\tilde{s}^S$.

10.4 Reduction to a Hierarchy of Near-Minimum Cuts

Let $\mathcal{L}$ be the almost laminar family of the η-mincuts that are not irrelevant (cf. Lemma 10.11). Define a graph with vertex set $\mathcal{L}$ and an edge for every pair of sets that cross. For every connected component of this graph that consists of more than one set, let P be the union of the sets in this connected component, and let $\{A_1, \ldots, A_k\}$ be the coarsest partition of P (numbered from left to right) such that every set in the connected component is $A_1 \cup \cdots \cup A_j$ for some $j \in \{2, \ldots, k-1\}$ or $A_i \cup \cdots \cup A_k$ for some $i \in \{1, \ldots, k-1\}$. This is possible because $\mathcal{L}$ is an almost laminar family of intervals. We call P a *nontrivial component* of $\mathcal{L}$ and call $A_1, \ldots, A_k$ the *atoms* of P. We call $A_2, \ldots, A_{k-1}$ the *inner atoms* of P.

The *hierarchy* $\mathcal{H}$ consists of all sets in $\mathcal{L}$ that do not cross any other set in $\mathcal{L}$ as well as all nontrivial components and their atoms and the irrelevant singletons. See Figure 10.2.

Proposition 10.13. *The hierarchy is a laminar family of intervals. It contains* $V \setminus \{u_0, v_0\}$.

Proof. By construction, the hierarchy is laminar. Since $x^*(\delta(V \setminus \{u_0, v_0\})) = 2$ and $V \setminus \{u_0, v_0\}$ is not irrelevant, it belongs to $\mathcal{H}$. □

The set $V \setminus \{u_0, v_0\}$ is the unique maximal element of the hierarchy $\mathcal{H}$. If A is another element of $\mathcal{H}$ and B is the minimal proper superset of A in $\mathcal{H}$, then B is called the *parent* of A and A is called a *child* of B. The children of a polygon are its atoms.

A *triangle* is a set in $\mathcal{H}$ with exactly two children. A *polygon* is a nontrivial component of $\mathcal{L}$ or a triangle. For a polygon P, let $A_1, \ldots, A_k$ be its children from left to right. We call the sets $A_1 \cup \cdots \cup A_j$ $(j = 1, \ldots, k-1)$ the *left-relevant* subsets of P and the sets $A_i \cup \cdots \cup A_k$ $(i = 2, \ldots, k)$ the *right-relevant* subsets of P.

Proposition 10.14. *Let P be a polygon with children $A_1, \ldots, A_k$ from left to right. Then all sets of the form $A_i \cup \cdots \cup A_j$ for $1 \le i \le j \le k$ are 4η-mincuts. The left-relevant and the right-relevant subsets of P are 2η-mincuts. Moreover,* $x^*(\delta(P) \setminus \delta(A_1 \cup A_k)) \le 4\eta$.

Proof. This is trivial for triangles, so let P be a nontrivial component. Elements of $\mathcal{L}$ are η-mincuts. If a left-relevant set $L = A_1 \cup \ldots \cup A_i$ is not in $\mathcal{L}$, then $P \setminus L$ is in $\mathcal{L}$ (and thus an η-mincut) and crosses a set $L' = A_1 \cup \cdots \cup A_j \in \mathcal{L}$ for some $j > i$. So $L = L' \setminus (P \setminus L)$ is a 2η-mincut by Proposition 10.9. By symmetry, every right-relevant set is a 2η-mincut. Every other set of the form $A_i \cup \cdots \cup A_j$ for $1 \le i \le j \le k$ is the union (if $i = 1$ and $j = k$) or the intersection of a left-relevant and a right-relevant set, and hence a 4η-mincut by Proposition 10.9.

For the last statement, first note that $P \setminus A_1$ and $P \setminus A_k$ are in $\mathcal{L}$ and are thus η-mincuts. By Proposition 10.8, $x^*(\delta(P \setminus A_1)) < 2+4\eta$ and $x^*(\delta(P \setminus A_k)) < 2+4\eta$. We conclude

$$
\begin{aligned}
&2x^*(\delta(P) \setminus \delta(A_1 \cup A_k)) \\
&\quad = x^*(\delta(P \setminus A_1)) + x^*(\delta(P \setminus A_k)) - x^*(\delta(A_1)) - x^*(\delta(A_k)) \\
&\quad < (2 + 4\eta) + (2 + 4\eta) - 2 - 2 \\
&\quad = 8\eta.
\end{aligned}
$$

□

Lemma 10.15. *Every η-mincut that is not irrelevant belongs to $\mathcal{H}$ or is a left-relevant or right-relevant subset of a polygon in $\mathcal{H}$.*

Proof. Let A be an η-mincut that is not irrelevant. Then A belongs to the almost laminar family $\mathcal{L}$. If $A \notin \mathcal{H}$, then A is part of a nontrivial component and thus left-relevant or right-relevant. □

Call a polygon P with children $A_1 \ldots, A_k$ (from left to right) *left-happy* if $|\delta_S(A_1) \cap \delta(P)|$ is odd and *right-happy* if $|\delta_S(A_k) \cap \delta(P)|$ is odd. Our next goal is to show that we can ignore left-relevant subsets of left-happy polygons (and right-relevant subsets of right-happy polygons). To this end, we need one more definition.

A left-relevant or right-relevant set A of a polygon P is called S-*normal* for a tree $S \in \mathcal{S}$ if $\delta_S(P) \subseteq \delta(A_1) \cup \delta(A_k)$ and $|\delta_S(A) \setminus \delta(P)| = 1$. If P is S-normal and left-happy, then $|\delta_S(A)|$ is even for every left-relevant subset A. The next theorem implies that we do not have to consider left-relevant and right-relevant sets of a polygon unless they are S-normal. Moreover, we can also ignore inner atoms of polygons. Again, we add some slack.

Theorem 10.16. *There are vectors $\hat{s}^S \in \mathbb{R}_{\geq 0}^{\binom{V}{2}}$ for $S \in \mathcal{S}$ such that*

- *$\mathbb{E}[\hat{s}_e^S] \leq 20\eta^2 o_e^*$ for all $e \in \binom{V}{2}$, and*
- *for every polygon P, we have $\hat{s}^S(\delta(A)) \geq \eta$ for every inner atom A with $|\delta_S(A)|$ odd and for every left-relevant or right-relevant set A that is not S-normal.*

Proof. For every edge $e \in O^* \setminus (\delta(u_0) \cup \delta(v_0))$, let P be the smallest set in $\mathcal{H}$ that contains both endpoints. If P is not a polygon, set $\hat{s}_e = 0$. Otherwise, let $A_1, \ldots, A_k$ be the children of P from left to right, and let $i \in \{1, \ldots, k-1\}$ such that $e \in \delta(A_i) \cap \delta(A_{i+1})$. Consider the left-relevant subset $L = A_1 \cup \cdots \cup A_i$ and the right-relevant subset $R = A_{i+1} \cup \cdots \cup A_k$. We set $\hat{s}_e^S = 0$ if $|\delta_S(L) \cap \delta_S(R)| = |\delta_S(A_i) \cap \delta_S(A_{i+1})| = 1$ and $\delta_S(P) \subseteq \delta(A_1) \cup \delta(A_k)$; otherwise, we set $\hat{s}_e^S = \eta$.

We show that the latter happens with probability at most 20η. Indeed, it implies one of the following events:

- at least one of the sets $A_i \cup A_{i+1}$, L, and R does not induce a tree, which happens with probability at most 16η by Propositions 10.8 and 10.14, or
- $\delta_S(P) \setminus (\delta(A_1) \cup \delta(A_k)) \neq \emptyset$, which happens with probability at most 4η by Proposition 10.14.

By construction, $\hat{s}^S(\delta(A)) \geq \eta$ for every inner atom A of P with $|\delta_S(A)| \neq 2$ and for every left-relevant or right-relevant set A of P that is not S-normal. □

Here is the main payment theorem for hierarchies of Karlin, Klein, and Oveis Gharan [2021]. Recall that $E = \{e \in \binom{V}{2} \setminus \{e_0\} : x^*_e > 0\}$.

Theorem 10.17. *Let μ be the maximum entropy distribution on $\mathcal{S}$ with marginals x^*. There exist constants $\varepsilon_0 > 0$ and $0 < \eta \leq \frac{\varepsilon_0}{174}$ and vectors $s^S \in \mathbb{R}^E$ for $S \in \mathcal{S}$ such that for the hierarchy $\mathcal{H}$ representing the relevant η-mincuts:*

(i) $s^S_e \geq -x^*_e$ *for all $e \in E$ and all $S \in \mathcal{S}$;*

(ii) *For every set $A \in \mathcal{H}$ whose parent is not a polygon and every $S \in \mathcal{S}$, we have $|\delta_S(A)|$ even or $s^S(\delta(A)) \geq 0$;*

(iii) *For every polygon P and every $S \in \mathcal{S}$:*

- *if P is not left-happy, then $s^S(\delta(A)) \geq 0$ for every left-relevant set A;*
- *if P is not right-happy, then $s^S(\delta(A)) \geq 0$ for every right-relevant set A;*

(iv) $\mathbb{E}[s^S_e] \leq -\varepsilon_0 x^*_e$.

The proof of this theorem is very long; see the next chapter. We can now show that it implies a better approximation ratio than $\frac{3}{2}$ rather easily:

Theorem 10.18 (Karlin, Klein, and Oveis Gharan [2021]). *There is a randomized α-approximation algorithm for* Symmetric *TSP for some $\alpha < \frac{3}{2}$.*

Proof. Let ε_0 and η be the constants from Theorem 10.17. Note that $\varepsilon_0 \leq 1$ and $\eta \leq \frac{\varepsilon_0}{174} < \frac{1}{8}$. Run Algorithm 10.5 with $\varepsilon_\mu = \eta^2$. Again, let μ^λ denote the λ-uniform distribution for λ computed by the algorithm, and let μ be the maximum entropy distribution (with marginals x^*). We have $\mathbb{P}_{S \sim \mu^\lambda}[e \in S] \leq \left(1 + \frac{\varepsilon_\mu^3}{n^4}\right) x^*_e$. By Theorem 5.24, $\sum_{S \in \mathcal{S}} |\mu^\lambda(S) - \mu(S)| \leq \varepsilon_\mu$.

For $S \in \mathcal{S}$ let

$$y^S := y^o + \tilde{s}^S + \hat{s}^S + (\tfrac{\eta}{2} - \eta^2) s^S \tag{10.2}$$

according to (10.1) and Theorems 10.12, 10.16, and 10.17, amended by $s^S_e = 0$ for $e \in \binom{V}{2} \setminus E$. Since $\tilde{s}^S \geq 0$ and $\hat{s}^S \geq 0$, we have

$$y^S_e \geq \tfrac{x^*_e}{4} + (\tfrac{\eta}{2} - \eta^2) s^S_e \geq \tfrac{x^*_e}{4} - \tfrac{\eta}{2} x^*_e \geq 0$$

for all $e \in \binom{V}{2}$ by Theorem 10.17 (i). We will show that y^S is a parity correction vector for S – that is,

$$y^S(\delta(A)) \geq 1 \qquad \text{for all } A \subseteq V \text{ with } |\delta_S(A)| \text{ odd.} \tag{10.3}$$

Then the expected cost of the tour computed by the algorithm is

$$\begin{aligned}
&\sum_{S\in\mathcal{S}} \mu^\lambda(S)\left(c(S) + \min\{c(J) : J \text{ is an odd}(S)\text{-join}\}\right)\\
&\leq \sum_{S\in\mathcal{S}} \mu^\lambda(S)c(S) + \sum_{S\in\mathcal{S}} \mu(S)c(y^S) + \sum_{S\in\mathcal{S}} \left|\mu^\lambda(S) - \mu(S)\right| c(\tfrac{1}{2}o^*)\\
&= \sum_{e\in E} c(e)\mathbb{P}_{S\sim\mu^\lambda}[e\in S] + \mathbb{E}_{S\sim\mu}\left[c(y^S)\right] + \sum_{S\in\mathcal{S}} \left|\mu^\lambda(S) - \mu(S)\right| \tfrac{1}{2}c(o^*)\\
&\leq \left(1 + \frac{\varepsilon_\mu^3}{n^4}\right) c(x^*) + \mathbb{E}_{S\sim\mu}\left[c(y^S)\right] + \frac{\varepsilon_\mu}{2}c(o^*)\\
&\leq (1+\varepsilon_\mu)c(o^*) + \mathbb{E}_{S\sim\mu}\left[c(y^S)\right]\\
&\leq (1+\eta^2)c(o^*) + \tfrac{1}{4}c(x^*+o^*) + 64\eta^2c(o^*) + 20\eta^2c(o^*)\\
&\quad - \frac{\varepsilon_0\eta}{2}c(x^*) + \eta^2\varepsilon_0 c(x^*)\\
&\leq \left(\tfrac{3}{2} + 86\eta^2 - \frac{\varepsilon_0\eta}{2}\right) c(o^*)\\
&\leq \left(\tfrac{3}{2} - \eta^2\right) c(o^*),
\end{aligned}$$

where the first inequality follows from (10.3) and Theorem 2.19; moreover, we used $c(x^*) \leq c(o^*)$ and $c(y^o) = \frac{1}{4}(c(x^*) + c(o^*))$ and the bounds from Theorems 10.12, 10.16, and 10.17 (iv). For $\alpha := \frac{3}{2} - \eta^2$, this shows the claimed expected performance ratio $\alpha < \frac{3}{2}$.

It remains to show (10.3). Let $S \in \mathcal{S}$. For sets A with $e_0 \in \delta(A)$, we have $y^S(\delta(A)) \geq y^o(\delta(A)) - \frac{\eta}{2}x^*(\delta(A)) \geq \chi^{e_0}(\delta(A)) = 1$ (using Theorem 10.17 (i)), so (10.3) holds for such sets.

Now let A be a (nonempty) subset of $V \setminus \{u_0, v_0\}$ with $|\delta_S(A)|$ odd. If A is not an η-mincut, then (using Theorem 10.17 (i) again)

$$\begin{aligned}
y^S(\delta(A)) &\geq y^o(\delta(A)) - \tfrac{\eta}{2}x^*(\delta(A)) + \eta^2x^*(\delta(A))\\
&= y^o(\delta(A)) - \tfrac{\eta}{2}\left(4y^o(\delta(A)) - o^*(\delta(A))\right) + \eta^2x^*(\delta(A))\\
&= y^o(\delta(A))(1-2\eta) + \tfrac{\eta}{2}o^*(\delta(A)) + \eta^2x^*(\delta(A))\\
&\geq (1+\eta)(1-2\eta) + \eta + 2\eta^2\\
&= 1.
\end{aligned}$$

Now suppose that A is an η-mincut. Then $x^*(\delta(A)) < 2 + 4\eta$ by Proposition 10.8. If A is irrelevant, then $\tilde{s}^S(\delta(A)) \geq \eta$ by Theorem 10.12. Then

$$\begin{aligned}
y^S(\delta(A)) &\geq 1 + \eta - \tfrac{\eta}{2}x^*(\delta(A)) + \eta^2x^*(\delta(A))\\
&> 1 + \eta - \tfrac{\eta}{2}(2+4\eta) + 2\eta^2 \qquad (10.4)\\
&= 1.
\end{aligned}$$

Otherwise, by Lemma 10.15, $A \in \mathcal{H}$ or A is a left-relevant or right-relevant subset of a polygon in $\mathcal{H}$. If A is an inner atom of a polygon, or a left-relevant or right-relevant set that is not S-normal, then $\hat{s}^S(\delta(A)) \geq \eta$ by Theorem 10.16, and (10.4) again holds.

If A is an S-normal left-relevant set of a polygon P, then P is not left-happy because $|\delta_S(A_1) \cap \delta(P)| = |\delta_S(A) \cap \delta(P)| = |\delta_S(A)| - 1$ is even. From Theorem 10.17 (iii), we conclude $s^S(\delta(A)) \geq 0$ and hence $y^S(\delta(A)) \geq y^o(\delta(A)) \geq 1$. The analogous conclusion holds for right-relevant sets.

Finally, if $A \in \mathcal{H}$ and the parent of A is not a polygon, then $s^S(\delta(A)) \geq 0$ by Theorem 10.17 (ii) and hence $y^S(\delta(A)) \geq y^o(\delta(A)) \geq 1$. □

Karlin, Klein, and Oveis Gharan [2022] obtained the explicit guarantee $\alpha = \frac{3}{2} - 10^{-36}$.

10.5 Bounding the Integrality Ratio

Theorem 10.18 does not imply an upper bound on the integrality ratio of the subtour LP (2.2) because o^* (the incidence vector of an optimum tour) is used in the basic parity correction vector y^o. This allowed for a relatively simple structure of the η-mincuts. In a more recent work, Karlin, Klein, and Oveis Gharan [2022] managed to also obtain an upper bound less than $\frac{3}{2}$ on the integrality ratio of the subtour LP. To this end, they redefined $y^o = \frac{1}{2}x^* + \chi^{e_0}$. If we were then simply considering the cuts $\delta(U)$ with $y^o(\delta(U)) = 1$, we could use the cactus representation by Dinits, Karzanov, and Lomonosov [1976] (see Exercise 10.6). However, one needs to consider η-mincuts for some constant $\eta > 0$, and their structure can be more complicated. Fortunately, for $\eta < \frac{1}{5}$, their structure can still be described, namely by the polygon representation due to Benczúr [1995] and Benczúr and Goemans [2008].

Karlin, Klein, and Oveis Gharan [2022] exploited this polygon representation, proved further properties, again obtained a hierarchy of η-mincuts, and showed that the main payment theorem for hierarchies also implies:

Theorem 10.19 (Karlin, Klein, and Oveis Gharan [2022]). *The integrality ratio of the subtour LP* (2.2) *is less than* $\frac{3}{2}$.

Again, Karlin, Klein, and Oveis Gharan [2022] obtained the explicit guarantee $\frac{3}{2} - 10^{-36}$. We conclude this chapter by summarizing again the state of the art for SYMMETRIC TSP in Table 10.1.

Table 10.1 *Approximation ratios and upper bounds on the integrality ratio of* (2.2) *for* Symmetric TSP with Triangle Inequality *in the order of their discovery. (R) means randomized; this algorithm computes a random tour, and the approximation ratio compares its expected cost to* OPT.

Approximation Ratio	Integrality Ratio	Year	Reference	Chapter
2	–	1974	Rosenkrantz, Stearns, and Lewis [1977]	–
$\frac{3}{2}$	–	1976	Christofides [1976]	1.4
$\frac{3}{2}$	–	1976	Serdyukov [1978]	1.4
$\frac{3}{2}$	$\frac{3}{2}$	1980	Wolsey [1980]	2.4
$\frac{3}{2} - 10^{-36}$ (R)	–	2020	Karlin, Klein, and Oveis Gharan [2021]	10–11
$\frac{3}{2} - 10^{-36}$ (R)	$\frac{3}{2} - 10^{-36}$	2021	Karlin, Klein, and Oveis Gharan [2022]	10.5
$\frac{3}{2} - 10^{-36}$	$\frac{3}{2} - 10^{-36}$	2022	Karlin, Klein, and Oveis Gharan [2023]	11.6

Exercises

10.1 Let (V, c) be an instance of the Symmetric TSP with Triangle Inequality, let x be an optimum solution to the subtour LP (2.2), and suppose that x is half-integral – that is, $x_e \in \{0, \frac{1}{2}, 1\}$ for all $e \in \binom{V}{2}$. Further suppose that $|V|$ is even and there are no non-singleton mincuts – that is, $x(\delta(U)) > 2$ for all $U \subseteq V$ with $|U| \geq 2$ and $|V \setminus U| \geq 2$. Let $G = (V, E)$ be the support graph of x. We will describe a randomized algorithm that computes a tour in G with expected cost much less than $\frac{3}{2}c(x)$.

(a) Show that G is 4-regular and 4-edge-connected.

(b) Show that the all-$\frac{1}{4}$ vector is in the perfect matching polytope of G, which is

$$\left\{x \in \mathbb{R}^E_{\geq 0} : x(\delta(v)) = 1 \ (v \in V),\ x(\delta(U)) \geq 1 \ (U \subseteq V,\ |U| \text{ odd})\right\}$$

(cf. Exercise 2.8).

(c) The algorithm begins by choosing a random perfect matching M in G such that $\mathbb{P}[e \in M] = \frac{1}{4}$ for all $e \in E$. This is possible by (b). Show that $y := \frac{1}{3}\chi^E + \frac{2}{3}\chi^M$ is in the subtour polytope.

(d) Show that the edges of M can be colored (greedily) with seven colors so that for all edges $e = \{v, w\} \in E \setminus M$, the matching edges incident to v and to w have different colors.

(e) Consider the subset $M_i \subseteq M$ of matching edges of a color $i \in \{1, \ldots, 7\}$ chosen uniformly at random. Show that the edge sets in $\mathcal{N} := \{\delta(u) \setminus M : u \in V,\ \delta(u) \cap M_i \neq \emptyset\}$ are pairwise disjoint, and hence the edge sets $S \subseteq E$ with $|S \cap N| \leq 1$ for all $N \in \mathcal{N}$ form a matroid (cf. Exercise 2.5).

(f) Recall the definition of 1-trees in Exercise 4.3. Suppose we can sample a 1-tree (V, S) with $|S \cap N| \leq 1$ for all $N \in \mathcal{N}$ such that $\mathbb{P}[e \in S] = y_e$. Conclude that the edges of M_i are not contained in any cut $\delta(U)$ with $x(\delta(U)) < 3$ and $|\delta(U) \cap S|$ odd.

(g) Finish the algorithm by adding an odd(S)-join. Show that this costs at most $c(z)$, where $z_e = \frac{1}{2}$ for all $e \in M \setminus M_i$ and $z_e = \frac{1}{6}$ for all other edges $e \in E$.

(h) Conclude that the expected cost of the resulting tour is at most $(\frac{3}{2} - \frac{1}{42})c(x)$.

Notes: Using the integrality of the matroid intersection polytope, sampling essentially as in (f) can be done by splitting one vertex as in Proposition 10.4. The assumptions that $|V|$ is even and that there are no non-singleton mincuts were removed by Gupta et al. [2022], who obtained a 1.499-approximation algorithm for instances in which x is half-integral (sometimes called half-integral TSP). We remark that Jin, Klein, and Williamson [2023a] obtained a $\frac{4}{3}$-approximation algorithm for a special case of half-integral TSP.
(Haddadan and Newman [2023], Gupta et al. [2022])

10.2 Let x be a feasible solution to the subtour LP (2.2) and let $\eta \geq 0$. Let $A, B \subsetneq V$ with $x(\delta(A)) \leq 2 + \eta$ and $x(\delta(B)) \leq 2 + \eta$ and suppose A and B are crossing. Show that $x(\delta(A \cap B) \cap \delta(A \setminus B)) \geq 1 - \frac{\eta}{2}$ and $x(\delta(A \setminus B) \cap \delta(A \cup B)) \geq 1 - \frac{\eta}{2}$.

10.3 Let P be a polygon with children $A_1, \ldots, A_k$ from left to right. Show that $1 - 8\eta \leq x^*(\delta(A_i) \cap \delta(A_{i+1})) \leq 1 + 16\eta$ for all $i = 1, \ldots, k-1$.

10.4 Show that the almost laminar family $\mathcal{L}$ of the η-mincuts that are not irrelevant contains less than $3|V|$ elements.

10.5 Let $G = (V, E)$ be an undirected graph. Call a set $X \subsetneq V$ *tight* if $|\delta(X)| = \min\{|\delta(U)| : \emptyset \neq U \subsetneq V\} =: \lambda(G)$ (in other words, $\delta(X)$ is a minimum-cardinality cut in G). The number $\lambda(G)$ is called the edge-connectivity of G. A tight set X is called *proper* if $|X| \geq 2$ and

$|V \setminus X| \geq 2$. Suppose that there is a proper tight set and every proper tight set is crossed by another tight set. Then prove:

(a) All vertices have degree $\lambda(G)$.
(b) Let $v \in V$ and X be a proper tight set containing v. Then there is a tight set Y with exactly two elements and $v \in Y \subseteq X$.
(c) Every vertex is contained in two two-element tight sets containing v.
(d) If $\{x, y\}$ is tight, then there are $\frac{1}{2}\lambda(G)$ parallel edges between x and y.
(e) G arises from a circuit by replacing every edge with $\frac{1}{2}\lambda(G)$ parallel edges.

(Fleiner and Frank [2009], Frank [2011])

10.6 A cactus is an undirected connected graph in which every edge belongs to exactly one circuit. Using Exercise 10.5, our goal is to show the following theorem:

For every connected undirected graph G, there is a cactus C and a map $\varphi : V(G) \to V(C)$ such that
(i) if Y is tight in C, then $\varphi^{-1}(Y)$ is tight in G, and
(ii) if X is tight in G, then there is a tight Y in C with $X = \varphi^{-1}(Y)$. (10.5)

The pair (C, φ) is called a cactus representation of the minimum-cardinality cuts of G. Prove:

(a) Every cactus with more than one vertex has edge-connectivity 2, and the minimum-cardinality cuts of a cactus are precisely those pairs of edges that belong to the same circuit.
(b) (10.5) holds if G has no proper tight set.
(c) (10.5) holds if every proper tight set in G is crossed by another proper tight set.
(d) (10.5) holds in general (use induction on $|V(G)|$).
(e) Show that the theorem extends to positive edge weights: For every connected undirected graph $G = (V, E)$ with weights $x : E \to \mathbb{R}_{>0}$, there is a cactus representation of the minimum-weight cuts.

(Dinits, Karzanov, and Lomonosov [1976])

10.7 Consider the GRAPH TSP instance shown in Figure 10.3: a "donut graph" with $n = 4k$ vertices. Note that these graphs are Hamiltonian. Consider the solution x^* to the subtour LP as shown in the figure. Show that based on this, $\frac{1}{n}$ times the expected cost of the tour produced by Algorithm 10.5 converges to $\frac{11}{8}$ as $k \to \infty$.
Hint: Assume $\varepsilon_\mu = 0$. Show that for almost every edge from the inside to the outside of the donut (except near e_0), exactly one vertex has

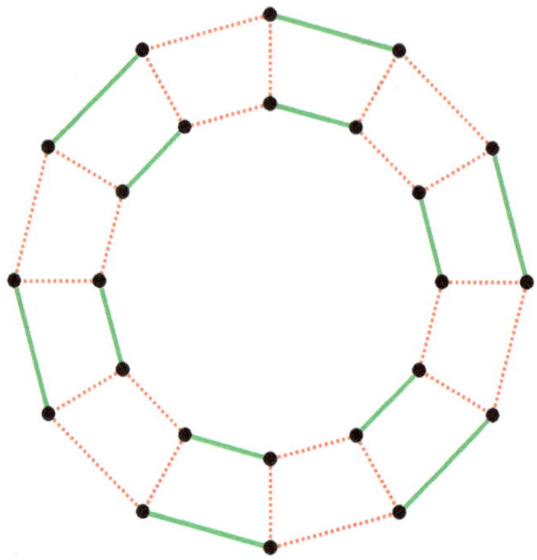

Figure 10.3 A donut graph (for $k = 6$). Consider the LP solution x^* with $x^*_e = 1$ for every solid green edge e and $x^*_e = \frac{1}{2}$ for every dotted red edge e.

odd degree in the sampled spanning tree, and the events which one are independent. To this end, apply Lemma 5.18 to sets U with $x^*(\delta(U)) = 2$. Use that the sum of Bernoulli random variables is strongly concentrated around its expectation (see, e.g., Lemma 5.6).
(Jin, Klein, and Williamson [2023b])

11

Proving the Main Payment Theorem for Hierarchies

This chapter is about the proof of the main payment theorem for hierarchies (Theorem 10.17) from Karlin, Klein, and Oveis Gharan [2021]. Because the proof is very long and technical, we will not give a complete proof here, but rather focus on explaining the key combinatorial ideas. This chapter is structured as follows. First, in Section 11.1, we describe the general proof strategy and prove the theorem in an idealized setting. Then, in Section 11.2, we discuss a few crucial properties of λ-uniform distributions. Sections 11.3–11.5 focus on the main ideas needed to address the hurdles we ignored in the idealized setting described in Section 11.1. Finally, in Section 11.6, we show how the Karlin–Klein–Oveis Gharan algorithm can be derandomized.

11.1 Outline of the Proof

Throughout this chapter, we let μ be a λ-uniform distribution that almost preserves the marginals x^* (up to an arbitrarily small error) and is hence almost identical to the maximum entropy distribution on $\mathcal{S}$ (with marginals x^*). The maximum entropy distribution itself may not be λ-uniform as shown in Exercise 5.12 (b), but we can get arbitrarily close (cf. Theorem 5.24). In the following, we ignore the arbitrarily small difference in the marginals.

Moreover, S will denote the edge set of a random spanning tree sampled from this distribution μ. We consider the hierarchy $\mathcal{H}$ representing the relevant η-mincuts, as constructed in Section 10.4, where η is a tiny positive constant that we choose later. In order to explain the main ideas of the proof of Theorem 10.17, let us assume for now that the hierarchy $\mathcal{H}$ contains no polygons. We will explain how the proof can be extended to handle polygons in Section 11.5.

We will use the following notation, which is illustrated in Figure 11.1:

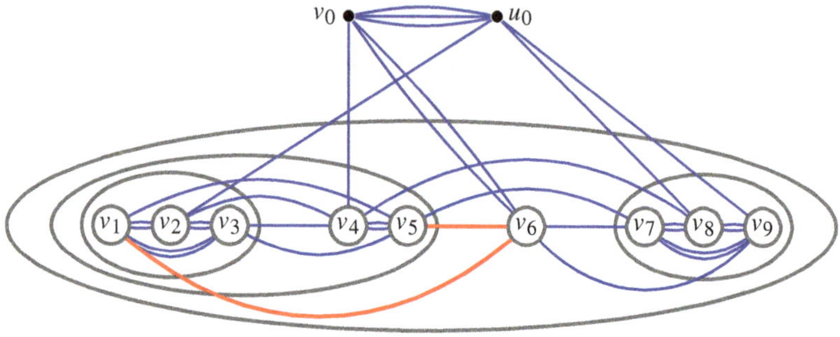

Figure 11.1 A hierarchy without polygons. Assume that the optimum tour visits $v_1, \ldots, v_9, u_0, v_0$ in this order and the LP solution x^* puts $x_e^* = \frac{1}{3}$ on every edge in the figure; when two or three parallel edges are drawn, the value is $\frac{2}{3}$ or 1, respectively. In this example, the family of η-mincuts is laminar, shown by the ellipses in the figure (including the singletons $\{v_1\}, \ldots, \{v_9\}$). One can see that here none of the η-mincuts is irrelevant, so this laminar family is identical to the hierarchy. The thick red edges $\{v_1, v_6\}$ and $\{v_5, v_6\}$ both have the top cuts $\{v_1, v_2, v_3, v_4, v_5\}$ and $\{v_6\}$.

Definition 11.1 (top cuts)**.** For an edge $e \in E$, we call the (up to) two maximal sets in $\mathcal{H}$ that contain exactly one endpoint of e the *top cuts* of the edge e. We say that an element U of the hierarchy is *even* if $|\delta_S(U)|$ is even (and *odd* otherwise).

We need to show that there exist slack vectors $s^{S'}$ for $S' \in \mathcal{S}$ fulfilling properties (i) – (iv) of Theorem 10.17. To this end, we will describe a random slack vector s, where the randomness depends not only on the random tree S (sampled from the distribution μ) but also on further random events that we will introduce along the way. To prove Theorem 10.17, we can obtain vectors $s^{S'}$ for $S' \in \mathcal{S}$ that depend deterministically only on the tree S' by setting $s_e^{S'} := \mathbb{E}\left[s_e \mid S = S'\right]$. Our random slack vector s and the random tree S are strongly correlated; they will always satisfy properties (i) – (iii) and $\mathbb{E}[s_e] \leq -\varepsilon_0 x_e^*$. Hence, we get properties (i) – (iv) for the deterministic vectors $s^{S'}$ ($S' \in \mathcal{S}$).

In order to define the random slack vector s, we will define a nonnegative reduction vector $r \in \mathbb{R}^E_{\geq 0}$ and a nonnegative increase vector $i \in \mathbb{R}^E_{\geq 0}$. Then we set $s := i - r$. The basic idea is to reduce the entries of the parity correction vector for an edge e (i.e., to set r_e to a positive value) when both of its top cuts are even. This does not lead to a feasible slack vector (i.e., it will often violate (ii) and (iii) of Theorem 10.17), which is why we need the vector i.

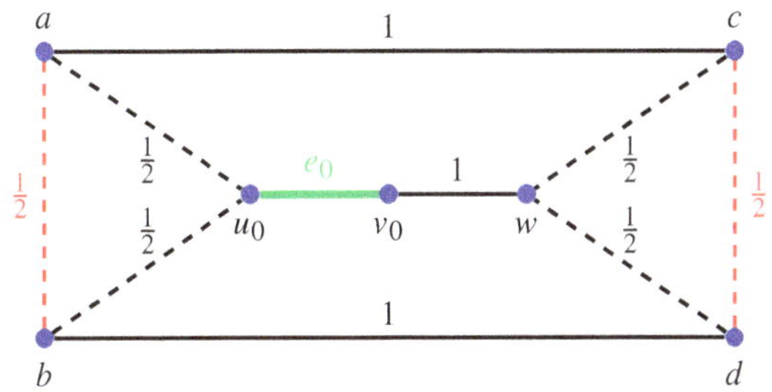

Figure 11.2 An example in which there exist edges e for which the probability that both top cuts of e are simultaneously even is zero. We have $x_e = \frac{1}{2}$ for each dashed edge e and $x_e = 1$ for each solid edge e. Suppose the optimum tour visits the vertices a, b, w, c, d, u_0, v_0 in this order. Then the hierarchy consists only of singleton cuts and the set $V \setminus \{u_0, v_0\}$. Every spanning tree that is sampled with positive probability contains the edges $\{a, c\}$, $\{v_0, w\}$, $\{b, d\}$. Moreover, it contains precisely one of the edges $\{a, u_0\}$ and $\{b, u_0\}$ (since it must connect u_0 and does not contain e_0) and precisely one of the edges $\{c, w\}$ and $\{d, w\}$. Thus, for the two red edges, the top cuts – which are $\{a\}$ and $\{b\}$ for the edge $\{a, b\}$ and $\{c\}$ and $\{d\}$ for the edge $\{c, d\}$ – are simultaneously even with probability zero.

Definition 11.2 (γ-good)**.** An edge in E is γ*-good* if the probability that all of its top cuts are even is at least γ.

Edges that have $V \setminus \{u_0, v_0\}$ as its only top cut are always γ-good (see Exercise 11.1). The interesting edges have exactly two top cuts. Note that $\delta_S(U)$ contains an even number of edges with constant probability for every η-mincut U (and thus every $U \in \mathcal{H}$) simply because $x^*(\delta(U)) \approx 2$ and because of concentration properties similar to Lemma 5.6. In fact, the probability is more than 40 % (but this requires stronger arguments; see Exercise 11.5). For some edges, however, its two top cuts might never be even simultaneously (Figure 11.2 shows an example). We will explain how to handle such *bad* edges in Section 11.3. Let us suppose for the remainder of Section 11.1 that for some constant $\gamma > 0$, all edges are γ-good. We may assume $\gamma \leq \frac{1}{5}$.

Assuming that all edges are γ-good, we can define for every edge e a reduction event R_e with probability exactly γ that is a subevent of the event that both top cuts of e are even. More precisely, let b_e be a Bernoulli variable with success probability $\frac{\gamma}{\mathbb{P}[\text{both top cuts of } e \text{ are even}]}$ that is independent of S and all other $b_{e'}$, and let R_e be the event that both top cuts of e are even and $b_e = 1$. We set

$$r_e := x_e^* \cdot \mathbb{1}_{R_e}, \tag{11.1}$$

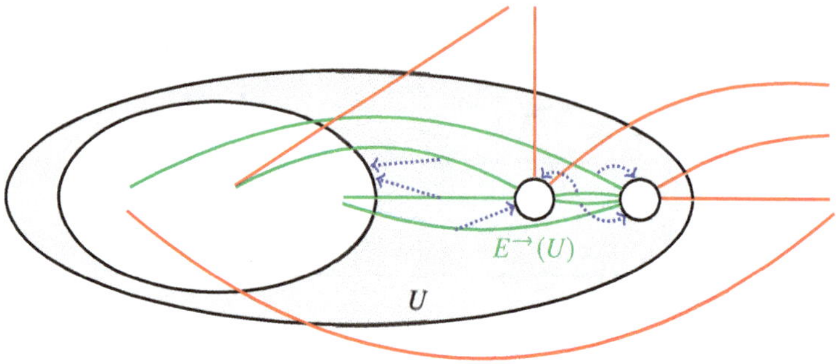

Figure 11.3 A set $U \in \mathcal{H}$ (the large ellipse, corresponding to $\{v_1, \ldots, v_5\}$ in Figure 11.1) with its three children. The edges in $E^{\rightarrow}(U)$ are shown in green; the edges in $\delta(U)$ are shown in red. The dotted blue arrows show a possible assignment of the green edges to the vertices: In this case, each green edge pays for exactly one incident red edge. In general, the assignment can be fractional.

where $\mathbb{1}_{R_e}$ denotes the indicator variable of the reduction event R_e. Then $r_e \leq x_e^*$ for all $S \in \mathcal{S}$, and by the choice of the reduction event R_e, we have $\mathbb{E}[r_e] = \gamma \cdot x_e^*$. We remark that we will choose the constant η much smaller than γ.

Simply setting $s = -r$ would not be feasible (i.e., it would violate condition (ii) of Theorem 10.17) because $u \in \mathcal{H}$ might be odd although r_e is negative for some edges $e \in \delta(u)$ for which u is not a top cut. To ensure $s(\delta(u)) \geq 0$ also in this case, we need the increase vector i. The basic idea is that we will make some edges for which u is a top cut "responsible for compensating the reduction $r(\delta(u))$" by ensuring $i(\delta(u)) \geq r(\delta(u))$ whenever $|\delta_S(u)|$ is odd. In order to decide which edges will be increased for the compensation, we use the responsibility assignment described in the next lemma (cf. Figure 11.3). This lemma does not apply to triangles (i.e., cuts with exactly two children in the hierarchy). However, recall that triangles are polygons and we assumed that our hierarchy does not contain polygons. (We will describe the main ideas needed to handle polygons and in particular triangles in Section 11.5.)

We use the following notation. For a set $U \in \mathcal{H}$, let $\mathscr{C}(U)$ denote the set of its children in $\mathcal{H}$ (i.e., the set of maximal proper subsets of U that belong to $\mathcal{H}$). For $u \in \mathcal{H}$ with parent U in $\mathcal{H}$ (i.e., $u \in \mathscr{C}(U)$), we define $\delta^{\uparrow}(u) := \delta(u) \cap \delta(U)$ and $\delta^{\rightarrow}(u) := \delta(u) \setminus \delta(U)$. Note that $\delta^{\rightarrow}(u)$ is the set of edges for which u is a top cut. Define $E^{\rightarrow}(U) := E[U] \setminus \bigcup_{u \in \mathscr{C}(U)} E[u] = \bigcup_{u \in \mathscr{C}(U)} \delta^{\rightarrow}(u)$.

Recall that we will choose $\eta > 0$ to be a tiny constant. In particular, we will choose $\eta \leq \frac{1}{10}$.

Lemma 11.3. *For every $U \in \mathcal{H}$ that is not a triangle, there exists a responsibility assignment $a : \mathscr{C}(U) \times E^{\rightarrow}(U) \rightarrow \mathbb{R}_{\geq 0}$ such that*

- $a(u, e) = 0$ *if e is not contained in $\delta_E(u)$,*
- $a(u, e) + a(v, e) \leq x^*_e \cdot (1 + 5\eta)$ *for every edge $e \in E^{\rightarrow}(U)$ with top cuts u and v, and*
- $\sum_{e \in \delta^{\rightarrow}(u)} a(u, e) = x^*(\delta^{\uparrow}(u))$ *for all $u \in \mathscr{C}(U)$.*

Proof. For a set $\mathcal{U} \subseteq \mathscr{C}(U)$, let $E^{\mathcal{U}} := \bigcup_{u \in \mathcal{U}} \delta^{\rightarrow}(u)$. By Theorem 3.13, a responsibility assignment a as claimed exists if and only if for all $\mathcal{U} \subseteq \mathscr{C}(U)$,

$$\sum_{e \in E^{\mathcal{U}}} (1 + 5\eta) \cdot x^*_e \; \geq \; \sum_{u \in \mathcal{U}} x^*(\delta^{\uparrow}(u)). \tag{11.2}$$

Recall that Proposition 10.8 implies $x^*(\delta(A)) \leq 2+4\eta$ for all $A \in \mathcal{H}$ (since there are no polygons, all elements of $\mathcal{H}$ are η-mincuts). Moreover, $x^*(\delta(A)) \geq 2$.

We first prove (11.2) for $\mathcal{U} = \{u\}$ with $u \in \mathscr{C}(U)$. Using

$$\begin{aligned} 2 \cdot x^*(\delta^{\rightarrow}(u)) &= x^*(\delta(u)) + x^*(\delta(U \setminus u)) - x^*(\delta(U)) \\ &\geq x^*(\delta(u)) + 2 - (2 + 4\eta) \end{aligned}$$

and

$$\begin{aligned} 2 \cdot x^*(\delta^{\uparrow}(u)) &= x^*(\delta(u)) + x^*(\delta(U)) - x^*(\delta(U \setminus u)) \\ &\leq x^*(\delta(u)) + (2 + 4\eta) - 2, \end{aligned} \tag{11.3}$$

we can bound the ratio by

$$\frac{x^*(\delta^{\uparrow}(u))}{x^*(E^{\mathcal{U}})} = \frac{x^*(\delta^{\uparrow}(u))}{x^*(\delta^{\rightarrow}(u))} \leq \frac{x^*(\delta(u)) + 4\eta}{x^*(\delta(u)) - 4\eta} \leq \frac{2 + 4\eta}{2 - 4\eta} \leq 1 + 5\eta,$$

which yields (11.2) in this case.

Let us now consider the remaining case $|\mathcal{U}| \geq 2$ and let $u, v \in \mathcal{U}$. Then

$$\begin{aligned} x^*(E^{\mathcal{U}}) &\geq \tfrac{1}{2}\Big(x^*(\delta(u)) + x^*(\delta(v)) + x^*(\delta(U \setminus (u \cup v)) - x^*(\delta(U))\Big) \\ &\geq \tfrac{1}{2}\big(2 + 2 + 2 - (2 + 4\eta)\big) \\ &= 2 - 2\eta. \end{aligned}$$

Because $\sum_{u \in \mathcal{U}} x^*(\delta^{\uparrow}(u)) \leq x^*(\delta(U)) \leq 2 + 4\eta$, this implies (11.2). □

Using the responsibility assignment from Lemma 11.3, we can now define the increase vector i. For an edge e with top cuts u and v, we set

$$\begin{aligned} i_e := \; & \frac{a(u, e)}{x^*(\delta^{\uparrow}(u))} \cdot \sum_{f \in \delta^{\uparrow}(u)} r_f \cdot \mathbb{1}[|\delta_S(u)| \text{ is odd}] \\ &+ \frac{a(v, e)}{x^*(\delta^{\uparrow}(v))} \cdot \sum_{f \in \delta^{\uparrow}(v)} r_f \cdot \mathbb{1}[|\delta_S(v)| \text{ is odd}], \end{aligned} \tag{11.4}$$

where $\mathbb{1}[|\delta_S(u)| \text{ is odd}]$ and $\mathbb{1}[|\delta_S(v)| \text{ is odd}]$ are the indicator functions of $|\delta_S(u)|$ and $|\delta_S(u)|$ being odd, respectively.

This ensures the properties (i) and (ii) of Theorem 10.17:

Lemma 11.4. *Let* $s := i - r$, *where* r *and* i *are defined by* (11.1) *and* (11.4). *Then the following always holds: For all* $e \in E$, *we have* $s_e \geq -x^*_e$, *and for every set* $A \in \mathcal{H}$, *we have* $|\delta_S(A)|$ *even or* $s(\delta(A)) \geq 0$.

Proof. The first property is guaranteed by $r_e \leq x^*_e$ and $i_e \geq 0$.

We show that for every set $u \in \mathcal{H}$, we have $s(\delta(u)) \geq 0$ or $|\delta_S(u)|$ even. Indeed, for $u \in \mathcal{H}$ with $|\delta_S(u)|$ odd, we have

$$r(\delta(u)) = \sum_{f \in \delta^{\uparrow}(u)} r_f$$

because u is a top cut for edges $f \in \delta(u) \setminus \delta^{\uparrow}(u)$ and hence $r_f = 0$ for such edges whenever $|\delta_S(u)|$ is odd. Moreover, because $|\delta_S(u)|$ is odd and by Lemma 11.3 we have $\sum_{e \in \delta^{\rightarrow}(u)} a(u,e) = x^*(\delta^{\uparrow}(u))$, we have

$$i(\delta(u)) \geq \sum_{e \in \delta^{\rightarrow}(u)} \frac{a(u,e)}{x^*(\delta^{\uparrow}(u))} \cdot \sum_{f \in \delta^{\uparrow}(u)} r_f = \sum_{f \in \delta^{\uparrow}(u)} r_f = r(\delta(u)),$$

implying $s(\delta(u)) = i(\delta(u)) - r(\delta(u)) \geq 0$. □

Property (iii) of Theorem 10.17 applies only to polygon cuts, and for the purpose of this overview, we assumed that the hierarchy $\mathcal{H}$ contains no polygon cuts.

Therefore, it remains to discuss the property (iv), which states that the expectation of s_e is negative for all $e \in E$. More precisely, we need $\mathbb{E}[s_e] \leq -\varepsilon_0 \cdot x^*_e$ for every $e \in E$ and some constant $\varepsilon_0 \geq 174\eta$ (for this inequality, we can choose η sufficiently small if $\varepsilon_0 > 0$).

For an edge e with top cuts u and v, we have $\mathbb{E}[s_e] = \mathbb{E}[i_e] - \mathbb{E}[r_e] = \mathbb{E}[i_e] - \gamma x^*_e$. Using Lemma 11.3 and upper bounding $\mathbb{1}[|\delta_S(u)| \text{ is odd}]$ and $\mathbb{1}[|\delta_S(v)| \text{ is odd}]$ in the definition of the increase vector by 1, we obtain the bound $\mathbb{E}[s_e] \leq \gamma(a(u,e) + a(v,e)) - \gamma x^*_e \leq 5\eta\gamma x^*_e$. To improve on this, we would like to show that for an edge e with top cut u, the cut size $|\delta_S(u)|$ is not always odd when edges $f \in \delta^{\uparrow}(u)$ are reduced (i.e., when $r_f > 0$).

Recall that for an edge f, we have $r_f = x^*_f \cdot \mathbb{1}_{R_e}$, where R_f denotes the reduction event for f, which occurs with probability γ. Thus,

$$\begin{aligned}\mathbb{E}[i_e] &= \frac{a(u,e)}{x^*(\delta^\uparrow(u))} \cdot \sum_{f \in \delta^\uparrow(u)} \gamma \cdot x^*_f \cdot \mathbb{P}\big[|\delta_S(u)| \text{ is odd} \bigm| R_f\big] \\ &\quad + \frac{a(v,e)}{x^*(\delta^\uparrow(v))} \cdot \sum_{f \in \delta^\uparrow(u)} \gamma \cdot x^*_f \cdot \mathbb{P}\big[|\delta_S(v)| \text{ is odd} \bigm| R_f\big].\end{aligned}$$

Hence, if we could bound $\mathbb{P}\big[|\delta_S(u)| \text{ is odd} \bigm| R_f\big]$ and $\mathbb{P}\big[|\delta_S(v)| \text{ is odd} \bigm| R_f\big]$ from above by some constant $p \le 1 - \frac{175\eta}{\gamma}$, we would get

$$\begin{aligned}\mathbb{E}[i_e] &= \frac{a(u,e)}{x^*(\delta^\uparrow(u))} \cdot x^*(\delta^\uparrow(u)) \cdot \gamma \cdot p + \frac{a(v,e)}{x^*(\delta^\uparrow(v))} \cdot x^*(\delta^\uparrow(v)) \cdot \gamma \cdot p \\ &= (a(u,e) + a(v,e)) \cdot \gamma \cdot p \\ &\le (1 + 5\eta) \cdot x^*_e \cdot \gamma \cdot p \\ &\le (1 + 5\eta) \cdot x^*_e \cdot (\gamma - 175\eta) \\ &\le (\gamma - 174\eta) \cdot x^*_e,\end{aligned}$$

implying $\mathbb{E}[s_e] = \mathbb{E}[i_e] - \gamma x^*_e \le -174\eta x^*_e$, which yields (iv) of Theorem 10.17 for $\varepsilon_0 := 174\eta$.

We now show that if $x^*(\delta^\uparrow(u))$ is "sufficiently fractional" for all sets $u \in \mathcal{H}$, then we can indeed show $\mathbb{P}\big[|\delta_S(u)| \text{ is odd} \bigm| R_f\big] < 1 - \frac{175\eta}{\gamma}$ for all $u \in \mathcal{H}$ and $f \in \delta^\uparrow(u)$. More precisely, we assume that there exists some constant $\varepsilon_F > 0$ such that

$$\varepsilon_F < x^*(\delta^\uparrow(u)) < 1 - \varepsilon_F \quad \text{for all sets } u \in \mathcal{H}, \tag{11.5}$$

and η is chosen small enough so that $\eta \le \frac{\varepsilon_F \cdot \gamma}{200}$. Note that we always have $0 \le x^*(\delta^\uparrow(u)) \le 1 + 4\eta$ (cf. (11.3)), and we will discuss how to get rid of the assumption (11.5) and deal with the cases $x^*(\delta^\uparrow(u)) \approx 0$ and $x^*(\delta^\uparrow(u)) \approx 1$ in Section 11.4.

We first argue that $|\delta^{\rightarrow}_S(u)|$ and the reduction event R_f for $f \in \delta^\uparrow(u)$ are approximately independent. Let U be the parent of u in the hierarchy $\mathcal{H}$. One of the top cuts of f is a superset of U (possibly U itself), and the other one is disjoint from U. The parity of these top cuts (and hence R_f if we choose the reduction event appropriately) depends only on the edges in S/U. These are approximately independent from the edges in $S[U]$, which contains $\delta^{\rightarrow}_S(u)$. More precisely, we recall that by Proposition 10.8, $\mathbb{P}\big[(U, S[U]) \text{ is a tree}\big] \ge 1 - 2\eta$. Conditioning on this event, the random variable r_f is independent of $S[U]$ due to Lemma 5.18.

Let $\tilde{\mu}$ be the distribution we obtain from μ by conditioning on $(U, S[U])$ being a tree. Then by Proposition 10.8 and Lemma 5.18, we have

$$\begin{aligned}
&\mathbb{P}_{S\sim\mu}\big[|\delta_S(u)| \text{ is odd} \bigm| R_f\big] \\
&\quad\le \mathbb{P}_{S\sim\tilde{\mu}}\big[|\delta_S(u)| \text{ is odd} \bigm| R_f\big]\cdot(1-2\eta)+2\eta \qquad (11.6)\\
&\quad\le \max\Big\{\mathbb{P}_{S\sim\tilde{\mu}}\big[|\delta_S^{\rightarrow}(u)| \text{ is odd}\big],\ \mathbb{P}_{S\sim\tilde{\mu}}\big[|\delta_S^{\rightarrow}(u)| \text{ is even}\big]\Big\}\cdot(1-2\eta)+2\eta,
\end{aligned}$$

where we used that the event R_f depends only on S/U and that $|\delta_S^{\rightarrow}(u)|$ depends only on $S[U]$.

Next we show that the expected value of $|\delta_S^{\rightarrow}(u)|$ does not differ much between μ and $\tilde{\mu}$:

Lemma 11.5. $\mathbb{E}_{S\sim\mu}\big[|\delta_S^{\rightarrow}(u)|\big] \le \mathbb{E}_{S\sim\tilde{\mu}}\big[|\delta_S^{\rightarrow}(u)|\big] \le \mathbb{E}_{S\sim\mu}\big[|\delta_S^{\rightarrow}(u)|\big]+2\eta.$

Proof. The lower bound follows from applying Corollary 5.17 to $A = \delta^{\rightarrow}(u)$. Applying it to $A = E[U] \setminus \delta^{\rightarrow}(u)$ yields the upper bound via

$$\begin{aligned}
\mathbb{E}_{S\sim\tilde{\mu}}\big[|\delta_S^{\rightarrow}(u)|\big] &= (|U|-1) - \mathbb{E}_{S\sim\tilde{\mu}}\big[|A\cap S|\big]\\
&\le (|U|-1) - \mathbb{E}_{S\sim\mu}\big[|A\cap S|\big]\\
&< x^*(E[U|) + 2\eta - \mathbb{E}_{S\sim\mu}\big[|A\cap S|\big]\\
&= 2\eta + \mathbb{E}_{S\sim\mu}\big[|\delta_S^{\rightarrow}(u)|\big].
\end{aligned}$$

In the strict inequality, we used $x^*(E[U]) = |U| - \frac{1}{2}x^*(\delta(U)) > |U| - 1 - 2\eta$ (cf. Proposition 10.8). □

Now we can prove:

Lemma 11.6. *Assuming* (11.5) *for some* $0 < \varepsilon_F \le \frac{1}{10}$ *and* $\eta \le \frac{\varepsilon_F\cdot\gamma}{200}$, *we have for all* $u \in \mathcal{H}$ *and all* $f \in \delta^{\uparrow}(u)$*:*

$$\mathbb{P}_{S\sim\mu}\big[|\delta_S(u)| \textit{ is odd} \bigm| R_f\big] < 1 - \frac{175\eta}{\gamma}.$$

Proof. Using our assumption $\varepsilon_F < x^*(\delta^{\uparrow}(u)) < 1 - \varepsilon_F$ together with the inequalities $2 \le x^*(\delta(u)) < 2 + 4\eta$ (cf. Proposition 10.8), we obtain the bounds $1 + \varepsilon_F < x^*(\delta^{\rightarrow}(u)) < 2 + 4\eta - \varepsilon_F$. Thus, using Lemma 11.5,

$$1+\varepsilon_F < \mathbb{E}_{S\sim\tilde{\mu}}\big[|\delta_S^{\rightarrow}(u)|\big] < 2 + 6\eta - \varepsilon_F.$$

This shows that $\mathbb{E}_{S\sim\tilde{\mu}}\big[|\delta_S^{\rightarrow}(u)|\big]$ has distance at least $\varepsilon_F - 6\eta \ge \frac{194\eta}{\gamma}$ to the nearest integer, where we used that we chose $\eta \le \frac{\varepsilon_F\cdot\gamma}{200}$. Using properties of the $\lambda|_{E[U]}$-uniform distribution on the spanning trees of $G[U]$ (cf. Lemma 5.18), this implies

$$\begin{aligned}\mathbb{P}_{S\sim\tilde{\mu}}\left[|\delta_S^{\rightarrow}(u)| \text{ is odd}\right] &\geq \tfrac{1}{2}\left(1-e^{-388\eta/\gamma}\right)\\ \mathbb{P}_{S\sim\tilde{\mu}}\left[|\delta_S^{\rightarrow}(u)| \text{ is even}\right] &\geq \tfrac{1}{2}\left(1-e^{-388\eta/\gamma}\right),\end{aligned} \tag{11.7}$$

as we will show in Section 11.2 (see Lemma 11.13). Then we conclude

$$\begin{aligned}&\max\left\{\mathbb{P}_{S\sim\tilde{\mu}}\left[|\delta_S^{\rightarrow}(u)| \text{ is odd}\right], \mathbb{P}_{S\sim\tilde{\mu}}\left[|\delta_S^{\rightarrow}(u)| \text{ is even}\right]\right\}\\ &= 1-\min\left\{\mathbb{P}_{S\sim\tilde{\mu}}\left[|\delta_S^{\rightarrow}(u)| \text{ is odd}\right], \mathbb{P}_{S\sim\tilde{\mu}}\left[|\delta_S^{\rightarrow}(u)| \text{ is even}\right]\right\}\\ &\leq 1-\tfrac{1}{2}\left(1-e^{-388\eta/\gamma}\right)\end{aligned}$$

and thus (using (11.6) and $e^{-388\eta/\gamma} < 1-\frac{352\eta}{\gamma}$ for $\frac{\eta}{\gamma} \leq \frac{1}{2000}$):

$$\begin{aligned}\mathbb{P}_{S\sim\mu}\left[|\delta_S(u)| \text{ is odd} \mid R_f\right] &\leq \left(1-\tfrac{1}{2}\left(1-e^{-388\eta/\gamma}\right)\right)\cdot(1-2\eta)+2\eta\\ &\leq \left(1-\tfrac{176\eta}{\gamma}\right)\cdot(1-2\eta)+2\eta\\ &= 1-\frac{176\eta}{\gamma}+\frac{352\eta^2}{\gamma}\\ &< 1-\frac{175\eta}{\gamma},\end{aligned}$$

as required. □

This completes our proof of Theorem 10.17 under several simplifying assumptions. The following sections will provide more details and sketch the main ideas needed to get rid of the assumptions we made. Specifically, in Section 11.2, we discuss some properties of λ-uniform distributions in more detail, in particular proving (11.7). Then we describe the main ideas used to get rid of the various simplifying assumptions we made in this overview. Section 11.3 describes how edges that are not γ-good can be handled. Section 11.4 addresses the case when $x^*(\delta^{\uparrow}(u))$ is not "sufficiently fractional." Finally, Section 11.5 discusses polygons (including triangles, i.e., cuts with exactly two children in the hierarchy).

11.2 Strongly Rayleigh Distributions

In this section, we show that λ-uniform distributions have much stronger properties than negative correlation, and these are crucial for the proof of the main payment theorem by Karlin, Klein, and Oveis Gharan [2021]. Borcea, Brändén, and Liggett [2009] defined strongly Rayleigh distributions and showed that λ-uniform distributions are strongly Rayleigh.

Definition 11.7 (generating polynomial, stable, strongly Rayleigh distribution). Let E be a finite set, and let μ be a probability distribution on 2^E. Then the *generating polynomial* of μ is

$$g(z) = \sum_{A \subseteq E} \mu(A) \cdot \prod_{e \in A} z_e.$$

A polynomial $g : \mathbb{C}^E \to \mathbb{C}$ is called *stable* if $g(z) \neq 0$ for every $z \in \mathbb{C}^E$ with $\mathrm{Im}(z_e) > 0$ for all $e \in E$. The probability distribution μ is called *strongly Rayleigh* if its generating polynomial is stable.

To show that λ-uniform distributions are strongly Rayleigh, we need a well-known lemma from complex analysis, which we cite without proof:

Lemma 11.8. *Let $(g_k)_{k \in \mathbb{N}}$ be a sequence of stable polynomials that converges uniformly on compact subsets to a polynomial g. Then g is stable or constant zero.*

Theorem 11.9 (Borcea, Brändén, and Liggett [2009]). *Let $G = (V, E)$ be a connected graph and $\lambda_e > 0$ for $e \in E$. Let μ^λ denote the λ-uniform distribution of the spanning trees of G (and $\mu^\lambda(S) := 0$ if S is not the edge set of a spanning tree). Then μ^λ is strongly Rayleigh.*

Proof. By Definition 11.7, the generating polynomial of μ^λ is

$$g(z) = \sum_{S \in \mathcal{S}} \mu^\lambda(S) \cdot \prod_{e \in S} z_e = \frac{1}{\Lambda} \sum_{S \in \mathcal{S}} \prod_{e \in S} \lambda_e z_e,$$

where $\Lambda = \sum_{S \in \mathcal{S}} \prod_{e \in S} \lambda_e$. Kirchhoff's matrix tree theorem (Theorem 5.19) implies that for all $z \in \mathbb{R}^E_{>0}$ we have

$$\Lambda \cdot g(z) = \det(L_z^{1..n-1}), \tag{11.8}$$

where L_z is the weighted Laplacian of G for edge weights $\lambda_e z_e$, and $L_z^{1..n-1}$ arises by deleting the n-th row and column. We claim that (11.8) holds for all $z \in \mathbb{C}^E$.

To prove this claim, we note that both sides of (11.8) are given by polynomials in z with real coefficients. Considering the difference of the two polynomials and substituting $z = v + \underline{1}$, where $\underline{1}$ is the all-1 vector, we obtain a polynomial p defined by

$$p(v) = \Lambda \cdot g(v + \underline{1}) - \det(L_{v+\underline{1}}^{1..n-1}) = \sum_{F \subseteq E} \alpha_F \prod_{e \in F} v_e$$

for some real coefficients α_F. We show that α_F is zero for all $F \subseteq E$. To this end, we interpret p as a function from $\mathbb{R}^E$ to $\mathbb{R}$. Because (11.8) holds for all

$z \in \mathbb{R}^E_{>0}$, this function p is constant zero in a neighborhood of the origin (here we use the shift by $\underline{1}$). Since $p(0) = 0$, we have $\alpha_\emptyset = 0$. To show that $\alpha_F = 0$ for a nonempty set $F \subseteq E$, we use that all partial derivatives are 0 at the origin. So $\alpha_F = \frac{\partial^{|F|} p}{\prod_{e \in F} \partial v_e}(0) = 0$. This proves the claim.

Having established (11.8), we note that for all $z \in \mathbb{C}^E$:

$$\Lambda \cdot g(z) = \det(L_z^{1..n-1}) = \det(L_z + e_n e_n^\top) = \det\left(e_n e_n^\top + \sum_{e \in E} z_e L_e\right), \quad (11.9)$$

where $e_n = (0, \ldots, 0, 1)^\top$ denotes the n-th unit vector and $L_e = \lambda_e(\chi^{\{v\}} - \chi^{\{w\}})(\chi^{\{v\}} - \chi^{\{w\}})^\top$ for $e = \{v, w\} \in E$. We used that $\det L_z = 0$ (every column sum of L_z is zero).

To show that the g is stable, we construct a sequence of stable polynomials converging uniformly on compact subsets to $\Lambda \cdot g$ and apply Lemma 11.8. For $k \in \mathbb{N}$, let

$$g_k(z) := \det\left(e_n e_n^\top + \sum_{e \in E} z_e L_e + \frac{1}{k} i I\right),$$

where I denotes the $n \times n$ identity matrix. The sequence $(g_k)_{k \in \mathbb{N}}$ converges uniformly on compact subsets to $\Lambda \cdot g$. Since g is not constant zero (e.g., $g(\underline{1}) = 1$), it remains to show that every g_k is stable.

Now fix $k \in \mathbb{N}$ and some $z = \mathrm{Re}(z) + i\,\mathrm{Im}(z) \in \mathbb{C}^E$ with $\mathrm{Im}(z) \in \mathbb{R}^E_{>0}$ (and $\mathrm{Re}(z) \in \mathbb{R}^E$). We show $g_k(z) \neq 0$.

Since $M := \frac{1}{k} I + \sum_{e \in E} \mathrm{Im}(z_e) L_e$ is positive definite, it has a square root $M^{1/2}$ (i.e., a symmetric nonsingular real matrix with $M^{1/2} M^{1/2} = M$). Define

$$H = -M^{-1/2}\left(e_n e_n^\top + \sum_{e \in E} \mathrm{Re}(z_e) L_e\right) M^{-1/2}$$

and note that H is a symmetric real matrix. Then (11.9) can be written as

$$g_k(z) = \det\left(M^{1/2}(iI - H) M^{1/2}\right) = (\det M) \cdot (\det(iI - H)).$$

Since $x \mapsto \det(xI - H)$ is the characteristic polynomial (in the single variable x) of the real symmetric matrix H, its roots are the eigenvalues of H, which are all real. Hence $\det(iI - H) \neq 0$. Since $\det M > 0$, we conclude $g_k(z) \neq 0$. □

In the rest of this section, we will explain how Theorem 11.9 can be used to derive good bounds on the probability of cuts having a certain parity. We will use the fact that the strongly Rayleigh property is maintained under projections. This allows us to restrict our attention to the probability distribution on a subset of the edges (e.g., the edges in a particular cut) while maintaining the property that the distribution is strongly Rayleigh.

Definition 11.10 (projection). Let E be a finite set, and let μ be a strongly Rayleigh probability distribution on 2^E. Let $F \subseteq E$. The *projection of μ onto F* is the probability distribution μ_F defined by

$$\mu_F(B) := \sum_{A \subseteq E : A \cap F = B} \mu(A).$$

Lemma 11.11. *Let E be a finite set and let μ be a strongly Rayleigh probability distribution on 2^E. Let $F \subseteq E$. Then the projection of μ onto F is strongly Rayleigh.*

Proof. Let g denote the generating polynomial of μ, and let g_F denote the generating polynomial of the projection μ_F of μ onto F. For $z \in \mathbb{C}^F$ and $c \in \mathbb{C}$, let $z|c$ be the vector in $\mathbb{C}^E$ for which $(z|c)_e = z_e$ for $e \in F$ and $(z|c)_e = c$ for $e \in E \setminus F$. Then $\prod_{e \in B} z_e = \prod_{e \in A} (z|1)_e$ whenever $A \cap F = B$ and hence

$$\begin{aligned} g_F(z) &= \sum_{B \subseteq F} \mu_F(B) \cdot \prod_{e \in B} z_e \\ &= \sum_{B \subseteq F} \sum_{A \subseteq E : A \cap F = B} \mu(A) \cdot \prod_{e \in B} z_e \\ &= \sum_{A \subseteq E} \mu(A) \cdot \prod_{e \in A} (z|1)_e \\ &= g(z|1). \end{aligned}$$

To show that g_F is stable, we again construct a sequence of stable polynomials $(g_k)_{k \in \mathbb{N}}$ that converges uniformly on compact subsets to g_F. For $k \in \mathbb{N}$, let $g_k : \mathbb{C}^F \to \mathbb{C}$ be defined by $g_k(z) = g\big(z|(1 + \frac{1}{k}i)\big)$ for $z \in \mathbb{C}^F$. Indeed, the sequence converges uniformly on compact subsets to g_F. Moreover, each polynomial g_k is stable because g is stable. By Lemma 11.8, the generating polynomial g_F is stable, and hence μ_F is strongly Rayleigh. □

Using Lemma 11.11, we now show that the number of elements sampled from a fixed subset of the edges (e.g., from a particular cut) obeys the law of a sum of independent Bernoulli random variables (random variables that take values 0 or 1 only), also known as Poisson binomial distribution. This allows us to prove strong bounds on the probability that this edge set contains an even or odd number of edges.

Lemma 11.12. *Let E be a finite set, and let μ be a strongly Rayleigh probability distribution on 2^E. Let $F \subseteq E$. Then there exist independent Bernoulli random variables $B_1, \ldots, B_{|F|}$ such that*

$$\mathbb{P}_{S \sim \mu}[|S \cap F| = k] = \mathbb{P}[B_1 + \cdots + B_{|F|} = k]$$

for all $k \in \mathbb{Z}_{\geq 0}$.

Proof. Consider the univariate polynomial

$$f(z) := \sum_{k=0}^{|F|} \mathbb{P}_{S\sim\mu}[|S \cap F| = k] \cdot z^k = g_F(z, z, \ldots, z),$$

where g_F is the generating polynomial of the projection of μ onto F. Because by Lemma 11.11 the projection of μ onto F is strongly Rayleigh, all (complex) roots of the polynomial f have a nonpositive imaginary part. Since f is a univariate polynomial with real coefficients, its complex zeros come in conjugate pairs. We conclude that f is real rooted. Thus, we can write

$$f(z) = b \cdot \prod_{j=1}^{l} (a_j + z)$$

for some numbers $b, a_1, \ldots, a_l \in \mathbb{R}$ and $l \leq |F|$. Because all coefficients of f are nonnegative by definition and $f \neq 0$, we have $b > 0$ and $a_j \geq 0$ for all $j \in \{1, \ldots, l\}$. Define Bernoulli random variables B_j with success probability $\mathbb{P}[B_j = 1] = p_j := \frac{1}{1+a_j}$ for $j = 1, \ldots, l$. Note that $0 < p_j \leq 1$. Then we have

$$1 = \sum_{k=0}^{|F|} \mathbb{P}_{S\sim\mu}[|S \cap F| = k] = f(1) = b \cdot \prod_{j=1}^{l} (a_j + 1) = b \cdot \prod_{j=1}^{l} \frac{1}{p_j},$$

implying $b = \prod_{j=1}^{l} p_j$. We can read off the probability that $|S \cap F| = k$ from the coefficient of z^k in $f(z)$, which is

$$\mathbb{P}_{S\sim\mu}[|S \cap F| = k] = \sum_{\substack{I \subseteq \{1,\ldots,l\} \\ |I|=k}} b \cdot \prod_{j \notin I} a_j = \sum_{\substack{I \subseteq \{1,\ldots,l\} \\ |I|=k}} \prod_{j \in I} p_j \cdot \prod_{j \notin I} (1 - p_j),$$

where we used $a_j = \frac{1-p_j}{p_j}$ in the last equation. □

Using this, we show the following:

Lemma 11.13. *Let E be a finite set, μ a strongly Rayleigh probability distribution on 2^E, and $F \subseteq E$. Let $q := \mathbb{E}_{S\sim\mu}[|S \cap F|]$, and let $\delta := \min\{q - \lfloor q \rfloor, \lceil q \rceil - q\}$ be the distance of the expectation to the nearest integer. Then*

$$\mathbb{P}_{S\sim\mu}[|S \cap F| \text{ odd}] \geq \frac{1}{2}\left(1 - e^{-2\delta}\right) \text{ and } \mathbb{P}_{S\sim\mu}[|S \cap F| \text{ even}] \geq \frac{1}{2}\left(1 - e^{-2\delta}\right).$$

Proof. By Lemma 11.12, it suffices to show that for any set of Bernoulli random variables $B_1, \ldots, B_n$ with $\mathbb{E}[B_1 + \cdots + B_n] = q$, the sum is even and odd each with probability at least $\frac{1}{2}\left(1 - e^{-2\delta}\right)$.

Fix $n \in \mathbb{N}$. By an observation of Hoeffding [1956], the expectation of any function of the number of successes, including $\mathbb{P}[B_1 + \cdots + B_n \text{ odd}]$ and

$\mathbb{P}[B_1+\cdots+B_n \text{ even}]$, is minimized when all the Bernoulli variables have success probabilities 0, 1, or p, for some $0 < p < 1$. Indeed, if two Bernoulli variables have success probabilities p and q with $0 < p < q < 1$, then $\mathbb{P}[B_1+\cdots+B_n \text{ odd}]$ is $\alpha(p(1-q)+q(1-p))+(1-\alpha)(1-p(1-q)-q(1-p))$ for some $0 \leq \alpha \leq 1$. This is $1-\alpha+(2\alpha-1)(p+q-2pq)$. If $\alpha \geq \frac{1}{2}$, this becomes no greater if we change the two success probabilities to $\frac{p+q}{2}$ and $\frac{p+q}{2}$, and it becomes no smaller if we change the two success probabilities to 0 and $p+q$ (if $p+q \leq 1$) or to $p+q-1$ and 1 (if $p+q \geq 1$). If $\alpha \leq \frac{1}{2}$, the opposite holds.

So let all the Bernoulli variables have success probabilities 0, 1, or p (for some $0 < p < 1$). Then it is sufficient to show the claim when all success probabilities are p. In this case,

$$\mathbb{P}[B_1+\cdots+B_n \text{ even}] = \frac{1}{2}\left(1+(1-2p)^n\right). \tag{11.10}$$

Indeed, adding $(p+(1-p))^n = \sum_{j=0}^{n}\binom{n}{j}p^j(1-p)^{n-j}$ and $(-p+(1-p))^n = \sum_{j=0}^{n}\binom{n}{j}(-p)^j(1-p)^{n-j}$ yields $1^n+(1-2p)^n = 2\sum_{j \text{ even}}\binom{n}{j}(p)^j(1-p)^{n-j}$. Dividing by 2 yields (11.10).

Hence $\mathbb{P}[B_1+\cdots+B_n \text{ odd}] = 1-\mathbb{P}[B_1+\cdots+B_n \text{ even}] = \frac{1}{2}\left(1-(1-2p)^n\right)$. So it suffices to show that $|(1-2p)^n| \leq e^{-2\delta}$. This is easy: If $p \leq \frac{1}{2}$, we have $0 \leq (1-2p)^n \leq e^{-2pn} = e^{-2q} \leq e^{-2\delta}$. If $p \geq \frac{1}{2}$, we have $|(1-2p)^n| = |(1-(2-2p))^n| \leq e^{-2(n-q)} \leq e^{-2\delta}$. □

Note that the Chernoff bound (Lemma 5.6) that we get from negative correlation is not sufficient to obtain this result. In fact, there exist negatively correlated Bernoulli random variables whose sum is always odd. See Exercise 11.6 for an example.

The proof of Lemma 11.13 also yields:

Lemma 11.14. *Let $B_1,\ldots,B_n$ be independent Bernoulli random variables with $\mathbb{E}[B_1+\cdots+B_n] = 1$. Then $\mathbb{P}[B_1+\cdots+B_n$ even$] < 0.5677$.*

Proof. As in the proof of Lemma 11.13, we may assume that all Bernoulli variables have success probability p. Then we have (cf. (11.10)):

$$\mathbb{P}[B_1+\cdots+B_n \text{ even}] = \frac{1}{2}\left(1+(1-2p)^n\right).$$

If $p > \frac{1}{2}$, then $n = 1$ and the assertion is trivial. Otherwise, $\frac{1}{2}\left(1+(1-2p)^n\right) \leq \frac{1}{2}(1+e^{-2pn}) = \frac{1}{2}(1+e^{-2}) < 0.5677$. □

11.3 Bad Edges

In this section, we discuss how one can address the issue that in general there does not exist a $\gamma > 0$ such that all edges are γ-good. An edge that is not γ-good is called *γ-bad*. (Recall Figure 11.2 for an example.)

The basic idea to deal with such γ-bad edges is to neither reduce nor increase the parity correction vector for them – that is, for every γ-bad edge e, we set $r_e := 0$ and $i_e := 0$ and thus $s_e = i_e - r_e = 0$. This choice of the slack vector does not have the properties required by the main payment theorem (Theorem 10.17) because we would need $\mathbb{E}[s_e] < -\varepsilon_0 x^*_e$ for every edge e, including the bad edges. Therefore, Karlin, Klein, and Oveis Gharan [2021] showed the following variant of Theorem 10.17, which implies Theorem 10.17 as we prove below.

Theorem 11.15 (Karlin, Klein, and Oveis Gharan [2021]). *Let μ be the maximum entropy distribution on $\mathcal{S}$. There exist constants $\varepsilon_0 > 0$ and $0 < \eta \leq \frac{\varepsilon_0}{522}$, a set $E_{\text{good}} \subseteq E$, and a random vector $s \in \mathbb{R}^E$ such that for the hierarchy $\mathcal{H}$ representing the relevant η-mincuts, the following always holds:*

(i) $s_e \geq -x^*_e$ *for all* $e \in E$;
(ii) *For every set* $A \in \mathcal{H}$ *whose parent is not a polygon, we have* $|\delta_S(A)|$ *even or* $s(\delta(A)) \geq 0$;
(iii) *For every polygon* P*, we have:*
  - *if* P *is not left-happy, then* $s(\delta(A)) \geq 0$ *for every left-relevant set* A;
  - *if* P *is not right-happy, then* $s(\delta(A)) \geq 0$ *for every right-relevant set* A;

(iv) $\mathbb{E}[s_e] \leq -\varepsilon_0 x^*_e$ *for every edge* $e \in E_{\text{good}}$;
(v) $s_e = 0$ *for all* $e \in E \setminus E_{\text{good}}$;
(vi) $x^*(\delta(A) \cap E_{\text{good}}) \geq \frac{3}{4}$ *if*
  - *A is contained in $\mathcal{H}$, or*
  - *A is a left-relevant or right-relevant set of a polygon $P \in \mathcal{H}$.*

We now show that Theorem 11.15 implies Theorem 10.17. The idea is to transfer part of the reduction from good to bad edges.

Proof of Theorem 10.17 We apply Theorem 11.15 to obtain a constant ε_0, an edge set E_{good}, and a random slack vector s. Then we define slack vectors $\bar{s}^{S'}$ for $S' \in \mathcal{S}$ by

$$\bar{s}^{S'}_e := \begin{cases} \mathbb{E}[s_e \mid S = S'] + \frac{2}{3}\varepsilon_0 \cdot x^*_e & \text{if } e \in E_{\text{good}} \\ \mathbb{E}[s_e \mid S = S'] - \frac{1}{3}\varepsilon_0 \cdot x^*_e & \text{if } e \in E \setminus E_{\text{good}}. \end{cases}$$

We claim that these slack vectors fulfill the conditions of Theorem 10.17 for the constant $\bar{\varepsilon}_0 := \frac{1}{3}\varepsilon_0$. Indeed, we have $\bar{s}^{S'}_e \geq -x^*_e$ for all $e \in E$ by Theorem 11.15 (i) and (v). Moreover, if $A \in \mathcal{H}$ or A is a left-relevant or right-relevant set of a polygon $P \in \mathcal{H}$, then by Theorem 11.15 (vi), we have

$$\begin{aligned}\bar{s}^{S'}(\delta(A)) &\geq \mathbb{E}[s(\delta(A)) \mid S = S'] \\ &\quad + \tfrac{2}{3}\varepsilon_0 \cdot x^*(\delta(A) \cap E_{\text{good}}) - \tfrac{1}{3}\varepsilon_0 \cdot x^*(\delta(A) \setminus E_{\text{good}}) \\ &\geq \mathbb{E}[s(\delta(A)) \mid S = S'] + \tfrac{2}{3}\varepsilon_0 \cdot \tfrac{3}{4} - \tfrac{1}{3}\varepsilon_0 \cdot (2 + 4\eta - \tfrac{3}{4}) \\ &\geq \mathbb{E}[s(\delta(A)) \mid S = S'],\end{aligned}$$

which together with (ii) and (iii) from Theorem 11.15 implies that the slack vectors $\bar{s}^{S'}$ fulfill (ii) and (iii) from Theorem 10.17. Finally, for $e \in E_{\text{good}}$, we have $\mathbb{E}_{S' \sim \mu}[\bar{s}_e^{S'}] \leq -\varepsilon_0 x_e^* + \frac{2}{3}\varepsilon_0 \cdot x_e^* = -\frac{1}{3}\varepsilon_0 \cdot x_e^* = -\bar{\varepsilon}_0 x_e^*$, and for $e \in E \setminus E_{\text{good}}$, we have $\mathbb{E}_{S' \sim \mu}[\bar{s}_e^{S'}] = -\frac{1}{3}\varepsilon_0 \cdot x_e^* = -\bar{\varepsilon}_0 x_e^*$ by Theorem 11.15 (v). □

To prove Theorem 11.15, one could fix some (small-enough) constant $\gamma > 0$ and choose E_{good} to be the set of γ-good edges. However, it is more convenient to strengthen the condition a little and choose E_{good} to be the set of *2-2-good* edges:

Definition 11.16 (2-2-happy, 2-2-good). For a tree $S \in \mathcal{S}$, we say that an edge e with top cuts v and w is *2-2-happy* if

- $|\delta_S(v)| = 2$ and $|\delta_S(w)| = 2$, and
- $(v, S[v])$ and $(w, S[w])$ are trees.

An edge e is *2-2-good* if the probability that e is 2-2-happy is at least γ. We call an edge *bad* if it is not 2-2-good.

Exploiting properties of strongly Rayleigh distributions, one can then prove Theorem 11.15 (vi). We remark that in contrast to the previous definition of γ-good edges, we now require the set $\delta_S(u)$ for a top cut u of a 2-2-happy edge to contain exactly two edges, rather than any even number. Moreover, we have the additional requirement that $(v, S[v])$ and $(w, S[w])$ are trees. Both of these differences are not crucial for the proof and are simply more convenient to work with.

In order to follow the proof approach described in Section 11.1 for good edges despite the fact that we set $s_e = 0$ for bad edges e, one needs a stronger version of Lemma 11.3 that assigns only good edges in $E^{\rightarrow}(U)$ to children of U. This makes it necessary to prove strong properties on the structure of bad edges. For example, let us consider a set $U \in \mathcal{H}$ with exactly three children; then $x^*(E^{\rightarrow}(U)) \approx 2$ and $\sum_{u \in \mathscr{C}(U)} x^*(\delta^{\uparrow}(u)) \approx 2$. If a significant fraction of the edges in $x^*(E^{\rightarrow}(U))$ were bad, the desired assignment would not exist because then $x^*(E^{\rightarrow}(U) \cap E_{\text{good}})$ would be much smaller than $\sum_{u \in \mathscr{C}(U)} x^*(\delta^{\uparrow}(u))$.

Karlin, Klein, and Oveis Gharan [2021] proved that in this case, where U has exactly three children, none of the edges in $E^{\rightarrow}(U)$ is bad.

However, $E^{\rightarrow}(U)$ can contain bad edges if U has more than three children. To describe the structure of bad edges, it is useful to partition the edges into bundles:

Definition 11.17 (edge bundle). For two distinct sets $U, W \in \mathcal{H}$, we say that the set of edges with top cuts U and W is an *edge bundle* and denote it by $\{U, W\}$.

Either all edges of a bundle are 2-2-good (then we also call the edge bundle 2-2-good) or all of them are bad (then we also call the edge bundle bad). For edges e in the same edge bundle, the corresponding entries in the slack vector can be chosen to be the same fraction of x^*_e. For an edge bundle f, we define x^*_f to be the sum of the values x^*_e over all edges e in the edge bundle f.

An edge bundle f with $x^*_f \approx \frac{1}{2}$ is also called a *half-edge bundle*. A key property is that every bad edge bundle is a half-edge bundle. This was essentially shown already by Oveis Gharan, Saberi, and Singh [2011].

Moreover, if a set $U \in \mathcal{H}$ has more than three children, there might be bad edges, but the bad edge bundles in $E^{\rightarrow}(U)$ form a matching, and for every top cut u of a bad edge we have $x^{\uparrow}(u) \lesssim \frac{1}{2}$. To prove this, Karlin, Klein, and Oveis Gharan [2021] used probabilistic arguments exploiting further properties of the λ-uniform distribution μ. These properties imply that the desired stronger version of Lemma 11.3 holds (see Exercise 11.7).

11.4 Insufficient Fractionality

Recall that in the analysis described in Section 11.1, we defined a reduction event R_f for every edge f. In Lemma 11.6, we considered $u \in \mathcal{H}$ with $f \in \delta^{\uparrow}(u)$ and bounded $\mathbb{P}_{S \sim \mu}\left[|\delta_S(u)| \text{ is odd } \mid R_f\right]$ by a constant less than 1 under the assumption that $x^*(\delta^{\uparrow}(u))$ is "sufficiently fractional." We used this to obtain a good upper bound on the increase i_e for edges $e \in \delta^{\rightarrow}(u)$.

In this section, we discuss how Karlin, Klein, and Oveis Gharan [2021] handled the fact that the assumption (11.5) of "sufficient fractionality" is not necessarily fulfilled. There are two ways how this assumption can be violated: either $x^*(\delta^{\uparrow}(u)) \leq \varepsilon_F$ or $x^*(\delta^{\uparrow}(u)) \geq 1 - \varepsilon_F$. (Karlin, Klein, and Oveis Gharan [2021] chose $\varepsilon_F = \frac{1}{10}$, but the precise value is irrelevant for our discussion.)

The first of these two cases ($x^*(\delta^{\uparrow}(u)) \leq \varepsilon_F$) is easier to deal with than the second one. Let us first assume the parent U of u has at least four children. Then one can prove a strengthening of the responsibility assignment from Lemma 11.3. More precisely, one ensures that compared to Lemma 11.3, twice as many edges

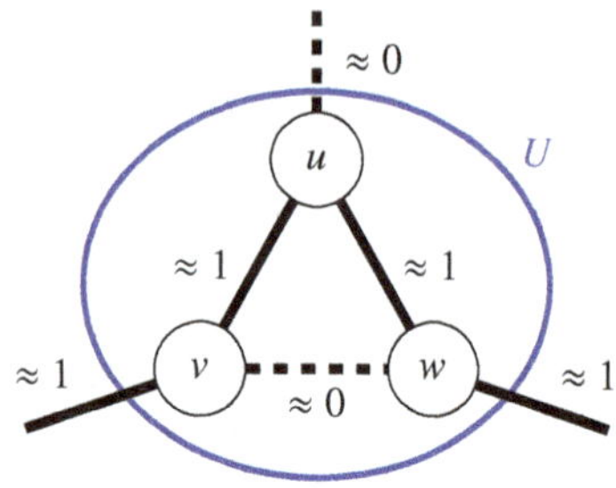

Figure 11.4 Suppose $U \in \mathcal{H}$ has exactly three children u, v, and w and $x^*(\delta^\uparrow(u)) \approx 0$. Then $x^*(\delta^\uparrow(v)) \approx 1$ and $x^*(\delta^\uparrow(w)) \approx 1$. Moreover, $x^*_{\{u,v\}} \approx 1$ and $x^*_{\{u,w\}} \approx 1$.

from $\delta^\rightarrow(u)$ are assigned to u, namely $2x^*(\delta^\uparrow(u))$ instead of just $x^*(\delta^\uparrow(u))$. This allows for replacing the term $\sum_{f\in\delta^\uparrow(u)} r_f \cdot \mathbb{1}[|\delta_S(u)| \text{ is odd}]$ in the definition of the increase vector (11.4) by $\sum_{f\in\delta^\uparrow(u)} \frac{1}{2} r_f$ whenever $x^*(\delta^\uparrow(u)) \le \varepsilon_F$.

Now consider the case where $x^*(\delta^\uparrow(u)) \le \varepsilon_F$ and the parent U of u has exactly three children, say u, v, and w. Then the above strengthening of Lemma 11.3 does not work, but this is actually an easier case. Indeed, the situation must then look as depicted in Figure 11.4. Here, we can ensure that every edge in $E^\rightarrow(U)$ is mostly responsible for v and w, and at most to a minor extent for u. Then we can afford to have a positive increase i_e *always* when the reduction event R_f for some $f \in \delta^\uparrow(u)$ occurs (instead of only with probability $1 - \frac{175\eta}{\gamma}$ as in Lemma 11.6).

Let us now consider the case where $x^*(\delta^\uparrow(u)) \approx 1$. Because $\mathcal{H}$ is a laminar family, the family of sets $U \in \mathcal{H}$ with $u \subsetneq U$ form a chain: We can number them as $U_1, \ldots, U_h$ with $u \subsetneq U_1 \subsetneq \ldots \subsetneq U_h$. We define

$$\delta^j(u) := \begin{cases} \delta^\uparrow(u) \cap \delta(U_j) \setminus \delta(U_{j+1}) & \text{if } j \in \{1, \ldots, h-1\} \\ \delta^\uparrow(u) \cap \delta(U_h) & \text{if } j = h. \end{cases}$$

See Figure 11.5. If $x^*(\delta^j(u))$ is much less than 1 for all $j \in \{1, \ldots, h\}$, then let k be an index such that $\sum_{j=1}^k x^*(\delta^j(u))$ and $\sum_{j=k+1}^h x^*(\delta^j(u))$ are both sufficiently far away from 0 and 1. Then we show that with constant probability, no increase is needed in the reduction event R_f for $f \in \bigcup_{j=k+1}^h \delta^j(u)$ because $\mathbb{P}_{S\sim\mu}\left[|\delta_S(u)| \text{ is odd} \mid R_f\right]$ is much smaller than 1 for such edges f. To see this, note that the parities of $|S \cap \bigcup_{j=k+1}^h \delta^j(u)|$ and $|S \cap \bigcup_{j=1}^k \delta^j(u)|$ and $|S \cap \delta^\rightarrow(u)|$ are approximately independent; this follows from applying Lemma 5.18 once to U_{k+1} and once to U_1. Since $\sum_{j=1}^k x^*(\delta^j(u))$ is far away from 0 and 1, $|S \cap \bigcup_{j=1}^k \delta^j(u)|$ is odd and even both with constant probability by

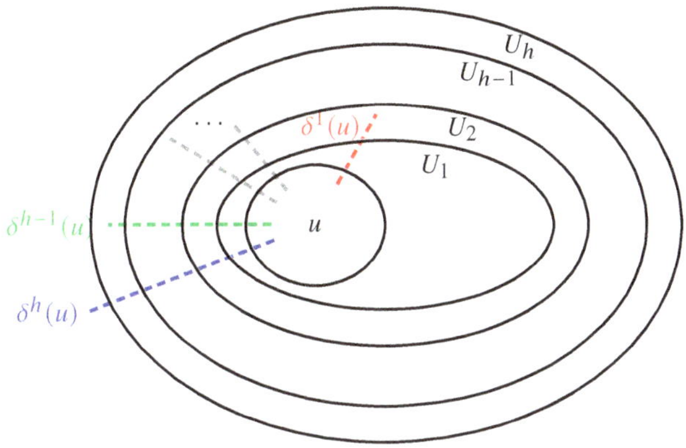

Figure 11.5 Illustration of the partition of $\delta^{\uparrow}(u)$, shown as dashed edges, into the sets $\delta^j(u)$ for $j \in \{1, \ldots, h\}$.

Lemma 11.13. Because $|S \cap \bigcup_{j=1}^{k} \delta^j(u)|$ is independent of R_f, this also holds when conditioning on R_f.

It remains to handle the case where there is an index $j \in \{1, \ldots, h\}$ with $x^*(\delta^j(u)) \approx 1$. Because $x^*(\delta^{\uparrow}(u)) \approx 1$, we have $x^*(\delta^{\rightarrow}(u)) \approx 1$. By Proposition 10.8, $(u, S[u])$ is a tree with probability at least $1 - 2\eta$ and thus the probability that $|S \cap \delta^{\rightarrow}(u)| = 1$ is close to 1. For this reason, it is desirable to have $r_f > 0$ for edge bundles $f \in \delta^j(u)$ when $|S \cap \delta^{\uparrow}(u)|$ is odd rather than when $|S \cap \delta^{\uparrow}(u)|$ is even. Karlin, Klein, and Oveis Gharan [2021] showed that it is possible to achieve that one of the following two things happens:

- at least for approximately half of the edge bundles $f \in \delta^j(u)$ (in terms of x^* value), we can achieve that at least approximately half of the times when the reduction event R_f happens, $|S \cap \delta^{\uparrow}(u)|$ is odd, or
- there is at least one bad half-edge bundle in $\delta^{\uparrow}(u)$.

Note that in the latter case, the bad half-edge bundle is never reduced and thus no compensation for this is needed. In order to prove that the first condition is fulfilled whenever $\delta^{\uparrow}(u)$ does not contain a bad half-edge bundle, a more careful definition of the reduction vector r is needed.

To define the reduction vector r, we now distinguish between different types of good edges. In particular, we will distinguish between so-called *2-1-1-good* edges and edges that are 2-2-good, but not 2-1-1-good. In order to define 2-1-1-good edges, we first fix a small constant $\varepsilon_{\text{partition}} > 0$ and construct a partition $A \,\dot{\cup}\, B \,\dot{\cup}\, C$ of every cut $\delta(W)$ with $W \in \mathcal{H}$ as in the following lemma. See Figure 11.6 for an illustration.

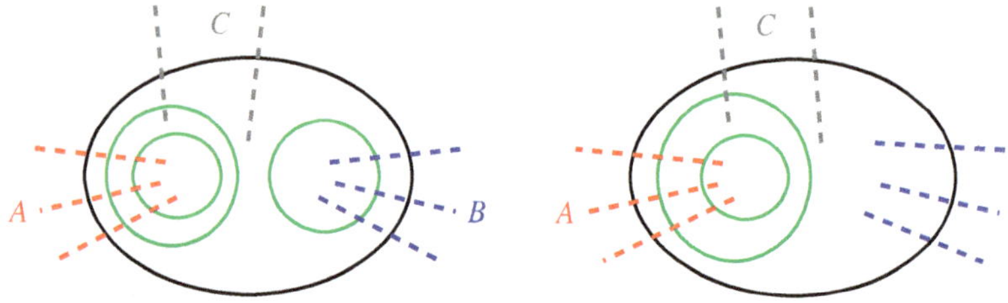

Figure 11.6 The figure shows two examples of a set W (black), the sets $w \in \mathcal{H}$ (green) with $x^*(\delta(w) \cap \delta(W)) \geq 1 - \varepsilon_{\text{partition}}$ and $w \subsetneq W$. The dashed lines show the edges in $\delta(W)$ partitioned into sets A (red), B (blue), and C (gray).

Lemma 11.18. *Let $6\eta < \varepsilon_{\text{partition}} < \frac{1-4\eta}{3}$. For every $W \in \mathcal{H}$, there exists a partition $\{A, B, C\}$ of $\delta(W)$ (possibly after splitting an edge into two copies) such that*

(i) $1 - \varepsilon_{\text{partition}} \leq x^*(A) \leq 1$,

(ii) $1 - \varepsilon_{\text{partition}} \leq x^*(B) \leq 1$, *and*

(iii) *for every $w \in \mathcal{H}$ with $w \subsetneq W$ and $x^*(\delta(w) \cap \delta(W)) \geq 1 - \varepsilon_{\text{partition}}$, we have either $A \subseteq \delta(w) \cap \delta(W) \subseteq A \cup C$ or $B \subseteq \delta(w) \cap \delta(W) \subseteq B \cup C$.*

Proof. We consider the minimal sets $w \in \mathcal{H}$ with $w \subsetneq W$ and $x^*(\delta(w) \cap \delta(W)) \geq 1 - \varepsilon_{\text{partition}}$. Because $x^*(\delta(W)) \leq 2 + 4\eta$ and $\mathcal{H}$ is laminar, there are at most two such sets $w \in \mathcal{H}$. If two such cuts exist, let $a, b \in \mathcal{H}$ be these sets, and let $A \subseteq \delta(a) \cap \delta(W)$ with (i), $B \subseteq \delta(b) \cap \delta(W)$ with (ii), and $C = \delta(W) \setminus (A \cup B)$.

If only one such set exists, let $a \in \mathcal{H}$ be this set, and let $A \subseteq \delta(a) \cap \delta(W)$ with (i). Moreover, let $a' \in \mathcal{H}$ be maximal with $a \subseteq a' \subsetneq W$. Then $x^*(\delta(a') \cap \delta(W)) \leq 1 + 6\eta$ by Proposition 10.8 and thus $x^*(\delta(W) \setminus \delta(a')) \geq 1 - 6\eta \geq 1 - \varepsilon_{\text{partition}}$. We choose $B \subseteq \delta(W) \setminus \delta(a')$ with (ii) and let $C = \delta(W) \setminus (A \cup B)$.

If no set $w \in \mathcal{H}$ with $w \subsetneq W$ and $x^*(\delta(w) \cap \delta(W)) \geq 1 - \varepsilon_{\text{partition}}$ exists, let $\{A, B, C\}$ be an arbitrary partition of $\delta(W)$ with (i) and (ii). This may require replacing an edge e by two copies e^1, e^2 with the same endpoints and $x^*_{e^1} + x^*_{e^2} = x^*_e$. □

A partition A, B, C of $\delta(W)$ as in Lemma 11.18 is called an $\varepsilon_{\text{partition}}$*-degree partition* of $\delta(W)$. Because $\delta(W)$ is an η-mincut, in an $\varepsilon_{\text{partition}}$-degree partition $\{A, B, C\}$, we have $x^*(C) \leq 4\eta + 2\varepsilon_{\text{partition}}$. In the following, we fix for every $W \in \mathcal{H}$ some $\varepsilon_{\text{partition}}$-degree partition A^W, B^W, C^W.

Definition 11.19 (2-1-1-happy, 2-1-1-good)**.** For a tree $S \in \mathcal{S}$, we say that an edge bundle $e = \{v, w\}$ (and every edge of this bundle) is *2-1-1-happy with respect to* w if

- $|\delta_S(v)| = 2$,
- $|S \cap A^w| = |S \cap B^w| = 1$ and $|S \cap C^w| = 0$, and
- $(v, S[v])$ and $(w, S[w])$ are trees.

An edge bundle $e = \{v, w\}$ (end every edge of this bundle) is *2-1-1-good with respect to* w if the probability that e is 2-1-1-happy with respect to w is at least γ.

Now we define the reduction r_f for an edge bundle $f = \{v, w\}$. If f is a bad edge bundle, we define $r_f = 0$ as before. Otherwise, we define events $R_{f,v}$ and $R_{f,w}$ of probability γ as follows. If f is 2-1-1-good with respect to v, then $R_{f,v}$ is a subevent of probability γ of the event that f is 2-1-1-happy with respect to v. If f is not 2-1-1-good with respect to v, then $R_{f,v}$ is a subevent of probability γ of the event that f is 2-2-happy. Such an event exists, because f is not a bad edge. $R_{f,w}$ is defined analogously. These subevents are defined by additional independent Bernoulli variables as in the original definition of the reduction events in Section 11.1. Finally, we define

$$r_f \ := \ x_e^* \cdot \tfrac{1}{2} \cdot \left(\mathbb{1}_{R_{f,v}} + \mathbb{1}_{R_{f,w}} \right). \tag{11.11}$$

Note that we still have $\mathbb{E}[r_f] = \gamma \cdot x_f^*$ for every 2-2-good edge bundle f and that the two top cuts of f are even whenever $r_f > 0$,

Let us now discuss how the refined definition (11.11) of the reduction vector r allows us to handle the remaining case where for some $j \in \{1, \dots, h\}$, we have $x^*(\delta^\uparrow(u)) \approx x^*(\delta^j(u)) \approx 1$. By Lemma 11.18, we have $A^{U_j} \subseteq \delta(u) \cap \delta(U_j) \subseteq A^{U_j} \cup C^{U_j}$ or $B^{U_j} \subseteq \delta(u) \cap \delta(U_j) \subseteq B^{U_j} \cup C^{U_j}$. See Figure 11.7.

Therefore, if an edge $f \in \delta^j(u)$ is 2-1-1-good with respect to U_j, then in the event R_{f,U_j}, we have $|A^{U_j} \cap S| = |B^{U_j} \cap S| = 1$ and $|C^{U_j} \cap S| = 0$ and thus $|\delta(u) \cap \delta(U_j) \cap S| = 1$. Moreover, conditioned on R_{f,U_j}, the graph $(U_j, S[U_j])$ is a spanning tree, and this tree is sampled independently from $S \setminus S[U_j]$ (Lemma 5.18). Using $x^*(\delta(u) \setminus \delta(U_j)) \approx 1$, one can conclude that conditioned on the event R_{f,U_j}, the probability that $|\delta_S(u)| = 2$ is close to 1. Therefore, it suffices to show that a large-enough fraction of the edges in $\delta^j(u)$ is bad or 2-1-1-good with respect to U_j. Karlin, Klein, and Oveis Gharan [2021] proved that this is indeed the case, relying again on the fact that the distribution μ is strongly Rayleigh.

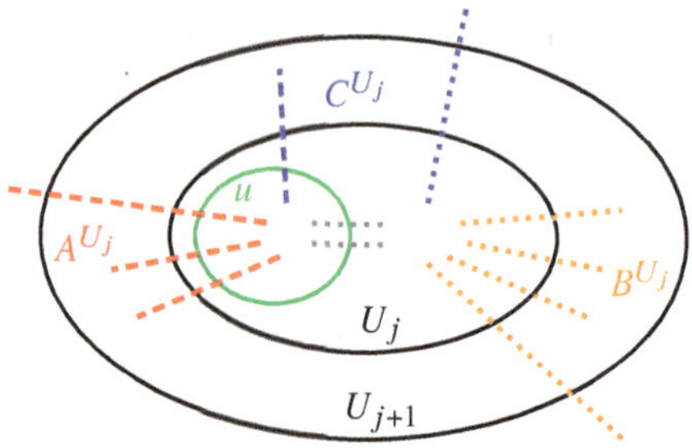

Figure 11.7 The figure shows an example of the sets u , U_j, and U_{j+1} together with the edges in $\delta(u)$ and in $\delta(U_j)$. The edges in $\delta(u) \cap \delta(U_j)$ are dashed, while other edges are dotted. The colors of the edges indicate the $\varepsilon_{\text{partition}}$-degree partition of $\delta(U_j)$ into A^{U_j} (red), B^{U_j} (orange), and C^{U_j} (blue).

11.5 Polygons

So far, we have assumed that the hierarchy $\mathcal{H}$ does not contain polygons. In this section, we discuss how Karlin, Klein, and Oveis Gharan [2021] removed this assumption.

Recall that a triangle is a set $U \in \mathcal{H}$ in the hierarchy with exactly two children $v, w \in \mathcal{H}$. Then we have $x^*(\delta(U)) \approx 2$, and for the edge bundle $\{u, w\}$, we have $x^*_{\{u,w\}} \approx 1$. Triangles are a special case of polygons, but it turns out that all polygons can be treated analogously to triangles. Recall that a polygon $P \in \mathcal{H}$ has children $A_1, \ldots, A_k$ (numbered from left to right), and for all $j \in \{1, \ldots, k-1\}$, the set $A_j \cup A_{j+1}$ is a 4η-mincut by Proposition 10.14. Thus, we have $x^*(\{A_j, A_{j+1}\}) \approx 1$ for all $j \in \{1, \ldots, k-1\}$ (see Figure 11.8 and Exercise 10.3).

We will define the slack vector s_e to be the same on all edge bundles $e = \{A_j, A_{j+1}\}$ for $j = 1, \ldots, k-1$. Then the slack of $\delta(W)$ for every left-relevant set W is almost equal to the slack of $\delta(A_1)$, and the slack of $\delta(W)$ for every right-relevant set W is almost equal to the slack of $\delta(A_k)$. In order to obtain (iii) of Theorem 11.15, it therefore essentially suffices to consider the case $k = 2$ (i.e., the case where P is a triangle). For general polygons, one has to take into account the existence of edges that connect two non-adjacent atoms or an inner atom with the complement of the polygon, but the total x^*-value of these edges is very small, and hence they can essentially be ignored. Therefore, in the following discussion, we will assume that all polygons are triangles.

The reason why we could not handle triangles with the methods described so far is that for triangles, a responsibility assignment with the desired properties (cf. Lemma 11.3) does not exist. Indeed, if $U \in \mathcal{H}$ is a triangle, we have

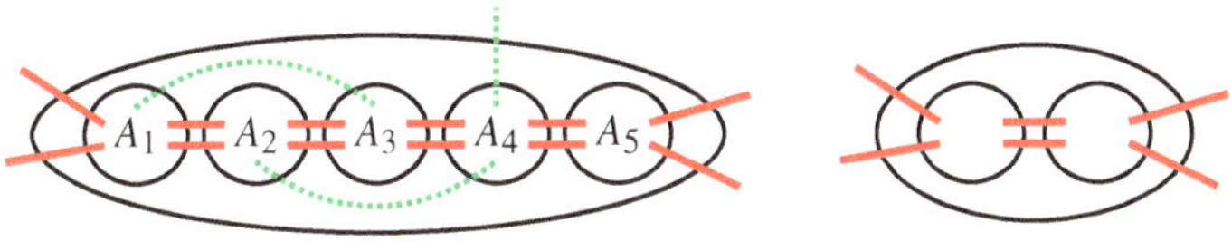

Figure 11.8 A polygon (here with atoms $A_1, \ldots, A_5$) looks like the picture on the left. Every pair of parallel red edges stands for a set of edges with total x^* value very close to 1. Other edges e (like those shown in green, dotted) may exist, but with x_e^* very small. Therefore, a polygon behaves like a triangle (shown on the right).

$\sum_{u \in \mathscr{C}(U)} x^*(\delta^\uparrow(u)) = x^*(\delta(U)) \approx 2 \cdot x^*(E^\to(U))$, and hence there does not exist a responsibility assignment as in Lemma 11.3. Because the edges in $E^\to(U)$ all belong to the same edge bundle, we define $a(u, e) = x^*(\delta^\uparrow(u))$ for both sets $u \in \mathscr{C}(U)$ and this edge bundle e. This results in $\sum_{u \in \mathscr{C}(U)} a(u, e) \approx 2 \approx 2 \cdot x_e^*$; in other words, we assign roughly twice as much to the edges $e \in E^\to(U)$ as for non-triangles.

Call an edge (bundle) e a *triangle edge (bundle)* if $e \in E^\to(U)$ for some triangle $U \in \mathcal{H}$ and a *non-triangle edge (bundle)*, otherwise. Because the condition (iii) in Theorem 11.15 asks for $s(\delta(u))$ to be nonnegative whenever $|\delta_S(u) \cap \delta_S(U)|$ is even – in contrast to condition (ii), which asked for nonnegative slack for sets w whose parent is not a polygon when $|\delta_S(w)|$ is odd – we adjust the definition of the increase event for triangle edge bundles accordingly. For a triangle edge bundle $e = \{u, v\}$, where U is the parent of u and v, this leads to the following choice of the increase vector:

$$
\begin{aligned}
i_e &= \frac{a(u,e)}{x^*(\delta^\uparrow(u))} \cdot \sum_{f \in \delta^\uparrow(u)} r_f \cdot \mathbb{1}[|\delta_S(u) \cap \delta_S(U)| \text{ is even}] \\
&\quad + \frac{a(v,e)}{x^*(\delta^\uparrow(v))} \cdot \sum_{f \in \delta^\uparrow(v)} r_f \cdot \mathbb{1}[|\delta_S(v) \cap \delta_S(U)| \text{ is even}] \\
&= \sum_{f \in \delta^\uparrow(u)} r_f \cdot \mathbb{1}[|\delta_S(u) \cap \delta_S(U)| \text{ is even}] \\
&\quad + \sum_{f \in \delta^\uparrow(v)} r_f \cdot \mathbb{1}[|\delta_S(v) \cap \delta_S(U)| \text{ is even}].
\end{aligned}
$$

In fact, it is sufficient to increase the slack on e by the maximum of these two summands – that is, we can define the increase for triangle edge bundles as

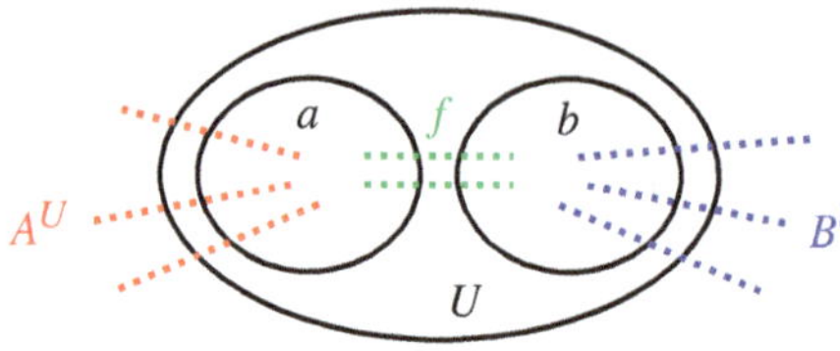

Figure 11.9 The figure illustrates the choice of the $\varepsilon_{\text{partition}}$-degree partition for a triangle U with children $a, b \in \mathcal{H}$. The green edges form the edge bundle $f = \{a, b\}$. We have $U = a \cup b$ and $x(A^U) \approx x(B^U) \approx x_f \approx 1$.

follows:

$$i_e := \max\Big\{ \textstyle\sum_{f\in\delta^\uparrow(u)} r_f \cdot \mathbb{1}[|\delta_S(u) \cap \delta_S(U)| \text{ is even}], \\ \textstyle\sum_{f\in\delta^\uparrow(v)} r_f \cdot \mathbb{1}[|\delta_S(v) \cap \delta_S(U)| \text{ is even}] \Big\}. \quad (11.12)$$

If only at most one of the two terms in the maximum is positive at each time, this does not change the increase vector at all. If both terms are positive at the same time, however, then the increase on e is smaller with this modified definition (11.12), which will turn out to be useful.

To handle the issue that triangle edge bundles need to compensate the reduction for twice as many edges (in terms of x^* value) than non-triangle ones, Karlin, Klein, and Oveis Gharan [2021] made several modifications of the approach discussed so far. The first modification is to reduce triangle edge bundles by a larger amount than non-triangle edge bundles (i.e., we simply scale up their reduction r_e by some constant factor). This will violate condition (i) of Theorem 11.15, but this can easily be fixed by scaling the final random slack vector s down (e.g., by $\frac{1}{2}$).

Of course, this change alone is not sufficient. If we reduce triangle edge bundles by too much, then a new issue arises when non-triangle edge bundles have to compensate for the reduction of triangle edge bundles. Moreover, this scaling of the reduction for triangle edge bundles does not help when triangle edge bundles have to compensate for the reduction of other triangle edge bundles.

To address these issues, Karlin, Klein, and Oveis Gharan [2021] defined the reduction event R_f for triangle edge bundles f in a very careful way. Let $U \in \mathcal{H}$ be a triangle with children $a, b \in \mathcal{H}$. We choose the $\varepsilon_{\text{partition}}$-degree partition of $\delta(U)$ (cf. Lemma 11.18) as $A^U = \delta^\uparrow(a)$, $B^U = \delta^\uparrow(b)$, and $C^U = \emptyset$. See Figure 11.9. Using properties of strongly Rayleigh distributions, Karlin, Klein, and Oveis Gharan [2021] proved that for a sufficiently small choice of the constants $\gamma > 0$ and $\eta > 0$ (where γ is much larger than η), there exists an event R_U such that

- the probability of R_U is γ,
- in the event R_U, we always have that $(U, S[U])$ is a tree and $|S \cap A^U| = |S \cap B^U| = 1$, implying that the triangle U is left-happy and right-happy,
- when conditioning on R_U, the marginals on the edges in A^U and B^U are approximately preserved – that is,

$$\sum_{e \in A^U} \left| \mathbb{P}[e] - \mathbb{P}[e|R_U] \right| \approx 0,$$

$$\sum_{e \in B^U} \left| \mathbb{P}[e] - \mathbb{P}[e|R_U] \right| \approx 0,$$

- when conditioning on R_U, the random variables $S[U]$ and S/U are independent, and the distribution of $S[U]$ when conditioning on R_U is the same as when conditioning on $(U, S[U])$ being a tree.

Except for the condition that $(U, S[U])$ is a tree, the first three properties only involve edges of S/U. Once we have these, the last condition can be obtained easily in addition because conditioned on $(U, S[U])$ being a tree, $S[U]$ is independent of S/U by Lemma 5.18. Obtaining the other properties is substantially more difficult.

We now show how to use the event R_U defined above. For the triangle edge bundle $f = \{a, b\}$, we choose $R_f = R_U$, where U is the parent of a and b. The fact that the marginals for edges in A^U and B^U are approximately preserved when conditioning on R_f is useful for two reasons.

First, consider a set $W \in \mathcal{H}$ with $W \subsetneq U$ and hence without loss of generality $W \subseteq a$. We now want to show that conditioned on R_U, the probability of $|\delta_S(W)|$ being odd is small. Recall that in the setting without triangles, we were just aiming to upper bound this probability by any constant strictly smaller than 1. Now we will aim for the much smaller bound of 0.5678.

Conditioned on the event R_U, we have $|\delta_S(W) \cap \delta_S(U)| \leq |\delta_S(a) \cap \delta_S(U)| = 1$, and thus $|\delta_S(W) \cap \delta_S(U)|$ is a Bernoulli random variable. Moreover, conditioned on the event R_U, the sampling of the tree $(U, S[U])$ is independent of the random variable $|\delta_S(W) \cap \delta_S(U)|$ by the last property of the event R_U. One can show that conditioning on $(U, S[U])$ being a tree yields another strongly Rayleigh distribution. Hence, by the last property of R_U and Lemma 11.12, $|\delta(W) \cap S[U]|$ obeys the law of a sum of independent Bernoulli random variables conditioned on R_U. Because $|\delta_S(W)| = |\delta_S(W) \cap \delta_S(U)| + |\delta(W) \cap S[U]|$, we conclude that there exist independent Bernoulli random variables $B_1, \ldots, B_q$ such that $\mathbb{P}_{S \sim \mu}\left[|\delta_S(W)| = k \;\middle|\; R_U\right] = \mathbb{P}[B_1 + \cdots + B_q = k]$ for all k. Because $\mathbb{P}_{S \sim \mu}\left[|\delta_S(W)| = 0 \;\middle|\; R_U\right] = 0$, at least one of the Bernoulli random variables $B_1, \ldots, B_q$, say B_1, is equal to 1 with probability one. Thus, the probability

$\mathbb{P}_{S\sim\mu}\big[|\delta_S(W)| \text{ is odd} \;\big|\; R_U\big]$ is equal to $\mathbb{P}[B_2 + \cdots + B_q \text{ is even}]$. Using that by the definition of R_U the marginals on the edges in $\delta_S(W)$ are approximately preserved when conditioning on R_U, we obtain

$$1+\mathbb{E}[B_2+\cdots+B_q] = \mathbb{E}_{S\sim\mu}\big[|\delta_S(W)| \;\big|\; R_U\big] \approx \mathbb{E}_{S\sim\mu}\big[|\delta_S(W)|\big] = x^*(\delta(W)) \approx 2$$

and thus $\mathbb{E}[B_2 + \cdots + B_q] \approx 1$. With Lemma 11.14, we get $\mathbb{P}[B_2 + \cdots + B_q \text{ is even}] < 0.5678$ and hence the desired good bound on the probability of $|\delta_S(W)|$ being odd.

This directly implies a better bound on the necessary increase for non-triangle edges compensating for the reduction of triangle edges as follows. Consider a non-triangle edge e with top cut W that needs to (partially) compensate for the reduction of a triangle edge $g \in \delta^{\uparrow}(W)$. We have shown $\mathbb{P}_{S\sim\mu}\big[|\delta_S(W)| \text{ is odd} \;\big|\; R_g\big] < 0.5678$. This improvement allows us to scale up the reduction vector for triangle edge bundles by almost $\frac{1}{0.5678}$ and maintain sufficiently negative expected slack on non-triangle edges. Karlin, Klein, and Oveis Gharan [2021] chose the scaling factor to be approximately 1.7513, which is strictly smaller than $\frac{1}{0.5678}$.

We still need to ensure sufficiently negative expected slack on triangle edges. Consider a triangle $W \in \mathcal{H}$, and denote the children of W by u and v. For the edges bundle $e = \{u, v\}$, we have

$$\mathbb{E}[s_e] \;=\; \mathbb{E}[i_e] - \mathbb{E}[r_e] \;=\; \mathbb{E}[i_e] - 1.7513 \cdot \gamma \cdot x_e^*. \tag{11.13}$$

Moreover, a similar argument as above can be used to show that for a triangle edge $f \in \delta(W)$, we have $\mathbb{P}[|\delta_S(u) \cap \delta_S(W)| \text{ even} \mid R_f] \le 0.5678$, implying

$$\mathbb{E}\Big[r_f \cdot \mathbb{1}\big[|\delta_S(v) \cap \delta_S(W)| \text{ is even}\big]\Big] \;\le\; 1.7513 \cdot x_f^* \cdot \gamma \cdot 0.5678 \;<\; \gamma \cdot x_f^* \tag{11.14}$$

because $\mathbb{P}[R_f] = \gamma$. On the other hand, for non-triangle edges f, we have $r_f \le x_f^*$ and $\mathbb{P}[r_f > 0] = \mathbb{P}[R_f] = \gamma$. Hence, we obtain $\mathbb{E}[i_e] \le x^*(\delta(W)) \cdot \gamma \approx 2\gamma \approx 2\gamma \cdot x_e^*$. Combining this with (11.13) is unfortunately not enough to prove that the slack vector is sufficiently negative as required by Theorem 11.15 (iv). To handle this, Karlin, Klein, and Oveis Gharan [2021] distinguished two cases.

First, consider a triangle u whose parent in the hierarchy is another triangle U. In this case, Karlin, Klein, and Oveis Gharan [2021] used again the marginal-preserving property of the reduction events for triangles. To explain this, let us assume for simplicity that $x^*(\delta(U)) = x^*(\delta(u)) = 2$ and $x^*(\delta(u) \cap \delta(U)) = 1$, which is in general only true approximately. Moreover, we will assume that for the children a and b of the triangle u, we have $x^*(\delta(a)) = x^*(\delta(b)) = 2$ and $x^*(\delta(a) \cap \delta(u)) = x^*(\delta(b) \cap \delta(u)) = 1$, which in general is also only true approximately. See Figure 11.10. These assumptions will simplify

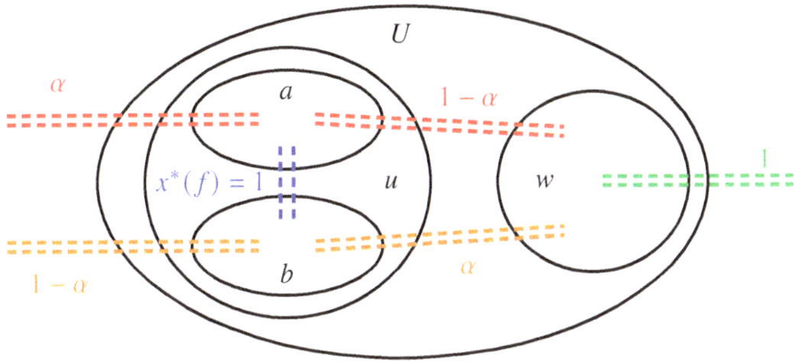

Figure 11.10 A triangle u with children a and b, where the parent U of u is also a triangle. The triangle edge bundle $f = \{a, b\}$ is shown in blue. The edges shown in red are the edges in $A^u = \delta(a) \cap \delta(u)$; the edges shown in orange are those in $B^u = \delta(b) \cap \delta(u)$. The edges leaving U shown on the left are those in $A^U = \delta(u) \cap \delta(U)$, and those edges leaving U on the right of this picture (shown in green) are those in $B^U = \delta(w) \cap \delta(U)$. We have $x^*(A^u) \approx x^*(B^u) \approx x^*(A^U) \approx x^*(B^U) \approx 1$. For simplicity, we assume in this example that these equations are fulfilled exactly and not just approximately. Then the numbers written next to the edges denote their x^* value.

our calculations, but essentially the same argument works in the general case. Let $\alpha := x^*(\delta(a) \cap \delta(U))$. Then $x^*(\delta(b) \cap \delta(U)) = 1 - \alpha$. Moreover, $x^*(\delta(a) \cap E^{\to}(U)) = 1 - \alpha$ and $x^*(\delta(b) \cap E^{\to}(U)) = \alpha$.

We calculate the expected slack $\mathbb{E}[s_f]$ for the triangle edge bundle $f = \{a, b\}$. The edges in $E^{\to}(U)$ belong to the same edge bundle $\{u, w\}$ and are thus all reduced simultaneously in the event R_U. Thus, the increase vector (11.12) here is

$$
\begin{aligned}
i_f &= \max\left\{ \sum_{g \in \delta^{\uparrow}(a)} r_g \cdot \mathbb{1}\big[|A^u| \text{ is even}\big], \sum_{g \in \delta^{\uparrow}(b)} r_g \cdot \mathbb{1}\big[|B^u| \text{ is even}\big] \right\} \\
&\le \max\left\{ \sum_{g \in E[U] \cap \delta(a)} r_g \cdot \mathbb{1}\big[|A^u| \text{ is even}\big], \sum_{g \in E[U] \cap \delta(b)} r_g \cdot \mathbb{1}\big[|B^u| \text{ is even}\big] \right\} \\
&\quad + \sum_{g \in \delta(U) \cap \delta(a)} r_g \cdot \mathbb{1}\big[|A^u| \text{ is even}\big] + \sum_{g \in \delta(U) \cap \delta(b)} r_g \cdot \mathbb{1}\big[|B^u| \text{ is even}\big] \\
&\le \max\Big\{ x^*(E[U] \cap \delta(a)) \cdot 1.7513 \cdot \mathbb{1}[R_U] \cdot \mathbb{1}\big[|A^u| \text{ is even}\big], \\
&\qquad x^*(E[U] \cap \delta(b)) \cdot 1.7513 \cdot \mathbb{1}[R_U] \cdot \mathbb{1}\big[|B^u| \text{ is even}\big] \Big\} \\
&\quad + \sum_{g \in \delta(U) \cap \delta(a)} r_g \cdot \mathbb{1}\big[|A^u| \text{ is even}\big] + \sum_{g \in \delta(U) \cap \delta(b)} r_g \cdot \mathbb{1}\big[|B^u| \text{ is even}\big].
\end{aligned}
$$

Hence, we can bound the expectation by

$$\begin{aligned}
\mathbb{E}[i_f] &\leq \max\{\alpha, 1-\alpha\} \cdot \mathbb{P}\big[\, |A^u| \text{ or } |B^u| \text{ is even} \,\big|\, R_U \,\big] \cdot 1.7513 \cdot \gamma \\
&\quad + \sum_{g \in \delta(U) \cap \delta(a)} r_g \cdot \mathbb{P}\big[\, |A^u| \text{ is even} \,\big|\, R_g \,\big] \\
&\quad + \sum_{g \in \delta(U) \cap \delta(b)} r_g \cdot \mathbb{P}\big[\, |B^u| \text{ is even} \,\big|\, R_g \,\big] \\
&\leq \max\{\alpha, 1-\alpha\} \cdot \mathbb{P}\big[\, |A^u| \text{ or } |B^u| \text{ is even} \,\big|\, R_U \,\big] \cdot 1.7513 \cdot \gamma \\
&\quad + x^*(\delta(U) \cap \delta(a)) \cdot \gamma + x^*(\delta(U) \cap \delta(b)) \cdot \gamma \\
&= \max\{\alpha, 1-\alpha\} \cdot \mathbb{P}\big[\, |A^u| \text{ or } |B^u| \text{ is even} \,\big|\, R_U \,\big] \cdot 1.7513 \cdot \gamma + \gamma,
\end{aligned}$$

where we used $\mathbb{E}[r_g] \leq \gamma \cdot x^*_g$ for non-triangle edges g and (11.14) for triangle edges $g \in \delta(u) \cap \delta(U)$.

Conditioned on the reduction event R_U for the triangle U, we have $|\delta_S(u) \cap \delta_S(U)| = 1$ by the choice of R_U. Moreover, the marginals of the edges in $\delta(u) \cap \delta(U)$ are approximately preserved, implying that conditioned on R_U, we have with probability approximately α that the one edge in $\delta_S(u) \cap \delta_S(U)$ is contained in $\delta(a)$, and with probability roughly $1-\alpha$, it is contained in $\delta(b)$. Because $x^*(\delta(U)) = 2$, Proposition 10.8 implies that $(U, S[U])$ is always a tree. Thus $x^*(E^{\rightarrow}(U)) = 1$ implies that we always have $|S \cap E^{\rightarrow}(U)| = 1$. Because the marginals of edges in $E^{\rightarrow}(U)$ are preserved when conditioning on R_U (by the last condition of the definition of R_U), the one edge in $S \cap E^{\rightarrow}(U)$ will be contained in $\delta(a)$ with probability $1-\alpha$ and in $\delta(b)$ with probability α. Moreover, conditioned on R_U, the edge in $\delta_S(u) \cap \delta_S(U)$ is independent of $S[U]$. Still conditioned on R_U, we have $|A^u|$ even if and only if $|B^u|$ even, and this happens with probability

$$\mathbb{P}\big[\, |A^u| \text{ or } |B^u| \text{ is even} \,\big|\, R_U \,\big] = 2 \cdot \alpha \cdot (1-\alpha),$$

implying (with (11.13))

$$\begin{aligned}
\mathbb{E}[s_f] &= \max\{\alpha, 1-\alpha\} \cdot \mathbb{P}\big[\, |A^u| \text{ or } |B^u| \text{ is even} \,\big|\, R_U \,\big] \cdot 1.7513 \cdot \gamma + \gamma \\
&\quad - 1.7513 \cdot \gamma \\
&\leq \max_{\alpha \in [0.5,1]} 2 \cdot \alpha^2 (1-\alpha) \cdot 1.7513 \cdot \gamma - 0.7513 \cdot \gamma \\
&\leq -0.23 \cdot \gamma.
\end{aligned}$$

We conclude that Theorem 11.15 (iv) is fulfilled for the triangle edge bundle f if the parent U of the triangle u with $f = E^{\rightarrow}(u)$ in the hierarchy $\mathcal{H}$ is also a triangle.

Finally, we have to consider triangles $u \in \mathcal{H}$ whose parent in $\mathcal{H}$ is not a triangle. To handle this case, Karlin, Klein, and Oveis Gharan [2021] introduced

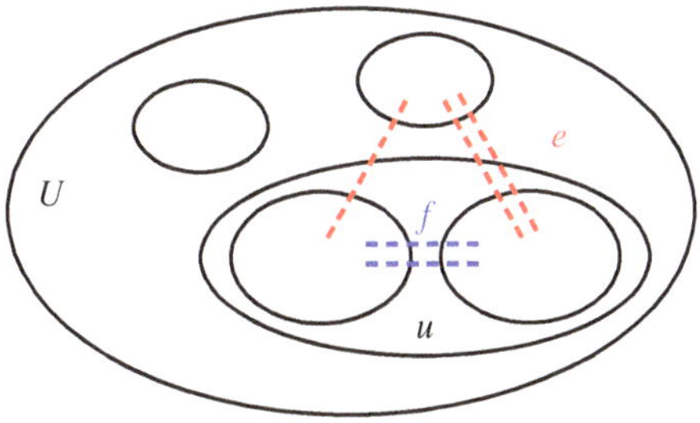

Figure 11.11 If the edge bundle e (shown in red) is bad, then e is a half-edge bundle. Because e is never reduced, the edge bundle $f \in E^{\rightarrow}(u)$ (shown in blue) does not need to be increased in order to compensate for reductions of e.

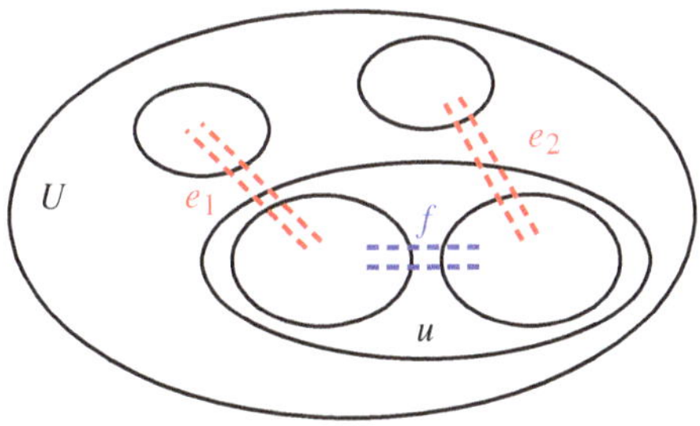

Figure 11.12 Illustration of the situation in which we choose the reduction events of the half-edge bundles e_1 and e_2 (shown in red) to be identical. By increasing the slack on the edge bundle $f = \delta^{\rightarrow}(u)$ (shown in blue), we can simultaneously compensate for the reduction on both e_1 and e_2.

another refinement of the definition of the reduction r_e for non-triangle edge bundles e. Consider a triangle u with parent $U \in \mathcal{H}$. If there is a bad edge bundle e with top cut u, then $r_e = 0$ and thus the edge bundle $f = E^{\rightarrow}(u)$ does not need to compensate for any reduction on e. See Figure 11.11. Because every bad edge bundle is a half-edge, this is sufficient to prove that the expectation of s_f is sufficiently negative (i.e., it fulfills Theorem 11.15 (iv)). Similarly, if there is a bundle e with top cut u that is 2-1-1-good with respect to u, then in half of the cases when e is reduced, the edge bundle f does not need to be increased to compensate for the reduction of e. If the sum of values x^*_e for all such edge bundles e is at least about $\frac{1}{2}$, this is again sufficient to prove that the expectation of s_f is sufficiently negative.

Karlin, Klein, and Oveis Gharan [2021] proved that if none of these two situations arises, then there are two half-edge bundles e_1, e_2 with top cut u

such that most of e_1 is a subset of A^u, most of e_2 is a subset of B^u, and with probability at least γ, the top cuts of both e_1 and e_2 are simultaneously even. See Figure 11.12. Then we can choose the reduction events $R_{e_1,u}$ and $R_{e_2,u}$ to be identical (see (11.11)). If we do this, we need to increase the slack on the edge bundle f only once to compensate simultaneously for the increase on both e_1 and e_2; see the definition (11.12) of the increase vector for triangle edge bundles. These savings again turn out to be sufficient to prove that s_f is sufficiently negative.

11.6 Derandomization

Karlin, Klein, and Oveis Gharan [2023] managed to derandomize their algorithm and finally settled the question of improving on Christofides' algorithm in a deterministic way:

Theorem 11.20 (Karlin, Klein, and Oveis Gharan [2023]). *There is a deterministic α-approximation algorithm for Symmetric TSP with Triangle Inequality for some $\alpha < \frac{3}{2}$.*

In this section, we sketch the main ideas that Karlin, Klein, and Oveis Gharan [2023] used to prove Theorem 11.20. It is based on the well-known method of conditional expectations. Recalling the proof of Theorem 10.18, we first note that the expected cost of the output of Algorithm 10.5 is at most

$$\mathbb{E}_{S\sim\mu^\lambda}\left[c(S) + c(y^S)\right], \tag{11.15}$$

where y^S is the parity correction vector defined in (10.2), and μ^λ is the λ-uniform distribution for the vector λ computed by the algorithm. Since the original definition of y^S involves the incidence vector o^* of an optimum Hamiltonian cycle (cf. (10.1)), which we of course do not know, we need to use the hierarchy construction and the variant of the parity correction vector y^S as in Section 10.5, which in particular replaces $\frac{1}{4}x^* + \frac{1}{4}o^*$ in (10.1) by $\frac{1}{2}x^*$. The η-mincuts can then be computed by Theorem 4.28, and they can be used to compute the desired hierarchy.

Now Step (4) of Algorithm 10.5 (sampling S from the distribution μ^λ) is replaced by the following deterministic procedure, which decides for the edges one by one whether to include them into S or not. Suppose we have decided already that $E_1 \subseteq S \subseteq E \setminus E_0$ for some disjoint subsets $E_0, E_1 \subseteq E$ such that (V, E_1) is a forest and $(V, E \setminus E_0)$ is connected. Then our current *pessimistic estimator* of the expected final cost is the conditional expectation of (11.15):

$$\mathrm{Est}(E_0, E_1) := \mathbb{E}_{S\sim\mu^\lambda}\left[c(S) + c(y^S) \,\middle|\, E_1 \subseteq S \subseteq E \setminus E_0\right]. \tag{11.16}$$

Let $e \in E \setminus (E_0 \cup E_1)$ be the next edge we consider, and let

$$p := \mathbb{P}_{S \sim \mu^\lambda}\left[e \in S \,\middle|\, E_1 \subseteq S \subseteq E \setminus E_0\right]$$

be the conditional probability that e is part of S. Note that

$$\mathrm{Est}(E_0, E_1) = p \cdot \mathrm{Est}(E_0, E_1 \cup \{e\}) + (1-p) \cdot \mathrm{Est}(E_0 \cup \{e\}, E_1). \quad (11.17)$$

If $p = 1$, we include e into E_1. If $p = 0$, we include e into E_0. If $0 < p < 1$, we include e into E_1 if $\mathrm{Est}(E_0, E_1 \cup \{e\}) \le \mathrm{Est}(E_0 \cup \{e\}, E_1)$ and include e into E_0 otherwise. By (11.17), this guarantees that the pessimistic estimator (11.16) never increases. In the end, there is no randomness anymore, so we are guaranteed to obtain a tour of cost at most (11.15), which is at most $\alpha \cdot \mathrm{OPT}$ for some $\alpha < \frac{3}{2}$.

The only remaining issue is how to compute the pessimistic estimator (11.16) in deterministic polynomial time for given E_0 and E_1.

First, it is relatively easy to compute the conditional expectation of the cost of the tree. To this end, contract an edge when including it into E_1, and delete an edge when including it into E_0. Then $\mathbb{P}_{S \sim \mu^\lambda}[e \in S \mid E_1 \subseteq S \subseteq E \setminus E_0]$ can be computed for any edge e by using Theorem 5.15 and Corollary 5.11. Hence, $\mathbb{E}_{S \sim \mu^\lambda}[c(S) \mid E_1 \subseteq S \subseteq E \setminus E_0]$ can be computed in polynomial time. The main difficulty is to compute the conditional expectation of the cost of y^S.

From the construction of the parity correction vector y^S in Chapter 10 and this chapter, one can see that each entry can be written as the weighted sum of indicators of events that depend on the size or on the parity of $|A \cap S|$ for a constant number of (not necessarily disjoint) edge sets A. Formally:

Lemma 11.21. *There is a universal constant k such that all entries of all parity correction vectors y^S ($S \in \mathcal{S}$) defined in (10.2) have the form*

$$y_e^S = \sum_{l \in \{1,\dots,L\} : S \in \mathcal{S}_l} w_e^l$$

where L is bounded by a polynomial in n, and for $l = 1, \dots, L$, we have $w_e^l \in \mathbb{Q}$ and

$$\mathcal{S}_l = \left\{S \in \mathcal{S} : |A_{l,j} \cap S| \equiv \sigma_{l,j} \pmod{\tau_{l,j}} \text{ for } j = 1, \dots, k_l\right\}$$

for some $k_l \in \{0, \dots, k\}$, $A_{l,j} \subseteq E$ and $\tau_{l,j} \in \{2, n\}$ and $\sigma_{l,j} \in \{0, 1, \dots, \tau_{l,j} - 1\}$ for all $j = 1, \dots, k_l$.

See Karlin, Klein, and Oveis Gharan [2023] for the proof of this lemma. Here is an example:

Example 11.22. Let e be an edge, and consider a component i_e of the random increase vector as defined in (11.4) (using (11.1)) in Section 11.1. Recall that we took expectations when going from the random slack vector s to individual deterministic slack vectors $s^{S'}$ ($S' \in \mathcal{S}$) via $s_e^{S'} = \mathbb{E}[s_e \mid S = S']$ for $e \in E$. Similarly, we consider the increase vectors defined by $i_e^{S'} = \mathbb{E}[i_e \mid S = S']$. Let $\mathcal{U}$ denote the (two-element) set of top cuts of e. For $S' \in \mathcal{S}$,

$$\begin{aligned} i_e^{S'} &= \mathbb{E}\left[\sum_{u\in\mathcal{U}} \sum_{f\in\delta^\uparrow(u)\ \gamma\text{-good}} \frac{a(u,e)}{x^*(\delta^\uparrow(u))} \cdot x_f^* \cdot \mathbb{1}_{R_f} \cdot \mathbb{1}[|\delta_S(u)| \text{ is odd}] \;\middle|\; S = S'\right] \\ &= \sum_{u\in\mathcal{U}} \sum_{f\in\delta^\uparrow(u)\ \gamma\text{-good}} \frac{a(u,e)}{x^*(\delta^\uparrow(u))} \cdot x_f^* \cdot \mathbb{1}[|\delta_{S'}(u)| \text{ is odd}] \cdot \mathbb{P}[R_f \mid S = S'], \end{aligned}$$

where R_f is the reduction event (both top cuts of f are even, total probability γ). Fix $u \in \mathcal{U}$ and a γ-good edge $f \in \delta^\uparrow(u)$. Write U_1 and U_2 for the top cuts of f and $A_1 = \delta(U_1)$ and $A_2 = \delta(U_2)$. Moreover, let $A_3 = \delta(u)$. Let $\tau_1 = \tau_2 = \tau_3 = 2$ and $\sigma_1 = \sigma_2 = 0$ and $\sigma_3 = 1$. This defines a subset $\mathcal{S}' = \{S \in \mathcal{S} : |A_j \cap S| \equiv \sigma_j \pmod{\tau_j} \text{ for } j = 1, \ldots, 3\}$. By definition of the reduction event R_f, we have

$$\mathbb{1}[|\delta_{S'}(u)| \text{ is odd}] \cdot \mathbb{P}[R_f \mid S = S'] = \begin{cases} \frac{\gamma}{\gamma^+} & \text{if } S' \in \mathcal{S}' \\ 0 & \text{if } S' \in \mathcal{S} \setminus \mathcal{S}', \end{cases}$$

where $\gamma^+ := \mathbb{P}\left[|A_1 \cap S| \text{ even and } |A_2 \cap S| \text{ even}\right] \geq \gamma$. Then we can set $w'_e = \frac{a(u,e)\cdot x_f^* \cdot \gamma}{x^*(\delta^\uparrow(u))\cdot\gamma^+}$. By taking the sum over all u and f, we can write $i_e^{S'}$ in the form of Lemma 11.21.

Given Lemma 11.21, all we need to do is compute the constants w_e^l and the probabilities $\mathbb{P}_{S\sim\mu^\lambda}\left[S \in \mathcal{S}_l \;\middle|\; E_1 \subseteq S \subseteq E \setminus E_0\right]$ for each $l = 1, \ldots, L$. To compute these probabilities, we consider the graph that arises from deleting edges in E_0 and contracting edges in E_1. Let λ' be the restriction of λ to the edges of this graph. After adapting $\mathcal{S}_l$ to $\mathcal{S}'_l$ by accounting for the edges in $A_{l,j} \cap E_1$, it suffices to compute $\mathbb{P}_{S\sim\mu^{\lambda'}}\left[S \in \mathcal{S}'_l\right]$, which can be done using Theorem 11.23. Most of the constants w_e^l are given directly by the construction, but for the reduction vector for triangle edges, it is more complicated. However, also this can be reduced to the following theorem:

Theorem 11.23. *Let k be a constant. Given a connected graph $G = (V, E)$ with n vertices, $\lambda \in \mathbb{R}^E_{>0}$, $\varepsilon > 0$, and $\tau_j \in \{2, \ldots, n\}$ and $\sigma_j \in \{0, 1, \ldots, \tau_j - 1\}$ and $A_j \subseteq E$ for all $j = 1, \ldots, k$. Let $\mathcal{S}$ denote the set of edge sets of spanning trees in G, and let*

$$\mathcal{S}' = \left\{S \in \mathcal{S} : |A_j \cap S| \equiv \sigma_j \pmod{\tau_j} \text{ for } j = 1, \ldots, k\right\}.$$

Then $\mathbb{P}_{S\sim\mu^\lambda}\left[S \in \mathcal{S}'\right]$ can be computed up to a factor $(1+\varepsilon)$ in polynomial time.

Proof. For a vector $x = (x_1, \ldots, x_k) \in \mathbb{C}^k$, we evaluate the generating polynomial g of μ^λ at $z_e = \prod_{j \in \{1,\ldots,k\}: e \in A_j} x_j$ $(e \in E)$ and obtain a new polynomial $h : \mathbb{C}^k \to \mathbb{C}$, where

$$h(x_1, \ldots, x_k) := \sum_{S \in \mathcal{S}} \mu^\lambda(S) \prod_{e \in S} \prod_{j: e \in A_j} x_j = \mathbb{E}_{S \sim \mu^\lambda}\left[\prod_{j=1}^{k} x_j^{|A_j \cap S|}\right]. \quad (11.18)$$

Note that evaluating h at any point $x \in \mathbb{C}^k$ can be done by computing $z_e = \prod_{j \in \{1,\ldots,k\}: e \in A_j} x_j$ $(e \in E)$ and evaluating $g(z)$. The latter can be done in polynomial time by Theorem 5.19.

Let $\omega_j = e^{2\pi i/\tau_j}$ denote the τ_j-th root of unity (here i denotes the imaginary unit). For positive integers s and t, we have

$$\sum_{r=0}^{t-1} e^{2\pi i r s/t} = \begin{cases} t & \text{if } \frac{s}{t} \in \mathbb{Z} \\ 0 & \text{if } \frac{s}{t} \notin \mathbb{Z} \end{cases} \quad (11.19)$$

(the second case follows from multiplying by $1 - e^{2\pi i s/t}$ and using $e^{2\pi i} = 1$). Using (11.19) in the second equation, we compute

$$\begin{aligned}
\mathbb{P}_{S \sim \mu^\lambda}\left[S \in \mathcal{S}'\right] &= \mathbb{P}_{S \sim \mu^\lambda}\left[|A_j \cap S| \equiv \sigma_j \pmod{\tau_j} \text{ for } j = 1, \ldots, k\right] \\
&= \mathbb{E}_{S \sim \mu^\lambda}\left[\prod_{j=1}^{k}\left(\frac{1}{\tau_j}\sum_{\rho_j=0}^{\tau_j-1} \omega_j^{\rho_j(|A_j \cap S| - \sigma_j)}\right)\right] \\
&= \frac{1}{\tau_1 \cdots \tau_k} \sum_{\rho_1=0}^{\tau_1-1} \cdots \sum_{\rho_k=0}^{\tau_k-1} \mathbb{E}_{S \sim \mu^\lambda}\left[\prod_{j=1}^{k} \omega_j^{\rho_j(|A_j \cap S| - \sigma_j)}\right] \\
&= \frac{1}{\tau_1 \cdots \tau_k} \sum_{\rho_1=0}^{\tau_1-1} \cdots \sum_{\rho_k=0}^{\tau_k-1} h(\omega_1^{\rho_1}, \ldots, \omega_k^{\rho_k}) \prod_{j=1}^{k} \omega_j^{-\rho_j \sigma_j}.
\end{aligned}$$

Hence, it suffices to evaluate $h(\omega_1^{\rho_1}, \ldots, \omega_k^{\rho_k})$ for all $\rho_j \in \{0, 1, \ldots, \tau_j - 1\}$ $(j = 1, \ldots, k)$. Note that the number of these evaluations can be bounded by n^k, which is a fixed polynomial. Exact evaluation with irrational numbers is impossible, but it can be done with the required precision (i.e., up to a factor $1 + \varepsilon$). □

As the final result, we obtain:

Theorem 11.24 (Karlin, Klein, and Oveis Gharan [2023]). *There is an $\alpha < \frac{3}{2}$ and a deterministic algorithm that computes for any given* Symmetric TSP *instance a solution of cost at most α times the value of the LP relaxation* (2.12).

Exercises

11.1 Show that an edge that has $V \setminus \{u_0, v_0\}$ as its only top cut is always γ-good.

11.2 Show that the responsibility assignment of Lemma 11.3 can be done so that for every edge bundle $\{u, v\}$, there is a constant $\varphi_{u,v}$ such that $a(u, e) = \varphi \cdot a(v, e)$ for every edge e in this edge bundle.

11.3 Show that Lemma 11.4 still holds if we redefine the increase vector i by taking the maximum instead of the sum of the two lines in (11.4).

11.4 Let E be a finite set and μ a probability distribution on 2^E in which the elements are picked independently of each other – that is,

$$\mathbb{P}_{S\sim\mu}[I \subseteq S] = \prod_{e\in I} \mathbb{P}_{S\sim\mu}[e \in S]$$

for all $I \subseteq E$. Show that then μ is strongly Rayleigh.

11.5 Let U be an η-mincut for $\eta \leq \frac{1}{400}$ and μ a strongly Rayleigh distribution of spanning trees. Prove that $\mathbb{P}_{S\sim\mu}[|S \cap \delta(U)| \text{ even}] \geq 0.43$.
Hint: Revisit the proof of Lemma 11.13.

11.6 Let b and b' be two independent Bernoulli random variables, each with success probability $\frac{1}{2}$. Let $b'' := |b - b'|$ be a third (not independent) Bernoulli random variable. Prove:

(a) The sum of the three Bernoulli random variables $b + b' + b''$ is always even.
(b) The expectation of the sum is $\frac{3}{2}$.
(c) The three random variables are negatively correlated.
(d) Compute the generating polynomial g of this distribution, and show that the distribution is not strongly Rayleigh.

11.7 Let E_{bad} be a set of edge bundles with the following properties:

- for any $u \in \mathcal{H}$, at most one edge bundle in E_{bad} has top cut u, and
- $x^*_e = \frac{1}{2}$ and $x^*(\delta^\uparrow(u)) \leq \frac{1}{2}$ and $x^*(\delta^\uparrow(v)) \leq \frac{1}{2}$ for all $e \in E_{\text{bad}}$, where u and v are the top cuts of e.

Show that then the strengthening of Lemma 11.3 holds where we impose the additional condition $a(u, e) = 0$ for every edge e that belongs to an edge bundle in E_{bad} with top cut u.
Note: Karlin, Klein, and Oveis Gharan [2021] showed that the above properties hold approximately for the set of bad edge bundles.

11.8 Consider a non-triangle edge f that is 2-1-1-good with respect to both of its top cuts. Show that the reductions $r_f^{S'} = \mathbb{E}[r_f \mid S = S']$ for $S' \in \mathcal{S}$ (cf. (11.11)) can be written in the form of Lemma 11.21.

12

Removable Pairings

So far, all algorithms for SYMMETRIC TSP began with a spanning tree and then added edges to make the graph Eulerian. Mömke and Svensson [2016] had a brilliant idea: If we begin with a 2-connected graph, we may also delete some edges for making it Eulerian, and this may be cheaper overall. They introduced the notion of removable pairings, which allow us to guarantee connectivity when deleting edges. This idea led to a substantial improvement over the result by Oveis Gharan, Saberi, and Singh [2011] (cf. Corollary 10.3) and is still used for the best algorithm for GRAPH TSP that we know today (cf. Chapter 13).

12.1 Graph TSP

We will mostly consider GRAPH TSP in this and the next chapter. This is the special case of SYMMETRIC TSP where $c(e) = 1$ for all edges e. The main motivation for studying GRAPH TSP is that the worst-known integrality ratio of the subtour LP (2.2) is attained by GRAPH TSP instances (cf. Proposition 2.24). So one could try to prove the $\frac{4}{3}$ conjecture first for this special case.

Let us restate the LP relaxation (2.12) for the special case GRAPH TSP:

$$\begin{array}{lrcll} \min & x(E) & & & \\ \text{subject to} & x(\delta(U)) & \geq & 2 & (\emptyset \neq U \subsetneq V) \\ & x_e & \geq & 0 & (e \in E). \end{array} \tag{12.1}$$

Here $G = (V, E)$ is an instance of GRAPH TSP (i.e., a connected undirected graph). We call this the Graph TSP LP, although its integral solutions are the incidence vectors of 2-edge-connected spanning multi-subgraphs. We denote the value of this linear program by $\text{LP}(G)$.

By Theorem 2.31, the value of (12.1) is identical to the value of the subtour LP (2.2) for the metric closure of G.

An undirected graph $G = (V, E)$ with at least three vertices is called *2-vertex-connected* (or *2-connected*) if $G - v$ is connected for all $v \in V$; here $G - v$ denotes the graph $G[V \setminus \{v\}]$, which results from G by deleting v and its incident edges. In SYMMETRIC TSP and in GRAPH TSP, we may restrict attention to instances where the given graph G is 2-connected:

Proposition 12.1. *Suppose there is an α-approximation algorithm for 2-connected instances of* SYMMETRIC *TSP or* GRAPH *TSP. Then the same holds for general instances.*

Suppose for every 2-connected instance of SYMMETRIC *TSP or* GRAPH *TSP, there is a tour of cost at most ρ times the value of* (2.12) *or* (12.1). *Then the same holds for general instances.*

Proof. By induction on $|V|$. For instances with $|V| \leq 2$, the statement is trivial. If $G - v$ is disconnected for an undirected graph $G = (V, E)$, let $G_1 = (V_1, E_1)$ and $G_2 = (V_2, E_2)$ be two graphs with $V_1 \cup V_2 = V$, $E_1 \cup E_2 = E$, and $V_1 \cap V_2 = \{v\}$. By induction, the assertion holds for G_1 and G_2, and the unions of tours in G_1 and G_2 are the tours in G, and the same holds for the LP solutions. □

12.2 The Mömke–Svensson Theorem

We now present the key concept of Mömke and Svensson [2016].

Definition 12.2 (removable pairing)**.** A *removable pairing* in a 2-connected graph (V, E) is a pair $(R, \mathcal{P})$ with the following properties:

(i) $R \subseteq E$;
(ii) for each $P \in \mathcal{P}$, there exists a vertex $v \in V$ and three distinct edges e_1, e_2, e_3 incident to v such that $P = \{e_1, e_2\} \subseteq R$;
(iii) the elements of $\mathcal{P}$ are pairwise disjoint;
(iv) for any set $F \subseteq R$ with $|F \cap P| \leq 1$ for all $P \in \mathcal{P}$, the graph $(V, E \setminus F)$ is connected.

The elements of R are called *removable edges* and the elements of $\mathcal{P}$ *pairs*. Figure 12.1 shows an example.

Now we can formulate and prove the main theorem of Mömke and Svensson [2016]. It works for general weights, although it has not yet been used for general SYMMETRIC TSP. We follow the proof of Sebő and Vygen [2014], a variant of the original proof.

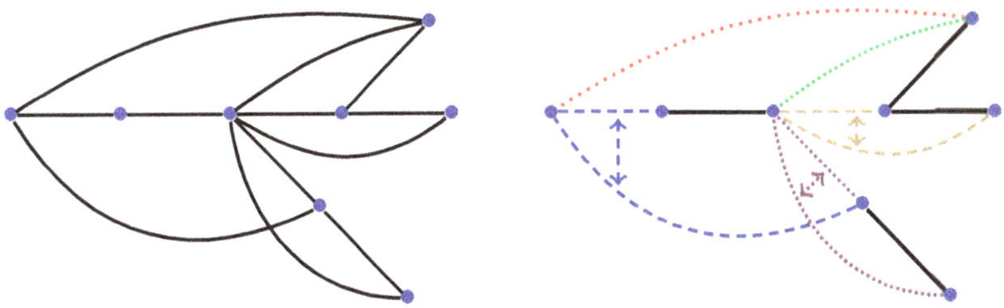

Figure 12.1 A 2-connected graph (left) and a removable pairing (right: dashed and dotted edges are removable; colors and arrows indicate pairs).

Theorem 12.3 (Mömke and Svensson [2016]). *Let $G = (V, E)$ be a 2-vertex-connected graph, $c : E \to \mathbb{R}$, and $(R, \mathcal{P})$ a removable pairing in G. Then one can find a tour in G of cost at most $\frac{4}{3}c(E) - \frac{2}{3}c(R)$ in $O((n + |\mathcal{P}|)^3)$ time. This tour contains each edge of $E \setminus R$ at least once.*

Proof. Let $\text{odd}(E)$ again denote the set of odd-degree vertices of G. Let $c'(e) = c(e)$ for $e \in E \setminus R$ and $c'(e) = -c(e)$ for $e \in R$. For any $\text{odd}(E)$-join J in G that intersects each pair $P \in \mathcal{P}$ in at most one edge, we construct a tour from E by doubling the edges in $J \setminus R$ and deleting the edges in $J \cap R$. This tour has cost $c(E) + c'(J)$.

To compute an $\text{odd}(E)$-join J with $c'(J) \le \frac{1}{3}c(E) - \frac{2}{3}c(R) = \frac{1}{3}c'(E)$, intersecting each pair at most once, we construct an auxiliary graph G' with weights c' from (G, c') as follows (cf. Figure 12.2). For each pair $P = \{\{v, w\}, \{v, w'\}\} \in \mathcal{P}$, we add a vertex v_P and an edge $\{v, v_P\}$ of weight zero, and we replace the two edges in P by $\{v_P, w\}$ and $\{v_P, w'\}$, keeping their weight. Note that every new vertex has degree three.

Let E' be the edge set of G', and let $\text{odd}(E')$ denote the set of odd-degree vertices of G'. Note that G' is 2-edge-connected because G is 2-vertex-connected. Hence, every $\text{odd}(E')$-cut contains at least three edges (by Lemma 2.20). Therefore, the vector with all components $\frac{1}{3}$ is in the $\text{odd}(E')$-join polytope of G' (cf. Theorem 2.21), and even in its face defined by $x(\delta(v_P)) = 1$ for all $P \in \mathcal{P}$. By Theorem 2.21 and Proposition 4.19, this face is integral. Hence, there is an $\text{odd}(E')$-join J' in G' with $|\delta_{J'}(v_P)| = 1$ for all $P \in \mathcal{P}$ and with weight at most $\frac{1}{3}c'(E)$. Such a J' can be found in $O(|V(G')|^3)$ time (by increasing the weight of the edges incident to v_P, for all $P \in \mathcal{P}$, by a large constant and applying Corollary 1.30). It corresponds to an $\text{odd}(E)$-join J in G that intersects each pair at most once and has weight at most $\frac{1}{3}c'(E)$. □

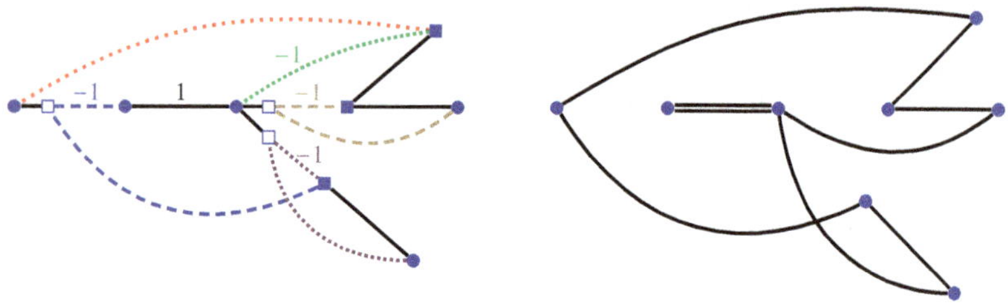

Figure 12.2 Proof of Theorem 12.3. The graph G' on the left results from G and $(R, \mathcal{P})$ in Figure 12.1. Squares denote odd-degree vertices. Here $|E| = 12$ and $|R| = 8$. As J' one could choose, for example, the five edges whose weight is shown (assuming that all edges of G had unit weight). This leads to the tour shown on the right.

So for a given instance G of Graph TSP, we should try to find a 2-connected spanning subgraph with few edges and a removable pairing with many removable edges. This is not easy: Even for the problem of finding a smallest 2-connected spanning subgraph, the best-known approximation ratio is $\frac{4}{3}$ (Bosch-Calvo, Grandoni, and Jabal Ameli [2023]).

We will see several methods to compute a removable pairing. First, Mömke and Svensson [2016] proposed to obtain a removable pairing via a DFS tree.

Definition 12.4 (DFS tree). Let $G = (V, E)$ be an undirected graph, $r \in V$, and (V, S) a spanning tree in G. Then (V, S) is called a *DFS tree* rooted at r if for every edge $e = \{v, w\} \in E$, either v lies on the path from r to w in (V, S) or w lies on the path from r to v in (V, S).

For any connected graph $G = (V, E)$ and any $r \in V$, one can find a DFS tree rooted at r in linear time by starting with $R = S = \emptyset$ and applying the following recursive function to r:

Visit(v): Add v to R.
While there exists an edge $e = \{v, w\} \in \delta(v)$ with $w \notin R$, add e to S and call Visit(w).

This is called *depth-first search*; hence the name DFS tree. Figure 12.3 (left-hand side) shows an example.

Lemma 12.5 (Mömke and Svensson [2016]). *Let $G = (V, E)$ be a 2-vertex-connected graph and (V, S) a DFS tree in G, rooted at $r \in V$. For each edge $e = \{v, w\} \in E \setminus S$, let without loss of generality be v on the r-w-path in (V, S), and let v' be the successor of v on this path. Add e to R; moreover, if $|\delta(v)| \geq 3$*

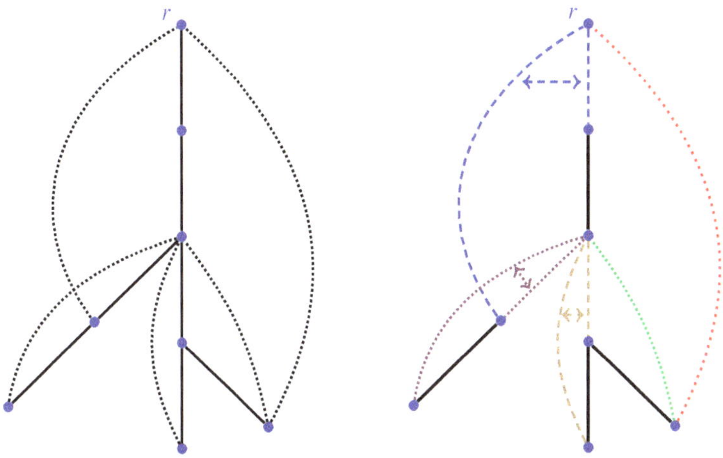

Figure 12.3 A 2-connected graph with a DFS tree rooted at r (left: solid edges) and a removable pairing resulting from the construction in Lemma 12.5 (right: dashed and dotted edges are removable; colors and arrows indicate pairs).

and $e' = \{v, v'\}$ has not yet been added to R, then also add e' to R and $\{e, e'\}$ to $\mathcal{P}$ (cf. Figure 12.3). Then $(R, \mathcal{P})$ is a removable pairing in G.

Proof. It is easy to see that conditions (i)–(iii) of Definition 12.2 hold. To show that condition (iv) holds, take $F \subseteq R$ with $|F \cap P| \leq 1$ for all $P \in \mathcal{P}$. For each $v \in V$, we consider the set W_v of vertices w for which v is on the r-w-path in (V, S). We show that for each $v \in V$, the vertex set W_v induces a connected subgraph of $(V, E \setminus F)$. Indeed, this follows by a straightforward induction on $|W_v|$. Since $W_r = V$, this yields (iv). □

As we will see now, this leads to short tours in some interesting cases.

12.3 Subcubic Graphs

Gamarnik, Lewenstein, and Sviridenko [2005] found the first approximation algorithm with approximation ratio (slightly) better than $\frac{3}{2}$ for *cubic* graphs (i.e., graphs in which every vertex has degree 3). Then Boyd, Sitters, van der Ster, and Stougie [2014] devised a $\frac{4}{3}$-approximation algorithm for cubic graphs.

Mömke and Svensson [2016] gave a simpler proof for this result and extended it to *subcubic* graphs (i.e., graphs with maximum degree 3):

Theorem 12.6 (Mömke and Svensson [2016]). *For any instance $G = (V, E)$ of* Graph TSP *where G is subcubic, one can compute a tour with less than $\frac{4}{3}n$ edges, where $n = |V|$.*

Proof. Let (V, S) be a DFS tree in G. Lemma 12.5 yields a removable pairing with $|R| \geq 2(|E| - |S|) - 1$ because all non-tree edges, except possibly one incident to the root, can be paired with tree edges if the graph is subcubic. Theorem 12.3 yields a tour with at most $\frac{4}{3}|E| - \frac{2}{3}|R| \leq \frac{4}{3}n - \frac{2}{3}$ edges. □

This is best possible, for example, for graphs that consist only of three internally vertex-disjoint paths of the same length and with the same endpoints. Note that the example from Figure 2.2 is also subcubic, which explains the interest in this special case.

Later, Correa, Larré, and Soto [2015], van Zuylen [2018], Duník and Lukot'ka [2018], Dvořák, Král', and Mohar [2017], and Wigal, Yoo, and Yu [2023] refined the techniques of Boyd et al. [2014] and obtained better approximation ratios for simple cubic graphs. The best-known ratio today is due to Wigal, Yoo, and Yu [2023], who proved that every 2-connected simple cubic graph has a tour of length less than $\frac{5}{4}n$. This was conjectured by Dvořák, Král', and Mohar [2017], who also showed that this bound is tight.

However, we will now focus on general graphs.

12.4 Removable Pairings via Circulation

Mömke and Svensson [2016] showed how to find a good removable pairing in general graphs using a network flow approach, somewhat similar to Lemma 5.1. The idea is to start with a DFS tree and include some of the non-tree edges to make the subgraph 2-vertex-connected but use as few non-pairable edges as possible.

First, the input graph G is transformed into a flow network (D, l) as follows (cf. Figure 12.4, left-hand side). Let (V, S) be again a DFS tree, rooted at r. Note that r has degree 1 because G is 2-connected. Orient all tree edges away from r (so that we get an arborescence rooted at r, which we also call (V, S) now) and subdivide each arc $e \in S$ by a vertex z_e. For each non-tree edge $\{v, w\}$, where v is on the r-w-path in (V, S), add an arc (w, z_e), where e is the first edge on the v-w-path in (V, S). This defines the digraph D. We define lower bounds on the flow along each arc by $l((v, z_e)) := 1$ for all $v \in V \setminus \{r\}$ and $e \in \delta^+_S(v)$, and $l(e) := 0$ for all other arcs in D. (The arcs that require at least one unit of flow are shown in green in Figure 12.4.) There are no upper bounds on the flow along any arc.

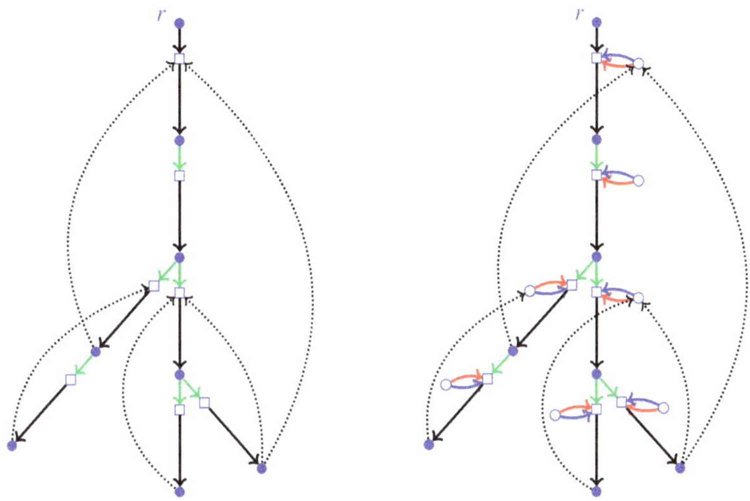

Figure 12.4 For the DFS tree in Figure 12.3, the left-hand side shows the constructed flow network (D, l) in which we look for a circulation f with minimum cost $\text{cost}(f)$. New vertices z_e $(e \in S)$ are shown as squares. Green arcs require at least one unit of flow. The right-hand side shows how to transform this to an equivalent standard minimum-cost circulation problem. Blue arcs can carry at most one unit of flow. Red arcs have cost 1; all other arcs have cost 0.

The cost of a circulation f in (D, l) is measured in an unusual way. We only pay for flow on non-tree edges, but the first unit entering a new vertex z_e is free. In other words,

$$\text{cost}(f) := \sum_{e=(v,w)\in S} \max\left\{0,\ f\left(\delta^-(z_e) \setminus \{(v, z_e)\}\right) - 1\right\}.$$

Finding a circulation f in (D, l) with minimum cost(f) can be reduced to a standard minimum-cost circulation problem. More precisely, we note:

Lemma 12.7. *There is an integral circulation $\tilde{f}$ in D such that $\tilde{f} \geq l$ and* $\text{cost}(\tilde{f}) \leq \text{cost}(f)$ *for every circulation f in D with $f \geq l$. Such an integral circulation $\tilde{f}$ can be computed in polynomial time.*

Proof. Add another vertex z'_e for each new vertex z_e, replace every non-tree edge (v, z_e) by (v, z'_e), and add two arcs from z'_e to z_e: one with capacity 1 and cost 0 and another one with infinite capacity and cost 1. (All other arcs have cost 0 and infinite capacity.) See Figure 12.4, right-hand side. There is a cost- and integrality-preserving one-to-one correspondence between circulations in

(D, l) and circulations in the new network. Hence, the result follows from Theorem 3.10 and Theorem 3.11. □

Mömke and Svensson [2016] proved:

Lemma 12.8. *Given an integral circulation $\tilde{f}$ in D with $\tilde{f} \geq l$, one can construct a tour in G with at most $\frac{4}{3}n + \frac{2}{3}\text{cost}(\tilde{f}) - \frac{2}{3}$ edges in $O(n^3)$ time.*

Proof. Let B be the set of non-tree edges that correspond to edges in D with positive flow. Then $(V, S \cup B)$ is 2-vertex-connected. Let

$$C := \left\{ e = (v, w) \in S : v \neq r, \, \tilde{f}(\delta^-(z_e) \setminus \{(v, z_e)\}) > 0 \right\}$$

be the set of tree edges that can be paired (with a non-tree edge). We do not include the edge leaving the root because the root might have degree 2 in $(V, S \cup B)$. Note that $|B| - |C| \leq \text{cost}(\tilde{f}) + 1$, where we used that the root has degree 1 in the DFS tree. Define a removable pairing in G by $R := B \cup C$ and by letting $\mathcal{P}$ contain a pair P for each element of C: For $e \in C$, choose an $e' \in B$ that corresponds to an edge in $\delta^-(z_e)$ and let $P = \{e, e'\}$. By Lemma 12.5, $(R, \mathcal{P})$ is indeed a removable pairing.

Now we apply Theorem 12.3 and obtain a tour with at most $\frac{4}{3}|S \cup B| - \frac{2}{3}|R| = \frac{4}{3}|S| + \frac{2}{3}|B| - \frac{2}{3}|C| \leq \frac{4}{3}(n-1) + \frac{2}{3}\text{cost}(\tilde{f}) + \frac{2}{3}$ edges. □

So all we need to do is find a cheap circulation in (D, l). Mömke and Svensson [2016] (and then also Mucha [2014]) proceeded as in Algorithm 12.9.

Algorithm 12.9: Mömke–Svensson Algorithm for Graph TSP

Input: a 2-vertex-connected graph $G = (V, E)$
Output: a tour in G

(1) Compute an optimum extreme point x of (12.1).
(2) Compute a DFS tree (V, S) in the support of x by choosing a root r arbitrarily and following always an edge e with maximum x_e to an unvisited vertex. Orient (V, S) as an arborescence rooted at r.
(3) Construct the associated flow network (D, l) and compute an integral circulation $\tilde{f}$ in D with $\tilde{f} \geq l$ that minimizes $\text{cost}(\tilde{f})$.
(4) Apply Lemma 12.8 to $\tilde{f}$ to obtain a tour in G.

This algorithm can be implemented to run in polynomial time: We use Corollary 4.21 for Step (1) and Lemma 12.7 for Step (3).

To bound the cost of $\tilde{f}$, Mömke and Svensson [2016] defined the fractional circulation $f = f' + f''$ in (D, l) as follows (see Figure 12.5):

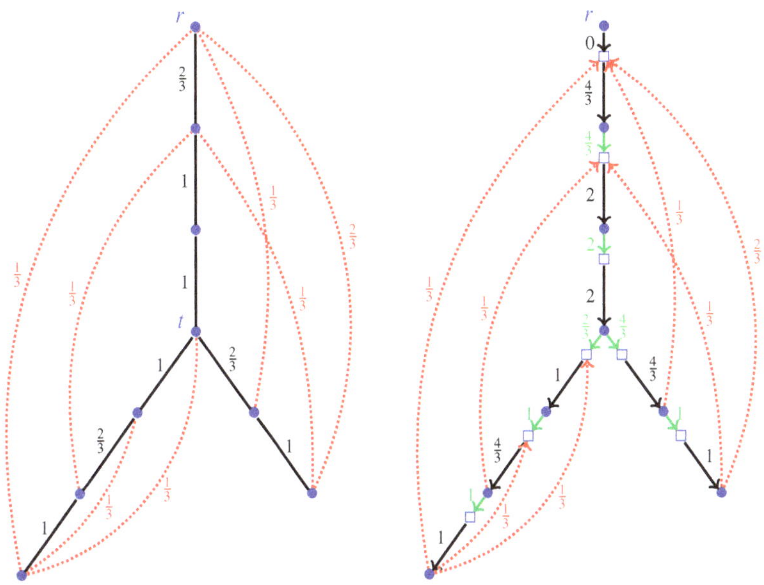

Figure 12.5 Example for constructing a cheap circulation. Left: an LP solution x and a possible DFS tree (rooted at r) as computed in Step (2) of Algorithm 12.9, shown by the black solid edges. Right: the circulation f'. In this case, there is one green edge carrying less than one unit of flow; this will be compensated by f''.

- First define f': For each $e \in E \setminus S$, send x_e units of flow along the fundamental cycle of e (the circuit in D corresponding to the unique circuit in $(V, S \cup \{e\})$).
- Then define f'': For each $v \in V \setminus r$ and $e \in \delta^+_S(v)$ with $f'((v, z_e)) < 1$, send $1 - f'((v, z_e))$ units of flow along any fundamental cycle containing (v, z_e). Such a fundamental cycle exists because G is 2-vertex-connected.

Then $f \geq l$. By Lemma 12.7, $\mathrm{cost}(\tilde{f}) \leq \mathrm{cost}(f)$. So using Lemma 12.8, all we need to bound is $\mathrm{cost}(f)$.

Mömke and Svensson [2016] proved $\mathrm{cost}(f) \leq (4\sqrt{2}-3)x(E) - (6\sqrt{2}-6)n$, which yields a tour with at most $\frac{4}{3}n + \frac{2}{3}(4\sqrt{2} - 3)x(E) - (4\sqrt{2} - 4)n$ edges. Combining this with Christofides' algorithm yields the approximation ratio 1.461 (Exercise 12.5).

For this bound on $\mathrm{cost}(f)$, and also for Mucha's [2014] improved bound that we prove in Section 12.5, it is essential that the LP solution x is an extreme point and hence has sparse support due to Theorem 4.23.

12.5 Mucha's Analysis

Mucha [2014] improved the Mömke–Svensson analysis and obtained $\text{cost}(f) \le \frac{5}{3}x(E) - \frac{3}{2}n$, which yields a $\frac{13}{9}$-approximation algorithm. Here we give a simpler proof of a slightly weaker bound, yielding the same approximation ratio. We continue to use the notation of Section 12.4.

Lemma 12.10 (Mucha [2014]).

$$\text{cost}(f) \;\le\; 2x(E) - \tfrac{11}{6}n + \tfrac{2}{3}, \tag{12.2}$$

and this holds even if the first edge of the DFS tree (the edge incident to the root r) is chosen arbitrarily (not necessarily in the support of x).

Proof. For $v \in V$, we consider the non-tree arcs entering z_e in D for all edges e that leave v in the DFS tree (V, S). This is the edge set

$$B_v \;:=\; \bigcup_{e \in \delta_S^+(v)} \left\{e' \in \delta_D^-(z_e) \setminus \{(v, z_e)\} : x_{e'} > 0\right\}.$$

Since x is an extreme point, the support graph G_x of x has at most $2n-3$ edges by Theorem 4.23. At least $n-2$ of these edges belong to $S \setminus \delta(r)$ (all edges of the DFS tree except possibly the edge incident to the root), and at least one of these edges belongs to $\delta(r)$. Hence,

$$\sum_{v \in V \setminus \{r\}} |B_v| \;\le\; |E(G_x)| - (n-2) - 1 \;\le\; (2n-3) - (n-1) \;=\; n-2. \tag{12.3}$$

We will bound, for each $v \in V$ separately, the number $\text{cost}_v(f)$, which is the contribution of the edges in B_v to $\text{cost}(f')$ plus the additional cost incurred by the f''-flow added for v. More precisely:

$$\text{cost}_v(f) \;:=\; \max\left\{0,\, f'(B_v) - 1\right\} + \sum_{e \in \delta_S^+(v)} \max\left\{0,\, 1 - f'((v, z_e))\right\}.$$

Note that $\text{cost}(f) \le \sum_{v \in V} \text{cost}_v(f)$.

For $v = r$, we have $\text{cost}_r(f) \le x(\delta(r)) - 1$. We show that for all $v \in V \setminus \{r\}$:

$$\text{cost}_v(f) \;\le\; x(\delta(v)) - 2 + \tfrac{1}{6}|B_v|. \tag{12.4}$$

Summing these inequalities and using (12.3) yields (12.2).

Let $v \in V \setminus \{r\}$, and let $w_1, \dots, w_l$ be the children of v in (V, S). Let $R(w)$ denote the set of descendants of a vertex w, including w itself (i.e., all vertices reachable from w in the arborescence (V, S)). See Figure 12.6. For $i = 1, \dots, l$, let $k_i := |\delta_{G_x}(R(w_i)) \cap \delta(v)| - 1$; so $|B_v| = \sum_{i=1}^{l} k_i$. Moreover, let $\gamma := x(\delta_G(R(v)) \cap \delta(v))$, let $\alpha_i := x(\delta_G(R(w_i)) \cap \delta(v))$, and let $\beta_i := x(\delta_G(R(w_i)) \setminus \delta(v))$ for $i = 1, \dots, l$.

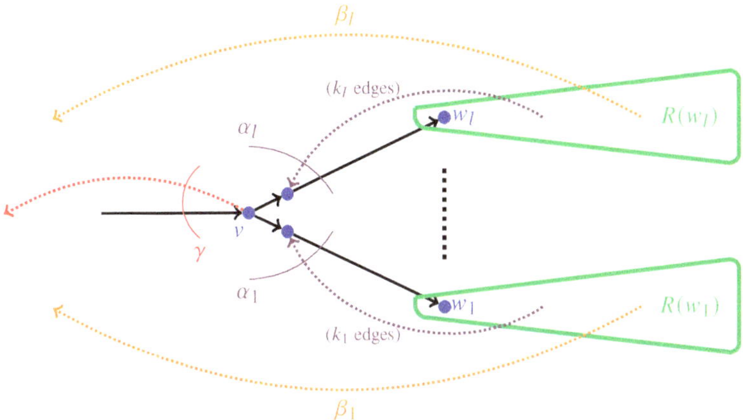

Figure 12.6 Illustrating α_i, β_i ($i = 1, \ldots, l$), and γ in the proof of Lemma 12.10. The edge set B_v consists of the violet dotted edges entering v: the k_i non-tree edges from $R(w_i)$ to v for $i = 1, \ldots, l$.

By the construction of the DFS tree in Step (2) of Algorithm 12.9, we have $x_e \le x_{\{v,w_i\}}$ for all $e \in \delta(R(w_i)) \cap \delta(v)$. Using this, we get

$$\begin{aligned}\text{cost}_v(f) &= \sum_{e\in\delta_S^+(v)} \Big(\max\big\{0,\ f'\big(\delta^-(z_e)\setminus\{(v,z_e)\}\big)-1\big\} \\ &\qquad\qquad + \max\big\{0,\ 1-f'\big((v,z_e)\big)\big\}\Big) \\ &= \sum_{i=1}^{l}\Big(\max\big\{0,\ \alpha_i - x_{\{v,w_i\}}-1\big\}+\max\big\{0,\ 1-\beta_i\big\}\Big) \\ &\le \sum_{i=1}^{l}\left(\max\left\{0,\ \frac{k_i}{k_i+1}\alpha_i-1\right\}+\max\{0,\ 1-\beta_i\}\right).\end{aligned}$$

Now we observe $\max\left\{0, \frac{k_i}{k_i+1}\alpha_i - 1\right\} \le \frac{k_i}{6} + \max\{0, \alpha_i - 2\}$: It is obviously sufficient to check this inequality for $\alpha_i = 2$; here it holds because $\frac{k_i-1}{k_i+1} \le \frac{k_i}{6}$ for all $k_i \in \mathbb{Z}_{\ge 0}$ (with equality for $k_i \in \{2, 3\}$). Hence,

$$\begin{aligned}\text{cost}_v(f) &\le \sum_{i=1}^{l}\left(\frac{k_i}{6}+\max\{0,\ \alpha_i-2\}+\max\{0,\ 1-\beta_i\}\right) \\ &= \frac{|B_v|}{6}+\sum_{i=1}^{l}\Big(\max\{0,\ \alpha_i-2\}+\max\{0,\ 1-\beta_i\}\Big).\end{aligned}$$

We claim

$$\sum_{i=1}^{l} \Big(\max\{0, \alpha_i - 2\} + \max\{0, 1 - \beta_i\} \Big) \leq \sum_{i=1}^{l} \alpha_i + \gamma - 2. \tag{12.5}$$

Since the right-hand side is $x(\delta(v)) - 2$, this implies (12.4).

We conclude the proof by showing (12.5). The right-hand side is nonnegative:

$$0 \leq \sum_{j=1}^{l} \alpha_j + \gamma - 2 \tag{12.6}$$

because $x(\delta(v)) \geq 2$. We use the following upper bounds on $1 - \beta_i$:

(i) $1 - \beta_i \leq 1$,
(ii) $1 - \beta_i \leq \alpha_i - 1$ (which follows from $x(\delta(R(w_i))) \geq 2$),
(iii) $1 - \beta_i \leq \sum_{j \neq i} \alpha_j + \gamma - 1$ (which follows from $x(\delta(\{v\} \cup R(w_i))) \geq 2$),
(iv) $1 - \beta_i \leq \sum_{j=1}^{l} \alpha_j + \gamma - 2$ (which follows from adding (ii), (iii), and (12.6)).

If none of the l summands in the left-hand side of (12.5) is positive, the inequality is given by (12.6). If only one of the l summands in the left-hand side of (12.5) is positive, we use (iii) or (iv) to bound it by $\sum_{j=1}^{l} \alpha_i + \gamma - 2$. If there are at least two positive summands, we use (i) or (ii) to bound each of them by $\alpha_i - 1$. In every case, (12.5) follows. □

Mucha [2014] proved the stronger bound $\text{cost}_v(f) \leq \frac{1}{6}|B_v| + \frac{5}{6}(x(\delta(v)) - 2)$, but this leads to the same worst-case bound. We conclude:

Theorem 12.11 (Mucha [2014]). *Algorithm 12.9 is a $\frac{13}{9}$-approximation algorithm for* Graph TSP.

Proof. By Proposition 12.1, we may assume that the instance G is 2-connected. Construct a flow network (D, l) and an integral circulation $\tilde{f}$ in D with $\tilde{f} \geq l$ as in Algorithm 12.9. By Lemma 12.10 and Lemma 12.7, we have $\text{cost}(\tilde{f}) \leq \text{cost}(f) \leq 2x(E) - \frac{11}{6}n + \frac{2}{3}$. By Lemma 12.8, we then get a tour in G with at most $\frac{4}{3}n + \frac{2}{3}\text{cost}(\tilde{f}) - \frac{2}{3} \leq \frac{4}{3}n + \frac{4}{3}x(E) - \frac{11}{9}n = \frac{1}{9}n + \frac{12}{9}x(E)$ edges. □

This proof also implies that the integrality ratio of the Graph TSP LP (12.1) is at most $\frac{13}{9}$. A stronger conclusion is that even the integrality ratio of (2.2) when restricted to metric closures of unweighted graphs is at most $\frac{13}{9}$. We will prove stronger upper bounds on both integrality ratios in Chapter 13.

We do not know whether Mucha's bound is tight. Newman [2020] proved a better bound for graphs with maximum degree 4 and noted that it is even possible that Algorithm 12.9 is a $\frac{4}{3}$-approximation algorithm for general graphs.

Open Problem 12.12. What is the approximation ratio of the Mömke–Svensson algorithm for GRAPH TSP?

We conclude this section with the following note for later use (in the proof of Theorem 14.20):

Lemma 12.13. *Let $G = (V, E)$ be a 2-connected graph and $\bar{e} = \{s, t\} \in E$ an edge. Then one can construct a tour in G that contains $\bar{e}$ exactly once and contains at most $\frac{13}{9}(\mathrm{LP}(G) - 1) + \frac{1}{3}\mathrm{dist}_{G-\bar{e}}(s, t)$ other edges in $O(n^3)$ time, where* $\mathrm{LP}(G)$ *is the value of* (12.1).

Proof. Let x be an extreme point solution to (12.1) with $x(E) = \mathrm{LP}(G)$. We choose s as the root of the DFS tree and $\bar{e}$ as its first edge; then we proceed with Algorithm 12.9. Construct a flow network (D, l) and a circulation $\tilde{f}$ as in Algorithm 12.9. By Lemma 12.10 and Lemma 12.7, $\mathrm{cost}(\tilde{f}) \leq \mathrm{cost}(f) \leq 2x(E) - \frac{11}{6}n + \frac{2}{3}$. We proceed as in the proof of Lemma 12.8, except that we set $c(\bar{e}) := \mathrm{dist}_{G-\bar{e}}(s, t)$ and $c(e) = 1$ for all $e \in E \setminus \{\bar{e}\}$. We obtain a tour $\bar{F}$ in G that contains $\bar{e}$ at least once (because $\bar{e}$ is not removable) and costs at most $\frac{4}{3}c(S) + \frac{2}{3}\mathrm{cost}(\tilde{f}) + \frac{2}{3} \leq \frac{4}{3}\left(c(\bar{e}) + n - 2\right) + \frac{2}{3}\left(2x(E) - \frac{11}{6}n + \frac{2}{3}\right) + \frac{2}{3} = \frac{4}{3}c(\bar{e}) + \frac{4}{3}x(E) + \frac{1}{9}n - \frac{14}{9}$.

If $\bar{F}$ contains a second copy of $\bar{e}$, we replace it by a shortest s-t-path in $G - \bar{e}$; this does not increase the cost, which thus remains at most $c(\bar{e}) + \frac{4}{3}x(E) + \frac{1}{9}n - \frac{14}{9} + \frac{1}{3}\mathrm{dist}_{G-\bar{e}}(s, t)$. Let F be the new tour, which now contains $\bar{e}$ exactly once and at most $\frac{4}{3}x(E) + \frac{1}{9}n - \frac{14}{9} + \frac{1}{3}\mathrm{dist}_{G-\bar{e}}(s, t)$ other edges. Since $n \leq x(E) = \mathrm{LP}(G)$, this yields the assertion. □

12.6 Removable Pairings via s-t-Numbering

Svensson [2013] suggested another way of constructing a removable pairing. For every 2-connected graph G and an edge $\{s, t\}$ of G, there is an *s-t-numbering*, which is a numbering $V = \{v_1, \ldots, v_n\}$ where $v_1 = s$, $v_n = t$, and every vertex v_i ($i = 2, \ldots, n-1$) has a left and a right neighbor (i.e., $\delta(v_i) \cap \delta(v_1, \ldots, v_{i-1}) \neq \emptyset$ and $\delta(v_i) \cap \delta(v_{i+1}, \ldots, v_n) \neq \emptyset$). In fact, this characterizes 2-connected graphs (cf. Exercise 12.8).

Now we can declare an edge $\{v_i, v_j\}$ (with $i < j$) removable if v_i has degree at least 3 and v_j is not the only right neighbor of v_i. For each such v_i, we take two edges between v_i and right neighbors of v_i and let them constitute a pair. The other removable edges do not participate in any pair. We also declare the edge $\{s, t\}$ as removable even if s has degree 2.

This is a removable pairing because t is still reachable from everywhere after removing removable edges, but at most one from each pair. Let R be this set of removable edges. Similarly, we can consider the symmetric construction (or the s-t-numbering $v_n, \ldots, v_1$) and get another removable pairing L. We have $|L| + |R| \geq 2m - n - n_2 + 2$, where n and m are the number of vertices and edges of G, and n_2 is the number of degree-2 vertices. (Note that the edge $\{s, t\}$ is always removable.)

The larger of the two removable pairings has at least $m - \frac{1}{2}(n + n_2 - 2)$ removable edges and yields a tour, via Theorem 12.3, that has length at most $\frac{1}{3}(2m + n + n_2 - 2)$. If G is subcubic, then $2m = 3n - n_2$, and so the tour has length at most $\frac{4}{3}n - \frac{2}{3}$. This gives another simple proof of Theorem 12.6.

Exercises

12.1 Show that adding the constraints $x_e \leq 1$ for all $e \in E$ does not change the value of the LP (12.1) for any 2-edge-connected graph G.
(Vygen [2012])

12.2 Prove that Theorem 12.3 does not hold in general if the graph G is only 2-edge-connected instead of 2-vertex-connected.
Note: A variant of Theorem 12.3 that works for 2-edge-connected graphs in a certain setting was used by Traub and Vygen [2023].

12.3 Show that Algorithm 12.9 has no better approximation ratio than $\frac{4}{3}$ for Graph TSP.

12.4 Show that f'' as defined in Section 12.4 sends at most $x(E) - n$ units of flow along fundamental cycles. In particular, $f'' = 0$ if $x(E) = n$.
(Mömke and Svensson [2016])

12.5 Show that the existence of a circulation f in (D, l) with

$$\text{cost}(f) \leq (4\sqrt{2} - 3)x(E) - (6\sqrt{2} - 6)n$$

for every instance yields a 1.461-approximation algorithm for Graph TSP.
Hint: If $x(E) \geq 1.041\, n$, use Christofides' algorithm.
(Mömke and Svensson [2016])

12.6 Show that Algorithm 12.9 computes a tour with at most $\frac{4}{3}x(E)$ edges if the LP solution x computed in Step (1) is half-integral. To this end, modify the analysis of Lemma 12.10 and observe that now $\text{cost}_v(f) \leq \sum_{i=1}^{l} \max\{0, \alpha_i - \frac{3}{2}\} + \max\{0, 1 - \beta_i\}$ and $\gamma \geq \frac{1}{2}$. Deduce from this that $\text{cost}_v(f) \leq x(\delta(v)) - 2$ for all $v \in V \setminus \{r\}$.
(Mömke and Svensson [2016])

12.7 Show that the bound (12.4) also holds for $v = r$ if the DFS tree is chosen as in Algorithm 12.9. Conclude that then $\text{cost}(f) \leq 2x(E) - \frac{11}{6}n$.

12.8 Show that a graph is 2-vertex-connected if and only if it has an s-t-numbering.
(Lempel, Even, and Cederbaum [1967])

13

Ear-Decompositions, Matchings, and Matroids

By combining the removable pairing technique presented in Chapter 12 with a new approach based on ear-decompositions and matroid intersection, Sebő and Vygen [2014] improved the approximation ratio for Graph TSP from $\frac{13}{9}$ to $\frac{7}{5}$. We will present this algorithm, which is still the best-known approximation algorithm for Graph TSP, in this chapter. An interesting feature of this algorithm is that it is purely combinatorial, does not need to solve a linear program, and runs in $O(n^3)$ time. A slight variant of the algorithm is a $\frac{4}{3}$-approximation algorithm for finding a smallest 2-edge-connected spanning subgraph, which was the best-known for many years. The proofs will also imply corresponding upper bounds on the integrality ratios.

13.1 Ear-Decompositions

Ear-decompositions are a classic concept to describe 2-connected or 2-edge-connected graphs, as well as strongly connected digraphs. We begin with the formal definition.

Definition 13.1 (ear-decomposition). An *ear-decomposition* of a graph $G = (V, E)$ (directed or undirected) is a sequence $P_0, P_1, \ldots, P_k$ of subgraphs of G such that P_0 consists of a single vertex, $\{E(P_1), \ldots, E(P_k)\}$ is a partition of E, and for $i = 1, \ldots, k$, either P_i is a path with exactly its endpoints in $V(P_0) \cup \cdots \cup V(P_{i-1})$ or P_i is a circuit with exactly one of its vertices (called its endpoint) in $V(P_0) \cup \cdots \cup V(P_{i-1})$.

The vertex of P_0 is called the *root*, and $P_1, \ldots, P_k$ are called *ears*. For an ear P, let $\text{in}(P)$ denote the set of *internal vertices* of P: those that are not endpoints. The length of an ear is the number of its edges; this is always the number of internal vertices plus one. Hence, the number of ears is always

$k = \sum_{i=1}^{k}(|E(P_i)| - |\text{in}(P_i)|) = |E| - |V| + 1$. An ear is called *trivial* if it has length 1; otherwise, it is *nontrivial*. We call an ear *short* if it has length 2 or 3 and *long* if it has length at least 4. An ear is called *odd* if its length is odd, otherwise *even*. See Figure 13.1, left-hand side, for an example.

An ear-decomposition is called *open* if $P_2, \ldots, P_k$ are paths, and it is *odd* if all ears are odd. Ear-decompositions can be used to characterize 2-connected or 2-edge-connected graphs or strongly connected digraphs:

Proposition 13.2 (Whitney [1932a]). *A digraph is strongly connected if and only if it has an ear-decomposition. An undirected graph is 2-edge-connected if and only if it has an ear-decomposition.*

Proof. We show the first statement. For the if-direction, an easy induction on i shows that for any $v \in \text{in}(P_i)$, there is a path from the root to v and a path from v to the root. For the only-if-direction, let G be a strongly connected digraph, and let G' be a maximal subgraph that has an ear-decomposition. If G' does not contain all vertices, there is an edge $e = (v, w) \in \delta_G^+(V(G'))$. Let P be a path from w to the root, and let p be the first vertex of P that belongs to $V(G')$. Then the edge e plus the subpath of P from w to p constitute a path or circuit that can be added as an ear, contradicting the maximality of G'. If G' contains all vertices but not all edges, any missing edge could be added as a trivial ear, which is again a contradiction. Hence $G' = G$.

The proof of the second statement is identical except that we work with $e = \{v, w\} \in \delta_G(V(G'))$ with $v \in V(G')$. □

Moreover, an undirected graph is 2-vertex-connected if and only if it has an open ear-decomposition (Exercise 13.1).

A relaxation of the GRAPH TSP that is closely related to ear-decompositions is the MINIMUM 2-EDGE-CONNECTED SPANNING SUBGRAPH PROBLEM: For a given 2-edge-connected graph G, find a 2-edge-connected spanning subgraph of G with minimum number of edges. Since the number of ears is always $|E| - |V| + 1$, this problem is equivalent to computing an ear-decomposition with minimum number of nontrivial ears. This problem is *NP*-hard because it includes the HAMILTONIAN CIRCUIT problem (cf. Theorem 1.21): Any 2-edge-connected spanning subgraph must contain at least $n = |V|$ edges, and it has exactly n edges if and only if it is a Hamiltonian circuit. In fact, Fernandes [1998] proved that the problem is *APX*-hard.

Computing an arbitrary ear-decomposition and deleting the trivial ears constitutes a 2-approximation algorithm for the MINIMUM 2-EDGE-CONNECTED SPANNING SUBGRAPH PROBLEM. Indeed, the number of edges in nontrivial ears is at most $2(n - 1)$ because every nontrivial ear has one more edge than it has internal vertices, so at most twice as many edges as internal vertices. We see that ears of length 2 are worst in this respect, and we will see better algorithms later. We do not know how to minimize the number of nontrivial ears or the number of short ears, but the number of even ears can be minimized in polynomial time, as we will see in Section 13.3.

A variant of the MINIMUM 2-EDGE-CONNECTED SPANNING SUBGRAPH PROBLEM allows us to pick some edges twice. However, this relaxation does not really change the problem:

Proposition 13.3. *Let G be a 2-edge-connected graph and H a 2-edge-connected spanning multi-subgraph of G. Then there exists a 2-edge-connected spanning subgraph H' of G with $|E(H')| \le |E(H)|$, and H' can be computed from H in polynomial time.*

Proof. If H contains two copies of an edge $e \in E(G)$ and none of these can be omitted, these copies are the only edges in $\delta_H(X)$ for some $\emptyset \neq X \subsetneq V(G)$. Since G is 2-edge-connected, it contains a different edge $f \in \delta_G(X) \setminus \{e\}$. Removing one of the copies of e and adding f maintains 2-edge-connectivity. □

Hence, a tour cannot have fewer edges than a minimum 2-edge-connected spanning subgraph.

13.2 Removable Pairings via Ear-Decompositions

Let us say that a nontrivial ear Q is *attached* to an ear P in an ear-decomposition if an endpoint of Q is an internal vertex of P. We call a nontrivial ear *pendant* if no nontrivial ear is attached to it. In Figure 13.1 (left-hand side), ears P_1 and P_2 are non-pendant, while ears P_3, P_4, P_5, and P_6 are pendant. By applying Theorem 12.3 to a removable pairing defined by an ear-decomposition, Sebő and Vygen [2014] observed:

Theorem 13.4. *Given a 2-vertex-connected graph $G = (V, E)$ with an ear-decomposition with π pendant ears and no trivial ears, one can construct a tour with at most $\frac{4}{3}(n-1) + \frac{2}{3}\pi$ edges in $O(n^3)$ time, where $n = |V|$.*

Proof. Define a removable pairing by taking an arbitrary edge of each pendant ear and taking for each non-pendant ear, a pair of its edges that are incident to a common vertex that is an endpoint of another ear. If $k = |E| - |V| + 1$ denotes the number of ears, we have $|R| = 2k - \pi$. Theorem 12.3 yields a tour with at most $\frac{4}{3}|E| - \frac{2}{3}|R| = \frac{4}{3}(n-1) + \frac{2}{3}\pi$ edges. □

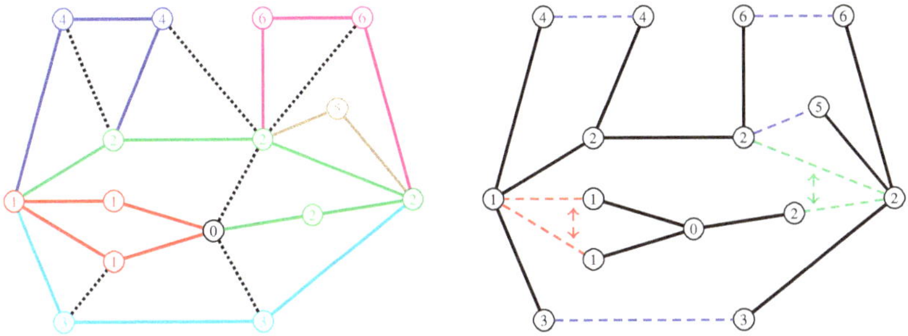

Figure 13.1 Left: A graph G with an open ear-decomposition. The ears $P_1, \ldots, P_6$ have distinct colors; moreover, the internal vertices of the i-th ear are labelled i. The ears $P_7, \ldots, P_{12}$ are trivial; each consists of a single dotted edge. The ears P_1 and P_5 are even; the other ears are odd. The ears P_3, P_4, P_5, and P_6 are short and pendant. Right: Trivial ears are deleted, and a removable pairing $(R, \mathcal{P})$ with $|R| = 8$ and $|\mathcal{P}| = 2$ as in the proof of Theorem 13.4 is shown; the dashed edges are removable, and the arrows indicate the two pairs.

See Figure 13.1 for an example. This bound is good if there are few pendant ears. Otherwise, we need something else. It turns out that long pendant ears are easy to deal with, but short pendant ears require care.

13.3 Frank's Theorem

In this section, we relate ear-decompositions to matching theory. A *matching* in an undirected graph G is a set of edges that have pairwise distinct endpoints. A matching *covers* a vertex v if it contains an edge incident to v. So a perfect matching is a matching that covers all vertices. A matching is called *near-perfect* if it covers all vertices but one. A graph $G = (V, E)$ is called *factor-critical* if $G - v$ contains a perfect matching for all $v \in V$. The graph in Figure 13.1 (left-hand side) is not factor-critical: Deleting the vertex marked 0 leaves a graph without a perfect matching.

The following classic theorem connects factor-critical graphs and odd ear-decompositions:

Theorem 13.5 (Lovász [1972]). *A graph is factor-critical if and only if it has an odd ear-decomposition. If it exists, such an ear-decomposition can be computed in polynomial time.*

Proof. Let $G = (V, E)$ be a graph with an odd ear-decomposition, and let $v \in V$. We show by induction on the number of ears that $G - v$ has a perfect matching. Let P be the last ear. If $v \notin \text{in}(P)$, we can pair up the vertices in $\text{in}(P)$ and apply the induction hypothesis to $(G - \text{in}(P), v)$. If $v \in \text{in}(P)$, we can pair up the vertices in $(\text{in}(P) \setminus \{v\}) \cup \{w\}$ for one endpoint w of P and apply induction to $(G - \text{in}(P), w)$.

For the other direction, let $G = (V, E)$ be a factor-critical graph, $r \in V$, and M_r a perfect matching of $G - r$. We construct an M_r-*alternating* (and hence odd) ear-decomposition of G (i.e., one in which the edges of every ear alternate between $E \setminus M_r$ and M_r). Let G' be a maximal subgraph of G that has an M_r-alternating ear-decomposition with root r. We show $G' = G$. Suppose G' is a proper subgraph of G. Since G is connected and edges with both endpoints in $V(G')$ could be added as trivial ears, there is an edge $e = \{v, w\} \in E$ with $v \in V(G')$ and $w \notin V(G')$. Let M_w be a perfect matching of $G - w$. Note that w and r are the only vertices of degree 1 in $(V, M_r \triangle M_w)$; all other vertices have degree 0 or 2. So this graph contains a path P from w to r, whose edges alternate between M_r and M_w. Let x be the first vertex of P (when traversed from w) that belongs to $V(G')$. Then e plus the subpath from w to x can be added as another M_r-alternating ear, contradicting the maximality of G'.

This proof can easily be turned into a polynomial-time algorithm, by extending G' (initially containing only r) and its ear-decomposition by one M_r-alternating ear at a time. A perfect matching M_v in $G - v$ can be found by Theorem 1.19. □

Factor-critical graphs also play a key role in the following Gallai–Edmonds structure theorem about maximum matchings. This structure results from Edmonds' [1965a] matching algorithm, but here we give a self-contained proof. The set of neighbors of a vertex set Y contains all vertices that do not belong to Y but are a neighbor of some vertex in Y. The partition $\{Y, X, V(G) \setminus (X \cup Y)\}$ in the following theorem is called the *Gallai–Edmonds decomposition* of G (see Figure 13.2 for an example).

Theorem 13.6 (Gallai [1964]). *Let $G = (V, E)$ be an undirected graph that does not have a perfect matching. Let Y be the set of all vertices y such that G has a maximum matching that does not cover y. Let X be the set of neighbors of Y. Then:*

(i) *every connected component of $G[Y]$ is factor-critical;*
(ii) *every maximum matching in G contains $|X|$ edges in $\delta(X) \cap \delta(Y)$;*
(iii) *every maximum matching in G contains a near-perfect matching in every connected component of $G[Y]$.*

Proof. We first prove (ii). Suppose there exists a maximum matching M in G that contains fewer than $|X|$ edges in $\delta(X) \cap \delta(Y)$. Let $x \in X$ be a vertex with $M \cap \delta(x) \cap \delta(Y) = \emptyset$, and let $e = \{x, y\}$ be an incident edge with $y \in Y$. Since

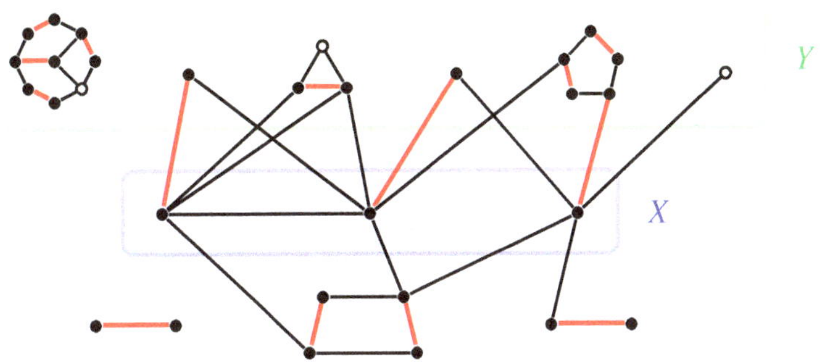

Figure 13.2 The Gallai–Edmonds decomposition of a graph G that has no perfect matching (cf. Theorem 13.6): Y contains every vertex that some maximum matching does not cover, and X is the set of neighbors of Y. The red edges form a maximum matching, covering all but three vertices (the unfilled circles in Y).

$x \notin Y$ and M is a maximum matching, there is an edge $\{x, w\} \in M \cap \delta(x)$. Note that $w \notin Y$. Let $M' = M \setminus \{\{x, w\}\}$. Let M_y be a maximum matching that does not cover y. Consider the maximal path P in $(V, M' \triangle M_y)$ that begins in y; its edges alternate between M' and M_y. This path P has an even number of edges (possibly zero), for otherwise $M_y \triangle E(P)$ would be a larger matching, contradicting the maximality of M_y. Moreover, P ends in Y because $M_y \triangle E(P)$ is a maximum matching that does not cover that endpoint of P. So neither x nor w is an endpoint of P, and in fact P does not contain x or w at all because each of these vertices has at most one incident edge in $M' \triangle M_y \supseteq E(P)$. Thus, $M' \triangle (E(P) \cup \{e\})$ is a maximum matching that does not cover w. This is a contradiction to $w \notin Y$. We have shown (ii).

Now we claim:

> Let H be a connected component of $G[Y]$, $r \in V(H)$, and
> M_r a maximum matching in G that does not cover r. (13.1)
> Then M_r contains a perfect matching in $H - r$.

Note that (13.1) directly implies (i) by the definition of Y. Moreover, (13.1) implies (iii) as follows. Let H be a connected component of $G[Y]$, and suppose there is a maximum matching M in G that does not contain a near-perfect matching in H. By (13.1), M covers all vertices of H. Let $y \in V(H)$, and let M_y be a maximum matching that does not cover y. Let P be the maximal path in $(V, M \triangle M_y)$ that begins in y; it alternates between M and M_y, and it ends at a vertex not covered by M (because M_y is a maximum matching), in particular outside H. Let w be the last vertex of P that belongs to H, and let P' be the subpath of P that begins in w. By (13.1), we have $M_y \cap \delta(V(H)) = \emptyset$; hence, this subpath P' must begin with an edge of M and thus has even length. Hence, $M \triangle E(P')$ is a maximum matching that is identical to M inside H and does not cover w, contradicting (13.1) by the choice of M.

We finally prove (13.1). Let $S \subseteq V(H)$ be a maximal set such that for all $s \in S$ there exists an M_r-alternating path of even length from s to r in $H[S]$. Note that the first edge of such a path (for any $s \in S \setminus \{r\}$) must be an edge of M_r. Therefore, $M_r[S]$ covers all vertices in $S \setminus \{r\}$, and hence $M_r \cap \delta(S) = \emptyset$. We show $S = V(H)$, which implies that $M_r \cap E(H)$ covers all vertices of $H \setminus \{r\}$. So suppose $S \subsetneq V(H)$. Since H is connected, there is an edge $e = \{v, w\}$ with $v \in S$ and $w \in V(H) \setminus S$. Let P_v be an M_r-alternating path of even length from v to r in $H[S]$.

Let M_w be a maximum matching that does not cover w, and consider the maximal path P in $(V, M_r \triangle M_w)$ that begins in w; its edges alternate between M_r and M_w, and it ends in Y. Again, P has an even number of edges for otherwise $M_w \triangle E(P)$ would be a larger matching, contradicting the maximality of M_w. If P did not contain any vertex of S, then $M_r \triangle (E(P_v) \cup \{e\} \cup E(P))$ would be a larger matching than M_r, which is impossible.

So let $f = \{y, z\}$ be the first edge of P (when traversed from w) with $y \in V(H) \setminus S$ and $z \in S \cup X$, and let P' be the subpath of P from w to z. If $z \in S$, then we could add the vertices of $V(P') \setminus \{z\}$ to S because $\{e\} \cup E(P')$ contains an M_r-alternating path of even length from every vertex of $V(P') \setminus \{z\}$ to v or z. Since S was chosen maximal, we conclude that $z \in X$. If f belongs to M_w, then $M_w \triangle E(P')$ is another maximum matching with fewer edges in $\delta(X) \cap \delta(Y)$, contradicting (ii). If f belongs to M_r, then $M_r \triangle (E(P_v) \cup \{e\} \cup E(P'))$ is another maximum matching with fewer edges in $\delta(X) \cap \delta(Y)$, again contradicting (ii). □

Given an algorithm for computing a maximum cardinality matching, it is easy to compute Y in polynomial time. In fact, Y is directly computed by Edmonds' [1965a] matching algorithm (see e.g., Schrijver [2003] or Korte and Vygen [2018]), which can be implemented to run in $O(n^3)$ time.

Let $\varphi(G)$ denote the minimum number of even ears in any ear-decomposition of G. This number yields a lower bound on the number of edges in any tour:

Proposition 13.7. *Let G be a 2-edge-connected graph with n vertices. Then any 2-edge-connected spanning subgraph, and any tour, has at least $n - 1 + \varphi(G)$ edges.*

Proof. Let H be a 2-edge-connected spanning subgraph of G. The number of edges of H is $n - 1$ plus the number of ears in any ear-decomposition of H.

Any ear-decomposition of H has at least $\varphi(G)$ ears because otherwise we could extend it to an ear-decomposition of G with fewer than $\varphi(G)$ even ears by adding the remaining edges as trivial ears. Hence, H has at least $n - 1 + \varphi(G)$ edges. By Proposition 13.3, no tour can have fewer edges. □

If G is a graph with n vertices, then $\varphi(G)$ is even if and only if n is odd (because even ears have an odd number of internal vertices). Since the graph G in Figure 13.1 (left-hand side) has 15 vertices and is not factor-critical, it has $\varphi(G) \geq 2$ by Theorem 13.5, and the ear-decomposition in this figure actually shows $\varphi(G) = 2$. Our next goal is to compute an ear-decomposition with $\varphi(G)$ even ears.

The following observation will be useful:

Lemma 13.8 (Schrijver [2003]). *If $G = (V, E)$ is 2-edge-connected and $\emptyset \neq U \subsetneq V$ such that $G[U]$ is 2-edge-connected, then $\varphi(G) \leq \varphi(G[U]) + \varphi(G/U)$.*

Proof. If $G[U]$ contains a Hamiltonian circuit C, then we can construct an ear-decomposition of G by taking an ear-decomposition of G/U with $\varphi(G/U)$ even ears, letting P be the first ear containing the vertex that resulted from contraction, and inserting (if possible, an even) part of C into P, as well as the rest of C as a new ear immediately behind P. The remaining edges of $G[U]$ can be added as trivial ears at the end. We obtain an ear-decomposition of G with $\varphi(G[U]) + \varphi(G/U)$ even ears.

Otherwise, let C be the first ear in an ear-decomposition of $G[U]$ with $\varphi(G[U])$ even ears. We have $\varphi(G[U]) = \varphi(C) + \varphi(G[U]/C)$. Then, by the first part, $\varphi(G) \leq \varphi(C) + \varphi(G/C)$. By induction on the number of vertices, $\varphi(G/C) \leq \varphi(G[U]/C) + \varphi(G/U)$. Combining this yields $\varphi(G) \leq \varphi(C) + \varphi(G[U]/C) + \varphi(G/U) = \varphi(G[U]) + \varphi(G/U)$. □

Now we can prove a fundamental result of Frank [1993]. Our proof largely follows Schrijver [2003].

Theorem 13.9 (Frank [1993]). *Let $G = (V, E)$ be a 2-edge-connected graph with n vertices.*

Then for any $T \subseteq V$ such that $|T|$ is even, there exists a T-join in G with at most $\frac{1}{2}(n + \varphi(G) - 1)$ edges. Moreover, there exists a $T \subseteq V$ such that $|T|$ is even and the minimum cardinality of a T-join in G is $\frac{1}{2}(n + \varphi(G) - 1)$. Such a T and an ear-decomposition with $\varphi(G)$ even ears can be found in polynomial time.

Proof. The first statement follows by induction on the number of ears. If P is the last ear in an ear-decomposition with $\varphi(G)$ even ears, P is the union of a T'-join J' and a T''-join J'', where $T' \cap \mathrm{in}(P) = T'' \cap \mathrm{in}(P) = T \cap \mathrm{in}(P)$. Take the smaller one, say J', with at most $\lfloor \frac{1}{2}|E(P)| \rfloor$ edges. By induction, there exists a $(T \triangle T')$-join J in $G - \text{in}(P)$ with at most $\frac{1}{2}(n - |\text{in}(P)| + \varphi(G - \text{in}(P)) - 1)$ edges. $J \cup J'$ does the job by Proposition 1.26.

The second statement is more difficult to prove. We proceed by induction on n. If G is factor-critical, we can find an odd ear-decomposition by Theorem 13.5, so $\varphi(G) = 0$, and any $T \subsetneq V$ with $|T| = n - 1$ proves the assertion. So we assume that G is not factor-critical.

If G has a perfect matching, let $G' = G - v$ for an arbitrary vertex v; otherwise, let $G' = G$. Let Y be the set of vertices $y \in V(G')$ for which there exists a maximum matching in G' that does not cover y. Let X be the set of neighbors of Y in G (as in the Gallai–Edmonds decomposition, but also including v if G has a perfect matching). Consider the bipartite graph that results from $G[Y \cup X]$ by deleting all edges in $G[X]$ and contracting all edges in $G[Y]$ (note that parallel edges may arise). See Figure 13.3. This bipartite graph has a matching M covering X (by Theorem 13.6 (ii) and the definition of Y, every maximum matching in G contains such a set M). Orient the edges of M towards X and the other edges towards Y. This digraph D (like every digraph, cf. Proposition 6.4) has a strongly connected component L such that no edge enters L.

Every vertex in D has an entering edge: for the vertices in X an edge of M, for the other vertices because only one edge (belonging to M) leaves such a vertex and G is 2-edge-connected. Because no edge enters L, we conclude $|L| \geq 2$. Moreover, no edge $e = (y, x)$ of M can leave L because otherwise y would have no outgoing edge in $D[L]$, but L is strongly connected.

Now $D[L]$ is strongly connected and thus has a directed ear-decomposition (by Proposition 13.2). All ears must alternate between edges in M and other edges. The first ear is even (the graph is bipartite), and a straightforward induction shows that for every k, the union of the first k ears contains a perfect matching within M, and all ears except the first one are odd. By Lemma 13.8, since the connected components of Y that we contracted are all factor-critical (by Theorem 13.6 (i)), we have $\varphi(G[U]) = 1$, where U arises from L by undoing the contraction. We also conclude $|V(L) \cap X| = |V(L) \cap Y|$.

So $\varphi(G[U]) = 1$, and hence, again by Lemma 13.8, $\varphi(G) \leq \varphi(G[U]) + \varphi(G/U) = 1 + \varphi(G/U)$. (Note that $U = V$ is possible, in which case G/U contains a single vertex and the above is also true.)

Let (by induction) $T \subseteq V(G/U)$ such that $|T|$ is even and the minimum size of a T-join is $\frac{1}{2}(|V(G/U)| + \varphi(G/U) - 1)$. Let $\bar{T}$ result from $(U \cap Y) \cup (T \setminus U)$ by adding an arbitrary vertex in $U \cap X$ if the set was odd before. Then we claim:

$$\text{Any } \bar{T}\text{-join in } G \text{ has at least } \tfrac{|U|}{2} \text{ edges inside } G[U]. \qquad (13.2)$$

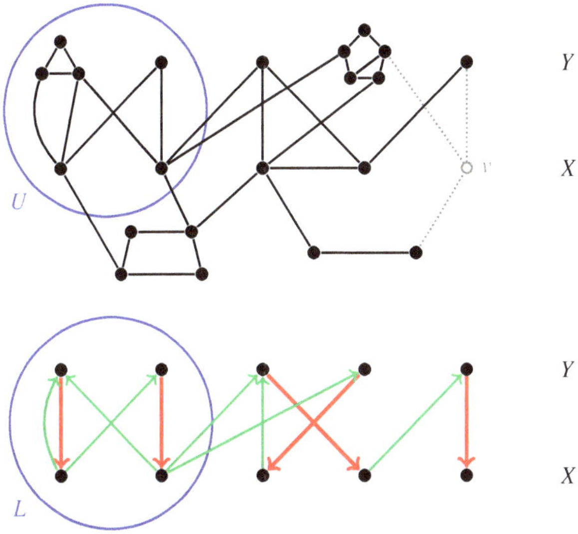

Figure 13.3 Top: The Gallai–Edmonds decomposition of $G' = G - v$ in the proof of Theorem 13.9. Bottom: The digraph D. The red, thick edges oriented from Y to X belong to M. The graph L is a strongly connected component of D without entering arcs, and U is the corresponding vertex set in G.

By (13.2), the minimum size of a $\bar{T}$-join in G is at least $\frac{1}{2}(|V(G/U)| + \varphi(G/U) - 1) + \frac{|U|}{2} = \frac{1}{2}(n + 1 + \varphi(G/U) - 1) \geq \frac{1}{2}(n + \varphi(G) - 1)$.

To prove (13.2), first observe that the vertices in $U \cap Y$ have no neighbors outside U. This holds by the choice of L and since no matching edge leaves L. We conclude that any $\bar{T}$-join in G contains at least $|U \cap Y| - \max\{|M| : M \text{ matching in } G[U \cap Y]\}$ edges inside $G[U]$. Because $|V(L) \cap X| = |V(L) \cap Y|$, the subgraph $G[U \cap Y]$ has $|U \cap X|$ connected components, each of which has an odd number of vertices. Hence, any matching in $G[U \cap Y]\}$ has at most $\frac{1}{2}(|U \cap Y| - |U \cap X|)$ edges. This proves (13.2).

All steps of this proof are constructive, implying a polynomial-time algorithm. Computing the Gallai–Edmonds decomposition can be done with any polynomial-time algorithm that computes a maximum matching (cf. Theorem 1.19). □

Note that this immediately implies a $\frac{3}{2}$-approximation algorithm for computing a smallest 2-edge-connected spanning subgraph: Compute an ear-decomposition with $\varphi(G)$ even ears and delete the trivial ears. The output has at most $\frac{3}{2}(n-1) + \frac{1}{2}\varphi(G)$ edges, which yields a $\frac{3}{2}$-approximation by Proposition 13.7. Cheriyan,

Sebő, and Szigeti [2001] improved the approximation ratio to $\frac{17}{12}$ as shown in Exercise 13.2.

Theorem 13.9 can also be used to strengthen Proposition 13.7. Let $\mathrm{LP}(G)$ denote the value of the Graph TSP LP (12.1), and let

$$L_\varphi(G) := n - 1 + \varphi(G).$$

Corollary 13.10 (Cheriyan, Sebő, and Szigeti [2001]). *For every 2-edge-connected graph G, we have*

$$L_\varphi(G) \le \mathrm{LP}(G).$$

Proof. By Theorem 13.9, there exists a set T of vertices such that $|T|$ is even and $\frac{1}{2}(n-1+\varphi(G))$ is the minimum cardinality of a T-join in G. Now use Theorem 2.19 and observe that the value of (2.9) (for $c(e) = 1$ for all edges e, and for every T) is at most half the value of (12.1). □

13.4 Nice and Nicer Ear-Decompositions

Frank's Theorem 13.9 was used by Cheriyan, Sebő, and Szigeti [2001] and Sebő and Vygen [2014] as a starting point to obtain an ear-decomposition with further properties. In the rest of this chapter, we assume G to be 2-vertex-connected, which is no loss of generality by Proposition 12.1.

Theorem 13.11 (Lovász and Plummer [1986]). *For any 2-connected factor-critical graph, an open odd ear-decomposition can be computed in polynomial time.*

Proof. Let G be a 2-connected factor-critical graph. By Theorem 13.5, we can compute an odd ear-decomposition $P_0, P_1, \ldots, P_k$ of G in polynomial time. Suppose this is not open; then let P_i be the first closed ear (i.e., $i \ge 2$ minimum such that P_i is a circuit). Let G' be the subgraph of G with ear-decomposition $P_0, P_1, \ldots, P_{i-1}$, and let v be the vertex that G' and P_i have in common.

Let P_j be the first ear after P_i so that the graph with ear-decomposition $P_0, P_1, \ldots, P_j$ contains a path from $\mathrm{in}(P_i)$ to the root that does not contain v. Let w_1 and w_2 be the two endpoints of P_j. For $i = 1, 2$, construct a path R_i from w_i to $V(G')$ by starting with $x = w_i$, letting Q be the ear with $x \in \mathrm{in}(Q)$ and y the endpoint of Q such that the x-y-path in Q has even length, setting $\bar{Q}$ to be the rest of Q (a path with odd length), and iterating with $x := y$ until $x \in V(G')$.

Exactly one of the two paths R_1, R_2 ends in v. Hence, concatenating R_1, P_j, and R_2 yields an open odd ear P_{new} that can be added to G', and

$$P_0, P_1, \ldots, P_{i-1}, P_{\mathrm{new}}, \bar{P}_i, \ldots, \bar{P}_{j-1}, P_{j+1}, \ldots, P_k$$

is an odd ear-decomposition of G. Iterating this procedure, we obtain an open odd ear-decomposition of G after less than n steps. □

Corollary 13.12 (Cheriyan, Sebő, and Szigeti [2001]). *For any 2-connected graph G, an open ear-decomposition with $\varphi(G)$ even ears can be computed in polynomial time.*

Proof. By Theorem 13.9, we can construct an ear-decomposition with $\varphi(G)$ even ears. Take an arbitrary edge in each even ear and subdivide it. We obtain an odd ear-decomposition, so the resulting graph is factor-critical. By Theorem 13.11, it has an open odd ear-decomposition. Take this and undo the $\varphi(G)$ subdivisions in order to obtain an open ear-decomposition of the original graph G, which then has at most (and hence exactly) $\varphi(G)$ even ears. □

Definition 13.13 (nice ear-decomposition). An ear-decomposition is called *nice* if it has $\varphi(G)$ even ears, all short ears are pendant, and there is no edge joining internal vertices of different short ears.

For example, Figure 13.1, left-hand side, displays a nice ear-decomposition. Let us call an ear of length 2 or 3 simply a 2-ear or 3-ear, respectively. The following lemma is essentially due to Cheriyan, Sebő, and Szigeti [2001], although they did not consider 2-ears. In the following form, the lemma appears in Sebő and Vygen [2014].

Lemma 13.14. *Given a 2-vertex-connected graph, one can compute a nice ear-decomposition in polynomial time.*

Proof. First, compute an open ear-decomposition with $\varphi(G)$ even ears, using Corollary 13.12. Then we perform a sequence of operations and maintain an ear-decomposition with $\varphi(G)$ even ears. Each step will increase the number of trivial ears. We will always assume that the trivial ears come last in the ear-decomposition. We proceed in three phases.

In the first phase, we make 2-ears pendant and maintain an open ear-decomposition. If a 2-ear P is not pendant, let Q be the first nontrivial ear attached to it. Remove P, extend Q by one edge of P so that it remains open, and make the other edge a trivial ear. Iterate this until all 2-ears are pendant.

In the second phase, we make 3-ears pendant and maintain the invariant that no closed ear is attached to any 3-ear. Let P be the first non-pendant 3-ear, and let Q be the first nontrivial ear attached to it. Note that Q is open. Remove P, extend Q by two edges of P (it may become closed), and make the third edge a trivial ear. Iterate this until all 3-ears are pendant.

Now all short ears are pendant (and will remain so). Suppose two short ears are adjacent – that is, there is an edge $\{v, w\}$ (a trivial ear) such that $v \in \mathrm{in}(P)$

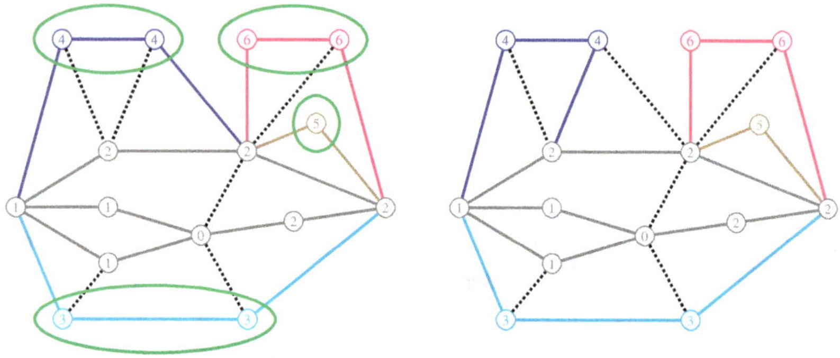

Figure 13.4 Left: A graph G with a nice ear-decomposition. The short ears are highlighted in color, and the elements of $\mathcal{I}$ are the green sets. Right: A nicer ear-decomposition, in which the spanning subgraph that contains only the edges of the short ears has fewer connected components (five instead of six). Only ear P_4 has been modified here.

and $w \in \text{in}(Q)$ for two short ears P and Q. Then we can extend this trivial ear by all but one edge from P and all but one edge from Q. The new ear is pendant and can be put at the end of the ear-decomposition, followed only by the trivial ears. One trivial ear vanishes, but two new trivial ears arise.

It is easy to see that the number of even ears never increases. □

Recall that Theorem 13.4 gives a poor bound if there are many pendant ears. In this case, we will start constructing a tour by taking all the edges of the pendant ears. Then their internal vertices have even degree, so we can proceed by considering only the subgraph induced by the remaining vertices. This is particularly efficient if the pendant ears already connect many of these vertices. This is what we try to achieve next, but we concentrate on the short ears.

Namely, a nice ear-decomposition allows for optimizing the short ears in the following sense. Let $\mathcal{I}$ contain for each short ear the set of its internal vertices (cf. Figure 13.4, left-hand side). For $I \in \mathcal{I}$, we denote by $\mathcal{P}_I$ the set of paths in G whose set of internal vertices is I. A key observation is that we can replace the ear $P_I \in \mathcal{P}_I$ by any other element of $\mathcal{P}_I$ (adapting the set of trivial ears accordingly). This maintains a nice ear-decomposition. Our goal is to choose an element of $\mathcal{P}_I$ for each $I \in \mathcal{I}$ so that the spanning subgraph that contains only the edges of these paths has as few connected components as possible. See Figure 13.4 for an example.

We will now show that this optimization problem can be solved in polynomial time. For this, we need the notion of matroids, a classic part of combinatorial optimization theory (see also Exercise 2.5):

Definition 13.15 (matroid)**.** Let U be a finite set and $\mathcal{M}$ a family of subsets of U. Then $(U, \mathcal{M})$ is called a *matroid* if

- $\emptyset \in \mathcal{M}$;
- for all $B \in \mathcal{M}$ and $A \subseteq B$, we have $A \in \mathcal{M}$; and
- for all $A, B \in \mathcal{M}$ with $|A| < |B|$, there exists an element $b \in B \setminus A$ with $A \cup \{b\} \in \mathcal{M}$.

Typically, the elements of $\mathcal{M}$ are given implicitly, and we assume that one can test in polynomial time whether a given subset of U belongs to $\mathcal{M}$. A simple example is a *partition matroid*: Here we have a partition $\mathcal{W}$ of U and a function $u : \mathcal{W} \to \mathbb{Z}_{>0}$ such that $\mathcal{M} = \{A \subseteq U : |A \cap W| \leq u(W) \text{ for all } W \in \mathcal{W}\}$. It is easy to see that then $(U, \mathcal{M})$ is indeed a matroid. We will consider the partition matroid $(U, \mathcal{M}_1)$ with $U = \bigcup_{I \in \mathcal{I}} \mathcal{P}_I$ and

$$\mathcal{M}_1 = \left\{A \subseteq U : |A \cap \mathcal{P}_I| \leq 1 \text{ for all } I \in \mathcal{I}\right\}.$$

Another example are *graphic matroids*: Here we have an undirected graph $G = (V, E)$ and let $\mathcal{M}$ contain all edge sets of forests in G. It is easy to show that $(E, \mathcal{M})$ is a matroid. For $P \in U = \bigcup_{I \in \mathcal{I}} \mathcal{P}_I$, let $e_P \in \binom{V}{2}$ denote an edge that connects the endpoints of P. We will consider the graphic matroid $(U, \mathcal{M}_2)$, where

$$\mathcal{M}_2 = \left\{A \subseteq U : \left(V, \dot{\bigcup}_{P \in A} \{e_P\}\right) \text{ is a forest}\right\}.$$

By the third property of Definition 13.15, it is trivial to find an element of $\mathcal{M}$ with maximum cardinality for any given matroid $(U, \mathcal{M})$. In fact, the greedy algorithm can also find an element A of $\mathcal{M}$ with maximum weight $w(A) = \sum_{a \in A} w(a)$ for given weights $w : U \to \mathbb{R}$ (see Exercise 2.5), but we do not need this here. A more difficult problem is to find a maximum-cardinality set that belongs to the intersection of two matroids. However, this is also well-solved due to Edmonds' matroid intersection theorem:

Theorem 13.16 (Edmonds [1970])**.** *Let U be a finite set and $\mathcal{M}_1, \mathcal{M}_2$ be families of subsets of U such that $(U, \mathcal{M}_1)$ and $(U, \mathcal{M}_2)$ are matroids. Suppose we can check in polynomial time for any $A \subseteq U$ whether $A \in \mathcal{M}_1$ and whether $A \in \mathcal{M}_2$. Then we can find a set $A \in \mathcal{M}_1 \cap \mathcal{M}_2$ of maximum cardinality in polynomial time, and this cardinality is* $\min\{r_1(X) + r_2(U \setminus X) : X \subseteq U\}$*, where* $r_i(X) = \max\{|Y| : Y \subseteq X,\ Y \in \mathcal{M}_i\}$ *is the rank function of $\mathcal{M}_i$ ($i = 1, 2$).*

To solve our optimization problem, we now apply this to the partition matroid $(U, \mathcal{M}_1)$ and the graphic matroid $(U, \mathcal{M}_2)$ defined above.

Theorem 13.17 (Sebő and Vygen [2014]). *Given a 2-connected graph $G = (V, E)$, we can compute in polynomial time a nice ear-decomposition of G such that the following holds. Let $\mathcal{I}$ contain the set of internal vertices for each short ear, and let E_{short} contain all edges of short ears. Then there exists a subset $\mathcal{B} \subseteq \mathcal{I}$ such that (V, E_{short}) has at most $|V| - |E_{\text{short}}| + |\mathcal{B}| - r_2(\bigcup_{I \in \mathcal{B}} \mathcal{P}_I)$ connected components.*

Proof. We begin with an arbitrary nice ear-decomposition of G, which we get from Lemma 13.14, and let $\mathcal{I}$ contain the set of internal vertices for each short ear. Then we find a set $A \in \mathcal{M}_1 \cap \mathcal{M}_2$ of maximum cardinality, where $(U, \mathcal{M}_1)$ and $(U, \mathcal{M}_2)$ are the two matroids defined earlier. By Theorem 13.16, A can be computed in polynomial time, and there exists a set $X \subseteq U$ with $|A| = r_1(X) + r_2(U \setminus X)$.

Since $A \in \mathcal{M}_1$, we have $A = \{P_I : I \in \mathcal{J}\}$ for some $\mathcal{J} \subseteq \mathcal{I}$ and $P_I \in \mathcal{P}_I$ $(I \in \mathcal{J})$. We change the ear-decomposition as follows. For each $I \in \mathcal{J}$, we replace the ear with internal vertices I by P_I (and adapt the set of trivial ears accordingly). We claim that this ear-decomposition does the job.

Let E_{short} be the set of edges of the short ears (after changing the ear-decomposition). Since $A \in \mathcal{M}_2$, omitting one edge from each ear with internal vertices $I \in \mathcal{I} \setminus \mathcal{J}$ yields a forest. Hence, the number of connected components of (V, E_{short}) is at most

$$|V| - \left(|E_{\text{short}}| - |\mathcal{I} \setminus \mathcal{J}|\right) = |V| - |E_{\text{short}}| + |\mathcal{I}| - |A|. \qquad (13.3)$$

Let $\mathcal{B} := \{I \in \mathcal{I} : \mathcal{P}_I \cap X = \emptyset\}$. Then we have

$$\begin{aligned} |A| &= r_1(X) + r_2(U \setminus X) \\ &= |\mathcal{I} \setminus \mathcal{B}| + r_2(U \setminus X) \\ &\geq |\mathcal{I} \setminus \mathcal{B}| + r_2\left(\bigcup_{I \in \mathcal{B}} \mathcal{P}_I\right). \end{aligned} \qquad (13.4)$$

Plugging this bound into (13.3) yields the result. □

In fact, we have equality because we need to remove at least $|\mathcal{B}| - r_2(\bigcup_{I \in \mathcal{B}} \mathcal{P}_I)$ edges from the short ears in order to obtain a forest.

In the example of Figure 13.4, $\mathcal{B}$ would consist of the sets of internal vertices of the ears P_5 and P_6 (see Figure 13.5, left-hand side); then $|\mathcal{B}| = 2$ and $r_2(\bigcup_{I \in \mathcal{B}} \mathcal{P}_I) = 1$, and indeed the edges of the short ears in the ear-decomposition shown on the right-hand side of Figure 13.4 form a spanning subgraph with $15 - 11 + 2 - 1 = 5$ connected components.

Although Theorem 13.17 yields a polynomial-time algorithm, one can exploit the fact that the two matroids have a special structure to obtain a faster running

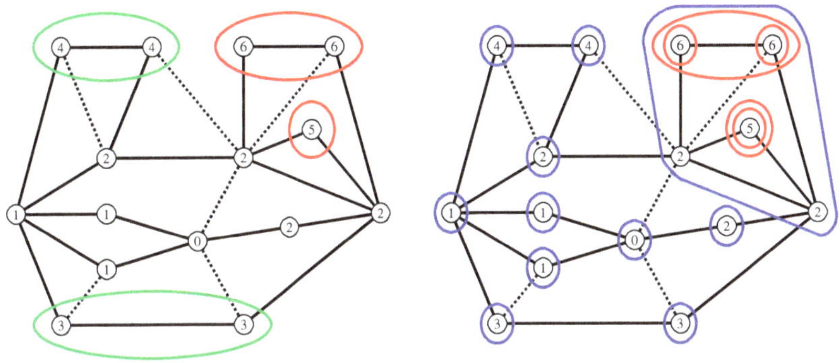

Figure 13.5 Left: The nicer ear-decomposition from Figure 13.4 (according to Theorem 13.17). The red sets are the elements of $\mathcal{B}$; the green sets are the elements of $\mathcal{I} \setminus \mathcal{B}$. Right: The cuts in the proof of Theorem 13.18. The partition $\mathcal{U}$ is shown in blue.

time. Already Rado [1942] considered the case when one matroid is a partition matroid. Sebő and Vygen [2014] reduced the problem to finding a maximum forest representative system of the hypergraph in which all endpoints of paths in $\mathcal{P}_I$ form a hyperedge (for each $I \in \mathcal{I}$). Using also a faster algorithm for Theorem 13.9 (Frank [1993]), one can compute an ear-decomposition as in Theorem 13.17 in $O(|V||E|)$ time.

From now on, we work with a "nicer" ear-decomposition as in Theorem 13.17. If it has many pendant ears and the short ears form a forest, this will help us because we simply take all edges of pendant ears (then the internal vertices of pendant ears have degree 2) and complete this edge set to a tour using only edges of the subgraph induced by the other vertices. If the short ears do not form a forest, we will obtain a stronger lower bound:

Theorem 13.18 (Sebő and Vygen [2014]). *Let $G = (V, E)$ be a graph with a nice ear-decomposition. Let $\mathcal{I}$ contain the set of internal vertices for each short ear. Then*

$$L_\mu(G, \mathcal{I}) := n - 1 + \max\left\{ |\mathcal{B}| - r_2\left(\bigcup_{I \in \mathcal{B}} \mathcal{P}_I\right) : \mathcal{B} \subseteq \mathcal{I} \right\} \le \mathrm{LP}(G),$$

where $n = |V|$ and r_2 is the rank function of the matroid $(U, \mathcal{M}_2)$.

Proof. Let $\mathcal{B} \subseteq \mathcal{I}$. Let $\mathcal{U}$ be the partition of V that contains the vertex sets of the connected components of $(V, \bigcup_{I \in \mathcal{B}, P_I \in \mathcal{P}_I} E(P_I))$. Then $|\mathcal{U}| = n - \sum_{I \in \mathcal{B}} |I| - r_2(\bigcup_{I \in \mathcal{B}} \mathcal{P}_I)$.

Consider the family of sets $\mathcal{F} := \mathcal{U} \cup \mathcal{B} \,\dot\cup\, \{\{v\} : v \in I \in \mathcal{B}\}$, taking singletons in $\mathcal{B}$ twice. See Figure 13.5 for an illustration. Let x be a feasible solution to (12.1). Summing over the inequalities $x(\delta(U)) \geq 2$ for the sets U in this family $\mathcal{F}$ (except for $U = V$) yields

$$2x(E) \;\geq \sum_{U \in \mathcal{F} \setminus \{V\}} x(\delta(U)) \;\geq\; 2(|\mathcal{F}| - 1) \;=\; 2\left(n - 1 + |\mathcal{B}| - r_2\left(\bigcup_{I \in \mathcal{B}} \mathcal{P}_I\right)\right)$$

because no edge is contained in more than two of these at least $|\mathcal{U}| - 1 + |\mathcal{B}| + \sum_{I \in \mathcal{B}} |I|$ cuts. □

Note that $L_\mu(G, \mathcal{I}) \geq n - 1$ because $\mathcal{B} = \emptyset$ is a possible choice.

13.5 The $\frac{7}{5}$-Approximation Algorithm

Now we can explain the $\frac{7}{5}$-approximation algorithm for Graph TSP by Sebő and Vygen [2014]. We will work with the lower bound

$$\Lambda(G, \mathcal{I}) := \tfrac{1}{3} L_\varphi(G) + \tfrac{2}{3} L_\mu(G, \mathcal{I}).$$

Proposition 13.19. *Let G be a 2-edge-connected graph, and let $\mathcal{I}$ contain the set of internal vertices for each short ear in a nice ear-decomposition. Then*

$$n - 1 \;\leq\; \Lambda(G, \mathcal{I}) \;\leq\; \mathrm{LP}(G).$$

Proof. $\Lambda(G, \mathcal{I}) \leq \mathrm{LP}(G)$ follows from Corollary 13.10 and Theorem 13.18. The inequality $\Lambda(G, \mathcal{I}) \geq n - 1$ follows from $L_\varphi(G) \geq n - 1$ and $L_\mu(G, \mathcal{I}) \geq n - 1$. □

The algorithm first computes an ear-decomposition as in Theorem 13.17 and then deletes all trivial ears. The following lemma shows what to do if there are many pendant ears (otherwise, we will use Theorem 13.4):

Lemma 13.20. *Let G be a 2-edge-connected graph with a nice ear-decomposition that has no trivial ears, let E_{short} be the set of edges of the short ears, let $\mathcal{I}$ contain the set of internal vertices for each short ear, and let k be the number of connected components of (V, E_{short}). Then one can compute a tour in G with at most $k - 1 + |E_{\text{short}}| + \frac{1}{2} L_\varphi(G) - \pi$ edges in $O(n^3)$ time, where π is the number of pendant ears.*

Proof. Let E_π be the set of all edges of pendant ears, and V_π the set of internal vertices of pendant ears. We have $|E_\pi| = |V_\pi| + \pi$. If λ denotes the number of long pendant ears and φ_π denotes the number of even pendant ears, we have $|V_\pi| \geq 2\pi + 2\lambda - \varphi_\pi$.

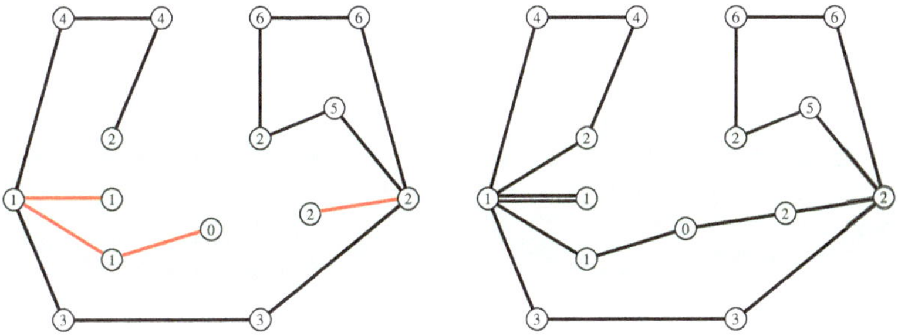

Figure 13.6 Illustrating the algorithm described in the proof of Lemma 13.20, continuing the example from Figure 13.5. Left: The edges of pendant ears after optimizing short ears (black) and a minimal set of edges of non-pendant ears (red) to make a connected spanning subgraph. Right: For correcting parities (in the green subgraph $G[V \setminus V_\pi]$), we need three more edges; a possible resulting tour is shown.

Our algorithm first takes all edges of pendant ears. Since (V, E_{short}) has k connected components and $E_{\text{short}} \subseteq E_\pi$, we conclude that (V, E_π) has at most $k - |E_\pi \setminus E_{\text{short}}| + \lambda$ connected components. Hence, we can add $k - |E_\pi \setminus E_{\text{short}}| + \lambda - 1$ edges to E_π to obtain a connected spanning subgraph in which all vertices in V_π have degree 2 (see Figure 13.6). Note that this subgraph has at most $k + |E_{\text{short}}| + \lambda - 1 \leq k + |E_{\text{short}}| + \frac{1}{2}|V_\pi| - \pi + \frac{1}{2}\varphi_\pi - 1$ edges.

Let $T \subseteq V \setminus V_\pi$ be the set of vertices with odd degree in this subgraph. Then add a minimum T-join in $G[V \setminus V_\pi]$ (cf. Theorem 1.29); note that this induced subgraph has an ear-decomposition with $\varphi(G) - \varphi_\pi$ even ears. By Theorem 13.9, this T-join has at most $\frac{1}{2}(n - |V_\pi| - 1 + \varphi(G) - \varphi_\pi)$ edges. Summing up, our tour has at most $k - 1 + |E_{\text{short}}| + \frac{1}{2}L_\varphi(G) - \pi$ edges. □

The overall algorithm is now easily described:

Theorem 13.21 (Sebő and Vygen [2014]). *There is a $\frac{7}{5}$-approximation algorithm for* Graph TSP. *For any given graph G, it computes a tour with at most $\frac{7}{5}$ LP(G) edges.*

Proof. By Proposition 12.1, we may assume that the given graph $G = (V, E)$ is 2-connected. Then we first compute an ear-decomposition as in Theorem 13.17. Next, we delete all trivial ears. Again, let $\mathcal{I}$ contain the set of internal vertices for each short ear.

Let $G' = (V, E')$ be the resulting spanning subgraph. G' is 2-edge-connected (since it has an ear-decomposition) but not necessarily 2-vertex-connected. Let $G_1, \ldots, G_p$ be the maximal 2-connected subgraphs of G'; we call them the *blocks* of G'. Several of them may share a vertex, but their edge sets form a partition of E'. Note that $\sum_{i=1}^{p}(|V(G_i)|-1) = |V|-1$ and $\sum_{i=1}^{p} L_\varphi(G_i) = L_\varphi(G)$.

Let π_i denote the number of pendant ears of G_i. Note that an ear can be pendant in some G_i but not pendant in G. Let k_i be the number of connected components in $(V(G_i), E_{\text{short}} \cap E(G_i))$ and k be the number of connected components in (V, E_{short}). By Theorem 13.17, we have (for some $\mathcal{B} \subseteq \mathcal{I}$)

$$\begin{aligned}\sum_{i=1}^{p}(k_i-1) &= k-1\\ &= n-1-|E_{\text{short}}|+|\mathcal{B}|-r_2\left(\bigcup_{I\in\mathcal{B}}\mathcal{P}_I\right)\\ &\le L_\mu(G,\mathcal{I})-|E_{\text{short}}|\end{aligned}$$

by definition of $L_\mu(G,\mathcal{I})$ (cf. Theorem 13.18).

We distinguish two cases. If $\sum_{i=1}^{p}\pi_i \le \frac{1}{10}\Lambda(G,\mathcal{I})$, apply Theorem 13.4 to each block and obtain a tour with at most $\frac{4}{3}(n-1)+\frac{2}{3}\sum_{i=1}^{p}\pi_i \le \frac{7}{5}\Lambda(G,\mathcal{I})$ edges. Otherwise, we apply Lemma 13.20 to each block to obtain a tour with at most

$$\begin{aligned}&\sum_{i=1}^{p}\left(k_i-1+|E_{\text{short}}\cap E(G_i)|+\tfrac{1}{2}L_\varphi(G_i)-\pi_i\right)\\ &=\sum_{i=1}^{p}(k_i-1)+|E_{\text{short}}|+\tfrac{1}{2}L_\varphi(G)-\sum_{i=1}^{p}\pi_i\\ &\le L_\mu(G,\mathcal{I})+\tfrac{1}{2}L_\varphi(G)-\sum_{i=1}^{p}\pi_i\\ &\le \tfrac{7}{5}\Lambda(G,\mathcal{I})\end{aligned}$$

edges. We can simply compute both tours and take the smaller one. We are done by Proposition 13.19. □

The algorithm can be implemented to run in $O(n^3)$ time. Theorem 13.21 still yields the best approximation ratio known for the GRAPH TSP. For some special cases, better approximation algorithms are known (see Section 12.3).

Theorem 13.21 also shows that the integrality ratio of the subtour LP (2.2) for metric closures of unweighted graphs is at most $\frac{7}{5}$. As of today, no better bound is known. Table 13.1 summarizes the history of GRAPH TSP.

Exercise 13.11 shows that the ratio $\frac{7}{5}$ is tight for this algorithm.

Open Problem 13.22. Devise a $\frac{4}{3}$-approximation algorithm for GRAPH TSP.

Table 13.1 *Approximation ratios and upper bounds on the integrality ratio for* Graph *TSP in the order of their discovery. The integrality ratio refers to the subtour LP (2.2) of the metric closure of* Graph *TSP instances. (R) means randomized; this algorithm computes a random tour. Only algorithms with ratio better than* $\frac{3}{2}$ *are shown.*

Approximation Ratio	Integrality Ratio	Year	Reference	Chapter
$\frac{3}{2} - 10^{-52}$ (R)	$\frac{3}{2} - 10^{-52}$	2011	Oveis Gharan, Saberi, and Singh [2011]	10.1
1.461	1.461	2011	Mömke and Svensson [2016]	12
$\frac{13}{9}$	$\frac{13}{9}$	2011	Mucha [2014]	12.5
$\frac{7}{5}$	$\frac{7}{5}$	2012	Sebő and Vygen [2014]	13

13.6 Two-Edge-Connected Spanning Subgraph Problem

The techniques described in this chapter also yield an approximation algorithm for the Minimum 2-Edge-Connected Spanning Subgraph Problem.

Theorem 13.23 (Sebő and Vygen [2014]). *There is a polynomial-time algorithm that computes, for any given 2-edge-connected graph G, a 2-edge-connected spanning subgraph with at most $\frac{4}{3}\,\mathrm{LP}(G)$ edges.*

Proof. Analogously to the proof of Proposition 12.1, we may assume that the given graph $G = (V, E)$ is 2-connected. Then we first compute an ear-decomposition as in Theorem 13.17. Next, we delete all trivial ears. Let G' be the resulting spanning subgraph. Again, G' is 2-edge-connected because it has an ear-decomposition.

Let π denote the number of pendant ears. Then G' has at most $\frac{5}{4}(n-1) + \frac{1}{4}\varphi(G) + \frac{1}{2}\pi$ edges because the number of edges in each ear P is at most $\frac{5}{4}|\mathrm{in}(P)|$ plus $\frac{1}{4}$ if it is even plus $\frac{1}{2}$ if it is short (and hence pendant). If $\pi \le \frac{1}{6}\,\mathrm{LP}(G)$, then $|E(G')| \le \frac{5}{4}L_\varphi(G) + \frac{1}{2}\pi \le \frac{4}{3}\,\mathrm{LP}(G)$ by Corollary 13.10.

If $\pi \ge \frac{1}{6}\,\mathrm{LP}(G)$, we apply Lemma 13.20 and obtain a tour F in G' with at most $k - 1 + |E_{\mathrm{short}}| + \frac{1}{2}L_\varphi(G) - \pi$ edges, which is at most $L_\mu(G, \mathcal{I}) + \frac{1}{2}L_\varphi(G) - \pi$ by Theorem 13.17 and the definition of $L_\mu(G, \mathcal{I})$. (Again, $\mathcal{I}$ contains the set of internal vertices of each short ear.) We get the bound

$$L_\mu(G, \mathcal{I}) + \tfrac{1}{2}L_\varphi(G) - \pi \;\le\; (1 + \tfrac{1}{2} - \tfrac{1}{6})\,\mathrm{LP}(G) \;=\; \tfrac{4}{3}\,\mathrm{LP}(G)$$

by Corollary 13.10 and Theorem 13.18.

Note that (V, F) is 2-edge-connected, but it may contain two copies of some edges of G. An application of Proposition 13.3 concludes the proof. □

This also shows that the integrality ratio of (12.1) is at most $\frac{4}{3}$. The best-known lower bound is $\frac{8}{7}$ (Boyd, Fu, and Sun [2016]).

Hunkenschröder, Vempala, and Vetta [2019] devised a completely different $\frac{4}{3}$-approximation algorithm for the minimum 2-edge-connected spanning subgraph problem. A better approximation ratio was achieved only recently by Garg, Grandoni, and Jabal Ameli [2023]. The currently best ratio is $1.3 + \varepsilon$ (for any $\varepsilon > 0$), obtained by Kobayashi and Noguchi [2023].

Heeger and Vygen [2017] found a $\frac{10}{7}$-approximation algorithm for the minimum 2-vertex-connected spanning subgraph problem, also based on ear-decompositions. This approximation ratio was recently improved to $\frac{4}{3}$ by Bosch-Calvo, Grandoni, and Jabal Ameli [2023].

The techniques described in this chapter also yield a $\frac{3}{2}$-approximation algorithm for a more general problem: finding a minimum T-tour (see Exercises 13.6–13.8). We will return to this problem in Chapter 14.

Exercises

13.1 Prove that an undirected graph is 2-vertex-connected if and only if it has an open ear-decomposition. Do not use Theorem 13.11 or Corollary 13.12.

13.2 Let G be a 2-vertex-connected graph. Consider a nice ear-decomposition of G. Show:

(a) If there are k_3 3-ears, then $\mathrm{LP}(G) \geq 3\,k_3$.
(b) The number of edges in nontrivial ears is at most $\frac{5}{4}\,L_\varphi(G) + \frac{1}{2}\,k_3$.
(c) Deleting the trivial ears yields a 2-edge-connected spanning subgraph with at most $\frac{17}{12}\,\mathrm{LP}(G)$ edges.

(Cheriyan, Sebő, and Szigeti [2001])

13.3 Show that the smallest 2-edge-connected spanning subgraph is at most a factor $\frac{4}{3}$ smaller than the shortest tour, and that this bound is tight.
(Monma, Munson, and Pulleyblank [1990])

13.4 For a graph G with edge weights $c : E(G) \to \mathbb{R}_{\geq 0}$, we define $\tau(G, c)$ as

$$\max\left\{\min\{c(J) : J \text{ is a } T\text{-join in } G\} : T \subseteq V(G),\ |T| \text{ even}\right\}.$$

Show that every unweighted 2-edge-connected graph G (i.e., $c(e) = 1$ for all e) has an ear-decomposition $P_0, P_1, \ldots, P_k$ such that $\tau(G, c) = \sum_{j=1}^{k} \tau(\bar{P}_j, c)$, where $\bar{P}_j$ arises from P_j by contracting the endpoints

of P_j. Show that this does not hold in general for weighted graphs.
Note: Computing $\tau(G, c)$ is *NP*-hard for general weights, even if the graph is a circuit, because this contains the well-known PARTITION problem. Iwata and Ravi [2013] devised a $\frac{3}{2}$-approximation algorithm.

13.5 Prove Theorem 13.16 in the special case when $(U, \mathcal{M}_1)$ and $(U, \mathcal{M}_2)$ are both partition matroids (given by the corresponding partitions).
Hint: Reduce this problem to a maximum flow problem and use Theorem 2.5.

13.6 Let G be a GRAPH TSP instance and $T \subseteq V$ such that $|T|$ is even. A *T-tour* is a multi-set F of edges such that (V, F) is connected and F is a T-join. Show that, given any ear-decomposition, one can construct a T-tour with at most $\frac{3}{2}(n-1) + \frac{1}{2}(k_2 - k_{\geq 4})$ edges, where k_2 and $k_{\geq 4}$ denote the number of 2-ears and long ears, respectively.
Hint: Use ear induction similar to the first part of the proof of Theorem 13.9.
(Sebő and Vygen [2014])

13.7 Continuing the previous exercise, let $\mathcal{I}$ contain the (one-element) set of internal vertices of each 2-ear, except those for which the internal vertex is in T. Now assume an ear-decomposition as in Theorem 13.17 (with the new definition of $\mathcal{I}$). Take all edges of 2-ears, but only one of the two edges if the middle vertex belongs to T. Augment this edge set minimally to a set S such that (V, S) is connected (without further edges from 2-ears) and add a minimum $(\text{odd}(S) \triangle T)$-join. Show that this yields a T-tour with at most $L_\mu(G, \mathcal{I}) + \frac{1}{2}(n - 1 - k_2 + k_{\geq 4})$ edges.
(Sebő and Vygen [2014])

13.8 Use Exercises 13.6 and 13.7 to obtain a $\frac{3}{2}$-approximation algorithm for finding a minimum T-tour in a graph. To this end, show that $L_\mu(G, \mathcal{I})$ is a lower bound on the number of edges in any T-tour.
(Sebő and Vygen [2014])

13.9 Show that in the setting of the proof of Theorem 13.21, we have

$$\sum_{i=1}^{p} L_\mu(G_i, \mathcal{I}_i) = L_\mu(G, \mathcal{I}).$$

13.10 Show that $(U, \mathcal{M})$ is a matroid if and only if $\emptyset \in \mathcal{M}$, subsets of elements of $\mathcal{M}$ are in $\mathcal{M}$, and the maximal elements of $\mathcal{M}$ (called the *bases*) satisfy the following basis exchange axiom: For any two distinct bases B_1 and B_2, there are elements $e_1 \in B_1 \setminus B_2$ and $e_2 \in B_2 \setminus B_1$ such that $B_1 \setminus \{e_1\} \cup \{e_2\}$ and $B_2 \setminus \{e_2\} \cup \{e_1\}$ are bases. (See Exercise 5.13 for a special case.)

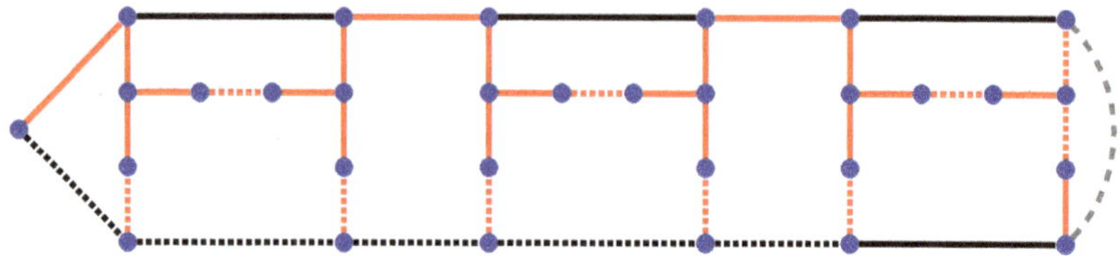

Figure 13.7 Example showing that the tour computed by the algorithm of Theorem 13.21 is not necessarily much shorter than $\frac{7}{5}$ times the optimum. For every $k \in \mathbb{N}$, we have a Hamiltonian graph with $10k + 1$ vertices and $13k + 1$ edges. The figure shows the case $k = 3$. A nice open ear-decomposition starts with $2k$ ears of length 5 from left to right, each with three vertical edges, and then proceeds with k horizontal pendant 3-ears and one trivial ear (the dashed edge on the right). Here $\pi = k = \frac{1}{10}\Lambda(G, I)$, but no matter whether we apply Theorem 13.4 (if dotted edges are removable) or Lemma 13.20 (if the red edges form the spanning tree), we can end up with $14k$ edges. This picture is adapted from Sebő and Vygen [2014] (with permission from Springer Nature).

13.11 Show with the example in Figure 13.7 that the bound $\frac{7}{5}$ on the approximation ratio of the algorithm of Theorem 13.21 is tight.

13.12 Let $\alpha > 1$. Prove: If there is an α-approximation algorithm for the minimum 2-edge-connected spanning subgraph problem, then there is a $\frac{2}{3}(\alpha + 1)$-approximation algorithm for the GRAPH TSP.
Hint: Use Theorem 13.4.
(Sebő and Vygen [2014])

14

Symmetric Path TSP and T-Tours

Like in the asymmetric case (cf. Chapter 9), one can consider the generalization of SYMMETRIC TSP where the start and end of the tour that we are looking for are not necessarily identical. Christofides' algorithm can be generalized to this problem but yields only a $\frac{5}{3}$-approximation here. This chapter contains basic results about this problem, as well as a further generalization called T-tours; these results will be used in the subsequent chapters where we will present better approximation algorithms. For unweighted graphs, a $\frac{3}{2}$-approximation algorithm can be obtained with the techniques of Chapter 13, or with a simple LP-based approach that we will present in this chapter.

14.1 Hoogeveen's Path Variants and T-Tours

Similar to Definition 9.2, we say that an *$\{s,t\}$-tour* in an undirected graph $G = (V, E)$ is a multi-subset F of E such that $(V, F \dot{\cup} \{\{t, s\}\})$ is connected and Eulerian. Again, the $\{s,t\}$-tours are precisely the footprints of the walks from s to t that visit every vertex at least once. Therefore, we call s and t the endpoints of the $\{s,t\}$-tour.

Before formally defining PATH TSP and the T-TOUR PROBLEM, let us consider the variant in which we look for an $\{s,t\}$-tour and at most one of the endpoints is given as input and at least one endpoint can be chosen freely. Then there is a simple $\frac{3}{2}$-approximation algorithm:

Theorem 14.1 (Hoogeveen [1991]). *There is a $\frac{3}{2}$-approximation algorithm for each of the following problems:*

(i) *Given an instance (G, c) of SYMMETRIC TSP with $G = (V, E)$, find a minimum-cost multi-subset F of E such that F is an $\{s,t\}$-tour for some $s, t \in V$ with $s \neq t$.*

(ii) *Given an instance (G, c) of* Symmetric *TSP with $G = (V, E)$ and $s \in V$, find a minimum-cost multi-subset F of E such that F is an $\{s,t\}$-tour for some $t \in V \setminus \{s\}$.*

Proof. The algorithm is similar to Christofides' algorithm (Algorithm 1.31): First, take a minimum-cost spanning tree (V, S) and compute a minimum-cost W-join J, where W can be chosen so that

(i) $|W \triangle \text{odd}(S)| \leq 2$ if no endpoints are given;

(ii) $|W \triangle \text{odd}(S) \triangle \{s\}| = 1$ if one endpoint s is given.

Such an edge set J of minimum cost can be found by complete enumeration on the possible sets W (or, better, by a simple direct reduction to weighted matching; cf. Exercise 14.1). Now $S \dot{\cup} J$ is a feasible solution or a tour. If it is a tour, we can delete any edge (incident to s in case (ii)) and obtain a feasible solution.

To show the approximation ratio, observe that any optimum solution F contains two disjoint sets J_1 and J_2 satisfying the above conditions: Traverse an s-t-walk with footprint F and color its edges red and blue, changing the color whenever we visit an element of $\text{odd}(S)$ for the first time. Then the two color classes are a W_1-join J_1 and a W_2-join J_2 for sets W_1 and W_2, where $W_1 \triangle \text{odd}(S)$ is empty and $W_2 \triangle \text{odd}(S)$ is the set $\{s, t\}$ of endpoints of F (or vice versa). So these sets W_1, W_2 satisfy (i) or (ii). □

The third variant studied by Hoogeveen [1991] is the one in which both endpoints are given as input. The two variants of Theorem 14.1 can be reduced easily to this third variant by enumerating all possible choices of (s and) t. Therefore, we focus on this problem from now on:

Problem 14.2 (Path TSP).

Instance: A connected undirected graph $G = (V, E)$, a cost function $c : E \to \mathbb{R}_{\geq 0}$, and two vertices $s, t \in V$.

Task: Compute an $\{s,t\}$-tour in G with minimum cost.

Note that this is equivalent to the special case of Asymmetric Path TSP where the cost function c is symmetric. We use the term Path TSP instead of Symmetric Path TSP or s-t-path TSP (a term that was used in several papers). The special case of Path TSP where $c(e) = 1$ for all $e \in E$ is called Graph Path TSP.

Several approaches to Path TSP also work for an interesting generalization:

Definition 14.3 (T-tour)**.** Given an undirected graph $G = (V, E)$ and a set $T \subseteq V$ of even cardinality, a *T-tour* in G is a multi-subset F of E such that (V, F) is connected and $\text{odd}(F) = T$.

So $\emptyset$-tours are tours in the sense of Definition 1.9, and in the case of $|T| = 2$ (say $T = \{s, t\}$), we get $\{s, t\}$-tours as already defined earlier. In some papers, T-tours have also been called connected T-joins. The most general problem we will study in this chapter is:

Problem 14.4 (T-Tour Problem)**.**

Instance: A connected undirected graph $G = (V, E)$, a cost function $c : E \to \mathbb{R}_{\geq 0}$, and a set $T \subseteq V$ of even cardinality.

Task: Compute a T-tour in G with minimum cost.

For $T = \emptyset$, this is Symmetric TSP; for $|T| = 2$, it is Path TSP. The double tree algorithm (Proposition 1.22) easily extends to the T-Tour Problem:

Proposition 14.5. *Let (G, c, T) be an instance of the T-Tour Problem with $G = (V, E)$. Let (V, S) be a minimum-cost spanning tree in (G, c) and $J \subseteq S$ be a T-join. Then $F := S \mathbin{\dot\cup} (S \setminus J)$ is a T-tour, and this yields a 2-approximation algorithm.*

Proof. The existence of a T-join $J \subseteq S$ follows from Proposition 1.27; actually, J is unique. We have $\text{odd}(F) = \text{odd}(J) = T$, so F is a T-tour. We have $c(F) = 2c(S) - c(J) \leq 2c(S) \leq 2\,\text{OPT}$ because no T-tour can be cheaper than a cheapest connector. □

The straightforward generalization of Christofides' algorithm also works for the T-Tour Problem (see Algorithm 14.6). Only the parity correction (Step (2)) needs to be adapted.

Algorithm 14.6: Christofides' Algorithm for T-Tours

Input: an instance (G, c, T) of the T-Tour Problem
Output: a T-tour F

(1) Compute a minimum-cost spanning tree (V, S) in (G, c).
(2) Let $W = T \mathbin{\triangle} \text{odd}(S)$, and let J be a minimum-cost W-join in (G, c).
(3) Output the T-tour $S \mathbin{\dot\cup} J$.

Hoogeveen [1991] (for $|T| = 2$) and Sebő and Vygen [2014] (for general T) proved:

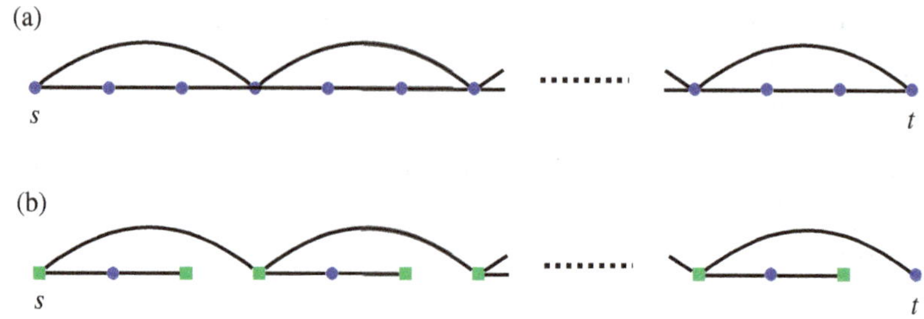

Figure 14.1 Picture (a) shows a family of examples due to Hoogeveen [1991] proving that the approximation ratio of $\frac{5}{3}$ of Christofides' algorithm for Graph Path TSP is tight. For every $k \geq 1$, we have a graph with $3k + 1$ vertices. Picture (b) shows a spanning tree (V, S) that might be chosen by Christofides' algorithm. The vertices with the wrong degree parity in the spanning tree (the elements of $T \triangle \text{odd}(S)$) are shown as green squares. Parity correction costs $2k$ here.

Theorem 14.7. *Algorithm 14.6 is a $\frac{5}{3}$-approximation algorithm for the T-Tour Problem, and this bound is tight even for $|T| = 2$ (and even for Graph Path TSP).*

Proof. Let (V, S) be a minimum-cost spanning tree. Let J be a minimum-cost $(T \triangle \text{odd}(S))$-join, and F^* an optimum solution (a minimum-cost T-tour). Then $S \dot{\cup} F^*$ is a $(T \triangle \text{odd}(S))$-join. Since both S and F^* are connected, each contains a $(T \triangle \text{odd}(S))$-join; so $S \dot{\cup} F^*$ can be partitioned into three $(T \triangle \text{odd}(S))$-joins. Hence, $3c(J) \leq c(S \dot{\cup} F^*)$, and we conclude $3c(S \dot{\cup} J) = 3c(S) + 3c(J) \leq 3c(S) + c(S \dot{\cup} F^*) = 4c(S) + c(F^*) \leq 5c(F^*)$.

The bound is tight even for $|T| = 2$ and unit weights: consider the graph (V, E) with $V = \{0, \ldots, 3\}$ and $E = \{\{0, 1\}, \{1, 2\}, \{2, 3\}, \{0, 3\}\}$ (a circuit of length 4) with $s = 0$ and $t = 3$. An optimum $\{s, t\}$-tour has three edges. If we choose $S = E \setminus \{2, 3\}$ as the spanning tree, we have $T \triangle \text{odd}(S) = \{0, 2\}$ and parity correction costs 2; hence, the resulting T-tour has five edges. See Figure 14.1 for an infinite set of examples. □

Table 14.1 gives an overview on the progress on Path TSP, starting with the seminal work by An, Kleinberg, and Shmoys [2015], who were the first to beat Christofides' algorithm. Table 14.2 summarizes the progress for the T-Tour Problem.

If we have an instance (V, c) of the Symmetric TSP with Triangle Inequality and $G = (V, E)$ is the complete graph on V, then we can turn any $\{s, t\}$-tour F in G into a Hamiltonian s-t-path without increasing the cost, simply

Table 14.1 *Approximation ratios for* Path TSP *in the order of their discovery. The ratio $\frac{3}{2} + \varepsilon$ holds for any $\varepsilon > 0$. The last result follows from the reduction from* Path TSP *to* Symmetric TSP *by Traub, Vygen, and Zenklusen [2022] (see Chapter 16).*

Approximation Ratio	Year	Reference	Chapter
$\frac{5}{3}$	1990	Hoogeveen [1991]	14.1
$\frac{1+\sqrt{5}}{2}$	2011	An, Kleinberg, and Shmoys [2015]	15.2
$\frac{8}{5}$	2012	Sebő [2013]	15.2
1.599	2015	Vygen [2016]	–
1.566	2015	Gottschalk and Vygen [2018]	15.3
$\frac{26}{17}$	2016	Sebő and van Zuylen [2019]	15.4
$\frac{3}{2} + \varepsilon$	2017	Traub and Vygen [2019a]	–
$\frac{3}{2}$	2018	Zenklusen [2019]	16.2
$\frac{3}{2} - 10^{-36}$	2022	Karlin, Klein, and Oveis Gharan [2023]	10–11, 16

by applying Lemma 1.7 to $F \dot{\cup} \{\{t, s\}\}$. Also for the T-Tour Problem, it is equivalent to work in the metric closure:

Proposition 14.8. *The T-Tour Problem is equivalent to its special case when G is complete and c satisfies the triangle inequality.*

Proof. Given a general instance (G, c, T) of the T-Tour Problem, let $(\bar{G}, \bar{c})$ be the metric closure of (G, c). Then a T-tour $\bar{F}$ in $(\bar{G}, \bar{c})$ can be transformed to a T-tour F in (G, c) of the same cost by replacing every edge $\{v, w\} \in \bar{F}$ by the edge set of a shortest v-w-path in G. □

Then optimum T-tours (with minimum number of edges) are trees:

Proposition 14.9 (Cheriyan, Friggstad, and Gao [2015]). *Let (G, c, T) be an instance of the T-Tour Problem where $G = (V, E)$ is a complete graph, c satisfies the triangle inequality, and $T \neq \emptyset$. Then there exists an optimum T-tour F such that (V, F) is a spanning tree of G.*

Proof. Let F be an optimum T-tour with minimum number of edges. By definition, (V, F) is connected. Suppose it contains a circuit C. Since $T \neq \emptyset$, the multi-graph (V, F) is connected but not a circuit itself, so there must be a vertex v that is incident to an edge $\{v, w\}$ of C and an edge $\{v, u\}$ of $F \setminus C$.

Table 14.2 *Approximation ratios and upper bounds on the integrality ratio for the T-Tour Problem in the order of their discovery. The integrality ratio refers to the LP* (14.2) *for instances that satisfy the triangle inequality.*

Approximation Ratio	Integrality Ratio	Year	Reference	Chapter
$\frac{5}{3}$	$\frac{5}{3}$	2012	Sebő and Vygen [2014]	14.1
$\frac{13}{8}$	$\frac{13}{8}$	2012	Cheriyan, Friggstad, and Gao [2015]	–
$\frac{8}{5}$	$\frac{8}{5}$	2012	Sebő [2013]	15.2
$\frac{11}{7}$	$\frac{11}{7}$	2020	Traub [2020b]	15.5

Replacing these two edges by an edge $\{u, w\}$ cannot increase the cost due to the triangle inequality, and it does not change the parity of any vertex degree. Hence, we obtain an optimum T-tour with one edge less. This contradicts our assumption. □

In particular, if $|T| = 2$, there is an optimum T-tour that is a Hamiltonian path (whose set of endpoints is T). This motivates the name Path TSP for this problem.

14.2 LP Relaxations of Path TSP and the T-Tour Problem

For Path TSP, we will mostly work in the metric closure, so (V, E) is a complete graph and c satisfies the triangle inequality. The natural LP relaxation then is:

$$
\begin{array}{llll}
\min\ c(x) & & & \\
\text{subject to} & x(\delta(U)) \geq 2 & (\emptyset \neq U \subsetneq V,\ |U \cap \{s,t\}| \text{ even}) & \\
& x(\delta(U)) \geq 1 & (\emptyset \neq U \subsetneq V,\ |U \cap \{s,t\}| \text{ odd}) & \\
& x(\delta(v)) = 2 & (v \in V \setminus \{s,t\}) & \\
& x(\delta(v)) = 1 & (v \in \{s,t\}) & \\
& x_e \geq 0 & (e \in E). &
\end{array}
\tag{14.1}
$$

The integral feasible solutions to this LP are the incidence vectors of Hamiltonian s-t-paths. Similar to Theorem 2.31, dropping the degree constraints does not change the LP value (see Exercise 14.2). The resulting relaxed LP is also applicable to instances when c does not satisfy the triangle inequality and is particularly useful when studying GRAPH PATH TSP (see Section 14.4).

Given Proposition 14.9, it is not surprising that feasible solutions to this LP are convex combinations of incidence vectors of spanning trees:

Proposition 14.10 (An, Kleinberg, and Shmoys [2015]). *Let x be a feasible solution to* (14.1). *Then x is a convex combination of incidence vectors of edge sets of spanning trees.*

Proof. For any $\emptyset \neq U \subseteq V$, we have

$$\begin{aligned} x(E[U]) &= \tfrac{1}{2}\left(\sum_{u\in U} x(\delta(u)) - x(\delta(U))\right) \\ &= |U| - \tfrac{1}{2}(|U \cap \{s,t\}| + x(\delta(U))) \\ &\leq |U| - 1, \end{aligned}$$

with equality for $U = V$, so x is in the spanning tree polytope (Theorem 2.16). □

We now consider a natural LP relaxation for the T-TOUR PROBLEM. If $G = (V, E)$ is a complete graph with $n = |V|$ vertices, c satisfies the triangle inequality, and $\emptyset \neq T \subseteq V$ with $|T|$ even, the following formulation is natural in view of Proposition 14.9:

$$\begin{array}{llll} \min & c(x) & & \\ \text{subject to} & x(\delta(U)) \geq 2 & (\emptyset \neq U \subsetneq V,\ |U \cap T| \text{ even}) & \\ & x(\delta(\mathcal{W})) \geq |\mathcal{W}|-1 & (\mathcal{W} \text{ partition of } V) & \qquad (14.2) \\ & x(E) = n - 1 & & \\ & x_e \geq 0 & (e \in E). & \end{array}$$

Here $\delta(\mathcal{W})$ again denotes the set of edges with endpoints in different classes of the partition $\mathcal{W}$. We first show that this LP is indeed a relaxation of the T-TOUR PROBLEM:

Proposition 14.11. *The integral feasible solutions of the T-Tour LP* (14.2) *are the incidence vectors of the T-tours F for which (V, F) is a tree.*

Proof. Let x be an integral feasible solution to (14.2). By Corollary 2.15, x is the incidence vector of (the edge set of) a spanning tree (V, F). By Proposition 1.27, this tree contains a T-join J. We show $J = F$. Let $e \in F$, and consider the corresponding fundamental cut of (V, F) – that is, the cut $\delta(U)$

with $\delta(U) \cap F = \{e\}$. We have $x(\delta(U)) = 1$, and hence $|U \cap T|$ is odd. Since every T-join must intersect every T-cut, we have $e \in J$. We conclude that F is a T-join and hence a T-tour.

Conversely, let x be the incidence vector of a T-tour F such that (V, F) is a tree. Then its fundamental cuts are T-cuts by Lemma 2.20, which implies $x(\delta(U)) \geq 2$ for $\emptyset \neq U \subsetneq V$ with $|U \cap T|$ even. The partition constraints $x(\delta(\mathcal{W})) \geq |\mathcal{W}| - 1$ are satisfied for every partition $\mathcal{W}$ of V because (V, F) is connected. □

By Propositions 14.9 and 14.11, the value of (14.2) is at most the minimum cost of a T-tour for every instance (G, c, T) where G is a complete graph and c satisfies the triangle inequality. Moreover, we note:

Proposition 14.12. *Any feasible solution to* (14.2) *is in the spanning tree polytope of G.*

Proof. This follows from Corollary 2.15 because connectors with $n - 1$ edges are spanning trees. □

With Theorem 2.16, this also shows that in (14.2), the partition constraints $x(\delta(\mathcal{W})) \geq |\mathcal{W}| - 1$ for all partitions $\mathcal{W}$ of V could be replaced by $x(E[U]) \leq |U| - 1$ for all $\emptyset \neq U \subsetneq V$ without changing the set of feasible solutions.

Proposition 14.13. *If $T = \{s, t\}$, then the sets of feasible solutions to* (14.1) *and* (14.2) *coincide.*

Proof. If x is a feasible solution to (14.1), then x is a convex combination of spanning trees by Proposition 14.10, which implies $x(E) = n-1$ and $x(\delta(\mathcal{W})) \geq |\mathcal{W}| - 1$ for every partition $\mathcal{W}$ of V. Conversely, if x is a feasible solution to (14.2), then x is a convex combination of spanning trees by Proposition 14.12, implying $x(\delta(U)) \geq 1$ for every $\emptyset \neq U \subsetneq V$. The degree constraints are implied by $x(\delta(v)) \geq 1$ for $v \in \{s, t\}$ and $x(\delta(v)) \geq 2$ for $v \in V \setminus \{s, t\}$ and $x(E) = n - 1$. □

The statement of Proposition 14.13 also holds for the two LPs that arise when we drop the degree constraints in (14.1) and the constraint $x(E) = n - 1$ in (14.2) (see Exercise 14.4).

Dropping the constraint $x(E) = n - 1$ makes the LP (14.2) applicable to instances when c does not satisfy the triangle inequality, in particular to the GRAPH T-TOUR PROBLEM (and also to $T = \emptyset$).

An, Kleinberg, and Shmoys [2015] (for $|T| = 2$) and Cheriyan, Friggstad, and Gao [2015] (for general T) observed the following strengthening of Theorem 14.7:

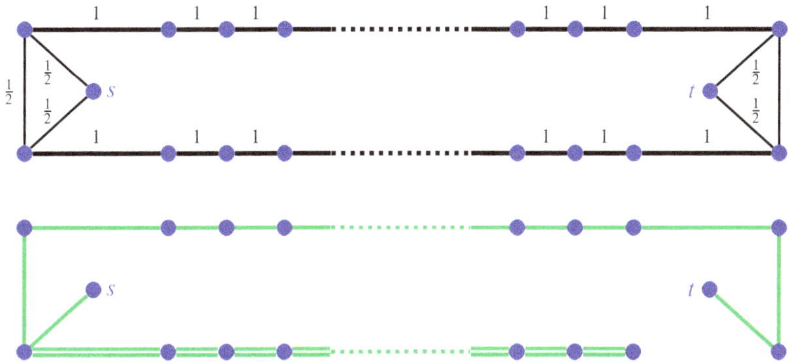

Figure 14.2 Example showing a lower bound of $\frac{3}{2}$ on the integrality ratio of the Path TSP LP (14.1) (and hence the T-Tour LP (14.2)). The top figure shows a graph $G = (V, E)$ with two vertices s and t. Together with $c(e) = 1$ for all $e \in E$, this defines an instance of PATH TSP. G is also the support graph of an optimum solution to the LP (14.1) in the metric closure, shown by the numbers x_e $(e \in E)$ in the top figure. The bottom figure shows an optimum $\{s, t\}$-tour.

Theorem 14.14. *Let (G, c, T) be an instance of the T-TOUR PROBLEM where G is a complete graph, c satisfies the triangle inequality, and $T \neq \emptyset$. Then Christofides' algorithm for T-tours (Algorithm 14.6) computes a solution of cost at most $\frac{5}{3}$ times the value of (14.2). In particular, the integrality ratio of (14.2) is at most $\frac{5}{3}$.*

Proof. Let x^* be an optimum solution to (14.2). By Proposition 14.12, x^* is contained in the spanning tree polytope (2.8). By Theorem 2.16, this implies that a cheapest spanning tree S has cost at most $c(x^*)$.

To bound the cost of the $(T \triangle \text{odd}(S))$-join J, we show that the vector $y := \frac{1}{3} \cdot x^* + \frac{1}{3} \cdot \chi^S$ is contained in the $(T \triangle \text{odd}(S))$-join polyhedron (see (2.9)). Then, by Theorem 2.19, the cost of the T-tour computed by Christofides' algorithm for T-tours is $c(S)+c(J) \leq c(S)+c(y) = \frac{4}{3}c(S)+\frac{1}{3}\cdot c(x^*) \leq \frac{5}{3}c(x^*)$.

It remains to show that the vector y is contained in the $(T \triangle \text{odd}(S))$-join polyhedron. To prove this, let $\emptyset \neq U \subsetneq V$ with $y(\delta(U)) < 1$. We show that $|(T \triangle \text{odd}(S)) \cap U|$ is even. Since $|S \cap \delta(U)| \geq 1$ and $x^*(\delta(U)) \geq 1$, we have $|S \cap \delta(U)| = 1$ and $x^*(\delta(U)) < 2$. Therefore, by Lemma 2.20, $|U \cap \text{odd}(S)|$ is odd. Moreover, $|U \cap T|$ is odd because $x^*(\delta(U)) < 2$. Hence, $|(T \triangle \text{odd}(S)) \cap U|$ is even. □

We will show better upper bounds on the integrality ratios of (14.1) and (14.2) in Chapter 15 (see Tables 14.2 and 14.3). The best-known lower bound is $\frac{3}{2}$, obtained for $|T| = 2$ and again for an unweighted graph instance:

Table 14.3 *Upper bounds on the integrality ratio for* Path TSP *in the order of their discovery. The integrality ratio refers to the LP* (14.1) *for instances that satisfy the triangle inequality.*

Integrality Ratio	Year	Reference	Chapter
$\frac{5}{3}$	2011	An, Kleinberg, and Shmoys [2015]	14.2
$\frac{1+\sqrt{5}}{2}$	2011	An, Kleinberg, and Shmoys [2015]	15.2
$\frac{8}{5}$	2012	Sebő [2013]	15.2
1.599	2015	Vygen [2016]	–
1.566	2015	Gottschalk and Vygen [2018]	15.3
$\frac{26}{17}$	2016	Sebő and van Zuylen [2019]	15.4
1.528	2018	Traub and Vygen [2019b], Zhong [2020]	–

Proposition 14.15. *The integrality ratio of* (14.1)*, and hence also the integrality ratio of* (14.2)*, is at least* $\frac{3}{2}$.

Proof. Figure 14.2 shows n-city instances for every even integer $n \geq 6$ in which the LP value is $n-1$, while an optimum $\{s,t\}$-tour has $\frac{3n}{2}-4$ edges. □

The LPs (14.1) and (14.2) can be solved in polynomial time using the equivalence of optimization and separation (Theorem 2.10): Separating over the partition constraints in (14.2) is possible since they define the connector polyhedron, over which we can optimize in polynomial time (Theorem 1.17). For the cut constraints, we could use Theorem 4.28 or the following:

Theorem 14.16 (Barahona and Conforti [1987]). *Given finite set* V *and* $T \subseteq V$ *with* $|T|$ *even and* $x : \binom{V}{2} \to \mathbb{R}_{\geq 0}$*, we can find a set attaining*

$$\min\left\{x(\delta(U)) : \emptyset \neq U \subsetneq V,\ |U \cap T| \text{ even}\right\} \tag{14.3}$$

in polynomial time.

Proof. Let U^* be an optimum solution.

We first find a set attaining $\min\{x(\delta(U)) : \emptyset \neq U \subsetneq V,\ |T \cap U| \in \{0,2\}\}$ by computing a minimum-weight cut (separating T' and $T \setminus T'$ if both sets are nonempty) for each $T' \subseteq T$ with $|T'| \in \{0,2\}$, using Corollary 2.9. If we do not have an optimal solution yet, U^* satisfies $|U^* \cap T| \geq 4$ and $|(V \setminus U^*) \cap T| \geq 4$. We assume this in the following.

Next, we find a set U attaining $\min\{x(\delta(U)) : |U \cap T| \geq 2,\ |(V \setminus U) \cap T| \geq 2\}$ by $|T|^4$ applications of a minimum s-t-cut algorithm (cf. Corollary 2.9). If $|U \cap T|$ is even, we are done. So far we have spent $O(n^8)$ time, where $n = |V|$. (This running time can be improved easily, but it is irrelevant for us.)

So assume that $|U \cap T|$ is odd. Then we claim:

$$\text{There is a set } W \text{ attaining (14.3) that does not cross } U. \tag{14.4}$$

Since $|U \cap T|$ is odd, we have

$$\begin{aligned} & |(U \cap U^*) \cap T| + |((V \setminus U) \cap (V \setminus U^*)) \cap T| \\ = \ & |(U \cap U^*) \cap T| + |T| - |(U \cup U^*) \cap T| \\ = \ & |U \cap T| + |U^* \cap T| + |T| - 2|(U \cup U^*) \cap T|, \end{aligned}$$

and this number is odd. Thus, $|(U \cap U^*) \cap T|$ is odd or $|((V \setminus U) \cap (V \setminus U^*)) \cap T|$ is odd.

Suppose U^* and U cross. We may assume that $|(U \cap U^*) \cap T|$ is odd; otherwise, we replace U by $V \setminus U$ and U^* by $V \setminus U^*$. Then $|(U \cup U^*) \cap T|$ and $|(U \setminus U^*) \cap T|$ are even. We show that one of them attains (14.3).

Case 1: $|(U \cap U^*) \cap T| \geq 3$. Then

$$x(\delta(U \cap U^*)) + x(\delta(U \cup U^*)) \ \leq \ x(\delta(U)) + x(\delta(U^*)).$$

Moreover, $x(\delta(U \cap U^*)) \geq x(\delta(U))$ by the choice of U, and hence $x(\delta(U \cup U^*)) \leq x(\delta(U^*))$.

Case 2: $|(U \cap U^*) \cap T| = 1$. Then

$$x(\delta(U^* \setminus U)) + x(\delta(U \setminus U^*)) \ \leq \ x(\delta(U)) + x(\delta(U^*)).$$

Moreover, $|(U^* \setminus U) \cap T| \geq 3$, implying $x(\delta(U^* \setminus U)) \geq x(\delta(U))$ by the choice of U, and hence $x(\delta(U \setminus U^*)) \leq x(\delta(U^*))$.

Claim (14.4) is proved. So we branch, once we contract U and once we contract $V \setminus U$, and apply the algorithm recursively. In the end, we output the best of all sets U that the algorithm computes and for which $|U \cap T|$ is even.

The subsets of V corresponding to sets considered as U in the course of the algorithm form a cross-free family. Hence, less than $4n$ applications of the above $O(n^8)$-time algorithm are needed (cf. Propositions 4.7 and 4.8). □

A variation of this proof (swapping even and odd, and adapting the initial step) also works for minimum-weight T-cuts (Exercise 14.6), but for this problem, Padberg and Rao [1982] found a more efficient algorithm (and here we do not need it anyway).

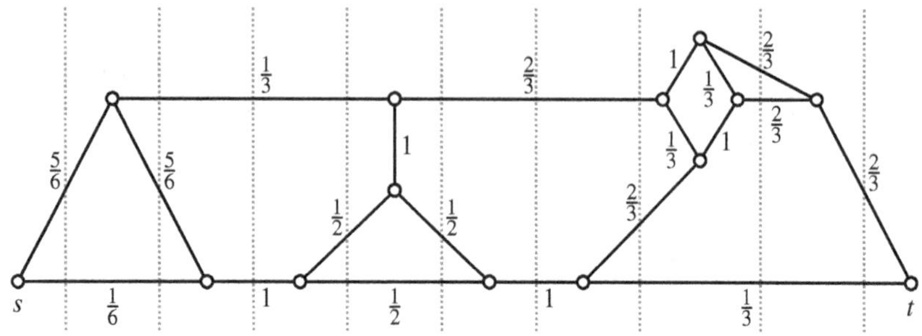

Figure 14.3 An LP solution (numbers in black) to a Path TSP instance and the narrow cuts (gray dotted lines), which form a chain.

14.3 Narrow Cuts

The following definition has played a key role in the development of approximation algorithms with a better guarantee than Christofides' algorithm for T-tours.

Definition 14.17 (narrow cut)**.** Let x be a feasible solution to the LP (14.2). We call a cut $\delta(U)$ *narrow* if $\emptyset \neq U \subsetneq V$ and $x(\delta(U)) < 2$.

See Figure 14.3 for an example. This notion is motivated by the fact that narrow cuts are the only reason that Wolsey's argument (Theorem 2.29) fails for T-tours. They therefore need special attention.

The following property, observed by An, Kleinberg, and Shmoys [2015] for $|T| = 2$ and by Cheriyan, Friggstad, and Gao [2015] for general T, is exploited by most algorithms that have been designed for $\{s,t\}$-tours or T-tours. If $V_1 \subsetneq \cdots \subsetneq V_k$, we say that $\{V_1, \ldots, V_k\}$ is a *chain*. If $\mathcal{U}$ is a chain, we say that the set of cuts $\{\delta(U) : U \in \mathcal{U}\}$ *forms a chain.*

Lemma 14.18. *Let x be a feasible solution to the LP* (14.2) *or* (14.1)*. Let $\mathcal{U} = \{U : \emptyset \neq U \subsetneq V,\ x(\delta(U)) < 2\}$. Then $\mathcal{U}$ is cross-free. If $|T| = 2$, then the set of narrow cuts forms a chain.*

Proof. Suppose there are sets $A, B \subsetneq V$ with $x(\delta(A)) < 2$, $x(\delta(B)) < 2$, and $A \cap B \neq \emptyset$ and $A \cup B \neq V$. Then

$$x(\delta(A \cap B)) + x(\delta(A \cup B)) \ \leq\ x(\delta(A)) + x(\delta(B)) \ <\ 2 + 2 \ =\ 4,$$

so at least one of $|T \cap (A \cap B)|$ and $|T \cap (A \cup B)|$ must be odd. Hence, both must be odd (as $|T \cap A|$ and $|T \cap B|$ are odd). But then $|T \cap (A \setminus B)|$ and $|T \cap (B \setminus A)|$

must be even. However,

$$x(\delta(A \setminus B)) + x(\delta(B \setminus A)) \le x(\delta(A)) + x(\delta(B)) < 2+2 = 4,$$

so $A \setminus B$ or $B \setminus A$ must be empty.

This shows the first statement. For the second statement, let $T = \{s,t\}$, and assume that $A, B \in \mathcal{U}$ and that A and B both contain s but not t (otherwise, take the complement). Then $A \cap B \neq \emptyset$, $A \cup B \neq V$, and the above yields $A \subsetneq B$ or $B \subseteq A$. □

With Theorem 4.28, one can enumerate all narrow cuts in polynomial time. For PATH TSP, there is a simpler way (see Exercise 14.7).

We have noted earlier that the narrow cuts are the reason why Wolsey's analysis fails for T-tours. More precisely, Wolsey's analysis fails only because of those narrow cuts that are $(\mathrm{odd}(S) \triangle T)$-cuts for the spanning tree (V, S). The following lemma implies that these are precisely the narrow cuts that contain an even number of edges of the spanning tree (V, S).

Lemma 14.19. *Let x be a feasible solution to LP* (14.2) *or* (14.1) *(and in this case, $T = \{s,t\}$). Let H be a multi-subset of E. Then a narrow cut C is an $(\mathrm{odd}(H) \triangle T)$-cut if and only if $|H \cap C|$ is even.*

Proof. The LP constraints imply that every narrow cut C is a T-cut. Hence, C is an $(\mathrm{odd}(H) \triangle T)$-cut if and only if C is not an $\mathrm{odd}(H)$-cut. Let U be a vertex set with $C = \delta(U)$. By Lemma 2.20, $|U \cap \mathrm{odd}(H)|$ is even if and only if $|\delta_H(U)|$ is even. Thus, C is not an $\mathrm{odd}(H)$-cut if and only if $|H \cap C|$ is even. □

14.4 *T*-Tours and Path TSP in Graphs

Within less than a year, Mömke and Svensson [2016], Mucha [2014], An, Kleinberg, and Shmoys [2015], and Sebő and Vygen [2014] proposed better and better approximation algorithms for GRAPH PATH TSP (see Table 14.4).

For GRAPH PATH TSP, it is natural to consider the following LP relaxation:

$$\begin{aligned} \min\ & x(E) \\ \text{subject to}\quad & x(\delta(U)) \ \ge\ 2 \qquad (\emptyset \neq U \subsetneq V,\ |U \cap \{s,t\}| \text{ even}) \\ & x(\delta(U)) \ \ge\ 1 \qquad (\emptyset \neq U \subsetneq V,\ |U \cap \{s,t\}| \text{ odd}) \\ & x_e \ \ge\ 0 \qquad (e \in E). \end{aligned} \tag{14.5}$$

The value of this LP equals the value of (14.1) in the metric closure of G (this follows from Exercise 14.2).

Table 14.4 *Approximation ratios and upper bounds on the integrality ratio for* Graph Path TSP *in the order of their discovery. The integrality ratio refers to the LP* (14.1) *of the metric closure of* Graph Path TSP *instances.*

Approximation Ratio	Integrality Ratio	Year	Reference	Chapter
1.586	–	2011	Mömke and Svensson [2016]	14.4
1.584	–	2011	Mucha [2014]	14.4
1.578	1.614	2011	An, Kleinberg, and Shmoys [2015]	–
1.5	1.5	2012	Sebő and Vygen [2014]	14.4
1.497	–	2018	Traub and Vygen [2023]	–
$1.4 + \varepsilon$	–	2019	Traub, Vygen, and Zenklusen [2022]	13, 16

First, we see that the removable pairing technique can be applied quite directly (as suggested by Mömke and Svensson [2016]). The following result is essentially due to Mucha [2014] (but he proved the approximation ratio $\frac{19}{12} + \varepsilon$ for any $\varepsilon > 0$ and did not obtain an upper bound on the integrality ratio).

Theorem 14.20. *There is a* $\frac{19}{12}$*-approximation algorithm for* Graph Path TSP*. For any given instance, it computes a solution with at most* $\frac{19}{12}$ LP *edges, where* LP *denotes the value of* (14.5).

Proof. Let (G, s, t) be a Graph Path TSP instance and x an optimum solution to the LP (14.5). Like in Proposition 12.1 we may assume that G is 2-connected. Adding an edge $\bar{e}$ from s to t and setting $x_{\bar{e}} = 1$ yields a solution to the Graph TSP LP (12.1) for the extended graph $G + \bar{e}$.

Applying Lemma 12.13 to $G+\bar{e}$ and removing $\bar{e}$ from the result yields an $\{s, t\}$-tour in G with at most $\frac{13}{9}(x(E \,\dot\cup\, \{\bar{e}\})-1)+\frac{1}{3}\text{dist}_G(s, t) = \frac{13}{9}x(E)+\frac{1}{3}\text{dist}_G(s, t)$ edges. This does the job if $\text{dist}_G(s, t) \le \frac{5}{12}x(E)$.

If $\text{dist}_G(s, t) > \frac{5}{12}x(E)$, we instead run the double tree algorithm (Proposition 14.5). This $\{s, t\}$-tour has at most $2(n - 1) - \text{dist}_G(s, t)$ edges. Using $n - 1 \le x(E)$ and $\text{dist}_G(s, t) > \frac{5}{12}x(E)$, this is at most $\frac{19}{12}x(E)$. □

Sebő and Vygen [2014] showed that their techniques (described in Chapter 13) also lead to a $\frac{3}{2}$-approximation algorithm for the Graph T-Tour Problem (the special case of the T-Tour Problem with $c(e) = 1$ for all $e \in E$). More precisely, they showed the following:

Theorem 14.21 (Sebő and Vygen [2014]). *For every connected graph $G = (V, E)$ and every $T \subseteq V$ with $|T|$ even, there is a T-tour with at most $\frac{3}{2}\,\mathrm{LP}(G,T)$ edges, where*

$$\begin{aligned}\mathrm{LP}(G,T) \;:=\; \min\bigl\{x(E) : \;& x(\delta(U)) \geq 2 \ (\emptyset \neq U \subsetneq V,\ |U \cap T| \text{ even}),\\ & x(\delta(\mathcal{W})) \geq |\mathcal{W}| - 1 \ (\mathcal{W} \text{ partition of } V),\\ & x_e \geq 0 \ (e \in E)\bigr\}\end{aligned}$$

denotes the value of the Graph T-Tour LP. This bound is tight. Such a T-tour can be computed in $O(n^3)$ time.

The algorithm is sketched in Exercises 13.6–13.8. See also Exercise 14.9. Proposition 14.15 shows that no better factor than $\frac{3}{2}$ is possible, even for $|T| = 2$. For $|T| = 2$, the linear program in Theorem 14.21 is equivalent to (14.5) (Exercise 14.4), and we obtain:

Corollary 14.22 (Sebő and Vygen [2014]). *For every connected graph G and every two vertices $s, t \in V$, there is an $\{s,t\}$-tour with at most $\frac{3}{2}\,\mathrm{LP}(G, \{s,t\})$ edges, where* $\mathrm{LP}(G, \{s,t\})$ *denotes the value of the Graph Path TSP LP* (14.5). *This bound is tight. Such an $\{s,t\}$-tour can be computed in polynomial time.*

We did not state the $O(n^3)$ running time that we immediately get from Theorem 14.21 because we now give a simpler proof of Corollary 14.22, due to Gao [2013], which is however based on a slower algorithm because it first solves the LP (14.5). The notion of narrow cuts extends to solutions to this LP without any change. Following Gottschalk and Vygen [2018], let us also introduce the following notion:

Definition 14.23 (Gao tree). Let x be a feasible solution to (14.5) and (V, S) a spanning tree. Then (V, S) is called a *Gao tree* (for x) if $x_e > 0$ for all $e \in S$ and $|S \cap N| = 1$ for every narrow cut N.

Figure 14.4 shows an example. Here is Gao's [2013] main lemma:

Lemma 14.24 (Gao [2013]). *For every feasible solution x of* (14.5), *there exists a Gao tree.*

Proof. Let $\{s\} \subseteq U_1 \subsetneq U_2 \subsetneq \cdots \subsetneq U_k \subseteq V \setminus \{t\}$ such that $\{\delta(U_i) : i = 1, \ldots, k\}$ is the set of narrow cuts (cf. Lemma 14.18). Let $U_0 := \emptyset$ and $U_{k+1} := V$. Let $G_x := (V, \{e \in E : x_e > 0\})$ denote the support graph of x. We claim:

$$\text{For all } 0 \leq i < j \leq k+1, \text{ the subgraph of } G_x \text{ induced by } U_j \setminus U_i \text{ is connected.} \tag{14.6}$$

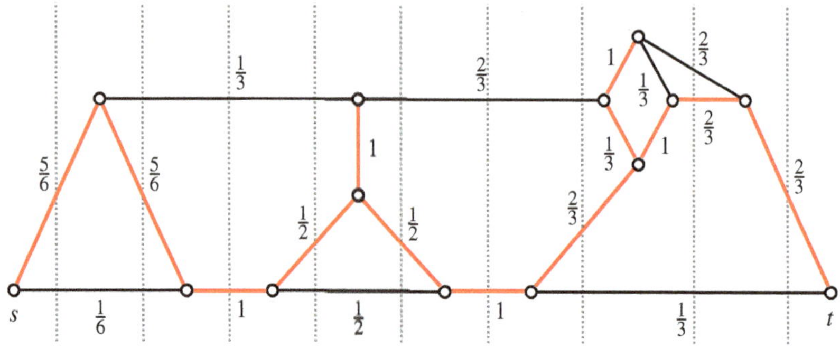

Figure 14.4 A Gao tree (red, thick) for the LP solution (numbers in black) of Figure 14.3; again, the narrow cuts are shown by gray dotted lines. The Gao tree contains exactly one edge in each narrow cut.

Suppose that (14.6) does not hold – that is, for some $0 \le i < j \le k+1$, there is a set $\emptyset \neq A \subsetneq U_j \setminus U_i$ such that $x(\delta(A) \cap \delta(B)) = 0$, where $B := (U_j \setminus U_i) \setminus A$. Then

$$x(\delta(A)) + x(\delta(B)) \;=\; x(\delta(U_j \setminus U_i)) \;\le\; x(\delta(U_j)) + x(\delta(U_i)).$$

Now there are two cases. If $i = 0$ or $j = k+1$, the right-hand side is less than 2, but the left-hand side is at least 2. If $i > 0$ and $j < k+1$, the right-hand side is less than 4, but the left-hand side is at least 4. This contradiction proves (14.6).

So we can design a Gao tree by taking a spanning tree of $G_x[U_{i+1} \setminus U_i]$ for $i = 0, \dots, k$ and then, for each $i = 1, \dots, k$, adding an edge of G_x that connects $U_i \setminus U_{i-1}$ and $U_{i+1} \setminus U_i$ (such an edge must exist because $G_x[U_{i+1} \setminus U_{i-1}]$ is connected). □

This immediately implies Corollary 14.22: Compute an optimum solution x to the LP (14.5), construct a Gao tree (V, S) for x, and do parity correction. The Gao tree has $n-1$ edges, and parity correction costs at most $c(\frac{x}{2}) = \frac{1}{2}\mathrm{LP}(G, \{s,t\})$ because $\frac{x}{2}$ is in the $(\mathrm{odd}(S) \triangle \{s,t\})$-join polyhedron: For all narrow cuts $N = \delta(U)$, we have $|N \cap S| = 1$, so $|U \cap (\mathrm{odd}(S) \triangle \{s,t\})|$ is even due to Lemma 14.19.

Gao [2015] showed that this analysis does not work for general weights: In the example of Figure 14.5, the only Gao tree costs more than the LP value (however, in this example, parity correction is very cheap).

Lemma 14.24 cannot be extended to T-tours: Let $G = (V, E)$ be the complete graph on four vertices, $T = V$, and $x_e = \frac{1}{2}$ for all $e \in E$; this is a feasible solution to (14.2) and all singletons induce narrow cuts, but of course there

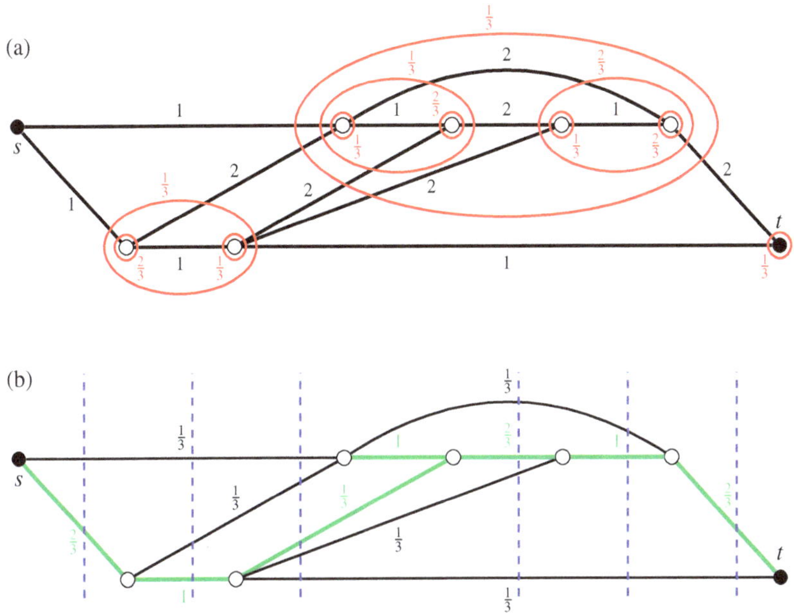

Figure 14.5 A Path TSP instance (from Gao [2015]) with edge costs and an optimum dual solution (red ellipses) shown in (a), and with an optimum primal solution and the narrow cuts (blue, dashed) shown in (b). The green tree in (b) is the only Gao tree, and its cost is 10, more than the LP value $\frac{29}{3}$.

is no spanning tree in which all four vertices have degree 1. The example of Gottschalk and Vygen [2018] in Figure 14.6 shows that even if x is an extreme point and we allow an odd number of edges instead of exactly one in every narrow cut, the existence of (the equivalent of) a Gao tree cannot be guaranteed for T-tours.

Corollary 14.22 implies that the Path TSP LP (14.1) restricted to metric closures of unweighted graphs has integrality ratio exactly $\frac{3}{2}$; this is the only case in this book where the LP is not integral and we know the exact integrality ratio. Although the approximation ratio matches this integrality ratio, Traub and Vygen [2023] improved on it by devising a 1.497-approximation algorithm. This proof also used ear-decompositions, refining the original Sebő–Vygen proof of Corollary 14.22. Finally, Traub, Vygen, and Zenklusen [2022] showed that Graph Path TSP is not much harder to approximate than Graph TSP (see Chapter 16).

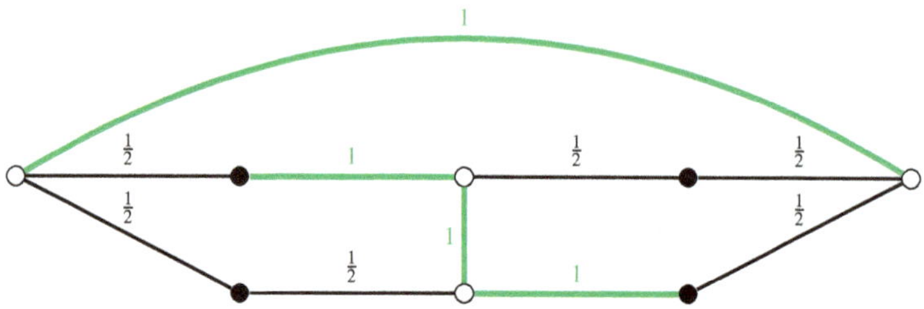

Figure 14.6 An instance of the T-Tour Problem (edge costs are not shown); T is the set of filled vertices. The numbers next to the edges show a feasible solution x^* to the LP (14.2): Thick green edges e have $x^*_e = 1$, thin edges e have $x^*_e = \frac{1}{2}$, other edges e (not shown) have $x^*_e = 0$. In fact, this is an extreme point (see Exercise 14.10). If G denotes the support graph of x^*, we see that $G - T$ is disconnected, so G contains no spanning tree in which all elements of T are leaves (and hence G contains no spanning tree in which all elements of T have odd degree). However, $\delta(t)$ is narrow for all $t \in T$, so G contains no "Gao tree." This picture is taken from Gottschalk and Vygen [2018] (with permission from Springer Nature).

14.5 A General Reduction

Let us conclude this chapter with a general reduction from Path TSP to Symmetric TSP.

Theorem 14.25. *Let $\varepsilon > 0$ and $\alpha, \rho \geq 1$ be fixed constants. If there is an α-approximation algorithm for Symmetric TSP, then there is a $(3 - \frac{2}{\alpha} + \varepsilon)$-approximation algorithm for Path TSP. If the integrality ratio of (2.2) is ρ, then the integrality ratio of (14.1) is at most $3 - \frac{2}{\rho}$.*

Proof. Let (G, c, s, t) be an instance of Path TSP. Let $K := \lceil \frac{2}{\varepsilon} \rceil$. Let P be the edge set of a shortest s-t-path in (G, c). We add to our graph a new s-t-path with $K - 1$ new vertices and K new edges $\bar{e}$, each with $c(\bar{e}) = \frac{1}{K} c(P)$. Let $\bar{E}$ be the set of these new edges, and $\bar{G}$ the resulting graph. Note that $c(\bar{E}) = c(P)$.

Let $\mathrm{OPT}(G, c, s, t)$ denote the minimum cost of an $\{s, t\}$-tour in (G, c), and let $\mathrm{OPT}(\bar{G}, c)$ be the minimum cost of a tour in $(\bar{G}, c)$. Since adding $\bar{E}$ to an $\{s, t\}$-tour in G yields a tour in $\bar{G}$, we have $\mathrm{OPT}(\bar{G}, c) \leq \mathrm{OPT}(G, c, s, t) + c(\bar{E}) = \mathrm{OPT}(G, c, s, t) + c(P)$.

We apply an α-approximation algorithm for Symmetric TSP to $(\bar{G}, c)$. Let the resulting tour be $\bar{F}$. We have

$$c(\bar{F}) \leq \alpha \cdot \mathrm{OPT}(\bar{G}, c) \leq \alpha \left(\mathrm{OPT}(G, c, s, t) + c(P)\right).$$

Because in $(V, \bar{F})$ every vertex has even degree, $\bar{F}$ must contain either each edge of $\bar{E}$ an odd number of times – then we simply remove all of them – or each edge of $\bar{E}$ an even number of times (and at least two for all but one edge of $\bar{E}$ due to connectivity) – then we remove all of them and add one copy of P. In both cases, we change the parity of the degree only at s and t and decrease the cost by at least $(1 - \frac{2}{K})c(P)$. Moreover, we do not destroy connectivity: In the "even case," this follows from the addition of P, and in the "odd case," this follows from the fact that in $(V, \bar{F})$ every cut contains an even number of edges.

We obtain an $\{s,t\}$-tour F in G with

$$\begin{aligned} c(F) &\leq c(\bar{F}) - \left(1 - \tfrac{2}{K}\right) c(P) \\ &\leq \alpha \, \mathrm{OPT}(G,c,s,t) + \left(\alpha - 1 + \tfrac{2}{K}\right) c(P) \\ &\leq \alpha \, \mathrm{OPT}(G,c,s,t) + (\alpha - 1 + \varepsilon) \; c(P) \\ &\leq (\alpha + \varepsilon) \, \mathrm{OPT}(G,c,s,t) + (\alpha - 1) \; c(P). \end{aligned}$$

If $c(P) \leq (\frac{2}{\alpha} - 1)\mathrm{OPT}(G,c,s,t)$, then this is at most

$$\left(\alpha + \varepsilon + (\alpha - 1)(\tfrac{2}{\alpha} - 1)\right) \mathrm{OPT}(G,c,s,t) \;=\; \left(3 - \tfrac{2}{\alpha} + \varepsilon\right) \mathrm{OPT}(G,c,s,t).$$

If $c(P) > (\frac{2}{\alpha} - 1)\,\mathrm{OPT}(G,c,s,t)$, the double tree algorithm (Proposition 14.5) yields a better result, whose cost is at most

$$2\,\mathrm{OPT}(G,c,s,t) - c(P) \;<\; \left(3 - \tfrac{2}{\alpha}\right) \mathrm{OPT}(G,c,s,t).$$

The proof of the second statement is almost the same. Let $\mathrm{LP}(G,c,s,t)$ denote the value of the LP (14.1). If $c(P) > (\frac{2}{\rho} - 1)\mathrm{LP}(G,c,s,t)$, the double tree algorithm yields an $\{s,t\}$-tour of cost at most $2\,\mathrm{LP}(G,c,s,t) - c(P) < (3 - \frac{2}{\rho})\mathrm{LP}(G,c,s,t)$ because the minimum cost of a spanning tree is at most $\mathrm{LP}(G,c,s,t)$ by Proposition 14.10.

If $c(P) \leq (\frac{2}{\rho} - 1)\mathrm{LP}(G,c,s,t)$, we construct $\bar{G}$ as above. From a solution x to the LP (14.1) for (G,c,s,t), we obtain a solution $\bar{x}$ to the LP (2.2) for $(\bar{G},c)$ by setting $\bar{x}_{\bar{e}} := 1$ for all $\bar{e} \in \bar{E}$ and $\bar{x}_e := x_e$ for all $e \in E$. Then $c(\bar{x}) = c(x) + c(\bar{E}) = c(x) + c(P)$. If the integrality ratio of (2.2) is ρ, then we get as above an $\{s,t\}$-tour F with

$$\begin{aligned} c(F) &\leq \rho \, \mathrm{LP}(G,c,s,t) + (\rho - 1 + \varepsilon)\, c(P) \\ &\leq (\rho + \varepsilon)\, \mathrm{LP}(G,c,s,t) + (\rho - 1)\, c(P) \\ &\leq \left(3 - \tfrac{2}{\rho} + \varepsilon\right) \mathrm{LP}(G,c,s,t), \end{aligned}$$

where we used Theorem 2.19 (applied to $T = \{s,t\}$) in the second inequality. Since we can choose $\varepsilon > 0$ arbitrarily small, this yields the claimed result. □

Corollary 14.26. *If the integrality ratio of* (2.2) *is* $\frac{4}{3}$, *then the integrality ratio of* (14.1) *is* $\frac{3}{2}$. □

At the moment, the reduction given by Theorem 14.25 does not really help. We will see a better general reduction in Chapter 16; however, this does not apply to the integrality ratio.

Exercises

14.1 Show that the algorithm in the proof of Theorem 14.1 can be implemented to run in $O(n^3)$ time (using Theorem 1.29).

14.2 Let (G, c, s, t) be a PATH TSP instance in which G is a complete graph and c satisfies the triangle inequality. Prove:

(a) The degree constraints $x(\delta(v)) = 2$ ($v \in V \setminus \{s, t\}$) and $x(\delta(v)) = 1$ ($v \in \{s, t\}$) can be dropped without changing the value of the Path TSP LP (14.1).

(b) The LP (14.1) has the same value as (9.1) on the digraph that arises from G by orienting every edge both ways (with the same cost).

14.3 Show that there always exists an optimum solution to (14.1) with less than $2n$ edges in the support.

14.4 Let (G, c, T) be an instance of the T-TOUR PROBLEM with $T = \{s, t\}$. Consider the LP that arises when we drop the degree constraints in (14.1) and the LP that arises when we drop the constraint $x(E) = n - 1$ in (14.2). Show that the sets of feasible solutions to these LPs coincide.

14.5 Show that replacing the partition constraints ($x(\delta(\mathcal{W})) \geq |\mathcal{W}| - 1$ for all partitions $\mathcal{W}$ of V) in the T-Tour LP (14.2) by cut constraints $x(\delta(U)) \geq 1$ for all $\emptyset \neq U \subsetneq V$ can change the LP value.
Hint: Modify the example in Figure 14.6.

14.6 Prove the variant of Theorem 14.16 with $|U \cap T|$ odd instead of even, once by a variation of that proof and once using the equivalence of optimization and separation (Theorem 2.10).

14.7 Let x be a solution to the LP (14.1) for a PATH TSP instance. Show how to compute all narrow cuts by computing a v-w-cut $\delta(U)$ with minimum $x(\delta(U))$ for all $v, w \in V$. Do not use Theorem 4.28.

14.8 Show that Proposition 12.1 applies also to PATH TSP, to the T-TOUR PROBLEM, and their special cases in unweighted graphs (with the corresponding LP relaxations): We may assume that the input graph G is 2-connected.

14.9 Let $G = (V, E)$ be a graph with a nice ear-decomposition, let $T \subseteq V$ with $|T|$ even, and let $\mathcal{I}$ contain the set of internal vertices of each short ear, except those for which T contains at least one of its internal vertices. Define L_μ as in Theorem 13.18 but with this definition of $\mathcal{I}$. Show that then $L_\mu(G, \mathcal{I}) \leq \text{LP}(G, T)$, where $\text{LP}(G, T)$ denotes the value of the LP (14.2).

14.10 Show that the LP solution shown in Figure 14.6 is an extreme point of the polytope described by (14.2).

15

Best-of-Many Christofides and Variants

An, Kleinberg, and Shmoys [2015] were the first to beat Christofides' algorithm for PATH TSP. Their algorithm, which they called *Best-of-Many Christofides*, is very natural: Since an LP solution can be written as a convex combination of spanning trees (Proposition 14.10), we can do parity correction on each of these trees and output the best of the resulting tours. It turns out that this yields a better guarantee than the $\frac{5}{3}$ (cf. Theorem 14.7) that Christofides' algorithm yields.

In this chapter, we analyze this algorithm and study various follow-up works that have yielded better and better approximation ratios; some of them also apply to general T-tours.

15.1 Decomposing an LP Solution into Spanning Trees

An, Kleinberg, and Shmoys [2015] proposed the Best-of-Many Christofides algorithm for PATH TSP and showed that it has approximation ratio at most $\frac{1+\sqrt{5}}{2} \approx 1.619$ (the golden ratio). This was the first improvement of Christofides' algorithm that is not restricted to unweighted graphs, and it opened the door for many further improvements. The Best-of-Many Christofides algorithm was generalized to T-tours by Cheriyan, Friggstad, and Gao [2015]; they proved an approximation ratio of 1.625 for the T-TOUR PROBLEM. Then Sebő [2013] showed that the same algorithm actually has an approximation ratio $\frac{8}{5}$ for the T-TOUR PROBLEM. We will see this in Section 15.2. Algorithm 15.1 shows a formal description of the algorithm. Recall that requiring that G is complete and c satisfies the triangle inequality is no restriction due to Proposition 14.8.

For SYMMETRIC TSP and even for GRAPH TSP, Best-of-Many Christofides does not have a better approximation ratio than $\frac{3}{2}$: If $G = (V, E)$ is a complete graph with an even number n of vertices and $x^*_e = \frac{2}{n-1}$ for all $e \in E$ and $\mu(S) = \frac{1}{n}$ for every star S (in which all vertices but one have degree 1), we

Algorithm 15.1: Best-of-Many Christofides

Input: an instance (G, c, T) of the T-TOUR PROBLEM, where G is a complete graph and c satisfies the triangle inequality

Output: a T-tour

(1) Let x^* be an optimum solution to the LP (14.2) (or (2.2) if $T = \emptyset$).
(2) Let $\mathcal{S}$ denote the set of edge sets of spanning trees in G. Find $\mu(S) \geq 0$ for $S \in \mathcal{S}$ such that $\sum_{S \in \mathcal{S}} \mu(S) = 1$ and $x^* \geq \sum_{S \in \mathcal{S}} \mu(S)\chi^S$.
(3) For each $S \in \mathcal{S}$ with $\mu(S) > 0$, compute a minimum-cost $\mathrm{odd}(S) \triangle T$-join J in (G, c) and consider the T-tour $S \dot{\cup} J$.
(4) Output the best of these T-tours.

obtain a tour with $\frac{3n}{2} - 1$ edges. A more interesting example due to Schalekamp and van Zuylen (cf. Genova and Williamson [2017]) is shown in Figure 15.1. We note:

Proposition 15.2. *The approximation ratio of the Best-of-Many Christofides algorithm is at least $\frac{3}{2}$, and exactly $\frac{3}{2}$ for* SYMMETRIC TSP.

Proof. Since $\sum_{S \in \mathcal{S}} \mu(S)c(S) \leq c(x^*)$, the cheapest tree S with $\mu(S) > 0$ costs at most $c(x^*)$, and for SYMMETRIC TSP ($T = \emptyset$), there is an $\mathrm{odd}(S)$-join J with $c(J) \leq \frac{1}{2}c(x^*)$ as in Wolsey's analysis (Theorem 2.29).

In the example in Figure 15.1, the algorithm might decompose x^* into the two shown spanning trees and then end up with a tour of cost 6, which is $\frac{3}{2}$ times the optimum. □

However, for PATH TSP and in fact the T-TOUR PROBLEM, Best-of-Many Christofides is better than Christofides' algorithm.

Note that for the special case $|T| = 2$ (PATH TSP), the LP (14.2) is equivalent to (14.1) by Proposition 14.13, and this is what An, Kleinberg, and Shmoys [2015] considered.

In the rest of this section, we discuss how to implement Step (2) of Best-of-Many Christofides. By Proposition 14.12 (for $T \neq \emptyset$) and Corollary 2.17 (for $T = \emptyset$), the LP solution x^* (scaled down by the factor $\frac{n-1}{n}$ if $T = \emptyset$) is in the spanning tree polytope of G. By Carathéodory's theorem, there exists a probability distribution μ as in Step (2) with less than n^2 nonzero entries, and such a distribution can be computed in polynomial time (see Theorem 4.22). One can also exploit the structure of the spanning tree polytope (see e.g., Cunningham [1984] or Exercise 15.1). Also the proof of Theorem 15.15 (see Section 15.3) can be turned into an algorithm.

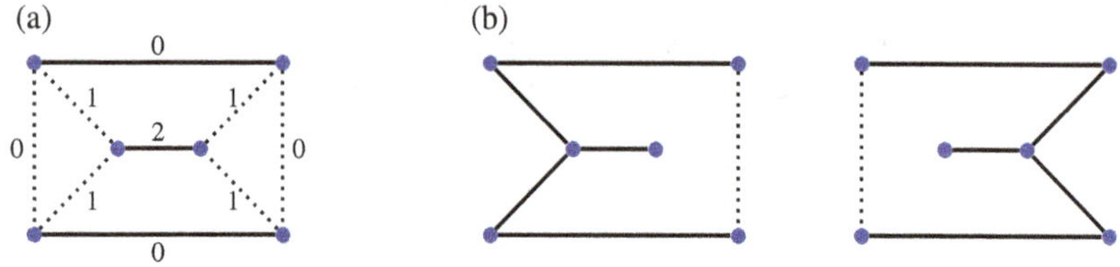

Figure 15.1 Example due to Schalekamp and van Zuylen, proving that for the Symmetric TSP, the Best-of-Many Christofides algorithm does not have a better approximation ratio than $\frac{3}{2}$. The numbers next to the edges in (a) denote their cost. In an optimum LP solution, the dotted edges in (a) have value $\frac{1}{2}$ and the solid edges have value 1. The LP value is 4. (b) shows a decomposition of the optimum LP solution into two spanning trees (each with weight $\frac{1}{2}$) plus the dotted edges. Both spanning trees cost 4, and for both of them, parity correction costs 2.

For Path TSP and Symmetric TSP, an elegant method was proposed by Genova and Williamson [2017]. We first describe it for a solution x to the subtour LP (2.2), which corresponds to the case $T = \emptyset$. The core of the proof is the following lemma:

Lemma 15.3. *Let $K \in \mathbb{N}$ and $G = (V, E)$ be an undirected $(2K)$-edge-connected multi-graph in which all vertices have degree $2K$. Then E can be partitioned into edge sets of K spanning trees of G plus K extra edges. If G contains at least K parallel edges between two vertices s and t, such K edges can be chosen as the extra edges.*

Proof. We proceed by induction on $|V|$. The assertion is trivial if G has only two vertices. Otherwise let z be an arbitrary vertex (but not s or t if we want to satisfy the extra requirement). Applying Theorem 2.30, we split off a pair of edges incident to z until z has degree zero. We obtain a multigraph $G' = (V \setminus \{z\}, E')$ in which all degrees are still $2K$ and in which still $|\delta_{G'}(U)| \geq 2K$ for all $\emptyset \neq U \subsetneq V \setminus \{z\}$. By the induction hypothesis, $E' = S_1 \,\dot{\cup}\, \cdots \,\dot{\cup}\, S_K \,\dot{\cup}\, \{e_1, \ldots, e_K\}$, where $(V \setminus \{z\}, S_i)$ is a tree for all $i = 1, \ldots, K$.

We now undo the splitting off operation. First, consider the extra edges $e_1, \ldots, e_K$. If an extra edge $e_i = \{x_i, y_i\}$ resulted from splitting off, we replace e_i by $\{x_i, z\}$ and put $\{z, y_i\}$ into a reserve list (which is initially empty).

If a tree S_i contains some edges $\{x_j, y_j\}$ $(j = 1, \ldots, l)$ that resulted from the splitting off operation, we first replace S_i by $S_i^1 := (S_i \setminus \{\{x_1, y_1\}\}) \cup \{\{x_1, z\}, \{z, y_1\}\}$ (now (V, S_i^1) is a tree). Then, for $j = 2, \ldots, l$, let without loss of generality x_j be closer to z than y_j in the current tree S_i^{j-1} (otherwise swap x_j

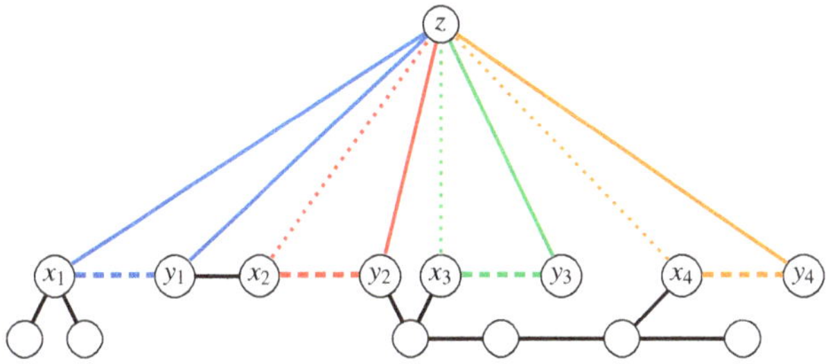

Figure 15.2 Undoing the splitting off operations at z in the proof of Lemma 15.3. Starting with the tree on the bottom (without z), we first replace the blue dashed edge $\{x_1, y_1\}$ by the two solid blue edges. Then, for $j = 2, 3, 4$, we replace $\{x_j, y_j\}$ by $\{z, y_j\}$ and put the dotted edge $\{x_j, z\}$ in the reserve list.

and y_j). Replace S_i^{j-1} by $S_i^j := (S_i^{j-1} \setminus \{\{x_j, y_j\}\}) \cup \{\{z, y_j\}\}$ and put $\{x_j, z\}$ into the reserve list. See Figure 15.2.

Now, for each tree S_i that contains no edge that resulted from splitting off, we take an arbitrary edge from the reserve list and add it to S_i. In the end, the reserve list is empty, and the induction step is complete. □

Theorem 15.4. *Let (V, c) be an instance of the* Symmetric TSP with Triangle Inequality, *and let x be a rational feasible solution to the subtour LP* (2.2). *Then we can compute in polynomial time an integer $K \in \mathbb{N}$, a list of spanning trees $(V, S_1), \ldots, (V, S_K)$ and numbers $\mu(S_1), \ldots, \mu(S_K) > 0$ such that $\sum_{i=1}^K \mu(S_i) = 1$ and $\sum_{i=1}^K \mu(S_i)\chi^{S_i} \le x$. If $x_e = 1$ for some $e = \{s, t\} \in \binom{V}{2}$, then the trees can be chosen so that $\sum_{i=1}^K \mu(S_i)\chi^{S_i} + \chi^e = x$.*

Proof. Let $K \in \mathbb{N}$ so that Kx_e is an integer for all e. Let G be the undirected multi-graph that contains Kx_e copies of every $e \in \binom{V}{2}$. Apply Lemma 15.3 to G, output the list of distinct trees, and set $\mu(S) = \frac{1}{K}|\{i \in \{1, \ldots, K\} : S = S_i\}|$ for each such tree (V, S).

For a polynomial-time algorithm, we proceed as in the proof of Lemma 15.3, but we do not store the edges of G explicitly but rather the number of parallel edges for every pair of vertices. Similarly, we store spanning trees with their multiplicities. Theorem 2.32 says that we can do all the splitting off steps at one vertex z in polynomial time. For every vertex pair $\{u, v\}$, we store the number of parallel edges with endpoints u and v that resulted from splitting off at z. When undoing the splitting off operation, we look at all spanning trees that contain an edge $\{u, v\}$ with their multiplicities. For at most one of these spanning trees, we create two copies, one for which we undo the splitting off operation for $\{u, v\}$ and one for which we do not change the edge $\{u, v\}$. Hence, each iteration increases the number of distinct spanning trees by at most $\binom{n-1}{2}$. □

Theorem 15.4 also works for the PATH TSP: Take a solution x of (14.1) and set $x_{\{s,t\}} := 1$ (this variable must have had value zero).

15.2 Analysis of Best-of-Many Christofides

Since the approximation ratio of the Best-of-Many Christofides algorithm is exactly $\frac{3}{2}$ for $T = \emptyset$, we assume $T \neq \emptyset$ from now on. Almost all works that analyze the Best-of-Many Christofides algorithm begin as follows:

Proposition 15.5 (An, Kleinberg, and Shmoys [2015]). *Best-of-Many Christofides (Algorithm 15.1) computes a T-tour of cost at most*

$$c(x^*) + \sum_{S \in \mathcal{S}} \mu(S) c(y^S)$$

for all vectors y^S in the $(T \triangle \mathrm{odd}(S))$-join polyhedron (cf. (2.9)) for $S \in \mathcal{S}$.

Proof. The cost of the T-tour computed by Best-of-Many Christofides is

$$\begin{aligned} &\min_{S \in \mathcal{S}:\, \mu(S)>0} \big(c(S) + \min\{c(J) : J \text{ is a } (T \triangle \mathrm{odd}(S))\text{-join}\}\big) \\ \le\ & \sum_{S \in \mathcal{S}} \mu(S) \big(c(S) + \min\{c(J) : J \text{ is a } (T \triangle \mathrm{odd}(S))\text{-join}\}\big) \\ \le\ & c(x^*) + \sum_{S \in \mathcal{S}} \mu(S) c(y^S), \end{aligned}$$

where the last inequality follows from $x^* \ge \sum_{S \in \mathcal{S}} \mu(S) \chi^S$ and Theorem 2.19. □

In other words, we can view μ as a probability distribution and analyze the expected cost if we sample S according to μ. The difficulty in the analysis lies in finding an appropriate set of vectors $(y^S)_{S \in \mathcal{S}}$ (which we call *parity correction vectors*). One can use $y^S = \frac{1}{3}x^* + \frac{1}{3}\chi^S$ as in Theorem 14.14 to obtain the approximation guarantee $\frac{5}{3}$ again, but one can do better.

We will again use that a narrow cut is a $(T \triangle \mathrm{odd}(S))$-cut if and only if the spanning tree (V, S) contains an even number of edges in that cut (cf. Lemma 14.19). To design a better parity correction vector, the first idea is to try $y^S = \beta x^* + (1 - 2\beta)\chi^S$ for some β slightly larger than $\frac{1}{3}$. This vector satisfies all constraints of the $(T \triangle \mathrm{odd}(S))$-join polyhedron except that $y^S(\delta(U)) \ge 1$ can be violated for narrow cuts C with $|S \cap C|$ even, but only if $x^*(C)$ is very close to 1 (less than $\frac{4\beta-1}{\beta}$). However, as we see now, if $x^*(C)$ is very close to 1, then the trees with more than one edge in C (in particular, trees S with $|S \cap C|$

even) cannot contribute much to the convex combination, so one might hope that correcting this is cheap.

Proposition 15.6. *For every narrow cut C, we have*

$$\sum_{S\in\mathcal{S}:|S\cap C|=1} \mu(S) \ \geq\ 2 - x^*(C) \tag{15.1}$$

and

$$\sum_{S\in\mathcal{S}:|S\cap C|\text{ even}} \mu(S) \ \leq\ x^*(C) - 1. \tag{15.2}$$

Proof. Since $|S \cap C| \geq 1$ for every cut C and $S \in \mathcal{S}$, we have $x^*(C) \geq \sum_{S\in\mathcal{S}} \mu(S)|S \cap C| \geq \sum_{S\in\mathcal{S}} 2\mu(S) - \sum_{S\in\mathcal{S}:|S\cap C|=1} \mu(S)$, which implies (15.1). Then (15.2) follows from $\sum_{S\in\mathcal{S}:|S\cap C|=1} \mu(S) + \sum_{S\in\mathcal{S}:|S\cap C|\text{ even}} \mu(S) \leq 1$ and (15.1). □

Let $\mathcal{N}$ denote the set of narrow cuts. We consider

$$\begin{aligned} y^S \ :=\ & \beta x^* + (1-2\beta)\chi^S \\ & + \sum_{C\in\mathcal{N}:|S\cap C|\text{ even}} \max\left\{0,\ 4\beta - 1 - \beta x^*(C)\right\} v^C \end{aligned} \tag{15.3}$$

for $S \in \mathcal{S}$, where $0 \leq \beta \leq \frac{1}{2}$ and $v^C \in \mathbb{R}^E_{\geq 0}$ are vectors with

$$v^C(C) \ \geq\ 1 \qquad \text{for all } C \in \mathcal{N}. \tag{15.4}$$

We now show that y^S is a parity correction vector for S.

Lemma 15.7 (An, Kleinberg, and Shmoys [2015]). *For every $S \in \mathcal{S}$, the vector y^S is in the $(T \,\triangle\, \mathrm{odd}(S))$-join polyhedron (cf. (2.9)).*

Proof. Clearly y^S is nonnegative. Let $S \in \mathcal{S}$ and C be a $(T \,\triangle\, \mathrm{odd}(S))$-cut. We need to show $y^S(C) \geq 1$.

If $C \notin \mathcal{N}$, then $x^*(C) \geq 2$ and hence

$$y^S(C) \ \geq\ \beta x^*(C) + (1-2\beta)|S \cap C| \ \geq\ 2\beta + 1 - 2\beta \ =\ 1.$$

If $C \in \mathcal{N}$, then by Lemma 14.19, $|S \cap C|$ is even and hence at least 2. Using (15.4), we get

$$\begin{aligned} y^S(C) \ &\geq\ \beta x^*(C) + (1-2\beta)|S\cap C| + 4\beta - 1 - \beta x^*(C) \\ &\geq\ \beta x^*(C) + 2(1-2\beta) + 4\beta - 1 - \beta x^*(C) \\ &=\ 1. \end{aligned}$$

□

This leads to the following analysis:

Lemma 15.8 (An, Kleinberg, and Shmoys [2015]). *Let (G, c, T) be an instance of the T-Tour Problem where G is a complete graph, c satisfies the triangle inequality, and $T \neq \emptyset$. Let x^* be a solution to the LP (14.2), and $0 \leq \beta \leq \frac{1}{2}$. Let $v^C \in \mathbb{R}^E_{\geq 0}$ be vectors with $v^C(C) \geq 1$ for all $C \in \mathcal{N}$. Then Best-of-Many Christofides computes a T-tour of cost at most*

$$(2-\beta)\,c(x^*) + \sum_{C \in \mathcal{N}} (x^*(C) - 1) \max\left\{0,\, 4\beta - 1 - \beta x^*(C)\right\} c(v^C).$$

Proof. By Proposition 15.5 and Lemma 15.7, the solution computed by Best-of-Many Christofides costs at most

$$\begin{aligned}
&(1+\beta)\,c(x^*) + \sum_{S \in \mathcal{S}} \mu(S)\Big((1-2\beta)c(S) \\
&\qquad\qquad + \sum_{C \in \mathcal{N}: |S \cap C| \text{ even}} \max\left\{0,\, 4\beta - 1 - \beta x^*(C)\right\} c(v^C)\Big) \\
&= (2-\beta)\,c(x^*) + \sum_{C \in \mathcal{N}} \left(\sum_{S \in \mathcal{S}: |S \cap C| \text{ even}} \mu(S) \right) \max\left\{0,\, 4\beta - 1 - \beta x^*(C)\right\} c(v^C) \\
&\leq (2-\beta)\,c(x^*) + \sum_{C \in \mathcal{N}} (x^*(C) - 1) \max\left\{0,\, 4\beta - 1 - \beta x^*(C)\right\} c(v^C),
\end{aligned}$$

where we used (15.2) in the inequality. □

Now, the next question is how to choose the vectors v^C. In the case $|T| = 2$, An, Kleinberg, and Shmoys [2015] showed:

Lemma 15.9 (An, Kleinberg, and Shmoys [2015]). *Let $|T| = 2$. Then there are vectors $v^C \in \mathbb{R}^E_{\geq 0}$ with $v^C(C) \geq 1$ for $C \in \mathcal{N}$ and $\sum_{C \in \mathcal{N}} v^C \leq x^*$.*

Proof. For any subset $\mathcal{N}' \subsetneq \mathcal{N}$, there is a partition $\mathcal{W}$ of V (the sets "in between" these narrow cuts) with $|\mathcal{W}| = |\mathcal{N}'| + 1$ and $\delta(\mathcal{W}) = \bigcup_{C \in \mathcal{N}'} C$. Then $x^*(\bigcup_{C \in \mathcal{N}'} C) = x^*(\delta(\mathcal{W})) \geq |\mathcal{W}| - 1 = |\mathcal{N}'|$. Hence, by Theorem 3.13, there is a function $f : E \times \mathcal{N} \to \mathbb{R}_{\geq 0}$ with $\sum_{C \in \mathcal{N}} f(e, C) \leq x^*_e$ $(e \in E)$ and $\sum_{e \in C} f(e, C) \geq 1$ $(C \in \mathcal{N})$. Set $v^C_e := f(e, C)$. □

This theorem does not hold for general T (see Exercise 15.2). Now it is easy to complete the analysis of the golden ratio approximation:

Theorem 15.10 (An, Kleinberg, and Shmoys [2015]). *Let (G, c, s, t) be a Path TSP instance where G is a complete graph and c satisfies the triangle inequality. Let x^* be a solution to the LP (14.1). Then Best-of-Many Christofides computes an $\{s, t\}$-tour of cost at most $\frac{1+\sqrt{5}}{2}\, c(x^*)$.*

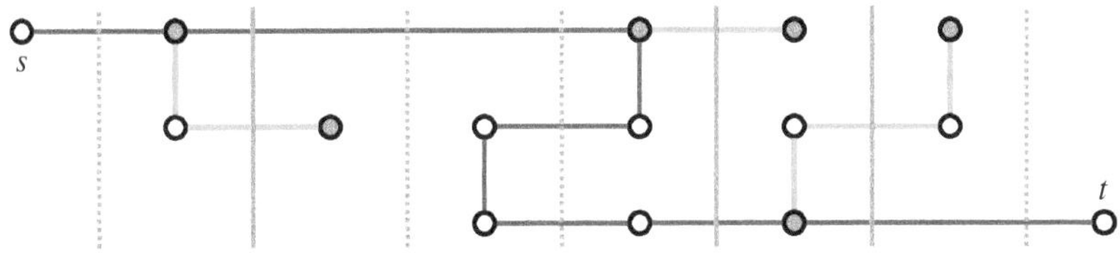

Figure 15.3 A spanning tree (V, S). Filled circles show vertices in $T \triangle \text{odd}(S)$, where $T = \{s, t\}$. The edge set S is partitioned into its s-t-path I_S (blue) and its $(\text{odd}(S) \triangle T)$-join J_S (green, thick). The narrow cuts are those shown in grey (dotted and solid). Those that have an even number of edges of S are shown in solid grey; each of them contains (at least) one green and (at least) one blue edge. Three of the blue edges are lonely.

Proof. Choose the vectors v^C as in Lemma 15.9. By Lemma 15.8, the solution computed by Best-of-Many Christofides costs at most

$$(2-\beta)\, c(x^*) + \sum_{C \in \mathcal{N}} (x^*(C) - 1) \max\left\{0,\, 4\beta - 1 - \beta x^*(C)\right\} c(v^C).$$

Using $(x-1)(4\beta - 1 - \beta x) \le \frac{(3\beta-1)^2}{4\beta}$ for all $x \in \mathbb{R}$, we get the upper bound

$$(2-\beta)\, c(x^*) + \sum_{C \in \mathcal{N}} \frac{(3\beta-1)^2}{4\beta}\, c(v^C).$$

Setting $\beta = \frac{1}{\sqrt{5}}$ and using Lemma 15.9 completes the proof. □

By choosing the vectors v^C differently, one can improve this bound, even for general T, as Sebő [2013] showed. To this end, we partition the edge set of a tree into two parts. By Proposition 1.27, every spanning tree contains a T-join. For $S \in \mathcal{S}$, let I_S denote the T-join in (V, S) and $J_S := S \setminus I_S$. See Figure 15.3 for an example. If J_S is cheap, we can use it for parity correction as in Proposition 14.5 because J_S is a $(T \triangle \text{odd}(S))$-join. Otherwise, I_S must be cheap, and this is also useful: I_S covers all narrow cuts and can be used for the vectors v^C $(C \in \mathcal{N})$, as we will see next.

From I_S, we actually only use the *lonely* edges – that is, those edges $e \in S$ for which there is a narrow cut C with $C \cap S = \{e\}$. Let L_S denote this set of lonely edges. We have $L_S \subseteq I_S$ because I_S (in fact every T-join) intersects every narrow cut (in fact every T-cut).

Theorem 15.11 (Sebő [2013]). *Let (G, c, T) be an instance of the* T-Tour Problem *where G is a complete graph, c satisfies the triangle inequality, and $T \neq \emptyset$. Let x^* be a solution to the LP* (14.2). *Then Best-of-Many Christofides computes a T-tour of cost at most $\frac{8}{5}c(x^*)$.*

Proof. Let

$$v^C := \frac{1}{2 - x^*(C)} \sum_{S \in \mathcal{S}: |S \cap C| = 1} \mu(S) \chi^{S \cap C} \tag{15.5}$$

for $C \in \mathcal{N}$. Then indeed $v^C(C) \geq 1$ for all $C \in \mathcal{N}$ due to (15.1). To bound the cost, we observe

$$\sum_{C \in \mathcal{N}} (2 - x^*(C))\, c(v^C) = \sum_{S \in \mathcal{S}} \mu(S) c(L_S). \tag{15.6}$$

Indeed, if e is the only edge of S in two narrow cuts, they are both the fundamental cut of e and S, and hence identical.

Using the new definition of v^C and (15.6), we get from Lemma 15.8 that the T-tour computed by Best-of-Many Christofides costs at most

$$(2 - \beta)\, c(x^*) + \max_{C \in \mathcal{N}} \left(\tfrac{x^*(C) - 1}{2 - x^*(C)} \max\left\{0,\, 4\beta - 1 - \beta x^*(C)\right\} \right) \sum_{S \in \mathcal{S}} \mu(S) c(L_S).$$

Observing $\frac{(x-1)(4\beta - 1 - \beta x)}{2 - x} \leq 1 - \beta - 2\sqrt{\beta - 2\beta^2}$ for all $x < 2$ and setting $\beta = \frac{4}{9}$, this bound is at most

$$\tfrac{14}{9} c(x^*) + \tfrac{1}{9} \sum_{S \in \mathcal{S}} \mu(S) c(L_S).$$

Using $y^S = \chi^{J_S}$ instead yields the bound

$$c(x^*) + \sum_{S \in \mathcal{S}} \mu(S) c(J_S). \tag{15.7}$$

Since $L_S \subseteq I_S = S \setminus J_S$, the better of the two bounds is at most

$$\begin{aligned} & \tfrac{9}{10} \left(\tfrac{14}{9} c(x^*) + \tfrac{1}{9} \sum_{S \in \mathcal{S}} \mu(S) c(I_S) \right) + \tfrac{1}{10} \left(c(x^*) + \sum_{S \in \mathcal{S}} \mu(S) c(J_S) \right) \\ = \; & \tfrac{15}{10} c(x^*) + \tfrac{1}{10} \sum_{S \in \mathcal{S}} \mu(S) c(S) \\ \leq \; & \tfrac{8}{5} c(x^*). \end{aligned}$$

□

Gao [2015] simply chose v^C to be the incidence vector of a cheapest edge in C, which is clearly best possible in this framework, but it does not lead to a better approximation ratio (see an example in the appendix of Vygen [2016]). The following question remains open:

Open Problem 15.12. What is the approximation ratio of the Best-of-Many Christofides algorithm for Path TSP and for the T-Tour Problem?

We only know that both answers are at least $\frac{3}{2}$ and at most $\frac{8}{5}$. Nevertheless, there have been several further improvements that work with variants of Best-of-Many Christofides, which we will discuss in the rest of this chapter. Most of them work only for PATH TSP, but in Section 15.5, we will also see an improvement for the general T-TOUR PROBLEM.

15.3 Working with a Better Distribution

Although no better upper bound than $\frac{8}{5}$ is known for Best-of-Many Christofides, Vygen [2016] showed that one can get a better approximation ratio for Path TSP by "reassembling trees" – that is, by changing the probability distribution μ so that certain bad configurations are avoided.

This result was improved by Gottschalk and Vygen [2018]. They showed that a distribution with certain global properties – containing, in a sense, many Gao trees – always exists. More precisely, the distribution consists of an ordered list of spanning trees such that the first one is a Gao tree, and for each narrow cut C with $x^*(C) = 2 - \alpha$, the first α fraction of the trees contains exactly one edge of C in each tree. More precisely:

Theorem 15.13 (Gottschalk and Vygen [2018]). *For every feasible solution x^* to (14.1), there are $S_1, \ldots, S_r \in \mathcal{S}$ and $\mu(S_1), \ldots, \mu(S_r) > 0$ with $\sum_{j=1}^{r} \mu(S_j) = 1$ such that $x^* = \sum_{j=1}^{r} \mu(S_j)\chi^{S_j}$, and for every $C \in \mathcal{N}$, there exists a $k \in \{1, \ldots, r\}$ with $\sum_{j=1}^{k} \mu(S_j) \geq 2 - x^*(C)$ and $|C \cap S_j| = 1$ for all $j = 1, \ldots, k$.*

Figure 15.4 shows an example. Theorem 15.13 gives us as many good trees in the beginning of our list as we can expect from (15.1). Gottschalk and Vygen [2018] also showed that such a list of trees can be computed in polynomial time, but we will not need this in the following. In fact, as observed by K. Pashkovich, a polynomial-time algorithm follows easily from Theorem 15.13 (see Exercise 15.5).

The general idea why Theorem 15.13 is useful is that Gao trees allow for cheap parity correction because they do not need any correction of narrow cuts. For a Gao tree S, the vector $\beta x^* + (1 - 2\beta)\chi^{J_S}$ is already a good parity correction vector, and the saved term $(1 - 2\beta)\chi^{I_S}$ in (15.3) can be used to help narrow cuts in other trees. The Gao trees in the distribution μ may have very little weight, but also the first trees with total weight of say 0.1 are nice: For each such tree S and each narrow cut C with $|S \cap C|$ even, we have $x^*(C) \geq 1.9$, and hence $\beta x^* + (1 - 2\beta)\chi^{J_S} + 0.1\beta\chi^{I_S}$ is a parity correction vector. The remaining $(1 - 2.1\beta)\chi^{I_S}$ can again be used to help later trees. Based on a parity correction vector proposed by Vygen [2016] (cf. Exercise 15.3), Gottschalk and Vygen

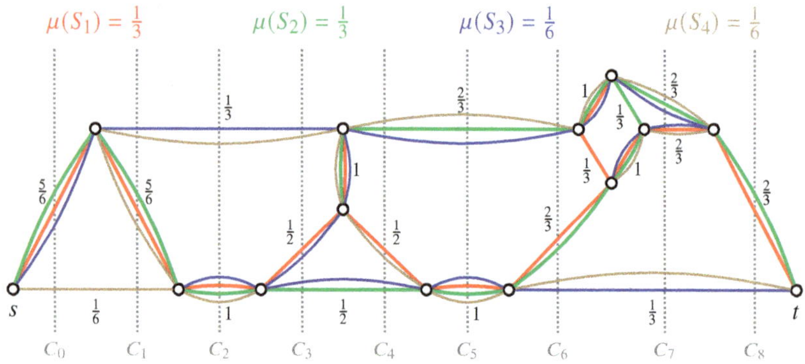

Figure 15.4 Continuing the example from Figure 14.4, the picture shows a decomposition of the LP solution x^* (black numbers) into spanning trees, satisfying the properties of Theorem 15.13. Here $x^* = \frac{1}{3}\chi^{S_1} + \frac{1}{3}\chi^{S_2} + \frac{1}{6}\chi^{S_3} + \frac{1}{6}\chi^{S_4}$. The trees S_1, S_2, S_3, S_4 are shown in red, green, blue, and brown, respectively. As required, S_1 is a Gao tree, and S_2 has only one edge in each of the cuts C_0, C_1, C_2, C_3, C_8 with value less than $\frac{5}{3}$.

[2018] obtained the approximation ratio 1.566 for any list of trees satisfying Theorem 15.13.

In the rest of this section, we prove Theorem 15.13, following the nicer proof of Schalekamp et al. [2018].

Definition 15.14 (chain-point). Let $G = (V, E)$ be an undirected graph, $\emptyset \subsetneq U_1 \subsetneq U_2 \subsetneq \ldots \subsetneq U_k \subsetneq V$, and $\mathscr{C} := \{\delta(U_1), \ldots, \delta(U_k)\}$. A *chain-point* for $\mathscr{C}$ is a vector $x \in \mathbb{R}^E_{\geq 0}$ such that

(a) x is contained in the spanning tree polytope (2.8) of G,
(b) $x(\delta(U_i)) < 2$ for $i = 1, \ldots, k$, and
(c) $\sum_{v \in U_j \setminus U_i} x(\delta(v)) = 2|U_j \setminus U_i|$ for $1 \leq i < j \leq k$.

Note that every feasible solution to the LP (14.1) is a chain-point for $\mathscr{C} = \mathcal{N}$ being the set of its narrow cuts. We will prove the following generalization of Theorem 15.13 to chain-points.

Theorem 15.15. *Let $G = (V, E)$ be an undirected graph, $\emptyset \subsetneq U_1 \subsetneq U_2 \subsetneq \ldots \subsetneq U_k \subsetneq V$, and $\mathscr{C} := \{\delta(U_1), \ldots, \delta(U_k)\}$. Let x^* be a chain-point for $\mathscr{C}$. Then there are $S_1, \ldots, S_r \in \mathcal{S}$ and $\mu(S_1), \ldots, \mu(S_r) > 0$ with $\sum_{j=1}^r \mu(S_j) = 1$ such that $x^* = \sum_{j=1}^r \mu(S_j)\chi^{S_j}$, and for every $C \in \mathscr{C}$, there exists a $k \in \{1, \ldots, r\}$ with $\sum_{j=1}^k \mu(S_j) \geq 2 - x^*(C)$ and $|C \cap S_j| = 1$ for all $j = 1, \ldots, k$.*

We call a spanning tree (V, S) of G a *Gao tree for* $\mathscr{C}$ if $|S \cap C| = 1$ for every $C \in \mathscr{C}$. In order to prove Theorem 15.15, we will repeatedly apply the following lemma:

Lemma 15.16. *Let x^* be a chain-point for $\mathscr{C}$. Then there are a Gao tree (V, S) for $\mathscr{C}$ and a vector x' in the spanning tree polytope* (2.8) *such that $x^* = \lambda \cdot \chi^S + (1 - \lambda) \cdot x'$ for some $0 < \lambda \le 1$.*

Proof. Let F be the minimal face of the spanning tree polytope (2.8) that contains x^*. We call a set $\emptyset \neq W \subseteq V$ *tight* if $x^*(E[W]) = |W| - 1$. Let $\mathcal{L}$ be a laminar family of tight sets defining the face F (i.e., a laminar family as in Lemma 4.16).

We will find a spanning tree (V, S) with $\chi^S \in F$ – that is, with $S \subseteq \{e \in E : x^*_e > 0\}$ and $|S[L]| = |L| - 1$ for every $L \in \mathcal{L}$. Then, by Lemma 4.16 (ii), for every tight set U we have, for suitable coefficients λ_L (for $L \in \mathcal{L}$) and λ_e (for $e \in E$ with $x^*_e = 0$),

$$\chi^{E[U]} = \sum_{L \in \mathcal{L}} \lambda_L \chi^{E[L]} + \sum_{e \in E : x^*_e = 0} \lambda_e \chi^{\{e\}}.$$

Because S contains only edges in the support of x^*, this implies

$$\begin{aligned} |S[U]| &= \sum_{L \in \mathcal{L}} \lambda_L |S[L]| \\ &= \sum_{L \in \mathcal{L}} \lambda_L (|L| - 1) \\ &= \sum_{L \in \mathcal{L}} \lambda_L x^*(E[L]) \\ &= x^*(E[U]) \\ &= |U| - 1. \end{aligned}$$

Hence, there is a vector x' in (2.8) such that $x^* = \lambda \cdot \chi^S + (1 - \lambda) \cdot x'$ for some $0 < \lambda \le 1$. The spanning tree (V, S) that we will find will in addition fulfill $|S \cap C| = 1$ for every $C \in \mathscr{C}$.

Let $\mathscr{C} = \{\delta(U_1), \ldots, \delta(U_k)\}$ with $\emptyset \subsetneq U_1 \subsetneq U_2 \subsetneq \ldots \subsetneq U_k \subsetneq V$. Let $U_0 := \emptyset$ and $U_{k+1} := V$, and let $G = (V, \{e : x^*_e > 0\})$ be the support graph of x^*. We claim that the following three statements hold.

(i) For every tight set W, the set $\{l \in \{0, 1, \ldots, k\} : W \cap (U_{l+1} \setminus U_l) \neq \emptyset\}$ is a set of consecutive indices – that is, of the form $\{i, i + 1, \ldots, j\}$ for some $i \le j$.

(ii) For every tight set W and for $0 \le i < j \le k + 1$, the induced subgraph $G[W \cap (U_j \setminus U_i)]$ of the support graph is connected.

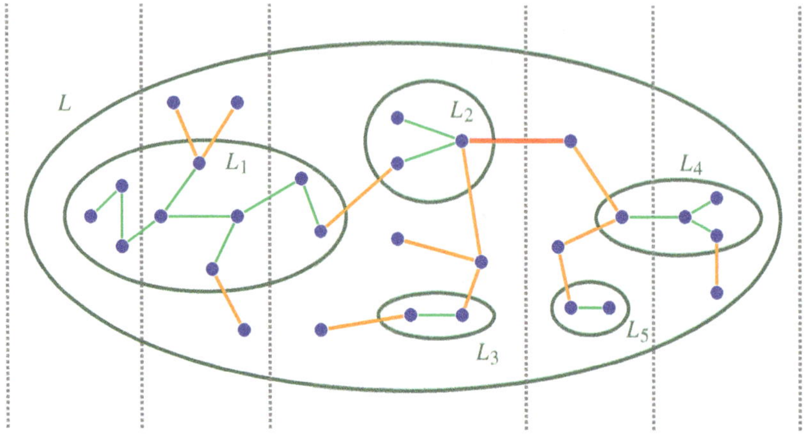

Figure 15.5 Illustration of the construction of S_L in the proof of Lemma 15.16. Narrow cuts are shown as gray dotted lines. The dark-green ellipses show a set L in the laminar family $V \cup \mathcal{L}$ and its maximal subsets $L_1, \ldots, L_5 \in \mathcal{L}$. The light-green edges are the edges of the trees $(L_1, S_{L_1}), \ldots, (L_5, S_{L_5})$. The edges added to $S_{L_1} \cup \cdots \cup S_{L_5}$ to obtain S'_L are shown in orange. Finally, we add the edge drawn in red, which completes S'_L to the tree S_L with the desired properties.

(iii) Let $1 \le i \le k$ and let W, W' be tight sets such that $W \setminus U_i$, $W \cap U_i$, $W' \setminus U_i$, and $W' \cap U_i$ are nonempty. Then $W \cap W' \neq \emptyset$.

Having (i)–(iii), we construct S as follows. We consider the sets $L \in \mathcal{L} \cup \{V\}$ in an order of non-decreasing cardinality and show that we can find a spanning tree (L, S_L) in the support graph G such that $|S_L \cap C| \le 1$ for every $C \in \mathscr{C}$. See Figure 15.5 for an illustration of the construction of S_L.

Let $L_1, \ldots, L_m \in \mathcal{L}$ be the maximal sets that are proper subsets of L. These are disjoint because $\mathcal{L}$ is laminar. By (iii), we have $|(S_{L_1} \cup \cdots \cup S_{L_m}) \cap C| \le 1$ for every $C \in \mathscr{C}$.

For $0 \le l \le k$ with $L \cap (U_{l+1} \setminus U_l) \neq \emptyset$, we now add edges of $G[U_{l+1} \setminus U_l]$ to $S_{L_1} \cup \cdots \cup S_{L_m}$ to obtain an edge set $S'_L \subseteq G[L]$ such that $S'_L[L \cap (U_{l+1} \setminus U_l)]$ is the edge set of a tree with vertex set $L \cap (U_{l+1} \setminus U_l)$; this is possible by (ii).

Let us now consider $0 \le l \le k$ where $L \cap U_l$ and $L \setminus U_l$ are nonempty. If there is a set L_i (with $1 \le i \le m$) such that both $L_i \cap U_l$ and $L_i \setminus U_l$ are nonempty, then by (i), L_i intersects both $U_l \setminus U_{l-1}$ and $U_{l+1} \setminus U_l$. Therefore, S_{L_i} contains the edge set of a path with one endpoint in each of these sets; in fact, it contains a single edge with this property. Otherwise, we add an edge from $U_l \setminus U_{l-1}$ to $U_{l+1} \setminus U_l$ to ensure connectivity; this results in the edge set of a spanning tree of $G[L]$ and maintains the property that every cut $C \in \mathscr{C}$ is intersected in at most one edge. The existence of such an edge follows from (i) and (ii) applied to $W = L$ and $i = l - 1$ and $j = l + 1$.

It remains to prove claims (i)–(iii).

We first show (i). Let W be a tight set. Suppose $W \cap (U_{l+1} \setminus U_l) = \emptyset$, but $W \cap U_l$ and $W \setminus U_{l+1}$ are both nonempty. By (a) of Definition 15.14, the chain-point x^* restricted to $E[W]$ is a convex combination of incidence vectors of (edge sets of) spanning trees of $G[W]$ because W is tight. Therefore $x^*(\delta(U_l) \cap \delta(U_{l+1})) \geq 1$. Moreover, we have

$$\begin{aligned} x^*(\delta(U_{l+1} \setminus U_l)) &= \sum_{v \in U_{l+1} \setminus U_l} x(\delta(v)) - 2x^*(E[U_{l+1} \setminus U_l]) \\ &\geq 2|U_{l+1} \setminus U_l| - 2(|U_{l+1} \setminus U_l| - 1) \\ &= 2 \end{aligned}$$

by (a) and (c). Therefore by (b),

$$\begin{aligned} 4 &> x^*(\delta(U_l)) + x^*(\delta(U_{l+1})) \\ &= x^*(\delta(U_{l+1} \setminus U_l)) + 2 \cdot x^*(\delta(U_l) \cap \delta(U_{l+1})) \\ &\geq 4. \end{aligned}$$

This contradiction implies (i).

Next we show (ii). Let $0 \leq i < j \leq k + 1$. We first show $x^*(E[U_j \setminus U_i]) > |U_j \setminus U_i| - 2$. For this, we distinguish three cases. If $i > 0$ and $j < k + 1$, we have

$$\begin{aligned} 4 &> x^*(\delta(U_i)) + x^*(\delta(U_j)) \\ &\geq \sum_{v \in U_j \setminus U_i} x^*(\delta(v)) - 2 \cdot x^*(E[U_j \setminus U_i]) \\ &= 2|U_j \setminus U_i| - 2 \cdot x^*(E[U_j \setminus U_i]). \end{aligned}$$

If $i = 0$ and $j = k + 1$, we have $|V| - 1 = x^*(E) = x^*(E[U_j \setminus U_i])$. Otherwise, we have either $U_j \setminus U_i = U_j$ or $U_j \setminus U_i = V \setminus U_i$. Without loss of generality $U_j \setminus U_i = U_j$ (the other case is symmetric). Then

$$\begin{aligned} |V| - 1 &= x^*(E) \\ &= x^*(E[U_j]) + x^*(\delta(U_j)) + x^*(E[V \setminus U_j]) \\ &< x^*(E[U_j]) + 2 + (|V| - |U_j| - 1). \end{aligned}$$

Hence, in all cases $x^*(E[U_j \setminus U_i]) > |U_j \setminus U_i| - 2$.

Now let W be a tight set. Then

$$\begin{aligned} & x^*(E[W \cap (U_j \setminus U_i)]) + x^*(E[W \cup (U_j \setminus U_i)]) \\ & \geq x^*(E[U_j \setminus U_i]) + x^*(E[W]) \\ & > |U_j \setminus U_i| - 2 + |W| - 1. \end{aligned}$$

Since $x^*(E[W \cup (U_j \setminus U_i)]) \leq |W \cup (U_j \setminus U_i)| - 1$, we have $x^*(E[W \cap (U_j \setminus U_i)]) > |W \cap (U_j \setminus U_i)| - 2$. This implies that $G[W \cap (U_j \setminus U_i)]$ is connected, because otherwise there is a partition of $W \cap (U_j \setminus U_i)$ into nonempty sets A and B such that $x^*(\delta(A) \cap \delta(B)) = \emptyset$, implying

$$\begin{aligned} x^*(E[W \cap (U_j \setminus U_i)]) &\leq x^*(E[A]) + x^*(E[B]) \\ &\leq |A| - 1 + |B| - 1 \\ &= |W \cap (U_j \setminus U_i)| - 2. \end{aligned}$$

Finally, we show (iii). Let $1 \leq i \leq k$, and suppose there are tight sets W and W' such that $W \setminus U_i$, $W \cap U_i$, $W' \setminus U_i$, and $W' \cap U_i$ are nonempty. Then $x^*(\delta(U_i) \cap E[W]) \geq 1$ because x^* restricted to $E[W]$ is a convex combination of incidence vectors of spanning trees of $G[W]$ (since W is tight). The same holds for W'. So if W and W' are disjoint, we have $x^*(\delta(U_i)) \geq x^*(\delta(U_i) \cap E[W]) + x^*(\delta(U_i) \cap E[W']) \geq 2$, contradicting (b). □

We now show how Lemma 15.16 implies Theorem 15.15.

Proof of Theorem 15.15. Let F be the minimal face of the spanning tree polytope that contains the chain-point x^*. We apply induction on the dimension of F. Let (V, S) be a Gao tree as in Lemma 15.16. Let $0 < \lambda \leq 1$ maximal such that $x^* = \lambda \cdot \chi^S + (1 - \lambda) \cdot x'$ for some x' in the spanning tree polytope (2.8). If $\lambda = 1$, we have $x^* = \chi^S$ and are done. Otherwise, x' is contained in a lower-dimensional face of the spanning tree polytope than x^*: By the maximality of λ, either the vector x' has smaller support, or there is a set U that is tight for x' but not for x^*, or both. Moreover, because $|S \cap C| = 1$ for every cut $C \in \mathscr{C}$, we have $(1 - \lambda) \cdot x'(C) = x^*(C) - \lambda$. Hence

$$\mathscr{C}' := \{C \in \mathscr{C} : x'(C) < 2\} = \{C \in \mathscr{C} : \lambda < 2 - x^*(C)\}.$$

We claim that x' is a chain-point for $\mathscr{C}'$. Properties (a) and (b) hold by construction. We now show (c).

Let U_i and U_j be two sets with $U_i \subsetneq U_j$ and $\delta(U_i), \delta(U_j) \in \mathscr{C}'$. We have $S \cap \delta(U_i) \cap \delta(U_j) = \emptyset$ because otherwise $S \cap \delta(U_j \setminus U_i) = \emptyset$ since S is a Gao tree for $\mathscr{C}$. Therefore, because $|S \cap \delta(U_i)| = |S \cap \delta(U_j)| = 1$, the graph $(V, S)[U_j \setminus U_i]$ is connected and hence $|S[U_j \setminus U_i]| = |U_j \setminus U_i| - 1$. This implies $\sum_{v \in U_j \setminus U_i} |\delta(v) \cap S| = 2 \cdot |S[U_j \setminus U_i]| + |S \cap \delta(U_i)| + |S \cap \delta(U_j)| = 2|U_j \setminus U_i|$.

Because x^* fulfills (c) for $\mathscr{C}$ and hence for $\mathscr{C}'$, this implies that x' fulfills (c) (for $\mathscr{C}'$) as well.

This shows that x' is a chain-point for $\mathscr{C}'$. Hence, we can apply the induction hypothesis to x' to obtain $x' = \sum_{j=1}^{r'} \mu'(S'_j)\chi^{S'_j}$. Then $x^* = \lambda \cdot \chi^S + (1-\lambda) \cdot \sum_{j=1}^{r'} \mu'(S'_j)\chi^{S'_j}$. We set $r := r'+1$, $\mu(S_1) := \lambda$, $S_1 := S$, and for $i = 2, \ldots, r'+1$, we set $S_i := S'_{i-1}$ and $\mu(S_i) := (1-\lambda)\mu'(S'_{i-1})$. Then $\mu(S_1), \ldots, \mu(S_r) > 0$ with $\sum_{j=1}^{r} \mu(S_j) = 1$ and $x^* = \sum_{j=1}^{r} \mu(S_j)\chi^{S_j}$.

Now consider a cut $C \in \mathscr{C}$. We claim that there exists a $k \in \{1, \ldots, r\}$ with $\sum_{j=1}^{k} \mu(S_j) \geq 2 - x^*(C)$ and $|C \cap S_j| = 1$ for all $j = 1, \ldots, k$. We always have $|C \cap S_1| = 1$ because (V, S) is a Gao tree for $\mathscr{C}$. If $C \notin \mathscr{C}'$ (i.e., $2 - x^*(C) \leq \lambda = \mu(S_1)$), the claim holds for $k = 1$. Otherwise, by the induction hypothesis, there exists a $k \in \{2, \ldots, r\}$ such that $\frac{1}{1-\lambda}\sum_{j=2}^{k} \mu(S_j) = \sum_{j=2}^{k} \mu'(S'_{j-1}) \geq 2 - x'(C)$ and $|C \cap S_j| = 1$ for all $j = 2, \ldots, k$. Then

$$\begin{aligned} \sum_{j=1}^{k} \mu(S_j) &\geq \lambda + (1-\lambda) \cdot (2 - x'(C)) \\ &= \lambda + 2(1-\lambda) - (x^*(C) - \lambda) \\ &= 2 - x^*(C). \end{aligned}$$

This completes the proof of the claim and of Theorem 15.15. □

As noted by Schalekamp et al. [2018], this proof can be turned into a polynomial-time algorithm; finding λ can be done with the technique of Exercise 5.14. Alternative algorithms were devised by Gottschalk and Vygen [2018] and by Pashkovich (see Exercise 15.5).

We remark that Theorem 15.13 cannot be generalized to T-tours: We saw in Figure 14.6 that there may not even be a single Gao tree in the support graph.

It is an interesting open question what the approximation ratio is for Best-of-Many Christofides for Path TSP with a distribution like in Theorem 15.13. We only know that it is between 1.5 (Proposition 15.2) and 1.566 (Gottschalk and Vygen [2018]), but it is not even clear whether it is really better than the generic Best-of-Many Christofides.

In Section 15.4, we show a better upper bound for a slightly different algorithm. There we will also need Theorem 15.13.

15.4 Parity Correction of Forests

The next improvement for Path TSP was found by Sebő and van Zuylen [2019]. They introduced the idea to delete some of the lonely edges in the spanning

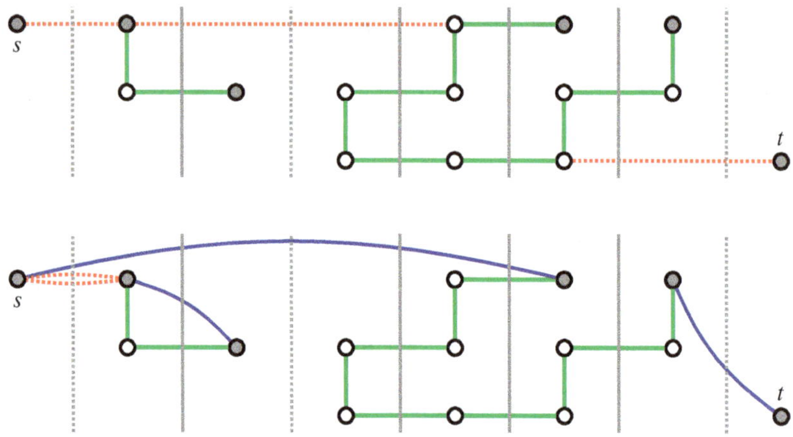

Figure 15.6 Top: The spanning tree (V, S) from Figure 15.3. Again, the narrow cuts are those shown in grey (dotted and solid). The lonely edges are dotted and red. For each lonely edge, there is one narrow cut (dotted) in which this is the only edge of S (other narrow cuts are shown as solid lines). Removing the lonely edges from S results in a forest F_S (green). Filled circles show vertices in $T \triangle \mathrm{odd}(F_S)$, where $T = \{s, t\}$.
Bottom: After removing the lonely edges, we do parity correction (blue, curved). The long blue edge is bad. Here we have to put back two copies of a lonely edge (red, dotted) to restore connectivity.

tree before parity correction, because parity correction will often reconnect the connected components of the forest anyway.

Deleting lonely edges from S results in a forest F_S (see Figure 15.6). We will do parity correction on this forest – that is, add a $(T \triangle \mathrm{odd}(F_S))$-join J. Deleting a lonely edge e from S generates two connected components X and $V \setminus X$, where $\delta(X)$ is a narrow cut. Hence, this is a T-cut and thus also a $(T \triangle \mathrm{odd}(F_S))$-cut because $F_S \cap \delta(X) = \emptyset$. Therefore, the $(T \triangle \mathrm{odd}(F_S))$-join J contains some edge (in fact an odd number of edges) in this cut. This makes us hope to restore connectivity often. Nevertheless, the resulting T-join $F_S \,\dot\cup\, J$ might be disconnected. In this case, we put back two copies of some of the edges from $S \setminus F_S$.

For a narrow cut C, we denote by $\mathcal{L}(C)$ the set of trees S for which $|S \cap C| = 1$ and $F_S \cap C = \emptyset$. For a tree $S \in \mathcal{L}(C)$, the forest F_S contains no edge in C because the only edge was deleted from S. Parity correction will add at least one edge in every such cut. The problem is, however, that an edge of J may belong to more than one of these cuts (such edges will be called *bad*). In this case, one way to ensure connectivity is to put back two copies of the lonely edge that we deleted in each of these cuts except one.

When choosing J, we will not just choose the cheapest $(T \triangle \mathrm{odd}(F_S))$-join but also account for this reconnection cost. If we choose J for parity correction, we will pay a total of $c^S(J)$, where

$$\begin{aligned} c^S(e) \;:=\; c(e) + \sum_{C \in \mathcal{N}: S \in \mathcal{L}(C),\, e \in C} 2c(C \cap S) \\ - \max\left\{0, \max_{C \in \mathcal{N}: S \in \mathcal{L}(C),\, e \in C} 2c(C \cap S)\right\}; \end{aligned} \tag{15.8}$$

here the second and third terms account for the reconnection cost as we will show in Lemma 15.19.

Definition 15.17 (bad edge). Call an edge $e \in E$ *bad* with respect to F_S if there are at least two narrow cuts C with $S \in \mathcal{L}(C)$ and $e \in C$.

If an edge is not bad, then $c^S(e) = c(e)$. We note that edges of $e \in S$ are not bad with respect to F_S; this is useful if we use such edges in a parity correction vector.

Algorithm 15.18: Best-of-Many Christofides with Lonely Edge Deletion

---

Input: an instance (G, c, T) of the T-TOUR PROBLEM with $T \neq \emptyset$, where G is a complete graph and c satisfies the triangle inequality

Output: a T-tour in G

(1) Let x^* be an optimum solution to the LP (14.2).

(2) Let $\mathcal{S}$ denote the set of edge sets of spanning trees in G. Find $\mu(S) \geq 0$ for $S \in \mathcal{S}$ such that $\sum_{S \in \mathcal{S}} \mu(S) = 1$ and $x^* = \sum_{S \in \mathcal{S}} \mu(S)\chi^S$. For each narrow cut C, let $\mathcal{L}(C) \subseteq \mathcal{S}$ such that $|S \cap C| = 1$ for all $S \in \mathcal{L}(C)$.

(3) For $S \in \mathcal{S}$ with $\mu(S) > 0$, let (V, F_S) be the forest that results from (V, S) by deleting every edge $e \in S$ for which $\{e\} = S \cap C$ for some $C \in \mathcal{N}$ with $S \in \mathcal{L}(C)$.

(4) For each $S \in \mathcal{S}$ with $\mu(S) > 0$, compute an $(T \triangle \mathrm{odd}(F_S))$-join J_S whose anticipated cost $c^S(J_S)$ is minimum. Select a minimum-cost set R_S of edges so that $F_S \cup J_S \cup R_S$ is connected, and consider the T-tour $F_S \,\dot{\cup}\, J_S \,\dot{\cup}\, R_S \,\dot{\cup}\, R_S$.

(5) Output the best of these T-tours.

Although Sebő and van Zuylen [2019] described their algorithm only for PATH TSP and with a specific distribution μ coming from Theorem 15.13, Algorithm 15.18 describes it in a slightly more general form, which we will

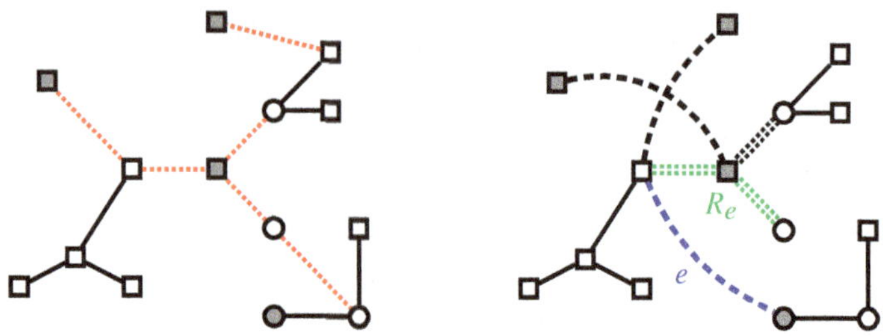

Figure 15.7 Illustration of the proof of Lemma 15.19. Vertices in T are shown as squares, other vertices as circles. On the left, we have the spanning tree (V, S) with the edges in $S \setminus F_S$ drawn dotted and red. Filled vertices are vertices in $T \triangle \text{odd}(F_S)$. On the right, we have a T-tour consisting of the forest F_S (solid), a $(T \triangle \text{odd}(F_S))$-join J_S (dashed), and a reconnection set $R \dot{\cup} R$ (dotted). For the blue edge $e \in J_S$, the two copies of the set R_e are highlighted in green.

need in Section 15.5. Note that $\mathcal{L}(C)$ does not necessarily contain all trees S with $|S \cap C| = 1$. We will analyze this algorithm for different choices of $\mathcal{L}(C)$.

Finding R_S as in Step (4) easily reduces to the Minimum Spanning Tree problem (Theorem 1.17) after contracting the connected components of $(V, F_S \cup J_S)$. The reconnection cost can indeed be bounded as anticipated.

Lemma 15.19. *For all $S \in \mathcal{S}$, we have $2c(R_S) \le c^S(J_S) - c(J_S)$.*

Proof. For each $e \in J_S$, let us denote by R_e the edge set that results from $\bigcup_{C \in \mathcal{N}: S \in \mathcal{L}(C), e \in C} (C \cap S)$ by removing a most expensive edge. See Figure 15.7 for an example. By the definition of c^S, we have $2c(R_e) = c^S(e) - c(e)$. Let $R := \bigcup_{e \in J_S} R_e$. We have $2c(R) \le c^S(J_S) - c(J_S)$. Hence, it suffices to show that $F_S \cup J_S \cup R$ is connected.

For $f = \{v, w\} \in S \setminus F_S$, we show that there is a path from v to w in $(V, F_S \cup J_S \cup R)$. This is obvious if $f \in R$. Hence, assume $f \notin R$ from now on. Consider the narrow cut C with $C \cap S = \{f\}$. Since this is a $(T \triangle \text{odd}(F_S))$-cut, J_S contains at least one edge e in C. The graph $(V, (S \setminus \{f\}) \cup \{e\})$ is a tree because $S \cap C = \{f\}$ and $e \in C$. This tree contains a path P from v to w. We claim that $E(P) \subseteq F_S \cup \{e\} \cup R_e$. Suppose there is an edge $f' \in S \setminus F_S$ that belongs to P. Then there is a narrow cut C' such that $S \in \mathcal{L}(C')$ and $C' \cap S = \{f'\}$. We have $e \in C'$ because P must contain an even number of edges in C'. We conclude $f' \in R_e$ (because $f \notin R_e$). $\square$

Just like Proposition 15.5, we continue by noting:

Lemma 15.20. *The cost of the T-tour computed by Algorithm 15.18 is at most $c(x^*) + \sum_{S \in \mathcal{S}} \mu(S)\big(c^S(\bar{y}^S) - c(S \setminus F_S)\big)$ for all vectors $\bar{y}^S$ in the $(T \triangle \mathrm{odd}(F_S))$-join polyhedron (cf. (2.9)) for $S \in \mathcal{S}$.*

Proof. The T-tour $F_S \,\dot\cup\, J_S \,\dot\cup\, R_S \,\dot\cup\, R_S$ costs at most

$$\begin{aligned} c(F_S) + c(J_S) + 2c(R_S) &\le c(S) + c^S(J_S) - c(S \setminus F_S) \\ &\le c(S) + c^S(\bar{y}^S) - c(S \setminus F_S), \end{aligned}$$

where the first inequality follows from Lemma 15.19 and the last inequality from Theorem 2.19. Taking the weighted sum yields the assertion. □

So we again use a parity correction vector to bound $c^S(J_S)$. This vector will again contain a fraction of x^*. Therefore, the following bound, shown by Sebő and van Zuylen [2019] for Path TSP and generalized by Traub [2020b] to T-tours, is useful:

Lemma 15.21. *Let $S \in \mathcal{S}$. Then*

$$c^S(x^*) - c(x^*) \le \sum_{C \in \mathcal{N}: S \in \mathcal{L}(C)} 2(x^*(C) - 1)\, c(C \cap S).$$

Proof. Let $\mathcal{N}^S$ denote the set of narrow cuts C for which $S \in \mathcal{L}(C)$. We claim:

$$\text{There are vectors } w^C \in \mathbb{R}^E_{\ge 0} \text{ for } C \in \mathcal{N}^S \text{ such that } w^C(C) \ge 1 \text{ for } C \in \mathcal{N}^S \text{ and } \textstyle\sum_{C \in \mathcal{N}^S} w^C \le x^*. \tag{15.9}$$

Then

$$\begin{aligned} &\sum_{e \in E} x^*_e \cdot \max\Big\{0, \max_{C \in \mathcal{N}^S: e \in C} 2c(C \cap S)\Big\} \\ &\ge \sum_{e \in E} \sum_{C' \in \mathcal{N}^S} w^{C'}_e \cdot \max\Big\{0, \max_{C \in \mathcal{N}^S: e \in C} 2c(C \cap S)\Big\} \\ &\ge \sum_{e \in E} \sum_{C' \in \mathcal{N}^S: e \in C'} w^{C'}_e \cdot 2c(C' \cap S) \\ &= \sum_{C \in \mathcal{N}^S} 2c(C \cap S) \cdot w^C(C) \\ &\ge \sum_{C \in \mathcal{N}^S} 2c(C \cap S). \end{aligned}$$

Together with the definition of c^S (cf. (15.8)) and $\mathcal{N}^S = \{C \in \mathcal{N} : S \in \mathcal{L}(C)\}$, this directly implies the assertion.

For $|T| = 2$, Lemma 15.9 directly implies (15.9), but we cannot use this for general T. To prove (15.9) for general T, we again use Theorem 3.13. Consider

$\mathcal{N}' \subseteq \mathcal{N}^S$. Removing the edge in $C \cap S$ from (V, S) for each $C \in \mathcal{N}'$ results in $1 + |\mathcal{N}'|$ connected components, whose vertex sets form a partition $\mathcal{W}$ of V. We have $\delta(\mathcal{W}) = \bigcup_{C \in \mathcal{N}'} C$ and hence $x^*(\bigcup_{C \in \mathcal{N}'} C) = x^*(\delta(\mathcal{W})) \geq |\mathcal{W}| - 1 = |\mathcal{N}'|$. Hence, by Theorem 3.13, there is a function $f : E \times \mathcal{N}^S \to \mathbb{R}_{\geq 0}$ with $\sum_{C \in \mathcal{N}^S} f(e, C) \leq x_e^*$ $(e \in E)$ and $\sum_{e \in C} f(e, C) \geq 1$ $(C \in \mathcal{N}^S)$. Set $w_e^C := f(e, C)$. □

In the following, we assume that μ satisfies the conditions of Theorem 15.13, which is feasible for PATH TSP. In addition, we assume that for each narrow cut C, the coefficients of the first trees sum up to exactly $2 - x^*(C)$, which can be obtained simply by duplicating some elements of $\mathcal{S}$.

Theorem 15.22 (Sebő and van Zuylen [2019]). *Let (G, c, s, t) be a PATH TSP instance and x^* an optimum solution to* (14.1). *Let $S_1, \ldots, S_r \in \mathcal{S}$ and $\mu(S_1), \ldots, \mu(S_r) > 0$ with $\sum_{j=1}^r \mu(S_j) = 1$ such that $x^* = \sum_{j=1}^r \mu(S_j)\chi^{S_j}$, and for every $C \in \mathcal{N}$, there exists a $k \in \{1, \ldots, r\}$ with $\sum_{j=1}^k \mu(S_j) = 2 - x^*(C)$ and $|C \cap S_j| = 1$ for all $j = 1, \ldots, k$; let $\mathcal{L}(C) = \{S_1, \ldots, S_k\}$ and $T = \{s, t\}$.*

Then Best-of-Many Christofides with lonely edge deletion (Algorithm 15.18) computes a solution of cost at most $\frac{26}{17}c(x^*)$.

Proof. Let $0 \leq \beta \leq \frac{1}{2}$ be a constant that we choose later. Similar to (15.3), let

$$\begin{aligned}\bar{y}^S &:= \beta x^* + (1 - 2\beta)\chi^S + \sum_{C \in \mathcal{N}: S \in \mathcal{L}(C)} \beta(2 - x^*(C))\chi^{S \cap C} \\ &\quad + \sum_{C \in \mathcal{N}: S \notin \mathcal{L}(C)} \max\left\{0,\ 4\beta - 1 - \beta x^*(C)\right\} v^C,\end{aligned}$$

where we now set $v^C := \frac{1}{2 - x^*(C)} \sum_{S \in \mathcal{L}(C)} \mu(S)\chi^{S \cap C}$ for $C \in \mathcal{N}$. Then $v^C(C) = 1$ for all $C \in \mathcal{N}$, and the equivalent of Lemma 15.7 still holds:

For every $S \in \mathcal{S}$ and every $(T \triangle \text{odd}(F_S))$-cut C, we have $\bar{y}^S(C) \geq 1$. (15.10)

To show (15.10), let $S \in \mathcal{S}$ and C be a $(T \triangle \text{odd}(F_S))$-cut. If $C \notin \mathcal{N}$, then $x^*(C) \geq 2$ and hence $\bar{y}^S(C) \geq \beta x^*(C) + (1 - 2\beta)|S \cap C| \geq 2\beta + (1 - 2\beta) = 1$.

If $C \in \mathcal{N}$, then $|F_S \cap C|$ is even by Lemma 14.19. If $F_S \cap C = \emptyset$, then $S \in \mathcal{L}(C)$ and thus $\bar{y}^S(C) \geq \beta x^*(C) + (1 - 2\beta) + \beta(2 - x^*(C))|S \cap C| \geq 1$.

If $|F_S \cap C| \geq 2$, then $S \notin \mathcal{L}(C)$ and thus

$$\begin{aligned}\bar{y}^S(C) &\geq \beta x^*(C) + (1 - 2\beta)|S \cap C| + 4\beta - 1 - \beta x^*(C) \\ &\geq \beta x^*(C) + 2(1 - 2\beta) + 4\beta - 1 - \beta x^*(C) \\ &= 1.\end{aligned}$$

We have shown (15.10); moreover, $\bar{y}^S$ is nonnegative. Hence $\bar{y}^S$ is in the $(T \,\triangle\, \mathrm{odd}(F_S))$-join polyhedron (cf. (2.9)), and by Theorem 2.19, we have $c^S(J_S) \le c^S(\bar{y}^S)$.

Recall that $c^S(e) = c(e)$ unless e is bad. The edges in S are never bad for S, nor are the edges that we deleted in trees that come earlier in the list of Theorem 15.13: This is because for every such edge e in an earlier tree S', there is only one narrow cut C with $e \in C$ and $S' \in \mathcal{L}(C)$, and $S' \in \mathcal{L}(C)$ whenever $S \in \mathcal{L}(C)$. Using the definition of $\bar{y}^S$, we conclude

$$c^S(J_S) \;\le\; c^S(\bar{y}^S) \;=\; c(\bar{y}^S) + \beta\big(c^S(x^*) - c(x^*)\big).$$

Using Lemmas 15.20 and 15.21, the total cost of the solution computed by Best-of-Many Christofides with lonely edge deletion is at most

$$(2-\beta)c(x^*) + \sum_{C\in\mathcal{N}} \sum_{S\in\mathcal{L}(C)} \mu(S)c(S\cap C)\cdot A,$$

where

$$\begin{aligned} A \;=\; & \max_{C\in\mathcal{N}} \Big(-1 + \beta\cdot 2(x^*(C)-1) + \beta(2-x^*(C)) \\ & \qquad\qquad + \max\left\{0,\, 4\beta - 1 - \beta x^*(C)\right\} \tfrac{x^*(C)-1}{2-x^*(C)} \Big) \\ \le\; & \max_{1\le x<2} \left(\beta x - 1 + \tfrac{x-1}{2-x}\max\{0, 4\beta - 1 - \beta x\} \right). \end{aligned}$$

For $\beta = \frac{8}{17}$, we get $A \le 0$ and hence total cost at most $(2-\beta)c(x^*) = \frac{26}{17}c(x^*)$. □

As said, the assumption on μ can be satisfied for Path TSP by Theorem 15.13. Hence, we immediately get:

Corollary 15.23. *The integrality ratio of* (14.1) *is at most* $\frac{26}{17} < 1.530$. □

It turns out that the algorithm can be implemented without actually computing μ (see Exercise 15.6). Although for general T, no μ as in Theorem 15.13 may exist and the above analysis does not work, we will return to Best-of-Many Christofides with lonely edge deletion for general T (and μ) in Section 15.5.

The analysis of Theorem 15.22 was improved by Traub and Vygen [2019b], who showed that Algorithm 15.18 (with $\mathcal{L}(C)$ as in Theorem 15.22) actually computes a solution of cost less than 1.5284 times the LP value for any given Path TSP instance. The main idea in this improved analysis is the following. For the first tree S_1, the vector $\frac{1}{2}x^*$ is a feasible parity correction vector since $|S_1 \cap C| = 1$ for all narrow cuts C. Hence, in the worst case, this tree must be more expensive than the average tree cost $c(x^*)$. Then, using a different parity correction vector for every tree – where for the earlier trees, the contribution of x^* to the parity correction vector is higher, and the contribution of the tree itself is lower than for later trees – yields a better bound on the integrality ratio. Traub and Vygen [2019b] conjectured that one can obtain the ratio 1.5273 with their approach, which was confirmed by Zhong [2020]:

Theorem 15.24 (Traub and Vygen [2019b], Zhong [2020]). *The integrality ratio of* (14.1) *is at most* 1.5273.

This is probably not the final word. The following conjecture is natural.

Open Problem 15.25. Prove that the integrality ratio of the LP relaxation (14.1) of PATH TSP is exactly $\frac{3}{2}$.

15.5 A Better Algorithm for *T*-Tours

Traub [2020b] devised an $\frac{11}{7}$-approximation algorithm for the T-TOUR PROBLEM. The algorithm is very simple: Run Best-of-Many Christofides (Algorithm 15.1) and Best-of-Many Christofides with lonely edge deletion (Algorithm 15.18) and output the best resulting T-tour. In contrast to the previous sections, we do not use a specific decomposition into spanning trees.

The idea why this is better than the plain Best-of-Many Christofides is as follows. Sebő's [2013] analysis (Theorem 15.11) is tight only if $I_S = L_S$, but in this case, deleting the lonely edges changes the parity in all the narrow cuts. The narrow cuts with exactly two edges were critical, but these end up with only one edge in the forest and no longer need parity correction.

Theorem 15.26 (Traub [2020b]). *There is a polynomial-time algorithm that computes a T-tour of cost at most $\frac{11}{7}$ times the value of the LP relaxation* (14.2) *for any given instance (G, c, T) of the T-TOUR PROBLEM where G is a complete graph, c satisfies the triangle inequality, and $T \neq \emptyset$.*

Proof. For a spanning tree $S \in \mathcal{S}$, let us again define

$$L_S := \bigcup_{C \in \mathcal{N}:|C \cap S|=1} (C \cap S)$$

as the set of its lonely edges. Let x^* denote an optimum solution to the LP relaxation (14.2). We run two algorithms and output the better result. The first algorithm is Best-of-Many Christofides. The second algorithm is Best-of-Many Christofides with lonely edge deletion with $\mathcal{L}(C) = \{S \in \mathcal{S} : |S \cap C| = 1\}$ (i.e., we delete all lonely edges). Both algorithms can be implemented in polynomial time by Theorem 4.22.

Recalling Lemma 15.8 and plugging in (15.5) and $\beta = \frac{7}{15}$, Best-of-Many Christofides yields a T-tour of cost at most

$$\frac{23}{15}c(x^*) + \sum_{C \in \mathcal{N}} \left(\frac{x^*(C)-1}{2-x^*(C)} \cdot \max\left\{0, \frac{13-7x^*(C)}{15}\right\}\right) \cdot \sum_{S \in \mathcal{L}(C)} \mu(S)c(S \cap C). \quad (15.11)$$

From (15.7), we also know that the cost is at most

$$c(x^*) + \sum_{S \in \mathcal{S}} \mu(S)c(J_S). \quad (15.12)$$

We will derive a third bound from the Best-of-Many Christofides with lonely edge deletion and then take a convex combination of the three bounds.

We will again use Lemma 15.20, but in contrast to Section 15.4, we now delete all lonely edges, we do not have a decomposition of x^* with any special properties, and we will not use v^C at all (because it might have contributions from bad edges). Therefore, we need a different parity correction vector. We use

$$\tilde{y}^S \ := \ \tfrac{2}{5}x^* + \tfrac{1}{5}\chi^S + \tfrac{1}{5}\chi^{I_S \cap F_S} + \sum_{C \in \mathcal{N}: S \in \mathcal{L}(C)} \tfrac{2}{5}(2 - x^*(C))\chi^{S \cap C}.$$

Again, we have to show:

For every $S \in \mathcal{S}$ and every $(T \,\triangle\, \mathrm{odd}(F_S))$-cut C, we have $\tilde{y}^S(C) \geq 1$. (15.13)

To show (15.13), let $S \in \mathcal{S}$, and let C be a $(T \,\triangle\, \mathrm{odd}(F_S))$-cut. If $C \notin \mathcal{N}$, then $x^*(C) \geq 2$ and hence $\tilde{y}^S(C) \geq \frac{2}{5}x^*(C) + \frac{1}{5}|S \cap C| \geq 1$. If $|S \cap C| \geq 3$, then again $\tilde{y}^S(C) \geq \frac{2}{5}x^*(C) + \frac{1}{5}|S \cap C| \geq 1$.

Now let $|S \cap C| \leq 2$ and $C \in \mathcal{N}$. Then C is a T-cut, and $|F_S \cap C|$ is even by Lemma 14.19.

If $|F_S \cap C| = 2$, then $S \cap C$ contains exactly these two edges. Moreover, the T-cut C contains an edge of the T-join I_S, and this edge also belongs to F_S, so $I_S \cap F_S \neq \emptyset$. We have $\tilde{y}^S(C) \geq \frac{2}{5}x^*(C) + \frac{2}{5} + \frac{1}{5} \geq 1$.

The final case is when $|S \cap C| \leq 2$ and $F_S \cap C = \emptyset$. We claim that then $|S \cap C| = 1$ and $S \in \mathcal{L}(C)$, which implies $\tilde{y}^S(C) \geq \frac{2}{5}x^*(C) + \frac{1}{5}|S \cap C| + \frac{2}{5}(2 - x^*(C))|S \cap C| \geq 1$. Suppose $S \notin \mathcal{L}(C)$. This is impossible if $|S \cap C| = 1$ because $F_S \cap C = \emptyset$. If $|S \cap C| = 2$, we deleted both edges in $S \cap C$, and hence $C = \delta(U_1 \cup U_2)$ for disjoint sets U_1 and U_2 with $S \in \mathcal{L}(\delta(U_1))$ and $S \in \mathcal{L}(\delta(U_2))$. This implies that $|U_1 \cap T|$ and $|U_2 \cap T|$ are both odd, which contradicts the fact that $C = \delta(U_1 \cup U_2)$ is a T-cut.

We have shown (15.13); moreover, $\tilde{y}^S$ is nonnegative. Using Lemma 15.21 and $c^S(e) = c(e)$ for all $e \in S$, we have

$$\begin{aligned}
c^S(\tilde{y}^S) &= c(\tilde{y}^S) + \tfrac{2}{5}(c^S(x^*) - c(x^*)) \\
&\le \tfrac{2}{5}c(x^*) + \tfrac{1}{5}c(S) + \tfrac{1}{5}c(I_S \cap F_S) + \sum_{C \in \mathcal{N}: S \in \mathcal{L}(C)} \tfrac{2}{5}(2 - x^*(C))\, c(S \cap C) \\
&\quad + \tfrac{4}{5} \sum_{C \in \mathcal{N}: S \in \mathcal{L}(C)} (x^*(C) - 1)\, c(C \cap S) \\
&= \tfrac{2}{5}c(x^*) + \tfrac{1}{5}c(S) + \tfrac{1}{5}c(I_S \cap F_S) + \tfrac{4}{5}c(S \setminus F_S) \\
&\quad - \sum_{C \in \mathcal{N}: S \in \mathcal{L}(C)} \tfrac{2}{5}(2 - x^*(C))\, c(S \cap C).
\end{aligned}$$

Hence, by Lemma 15.20 and using $L_S = S \setminus F_S = I_S \setminus F_S$, the result of Best-of-Many Christofides with lonely edge deletion costs at most

$$\begin{aligned}
& c(x^*) + \sum_{S \in \mathcal{S}} \mu(S) \left(c^S(\tilde{y}^S) - c(L_S)\right) \\
&\le \tfrac{7}{5}c(x^*) + \sum_{S \in \mathcal{S}} \mu(S) \Bigg(\tfrac{1}{5}c(S) + \tfrac{1}{5}c(I_S) - \tfrac{1}{5}c(I_S \setminus F_S) + \tfrac{4}{5}c(I_S \setminus F_S) \\
&\qquad - \sum_{C \in \mathcal{N}: S \in \mathcal{L}(C)} \tfrac{2}{5}(2 - x^*(C))c(C \cap S) - c(L_S)\Bigg) \\
&= \tfrac{8}{5}c(x^*) + \tfrac{1}{5} \sum_{S \in \mathcal{S}} \mu(S)c(I_S) - \tfrac{2}{5} \sum_{S \in \mathcal{S}} \mu(S)c(L_S) \\
&\quad - \tfrac{2}{5} \sum_{C \in \mathcal{N}} (2 - x^*(C)) \sum_{S \in \mathcal{L}(C)} \mu(S)c(S \cap C).
\end{aligned} \tag{15.14}$$

Taking $\frac{15}{21}$ times (15.11), $\frac{1}{21}$ times (15.12), and $\frac{5}{21}$ times (15.14) (and using (15.6)) yields the upper bound

$$\tfrac{32}{21}c(x^*) + \tfrac{1}{21} \sum_{S \in \mathcal{S}} \mu(S)c(S) + \sum_{S \in \mathcal{S}} \mu(S)c(L_S) \cdot A,$$

where

$$\begin{aligned}
A &= \max_{C \in \mathcal{N}} \left(\tfrac{x^*(C)-1}{2-x^*(C)} \max\left\{0, \tfrac{13}{21} - \tfrac{7}{21}x^*(C)\right\} - \tfrac{2}{21}(2 - x^*(C))\right) - \tfrac{2}{21} \\
&\le \max_{x<2} \left(\tfrac{x-1}{2-x} \max\left\{0, \tfrac{13}{21} - \tfrac{7}{21}x\right\} - \tfrac{2}{21}(2 - x)\right) - \tfrac{2}{21} \\
&= 0.
\end{aligned}$$

This yields the upper bound $\frac{32}{21}c(x^*) + \frac{1}{21}c(x^*) + 0 = \frac{11}{7}c(x^*)$ as asserted. □

This is the best-known ratio today. Given the progress on Path TSP (Chapter 16), the following question is natural:

Open Problem 15.27. Is there a $\frac{3}{2}$-approximation algorithm for the T-tour problem?

Exercises

15.1 Let x be a vector in the connector polyhedron of a graph $G = (V, E)$, and let $L \in \mathbb{Z}_{>0}$ such that there exists a distribution μ of spanning trees with $x \geq \sum_{S \in \mathcal{S}} \mu(S)\chi^S$ such that $L\mu(S)$ is integral for all $S \in \mathcal{S}$. Let $U := E \times \{1, \ldots, L\}$, and consider the subsets

$$\mathcal{M}_1 := \big\{W \subseteq U : (V, \{e \in E : (e,i) \in W\}) \text{ is a forest for all } i = 1, \ldots, L\big\}$$

and

$$\mathcal{M}_2 := \big\{W \subseteq U : |\{i : (e,i) \in W\}| \leq Lx_e \text{ for all } e \in E\big\}.$$

(a) Show that $(U, \mathcal{M}_1)$ and $(U, \mathcal{M}_2)$ are matroids.
(b) Show that ($W \in \mathcal{M}_1 \cap \mathcal{M}_2$ and $|W| = L(n-1)$) if and only if $x \geq \sum_{S \in \mathcal{S}} \mu(S)\chi^S$, where $\mu(S) := \frac{1}{L}|\{i : (e,i) \in W \text{ for all } e \in S\}|$.
(c) Conclude that for any vector x in the spanning tree polytope and any $\varepsilon > 0$, we can find a distribution μ of spanning trees with $\sum_{S \in \mathcal{S}} \mu(S)\chi^S \leq (1+\varepsilon)x$ in polynomial time using matroid intersection (Theorem 13.16).

Note: The $1+\varepsilon$ factor can be removed (i.e., the limitation that L is polynomially bounded can be overcome). See Corollary 40.4a in Schrijver [2003].

15.2 Show that Lemma 15.9 does not hold for general T.

15.3 This exercise analyzes Best-of-Many Christofides (for T-tours) slightly differently. Let $0 \leq \beta \leq \frac{1}{2}$. Let $z^S \in \mathbb{R}^E_{\geq 0}$ and $y^S = \beta x^* + (1-2\beta)\chi^{J_S} + z^S$ for $S \in \mathcal{S}$.

(a) Show that y^S is a parity correction vector for $S \in \mathcal{S}$ if $z^S(C) \geq \beta(2 - x^*(C))$ for all narrow cuts C with $|C \cap S|$ even.
(b) Suppose $\sum_{S \in \mathcal{S}} \mu(S) z^S \leq (1-2\beta) \sum_{S \in \mathcal{S}} \mu(S)\chi^{I_S}$. Show that then we have a $(2-\beta)$-approximation algorithm for the T-Tour Problem.
(c) Show that one can choose $\beta = \frac{2}{5}$ and

$$z^S = \tfrac{1}{10}\chi^{I_S} + \sum_{C \in \mathcal{N}: |S \cap C| \text{ even}} \max\left\{0, \tfrac{7-4x^*(C)}{10}\right\} v^C,$$

where v^C is defined as in (15.5).
Note: This yields another proof of Theorem 15.11.

(d) Show that for Path TSP ($T = \{s, t\}$), one can obtain an approximation ratio better than $\frac{8}{5}$ with the distribution from Theorem 15.13.

(Vygen [2016], Gottschalk and Vygen [2018])

15.4 Show that for every extreme point of the Path TSP LP (14.1), there always exists a distribution like in Theorem 15.13 with fewer than $2n$ trees. Use Exercise 14.3.

15.5 Let x^* be an optimum solution to the Path TSP LP (14.1). Let $1 = \xi_1 < \cdots < \xi_k < 2$ be the distinct values $x^*(C)$ of the narrow cuts $C \in \mathcal{N}$.

(a) Show that the separation problem for the convex hull of incidence vectors of spanning trees (V, S) with $|S \cap C| = 1$ for all narrow cuts C with $x^*(C) \le \xi_i$ can be solved in polynomial time for all $i = 1, \ldots, k$.

(b) Use (a) to devise a polynomial-time algorithm to compute an optimum LP solution x^* and a distribution as in Theorem 15.13 (without using the proof of this theorem).
Hint: Set up a linear program and use Theorems 2.10 and 4.22.

(K. Pashkovich; see Gottschalk and Vygen [2018])

15.6 Let x^* be an optimum solution to the Path TSP LP (14.1). Let $1 = \xi_1 < \cdots < \xi_k < 2$ be the distinct values $x^*(C)$ of the narrow cuts $C \in \mathcal{N}$. For $i = 1, \ldots, k$, let S_i be a minimum-cost spanning tree such that $|S_i \cap C| = 1$ for all narrow cuts C with $x^*(C) \le \xi_i$. Computing such spanning trees S_i obviously reduces to the Minimum Spanning Tree problem. For a narrow cut C with $x^*(C) = \xi_i$, let $\mathcal{L}(C) = \{S_i, \ldots, S_k\}$. Show that continuing Best-of-Many Christofides with lonely edge deletion (Algorithm 15.18) with the trees $S_1, \ldots, S_k$ yields a $\frac{26}{17}$-approximation algorithm.
(Sebő and van Zuylen [2019])

15.7 Show that if there is no cut C with $\frac{3}{2} < x^*(C) < 2$ for a solution x^* to the Path TSP LP (14.1), then an $\{s, t\}$-tour of cost at most $\frac{3}{2}c(x^*)$ can be computed in polynomial time.
Hint: Consider the analysis in the proof of Theorem 15.22.
(Sebő and van Zuylen [2019])

16

Path TSP by Dynamic Programming

Traub and Vygen [2019a] used recursive dynamic programming to obtain a $(\frac{3}{2} + \varepsilon)$-approximation algorithm for PATH TSP for any $\varepsilon > 0$. This approach was then improved and simplified by Zenklusen [2019], who obtained a $\frac{3}{2}$-approximation for PATH TSP. After discussing the dynamic programming approach in a simple context in Section 16.1, we present Zenklusen's algorithm in Section 16.2.

In Sections 16.3–16.7, we present a black-box reduction from PATH TSP to SYMMETRIC TSP, similar to the one proposed by Traub, Vygen, and Zenklusen [2022]. This shows that the former is not much harder to approximate than the latter. This implies the currently best-known approximation guarantees for PATH TSP and the special case GRAPH PATH TSP. Our new proof actually yields the same result even for a more general problem, which we call MULTI-PATH TSP.

Although the LP relaxation is still used, none of the dynamic programming approaches presented in this chapter imply upper bounds on the integrality ratio.

16.1 Reducing Path TSP to Instances with Near Endpoints

In this section, we first explain the dynamic programming idea in a simpler setting. Instances of PATH TSP where the distance from s to t is small are similar to the corresponding SYMMETRIC TSP instance. One could therefore think that the larger the distance from s to t is, the more difficult such an instance would be, and indeed the instances with the worst-known integrality ratio (cf. Figure 14.2) do have a large distance from s to t. However, Traub [2017] (inspired by the work of Blum et al. [2007]) showed how to use dynamic programming to deal with such instances.

Theorem 16.1 (Traub [2017]). *Let $\varepsilon > 0$ and $\alpha > 1$ be constants. If there is an α-approximation algorithm for* Path TSP *restricted to instances (G, c, s, t) in which $\mathrm{dist}_{(G,c)}(s,t) \leq (\frac{1}{3}+\varepsilon)\,\mathrm{OPT}$, then there is an α-approximation algorithm for general* Path TSP. *Here* OPT *denotes the minimum cost of an $\{s,t\}$-tour in G.*

Proof. Consider an instance (G, c, s, t) in which $\mathrm{dist}_{(G,c)}(s,t) > (\frac{1}{3}+\varepsilon)\,\mathrm{OPT}$. Let $G = (V, E)$ and

$$f(v) \;:=\; \min\{\mathrm{dist}_{(G,c)}(s,v), \mathrm{dist}_{(G,c)}(s,t)\}$$

for $v \in V$, and order the vertices so that $v_1 = s$, $v_n = t$, and $f(v_1) \leq f(v_2) \leq \cdots \leq f(v_n)$. Consider the chain of cuts $\delta(U_j)$ for $j = 1, \ldots, n-1$, where $U_j = \{v_1, \ldots, v_j\}$, and set $y(U_j) := f(v_{j+1}) - f(v_j)$. Note that

$$\sum_{j=1}^{n-1} y(U_j) \;=\; \sum_{j=1}^{n-1}\bigl(f(v_{j+1}) - f(v_j)\bigr) \;=\; f(t) - f(s) \;=\; \mathrm{dist}_{(G,c)}(s,t)$$

and

$$\begin{aligned} c(e) \;&\geq\; |\mathrm{dist}_{(G,c)}(s,v) - \mathrm{dist}_{(G,c)}(s,w)| \\ &\geq\; |f(v) - f(w)| \\ &=\; \sum_{j:e\in\delta(U_j)} y(U_j) \end{aligned} \tag{16.1}$$

for all $e = \{v, w\} \in E$. (This implies that y can be interpreted as an optimum dual solution to a shortest path LP; see Exercise 16.1.)

By Lemma 2.20, every $\{s,t\}$-tour F contains an odd number of edges in each of the cuts $\delta(U_j)$. Our goal is to show that some of these cuts contain only one edge of F and to "guess" those edges.

Let F^* be an optimum $\{s,t\}$-tour, and let L^* be the set of edges $e \in F^*$ that are the only edge in one of these cuts – that is,

$$L^* \;=\; \bigl\{e \in F^* : F^* \cap \delta(U_j) = \{e\} \text{ for some } j \in \{1, \ldots, n-1\}\bigr\}.$$

See Figure 16.1 for an example. Note that no edge can be the only edge in two of these cuts, say $\delta(U_i)$ and $\delta(U_j)$ for $i < j$, because otherwise $F^* \cap \delta(\{v_{i+1}, \ldots, v_j\}) = \emptyset$, contradicting the fact that (V, F^*) is connected. We claim that

$$c(L^*) \;\geq\; \tfrac{3}{2}\varepsilon \cdot \mathrm{OPT}. \tag{16.2}$$

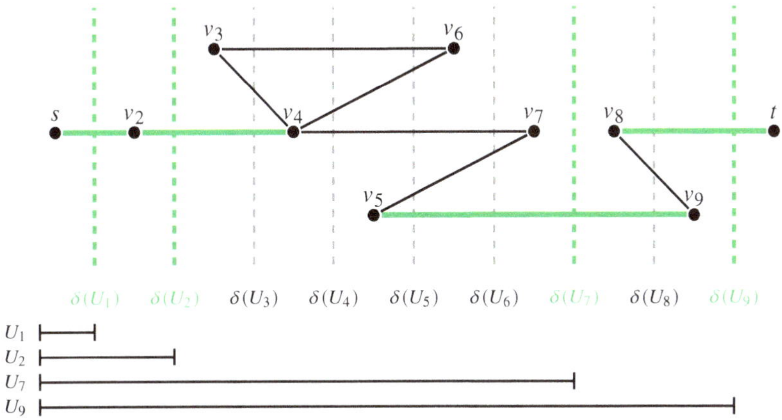

Figure 16.1 The ten vertices of a PATH TSP instance, ordered by the distance from s as in the proof of Theorem 16.1. Not all edges of the instance are shown. The dashed vertical lines indicate the cuts $\delta(U_j)$ for $j = 1, \ldots, n-1$. The solid lines show the optimum $\{s,t\}$-tour F^*. The thick green edges are those in L^*. The $\{s,t\}$-tour F^* corresponds to the path with vertices $(1, s, v_2)$, $(2, v_2, v_4)$, $(7, v_5, v_9)$, $(9, v_8, t)$ in the digraph (D, A). This picture is taken (in modified form) from Traub and Vygen [2023], with permission from SIAM.

To show this, first note that (16.1) implies that for any edge set F,

$$c(F) \geq \sum_{j=1}^{n-1} |F \cap \delta(U_j)| \cdot y(U_j). \tag{16.3}$$

Applying the inequality (16.3) first to F^* and then to L^*, we get

$$\begin{aligned} \mathrm{OPT} = c(F^*) &\geq \sum_{j=1}^{n-1} |F^* \cap \delta(U_j)| \cdot y(U_j) \\ &\geq 3\sum_{j=1}^{n-1} y(U_j) - 2 \sum_{j:|F^*\cap\delta(U_j)|=1} y(U_j) \\ &\geq 3\sum_{j=1}^{n-1} y(U_j) - 2c(L^*). \end{aligned}$$

Combining this with $\sum_{j=1}^{n-1} y(U_j) = \mathrm{dist}_{(G,c)}(s,t) > (\frac{1}{3} + \varepsilon)\mathrm{OPT}$ yields (16.2).

Our goal is to "guess" the cuts $\delta(U_j)$ that contain only one edge of F^* and these edges (i.e., the set L^*). For simplicity, assume that F^* has only one edge incident to s and only one edge incident to t. This is no loss of generality, either by working in the metric closure or by introducing new vertices s' and t' and edges $\{s', s\}$ and $\{t, t'\}$ of zero cost and asking for an $\{s', t'\}$-tour.

We now construct an acyclic digraph to set up our dynamic program. Let

$$D = \{(j,u,v) : j \in \{1,\ldots,n-1\},\, u \in U_j,\, v \in V \setminus U_j,\, \{u,v\} \in E\}$$

represent the candidates of the dynamic program, each consisting of a set U_j and an edge $e = \{u,v\} \in \delta(U_j)$ with $u \in U_j$ and $v \notin U_j$. Our acyclic digraph has vertex set D and weights both on vertices and arcs. The weight of a vertex (j,u,v) is $d((j,u,v)) := c(\{u,v\})$.

For two candidates $(j',u',v'), (j,u,v) \in D$, we introduce an arc a from (j',u',v') to (j,u,v) if $j' < j$ and $v', u \in U_j \setminus U_{j'}$. We consider the Path TSP instance corresponding to this arc a, which has vertex set $U_j \setminus U_{j'}$ and endpoints $v'$and u (so we ask for a minimum-cost $\{v', u\}$-tour in $G[U_j \setminus U_{j'}]$). We will call a β-approximation algorithm (for some $\beta > \alpha$) on this instance to compute a $\{v', u\}$-tour F_a with vertex set $U_j \setminus U_{j'}$ and set $d(a) := c(F_a)$.

If we denote by A the set of those arcs, (D, A) is an acyclic digraph. The total weight of a path in (D, A) is the total weight of its vertices and arcs.

By construction, any path P of total weight $d(P)$ from $D_s = \{(1,s,v) : \{s,v\} \in \delta_G(s)\}$ to $D_t = \{(n-1,v,t) : \{v,t\} \in \delta_G(t)\}$ corresponds to an $\{s,t\}$-tour of cost $d(P)$. Conversely, F^* corresponds to a path P^* from D_s to D_t of total weight $d(P^*) \le c(L^*) + \beta c(F^* \setminus L^*)$. See Figure 16.1 for an example. Hence, computing a minimum-weight path from D_s to D_t yields an $\{s,t\}$-tour of cost at most

$$\begin{aligned} c(L^*) + \beta c(F^* \setminus L^*) &= \beta \mathrm{OPT} - (\beta - 1) c(L^*) \\ &\le \left(\beta - (\beta - 1)\tfrac{3}{2}\varepsilon\right) \cdot \mathrm{OPT} \\ &\le \left(\beta - (\alpha - 1)\tfrac{3}{2}\varepsilon\right) \cdot \mathrm{OPT}, \end{aligned}$$

where we used (16.2) in the first inequality.

Constructing this digraph with $O(n^3)$ nodes and $O(n^6)$ arcs can be done in polynomial time (by calling the β-approximation algorithm for every arc). Computing a minimum-weight path from D_s to D_t is a simple dynamic program: The minimum total weight of a path from D_s to a vertex $(j,u,v) \in D$ is $\ell(j,u,v) := d((j,u,v))$ if $j = 1$ and

$$\begin{aligned} \ell(j,u,v) := \; & d((j,u,v)) \\ & + \min \Big\{ \ell(j',u',v') + d\left((j',u',v'),(j,u,v)\right) : \\ & \qquad (j',u',v') \in D,\ j' < j, \text{ and } v', u \in U_j \setminus U_{j'} \Big\} \end{aligned}$$

if $j > 1$.

For a given instance (G, c, s, t), we do not know whether we are in the case $\mathrm{dist}_{(G,c)}(s,t) \le (\frac{1}{3} + \varepsilon)\,\mathrm{OPT}$. However, we can simply run both algorithms: the dynamic programming algorithm and the given algorithm that does the job if $\mathrm{dist}_{(G,c)}(s,t) \le (\frac{1}{3} + \varepsilon)\,\mathrm{OPT}$. We return the better of the two resulting solutions. We conclude that an α-approximation algorithm for PATH TSP instances (G, c, s, t) with $\mathrm{dist}_{(G,c)}(s,t) \le (\frac{1}{3} + \varepsilon)\,\mathrm{OPT}$ plus a β-approximation algorithm for general PATH TSP imply a $\big(\max\{\alpha,\ \beta-(\alpha-1)\frac{3}{2}\varepsilon\}\big)$-approximation algorithm for general PATH TSP.

Starting with $\beta = 2$ (we can take the double tree algorithm; cf. Proposition 14.5) and applying this "booster" statement $\left\lceil \frac{2-\alpha}{(\alpha-1)\frac{3}{2}\varepsilon} \right\rceil$ times completes the proof. Note that the running time increases with each application of the booster by a factor $O(n^6)$, but we apply it only a constant number of times. □

The theorem also works for GRAPH PATH TSP (here the resulting subinstances are also instances of GRAPH PATH TSP), for which it was used by Traub and Vygen [2023] to beat the integrality ratio (i.e., to obtain an approximation ratio less than $\frac{3}{2}$). The dynamic programming technique was developed further by Traub and Vygen [2019a], Zenklusen [2019], and Traub, Vygen, and Zenklusen [2022].

Instead of computing a tour directly in the dynamic program, Traub and Vygen [2019a] computed a spanning tree and a parity correction vector. Their algorithm starts with a solution x_1 to the LP (14.1), which already yields a cheap parity correction vector $\frac{1}{2}x_1$ (for an arbitrary spanning tree) except for the deficiency on the narrow cuts (cf. Definition 14.17). The new dynamic program now guesses which of the narrow cuts contain only one edge of F^* (and guesses these edges and includes them in the spanning tree) and strengthens the LP relaxation of the subinstance in between two narrow cuts by requiring $x(\delta(U)) \ge 3$ for each narrow cut $\delta(U)$ in between. Combining the LP solutions of the subinstances with the incidence vector of the guessed edges yields an LP solution x_2, and $\frac{1}{4}(x_1 + x_2)$ is now good on all original narrow cuts. However, new narrow cuts show up in x_2. By iterating this process and giving x_i weight approximately 2^{-i-1}, one approaches a valid parity correction vector of cost at most $\frac{1}{2}\mathrm{OPT}$ and a spanning tree of cost at most OPT. One will not always make the correct guesses, but the output of the (recursive) dynamic program will not be worse than with correct guesses. In this way, Traub and Vygen [2019a] obtained a $(\frac{3}{2} + \varepsilon)$-approximation algorithm for PATH TSP for any constant $\varepsilon > 0$.

We omit the details because this approach was simplified and improved by Zenklusen [2019]. This is the subject of the next section.

16.2 Zenklusen's $\frac{3}{2}$-Approximation Algorithm for Path TSP

Zenklusen's [2019] Path TSP algorithm uses a similar dynamic program to the one proposed by Traub and Vygen [2019a], but it does not require recursion. Throughout this section, we will work with an instance (G, c, s, t) where G is a complete graph and c satisfies the triangle inequality. This is no restriction due to Proposition 14.8.

Lemma 16.2. *Let x^* be an optimum solution to* (14.1)*, and let x' be a feasible solution to* (14.1) *such that for every narrow cut C of $\frac{1}{2}x^* + \frac{1}{2}x'$, we have $x'(C) = 1$. Then we can compute an $\{s,t\}$-tour of cost at most $\frac{3}{2}c(x')$ in polynomial time.*

Proof. By Proposition 14.10, the vector x' is a convex combination of incidence vectors of (edge sets of) spanning trees. Each of these intersects every narrow cut of $\frac{1}{2}x^* + \frac{1}{2}x'$ in exactly one edge. For every such tree (V, S), the narrow cuts of $\frac{1}{2}x^* + \frac{1}{2}x'$ are no $(\text{odd}(S) \triangle \{s,t\})$-cuts (by Lemma 14.19). Hence, $\frac{1}{4}x^* + \frac{1}{4}x'$ is a feasible parity correction vector for these trees.

Let (V, S) be a cheapest spanning tree that has exactly one edge in each narrow cut of $\frac{1}{2}x^* + \frac{1}{2}x'$ (such a tree can be computed easily because the narrow cuts form a chain by Lemma 14.18). We have $c(S) \leq c(x')$. By Theorem 2.19, applying parity correction to S (i.e., adding a cheapest $(\text{odd}(S) \triangle \{s,t\})$-join) yields an $\{s,t\}$-tour of cost at most $\frac{1}{4}c(x^*) + \frac{5}{4}c(x') \leq \frac{3}{2}c(x')$. □

We will continue to denote by x^* an optimum solution to (14.1) throughout this section. We will show how to compute a vector x' as in Lemma 16.2 with $c(x') \leq \text{OPT}$. This immediately implies a $\frac{3}{2}$-approximation for Path TSP.

Now, let x be any feasible solution to (14.1), and let C be a narrow cut of $\frac{1}{2}x^* + \frac{1}{2}x$. Since the vector $\frac{1}{2}x^* + \frac{1}{2}x$ is a feasible solution to (14.1), C must be an s-t-cut. Moreover, because C is narrow, $\frac{1}{2}x^*(C) + \frac{1}{2}x(C) < 2$, and hence $x^*(C) < 3$ and $x(C) < 3$. Define

$$\mathcal{Q} := \{Q : Q \text{ is an } s\text{-}t\text{-cut with } x^*(Q) < 3\}.$$

By these observations, it suffices to compute a solution x' to (14.1) with $c(x') \leq \text{OPT}$ such that for every cut $Q \in \mathcal{Q}$, we have either $x'(Q) \geq 3$ or $x'(Q) = 1$.

The cuts in $\mathcal{Q}$ are the s-t-cuts Q with $(x^* + \chi^{\{s,t\}})(Q) < 4$. Because $(x^* + \chi^{\{s,t\}})(Q) \geq 2$ for all cuts, we have $|\mathcal{Q}| \leq n^4$ due to Theorem 4.25. Furthermore, the set $\mathcal{Q}$ can be computed deterministically in polynomial time by Theorem 4.28.

Definition 16.3 ($\mathcal{Q}$-good solution). A *$\mathcal{Q}$-good solution* is a solution x to the LP (14.1) such that there exists an $\mathcal{L} \subseteq \mathcal{Q}$ with the following two properties:

- For every cut $Q \in \mathcal{L}$, we have $x(Q) = 1$ and x restricted to Q is integral.
- For every cut $Q \in \mathcal{Q} \setminus \mathcal{L}$, we have $x(Q) \geq 3$.

We call $\mathcal{L}$ the set of *lonely cuts* of the LP solution x. Moreover, we call an edge $e \in Q \in \mathcal{L}$ with $x_e = 1$ a *lonely edge*.

The set $\mathcal{L}$ always contains $\delta(s)$ and $\delta(t)$. By Lemma 14.18, the lonely cuts form a chain. The reason for requiring integrality of x on the cuts in $\mathcal{L}$ is that this will allow us to compute a minimum-cost $\mathcal{Q}$-good solution in polynomial time.

Lemma 16.4 (Zenklusen [2019]). *A minimum-cost $\mathcal{Q}$-good solution can be computed in polynomial time.*

Before proving this lemma, we show that it implies the following theorem.

Theorem 16.5 (Zenklusen [2019]). *There exists a $\frac{3}{2}$-approximation algorithm for* Path TSP.

Proof. First compute an optimum solution x^* to (14.1) and compute $\mathcal{Q}$, using Theorem 4.28. Now compute a minimum-cost $\mathcal{Q}$-good solution x' (by Lemma 16.4). We have $x'(Q) = 1$ for every cut Q with $\frac{1}{2}x^*(Q) + \frac{1}{2}x'(Q) < 2$. We need to show $c(x') \leq \mathrm{OPT}$; the result then follows from Lemma 16.2.

Fix an optimum $\{s,t\}$-tour F. Let $\mathcal{L} = \{Q \in \mathcal{Q} : |F \cap Q| = 1\}$. Then $|F \cap Q| \geq 3$ for all $Q \in \mathcal{Q} \setminus \mathcal{L}$ because all cuts in $\mathcal{Q}$ are s-t-cuts and F is the footprint of a walk from s to t. Thus, the incidence vector of F is a $\mathcal{Q}$-good solution of cost OPT. In particular, our minimum-cost $\mathcal{Q}$-good solution x' has cost at most OPT. □

It remains to prove Lemma 16.4. If we fix $\mathcal{L}$ and the set of lonely edges, it is easy to compute a cheapest $\mathcal{Q}$-good solution with these given lonely cuts and edges: We can simply add constraints $x(Q) = 1$ for all $Q \in \mathcal{L}$, $x(Q) \geq 3$ for all $Q \in \mathcal{Q} \setminus \mathcal{L}$, and $x_e = 1$ for all lonely edges e to the LP (14.1) and compute an optimum solution to this strengthened LP. Since $|\mathcal{Q}| \leq n^4$, this is possible in polynomial time. To obtain a minimum-cost $\mathcal{Q}$-good solution x' even without knowing the lonely cuts and edges, we will use dynamic programming (similar to Section 16.1) to "guess" them.

As a first step towards the dynamic program, we prove that for fixed lonely cuts and lonely edges, we can partition the instance at the lonely cuts into several independent subinstances.

For a set $W \subseteq V$ and $s', t' \in W$ with $s' \neq t'$, let $\mathrm{LP}[W, s', t']$ denote the following linear program:

$$
\begin{array}{lrcll}
\min & c(x) & & & \\
\text{subject to} & x(\delta(U)) & \geq & 2 & (\emptyset \neq U \subsetneq W,\ |U \cap \{s', t'\}| \text{ even}) \\
 & x(\delta(U)) & \geq & 1 & (\emptyset \neq U \subsetneq W,\ |U \cap \{s', t'\}| \text{ odd}) \\
 & x(\delta(v)) & = & 2 & (v \in W \setminus \{s', t'\}) \\
 & x(\delta(v)) & = & 1 & (v \in \{s', t'\}) \\
 & x_e & \geq & 0 & (e \in E) \\
 & x_e & = & 0 & (e \in E \setminus E[W]).
\end{array}
$$

Note that $\mathrm{LP}[V, s, t]$ is the Path TSP LP (14.1).

Lemma 16.6. *A vector x with support in $E[W]$ is a feasible solution to $\mathrm{LP}[W, s', t']$ if and only if it is a convex combination of incidence vectors of trees spanning W and fulfills the degree constraints $x(\delta(v)) = 2$ for $v \in W \setminus \{s', t'\}$ and $x(\delta(s')) = x(\delta(t')) = 1$.*

Proof. For a vector x in the spanning tree polytope (2.8) of $G[W]$ that satisfies the degree constraints, we have $x(\delta(U)) \geq 1$ for all nonempty proper subsets U of W and $x(\delta(U)) = \sum_{v \in U} x(\delta(v)) - 2x(E[U]) = 2|U| - 2x(E[U]) \geq 2$ for all $\emptyset \neq U \subseteq W \setminus \{s', t'\}$. Conversely, every feasible solution to $\mathrm{LP}[W, s', t']$ is a convex combination of incidence vectors of trees spanning W by Proposition 14.10. □

For a vertex v, denote by $\mathrm{LP}[\{v\}, v, v]$ the linear program $\min\{0 : x_e = 0\ (e \in E)\}$ and by $\mathrm{LP}[W, v, v]$ with $W \neq \{v\}$ a linear program that has no feasible solution.

We now show how LP solutions can be decomposed into solutions for independent subinstances (see Figure 16.2):

Lemma 16.7. *Let $\{s\} = L_1 \subsetneq L_2 \subsetneq \ldots \subsetneq L_k = V \setminus \{t\}$. Let $w_i, v_{i+1} \in L_{i+1} \setminus L_i$ for $i = 1, \ldots, k-1$, and let $v_1 := s$ and $w_k := t$. Consider the linear program*

$$
\begin{array}{ll}
\min c(x) & \\
\textit{subject to} & x \textit{ feasible solution to } \mathrm{LP}[V, s, t] \\
 & x(\delta(L_i)) = 1 \textit{ for } i = 1, \ldots, k \\
 & x_{\{v_i, w_i\}} = 1 \textit{ for } i = 1, \ldots, k.
\end{array}
\tag{16.4}
$$

Then any feasible solution x to (16.4) can be written as $x = \sum_{i=1}^{k} \chi^{\{v_i, w_i\}} + \sum_{i=1}^{k-1} x^i$, where x^i is a feasible solution to $\mathrm{LP}[L_{i+1} \setminus L_i, w_i, v_{i+1}]$. Moreover,

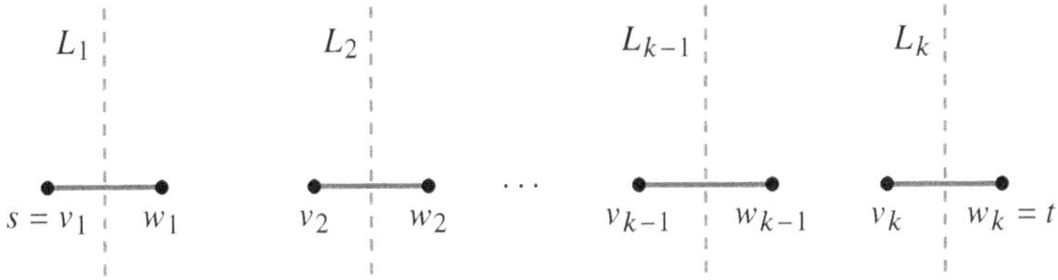

Figure 16.2 Illustration of Lemma 16.7. The dashed lines show the cuts $\delta(L_i)$ for $i = 1, \dots, k$, where the sets L_i are the sets left of the dashed lines.

for any feasible solutions x^i to $\text{LP}[L_{i+1} \setminus L_i, w_i, v_{i+1}]$ *for $i = 1, \dots, k-1$, the vector $x = \sum_{i=1}^{k} \chi^{\{v_i,w_i\}} + \sum_{i=1}^{k-1} x^i$ is a feasible solution to* (16.4).

Proof. Let x be a feasible solution to (16.4). By Lemma 16.6, x is a convex combination of incidence vectors of spanning trees. Because $x(\delta(L_i)) = x_{\{v_i,w_i\}} = 1$, all these spanning trees intersect the cut $\delta(L_i)$ in exactly the edge $\{v_i, w_i\}$. If (V, S) is a spanning tree with $S \cap \delta(L_i) = \{\{v_i, w_i\}\}$ for all $i = 1, \dots, k$, then $(L_{i+1} \setminus L_i, S[L_{i+1} \setminus L_i])$ is a spanning tree for all $i = 1, \dots, k-1$. Thus, the vector x^i that equals x for edges in $E[L_{i+1} \setminus L_i]$ and is zero for all other edges, is a convex combination of incidence vectors of trees spanning $L_{i+1} \setminus L_i$. We have $x = \sum_{i=1}^{k} \chi^{\{v_i,w_i\}} + \sum_{i=1}^{k-1} x^i$. If $w_i = v_{i+1}$, then this vertex is the only one in $L_{i+1} \setminus L_i$ because otherwise $x(\delta((L_{i+1} \setminus L_i) \setminus \{w_i\})) = 0$ would contradict a constraint of (16.4). If $w_i \neq v_{i+1}$, the degree constraints for x imply $x^i(\delta(v)) = 2$ for $v \in (L_{i+1} \setminus L_i) \setminus \{w_i, v_{i+1}\}$ and $x^i(\delta(w_i)) = x^i(\delta(v_{i+1})) = 1$. Thus by Lemma 16.6, the vector x^i is a feasible solution to $\text{LP}[L_{i+1} \setminus L_i, w_i, v_{i+1}]$.

Now let x^i $(i = 1, \dots, k-1)$ be feasible solutions to the linear programs $\text{LP}[L_{i+1} \setminus L_i, w_i, v_{i+1}]$ and $x = \sum_{i=1}^{k} \chi^{\{v_i,w_i\}} + \sum_{i=1}^{k-1} x^i$. Then $w_i = v_{i+1}$ only if $|L_{i+1} \setminus L_i| = 1$. Since the support of a vector x^i is contained in $E[L_{i+1} \setminus L_i]$, we have $x(\delta(L_i)) = 1$ and $x_{\{v_i,w_i\}} = 1$ for all $i = 1, \dots, k$. By Lemma 16.6, for every $i = 1, \dots, k-1$, the vector x^i is a convex combination of incidence vectors of trees spanning $L_{i+1} \setminus L_i$. Since the union of the edge sets of spanning trees for all sets $L_{i+1} \setminus L_i$ $(i = 1, \dots, k-1)$ together with the edges $\{v_i, w_i\}$ $(i = 1, \dots, k)$ is a spanning tree of G, the vector x is a convex combination of incidence vectors of spanning trees. Moreover, the degree constraints of $\text{LP}[L_{i+1} \setminus L_i, w_i, v_{i+1}]$ imply the degree constraints of $\text{LP}[V, s, t]$. Hence, by Lemma 16.6, x is a feasible solution to $\text{LP}[V, s, t]$. □

In order to prove that also any optimum solution to (16.4) with additional constraints $x(Q) \geq 3$ for all $Q \in \mathcal{Q} \setminus \{\delta(L_1), \dots, \delta(L_k)\}$ can be obtained as a combination of (independently computed) solutions for LPs for all the

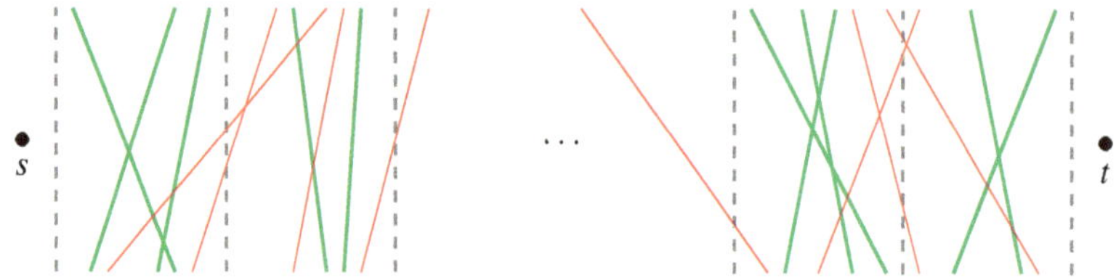

Figure 16.3 Illustration of Lemma 16.8. The dashed lines show the cuts in $\mathcal{L}$, which form a chain. If $\mathcal{Q} \setminus \mathcal{L}$ consists of the cuts sketched by the solid lines, $\mathcal{B}$ contains the green cuts, and $\mathcal{Q} \setminus (\mathcal{B} \cup \mathcal{L})$ contains the red cuts (which can be ignored if the cuts in $\mathcal{L}$ are lonely).

subinstances, we need to get rid of such constraints that affect several subinstances. See Figure 16.3. The following lemma shows that these can be simply omitted, since they are implied by other constraints:

Lemma 16.8 (Traub and Vygen [2019a]). *Let $\mathcal{L} \subseteq \mathcal{Q}$ such that $\mathcal{L}$ forms a chain. Let*

$$\mathcal{B} := \{Q \in \mathcal{Q} \setminus \mathcal{L} : \mathcal{L} \,\dot{\cup}\, \{Q\} \text{ forms a chain}\}.$$

Then any feasible solution x to (14.1) *with $x(Q) \geq 3$ for all $Q \in \mathcal{B}$ and $x(Q) = 1$ for $Q \in \mathcal{L}$ fulfills $x(Q) \geq 3$ for all cuts $Q \in \mathcal{Q} \setminus \mathcal{L}$.*

Proof. Let $\{s\} \subseteq U \subseteq V \setminus \{t\}$ such that $\delta(U) \in \mathcal{Q} \setminus \mathcal{L}$. If $\delta(U) \in \mathcal{B}$, we have $x(\delta(U)) \geq 3$ by assumption. Otherwise, there is a set L with $\{s\} \subseteq L \subseteq V \setminus \{t\}$ such that $\delta(L) \in \mathcal{L}$ and both $L \setminus U$ and $U \setminus L$ are nonempty. But then

$$4 \leq x(\delta(L \setminus U)) + x(\delta(U \setminus L)) \leq x(\delta(U)) + x(\delta(L)) = x(\delta(U)) + 1,$$

implying $x(\delta(U)) \geq 3$. □

We now describe the dynamic programming algorithm to compute a $\mathcal{Q}$-good solution x' with minimum cost. This will be very similar to the dynamic program in the proof of Theorem 16.1. We construct an auxiliary digraph (D, A) with vertices corresponding to possible pairs of a lonely cut and a lonely edge and arcs corresponding to subinstances. We define the vertex set to be

$$D := \big\{(U, v, w) : \delta(U) \in \mathcal{Q},\ \{s, v\} \subseteq U \subseteq V \setminus \{w, t\}\big\}$$

and the arc set to be

$$A := \big\{((U_1, v_1, w_1), (U_2, v_2, w_2)) \in D \times D : \\ U_1 \subsetneq U_2 \text{ and } w_1, v_2 \in U_2 \setminus U_1 \text{ and } (w_1 \neq v_2 \text{ or } |U_2 \setminus U_1| = 1)\big\}.$$

We now define weights d for both arcs and vertices of this digraph (D, A). For every vertex $(U, v, w) \in D$, define its weight as $d((U, v, w)) := c(v, w)$. For an arc $a = ((U_1, v_1, w_1), (U_2, v_2, w_2))$, let

$$\mathcal{B}^a := \{\delta(W) \in \mathcal{Q} : U_1 \subsetneq W \subsetneq U_2\}$$

and denote by x^a an optimum solution to the following linear program:

$$\begin{aligned} &\min c(x) \\ &\text{subject to} \quad x \text{ feasible solution to LP}[U_2 \setminus U_1, w_1, v_2] \\ &\qquad\qquad\quad x(Q) + \chi^{\{v_1,w_1\}}(Q) + \chi^{\{v_2,w_2\}}(Q) \geq 3 \text{ for all } Q \in \mathcal{B}^a. \end{aligned} \tag{16.5}$$

Then we define the weight of a to be $d(a) := c(x^a)$. Let

$$D_s := \Big\{(\{s\}, s, w) : w \in V \setminus \{s\}\Big\} \quad \text{and} \quad D_t := \Big\{(V \setminus \{t\}, v, t) : v \in V \setminus \{t\}\Big\}.$$

Now we compute a path P^* in (D, A) from D_s to D_t that minimizes the total weight $d(P^*)$, which is the total weight of its vertices plus the total weight of its arcs. Computing P^* can be done by a simple dynamic program, as in the proof of Theorem 16.1. Let $D(P^*)$ and $A(P^*)$ denote the vertex and arc set of this path P^*. Then we set x' to be

$$x' := \sum_{a \in A(P^*)} x^a + \sum_{(U,v,w) \in D(P^*)} \chi^{\{v,w\}}.$$

Lemma 16.9. *The vector x' is a $\mathcal{Q}$-good solution.*

Proof. Let (L_i, v_i, w_i) for $i = 1, \ldots, k$ be the vertices of the path P^* visited in this order. Let $x^i := x^{a_i}$ where $a_i = ((L_i, v_i, w_i), (L_{i+1}, v_{i+1}, w_{i+1}))$ for $i = 1, \ldots, k-1$. Then by construction of the vectors x^i and by Lemma 16.7, the vector x' is a feasible solution to LP$[V, s, t]$, fulfills $x(\delta(L_i)) = 1$ for all $i = 1, \ldots, k$, and is integral for edges in $\delta(L_i)$. To show that x' is a $\mathcal{Q}$-good solution, it remains to prove $x(Q) \geq 3$ for all $Q \in \mathcal{Q} \setminus \mathcal{L}$, where $\mathcal{L} = \{\delta(L_i) : i = 1, \ldots, k\}$. By Lemma 16.8, it suffices to prove this for the set $\mathcal{B}$ of cuts $Q \in \mathcal{Q} \setminus \mathcal{L}$ for which $\mathcal{L} \cup \{Q\}$ forms a chain. Since $\mathcal{B} = \bigcup_{a \in A(P^*)} \mathcal{B}^a$, every cut $Q \in \mathcal{B}$ is contained in some set $\mathcal{B}^a$ with $a \in A(P^*)$. But then $x(Q) \geq 3$ by construction of x^a. This proves that x' is a $\mathcal{Q}$-good solution. □

Lemma 16.10. *For every $\mathcal{Q}$-good solution x, there exists a path P in (D, A) from D_s to D_t with weight $d(P) \leq c(x)$.*

Proof. Let x be a $\mathcal{Q}$-good solution. Let $\{s\} = L_1 \subsetneq L_2 \subsetneq \ldots \subsetneq L_k = V \setminus \{t\}$ be the chain such that $\delta(L_1), \ldots, \delta(L_k)$ are the lonely cuts of x. Let $e_1, \ldots, e_k$ be the lonely edges of x with $e_i \in \delta(L_i)$. Then e_i is the unique edge in $\delta(L_i)$ with $x_{e_i} = 1$, and all other edges $e \in \delta(L_i)$ have $x_e = 0$. We claim that for

$i, j \in \{1, \ldots, k\}$ with $i < j$, the edges e_i and e_j are different. Suppose not, then $e_i = e_j \in \delta(L_i) \cap \delta(L_j)$ and

$$\begin{aligned} 2 \le x(\delta(L_j \setminus L_i)) &= x(\delta(L_j)) + x(\delta(L_i)) - 2x(\delta(L_i) \cap \delta(L_j)) \\ &\le 1 + 1 - 2x(e_i) = 0, \end{aligned}$$

a contradiction. Let $e_i = \{v_i, w_i\}$ with $v_i \in L_i$ for $i = 1, \ldots, k$. Consider $i \in \{1, \ldots, k-1\}$. Because $x(\delta(w_i)) = x(\delta(v_{i+1})) = 2$ and $x(\delta(L_{i+1} \setminus L_i)) = x_{e_i} + x_{e_{i+1}}$, we can have $w_i = v_{i+1}$ only if $|L_{i+1} \setminus L_i| = 1$. Thus the digraph (D, A) contains a path P with vertices

$$(\{s\}, s, w_1) = (L_1, v_1, w_1),\ (L_2, v_2, w_2),\ \ldots,\ (L_k, v_k, w_k) = (V \setminus \{t\}, v_k, t)$$

visited in this order. By Lemma 16.7, $x = \sum_{i=1}^{k} \chi^{\{v_i, w_i\}} + \sum_{i=1}^{k-1} x^i$, where x^i is a feasible solution to the linear program $\mathrm{LP}[L_{i+1} \setminus L_i, w_i, v_{i+1}]$. Let $a_i = ((L_i, v_i, w_i), (L_{i+1}, v_{i+1}, w_{i+1})) \in A(P)$. Then for every cut $Q \in \mathcal{B}^{a_i} \subseteq Q \setminus \mathcal{L}$, we have $3 \le x(Q) = x^i(Q) + \chi^{\{v_i, w_i\}}(Q) + \chi^{\{v_{i+1}, w_{i+1}\}}(Q)$, implying that x^i is a feasible solution to the linear program defining x^{a_i}. Thus, $d(a_i) = c(x^{a_i}) \le c(x^i)$. This proves

$$d(P) = \sum_{p \in D(P)} d(p) + \sum_{a \in A(P)} d(a) \le \sum_{i=1}^{k} c(\{v_i, w_i\}) + \sum_{i=1}^{k-1} c(x^i) = c(x).$$

□

We have $c(x') = d(P^*) \le d(P)$ for every path P from D_s to D_t in (D, A). Thus, Lemma 16.9 and Lemma 16.10 imply Lemma 16.4. This concludes the proof of Theorem 16.5.

16.3 Reducing Path TSP to TSP: Outline

Traub, Vygen, and Zenklusen [2022] devised a general reduction of Path TSP to Symmetric TSP: Any α-approximation algorithm for Symmetric TSP implies an $(\alpha + \varepsilon)$-approximation algorithm for Path TSP, for every constant $\varepsilon > 0$. The rest of this chapter is devoted to this reduction; however, we will give a slightly different proof.

The first idea is the following. If a Path TSP instance (G, c, s, t) satisfies $\mathrm{dist}_{(G,c)}(s, t) \le \frac{\varepsilon}{\alpha+1} \cdot \mathrm{OPT}(G, c, s, t)$, then there is not a big difference between the cost of $\{s, t\}$-tours and tours: Any $\{s, t\}$-tour can be extended to a tour by adding a shortest s-t-path and vice versa. Hence, applying the given α-approximation algorithm to the Symmetric TSP instance (G, c) and adding a

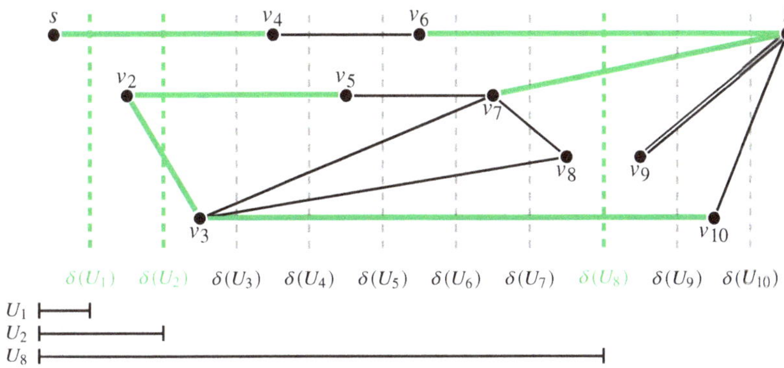

Figure 16.4 A Path TSP instance with 10 vertices, ordered by the distance from s as in Figure 16.1. The dashed vertical lines again indicate the cuts $\delta(U_j)$ for $j = 1, \dots, n-1$. Suppose the solid lines show the optimum $\{s,t\}$-tour F^*, and consider the cuts that contain at most three edges of F^*. These are the green cuts $\delta(U_1)$, $\delta(U_2)$, $\delta(U_8)$ here, so we would try to "guess" these cuts and the bold green edges. Note that in between $\delta(U_2)$ and $\delta(U_8)$, we do not have a Path TSP instance (but a Multi-Path TSP instance).

shortest s-t-path to the result yields an $\{s,t\}$-tour of total cost at most

$$\begin{aligned}
&\alpha \cdot \mathrm{OPT}(G,c) + \mathrm{dist}_{(G,c)}(s,t) \\
&\le \alpha \cdot (\mathrm{OPT}(G,c,s,t) + \mathrm{dist}_{(G,c)}(s,t)) + \mathrm{dist}_{(G,c)}(s,t) \\
&= \alpha \cdot \mathrm{OPT}(G,c,s,t) + (\alpha+1) \cdot \mathrm{dist}_{(G,c)}(s,t) \\
&\le (\alpha+\varepsilon) \cdot \mathrm{OPT}(G,c,s,t).
\end{aligned}$$

Otherwise (i.e., if the distance from s to t is large), we will give a reduction inspired by the one in Section 16.1. Like in the proof of Theorem 16.1, we set up a dynamic program. We will exploit the fact that some cuts $\delta(U_i)$ contain at most $\frac{\alpha+1}{\varepsilon}$ edges of any optimum $\{s,t\}$-tour. This is still a constant number of edges, so we can still hope to guess them, and we can still list all candidates of the dynamic program in polynomial time. The difficulty is that the subproblems in between two candidates are no longer Path TSP instances (see Figure 16.4 for an example). Therefore we generalize the entire approach to Multi-Path TSP.

Originally, Traub, Vygen, and Zenklusen [2022] generalized the approach to a different problem (called Φ-TSP, cf. Exercises 16.5–16.7), but here we present a different proof.

In the rest of this chapter, we will not work in the metric closure because we will also apply our reduction to Graph Path TSP.

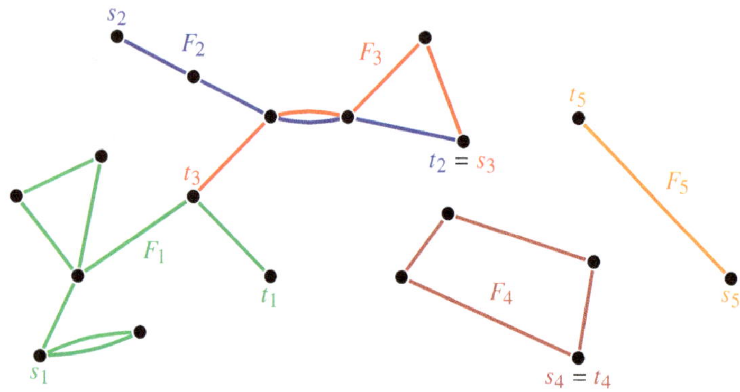

Figure 16.5 Example of a solution to a Multi-Path TSP instance. The different multi-edge sets F_i are shown in different colors.

16.4 Multi-Path TSP

We now present the problem that plays the principal role in the rest of this chapter. It generalizes Path TSP, but instead of asking for a single walk from s to t (whose footprint is an $\{s,t\}$-tour), we ask for walks from s_i to t_i for given terminal pairs $\{s_i, t_i\}$ such that every vertex belongs to at least one of these walks:

Problem 16.11 (Multi-Path TSP).

Instance: An undirected graph $G = (V, E)$ with weights $c : E \to \mathbb{R}_{\geq 0}$. Terminal pairs $\{s_1, t_1\}, \ldots, \{s_k, t_k\}$ with $s_i, t_i \in V$ for $i = 1, \ldots, k$.

Task: Compute multi-subgraphs (V_i, F_i) of G such that

- $s_i, t_i \in V_i$ for $i = 1, \ldots, k$ and $V_1 \cup \cdots \cup V_k = V$, and
- F_i is an $(\{s_i\} \triangle \{t_i\})$-tour in $G[V_i]$ for $i = 1, \ldots, k$,

and the total cost $\sum_{i=1}^{k} c(F_i)$ is minimum (or declare that no such solution exists).

Note that $s_i = t_i$ is allowed, in which case $\{s_i, t_i\}$ is a multi-set containing one vertex twice and F_i must be a tour in $G[V_i]$ with $s_i \in V_i$. Also other terminals can coincide. We do not require that the sets V_i are pairwise disjoint, and even an edge of G can appear in several $(\{s_i\} \triangle \{t_i\})$-tours. See Figure 16.5 for an example. The special case when $k = 1$ is the regular Path TSP.

For a Multi-Path TSP instance, we denote by $\sum_{i=1}^{k} \mathrm{dist}_{(G,c)}(s_i, t_i)$ the *total distance*. This is an obvious lower bound on the cost of an optimum solution, which we again denote by OPT. Let us first note that there is a simple constant-factor approximation algorithm:

Proposition 16.12. *There is a 3-approximation algorithm for Multi-Path TSP.*

Proof. Let $(G, c, k, \{s_1, t_1\}, \dots, \{s_k, t_k\})$ be a Multi-Path TSP instance. Consider the weighted graph (G', c') that results from (G, c) by contracting all terminals into a single vertex t_0. Contracting the terminals in an arbitrary solution to the Multi-Path TSP instance yields a connected graph; hence, the instance has no solution if G' is disconnected. If G' is connected, (G', c') contains a spanning tree of cost at most OPT. Taking a minimum-cost spanning tree in (G', c') and uncontracting yields a set F of edges that connects every vertex to a terminal. We double all edges of F.

If there is no s_i-t_i-path in (G, c) for some $i \in \{1, \dots, k\}$, then the instance has no solution. Otherwise, we start by letting (V_i, F_i) be a shortest s_i-t_i-path in (G, c) for all $i = 1, \dots, k$. Then we add each connected component X of $F \dot{\cup} F$ to a graph (V_i, F_i) that contains a vertex of X. The total cost of the resulting solution is at most $\sum_{i=1}^{k} \mathrm{dist}_{(G,c)}(s_i, t_i) + 2\,\mathrm{OPT} \le 3\,\mathrm{OPT}$. □

In the following, we will assume that k is bounded by a constant (recall that our final goal is to solve Path TSP, the special case $k = 1$). All subinstances in between two cuts that we will consider in our dynamic program will be of this kind.

Consider, for example, Figure 16.4. The solution F^* (more precisely, a walk traversing F^*) induces a nontrivial Multi-Path TSP instance in between $\delta(U_2)$ and $\delta(U_8)$, with vertex set $U_8 \setminus U_2 = \{v_3, \dots, v_8\}$. The terminal pairs of the induced instance depend on the order in which we traverse F^*. If we traverse F^* in the order $s, v_4, v_6, t, v_9, t, v_7, v_8, v_3, v_2, v_5, v_7, v_3, v_{10}, t$, the resulting terminal pairs are $\{s_1, t_1\} = \{v_4, v_6\}$, $\{s_2, t_2\} = \{v_7, v_3\}$, and $\{s_3, t_3\} = \{v_5, v_3\}$. We will formalize the notion of an induced subinstance in Definition 16.14.

16.5 The Case of Short Total Distance

In this section, we reduce Multi-Path TSP instances with small total distance to Symmetric TSP.

Lemma 16.13. *Let $k \in \mathbb{N}$ and $\varepsilon > 0$ and $\alpha > 1$ be constants such that there is an α-approximation algorithm for Symmetric TSP. Then there is an $(\alpha + \varepsilon)$-approximation algorithm for Multi-Path TSP restricted to instances with at most k terminal pairs and total distance at most $\frac{\varepsilon}{3\alpha} \cdot \mathrm{OPT}$.*

Proof. Let $(G, c, k, \{s_1, t_1\}, \ldots, \{s_k, t_k\})$ be a MULTI-PATH TSP instance with total distance at most $\frac{\varepsilon}{3\alpha} \cdot \mathrm{OPT}$. Let $G = (V, E)$. Let $(V_1, F_1^*), \ldots, (V_k, F_k^*)$ be an optimum solution, and let F^* be the disjoint union of $F_1^*, \ldots, F_k^*$. Let P_i be the edge set of a shortest s_i-t_i-path ($i = 1, \ldots, k$).

Call an edge $e \in E$ *heavy* if $c(e) \geq \frac{\varepsilon}{6k\alpha} \cdot \mathrm{OPT}$. We first "guess" the heavy edges in F^*. We do not know this threshold, but there are at most $|E| + 1$ choices of the set H of heavy edges in E. Moreover, there are at most $(|E| + 1)^{\frac{6k\alpha}{\varepsilon}}$ possibilities for the set H^* of heavy edges in F^*, because $|H^*| \leq \frac{6k\alpha}{\varepsilon}$. Hence, we can enumerate all possibilities for H and H^*. In one of the enumerated cases, we have the correct H and H^*, and we show that in this case we get a sufficiently cheap solution.

Given H and H^*, we delete all edges in $D := H \setminus \left(H^* \cup \bigcup_{i=1}^k P_i\right)$ and compute a tour in each connected component of the resulting graph $G - D$. Since s_i and t_i are in the same connected component of $G - D$ for all $i = 1, \ldots, k$, such tours can be obtained from F^* by adding P_i for each i (to correct the parity of the degree of the terminals) and by adding at most $k - 1$ pairs of parallel edges (so that the connected components are the same as in $G - D$). The latter edges are not heavy because the heavy edges that do not belong to $F^* \cup \bigcup_{i=1}^k P_i$ have been deleted. Hence, there exist tours in the connected components of $G - D$ that have total cost at most $\mathrm{OPT} + \sum_{i=1}^k \mathrm{dist}_{(G,c)}(s_i, t_i) + 2(k-1)\frac{\varepsilon}{6k\alpha} \cdot \mathrm{OPT}$. Since the total distance is at most $\frac{\varepsilon}{3\alpha} \cdot \mathrm{OPT}$, this is at most $(1 + \frac{2\varepsilon}{3\alpha})\mathrm{OPT}$.

We call an α-approximation algorithm for SYMMETRIC TSP on each connected component of $G - D$ and obtain tours of total cost $(\alpha + \frac{2\varepsilon}{3})\mathrm{OPT}$. From these tours, we obtain a solution to our MULTI-PATH TSP instance by adding a shortest s_i-t_i-path for $i = 1, \ldots, k$, which increases the total cost by at most $\frac{\varepsilon}{3\alpha} \cdot \mathrm{OPT}$. The cost of the final solution is at most $(\alpha + \varepsilon)\mathrm{OPT}$. □

Note that if the original instance has unit weights, all SYMMETRIC TSP instances in this proof have unit weights too.

16.6 Induced Instances on Subsets

In Section 16.4, we already indicated how we construct instances on subsets of V; let us formalize this now. First, we need to consider solutions to MULTI-PATH TSP instances as walks: A solution (V_i, F_i) for $i = 1, \ldots, k$ can also be represented by the sequence of vertices W_i in a walk from s_i to t_i whose footprint is F_i (for each $i \in \{1, \ldots, k\}$); then we call $W = (W_1, \ldots, W_k)$ a *walk solution* to the given instance.

Let U be a subset of the vertex set V. Given a walk solution, the first and the last vertex in each maximal subwalk inside U induce a terminal pair. See Figure 16.6 for an example. Later, we will apply the following definition not only to walk solutions but also to other sequences of vertices.

Definition 16.14 (induced terminal pairs, induced instance). Let $G = (V, E)$ be an undirected graph and $U \subseteq V$ a subset of vertices. Let $\mathcal{S}$ be a multi-set of sequences of vertices. A subsequence S of consecutive elements of a sequence in $\mathcal{S}$ is called a *U-sequence* in $\mathcal{S}$ if all of its vertices belong to U and S is maximal with this property. The *induced terminal pairs* $T[U, \mathcal{S}]$ contain a pair $\{s, t\}$ for each U-sequence S of $\mathcal{S}$, where s is the first and t is the last element of S.

For a Multi-Path TSP instance $\mathcal{I}$ with graph G and edge weights c, the *induced instance* $\mathcal{I}[U, \mathcal{S}] = (G[U], c, T[U, \mathcal{S}])$ consists of the subgraph of (G, c) induced by U and the induced terminal pairs $T[U, \mathcal{S}]$.

In fact, a walk solution does not only induce an instance, but also a solution to that instance:

Lemma 16.15. *Let $\mathcal{I}$ be a Multi-Path TSP instance with graph $G = (V, E)$. Let W be a walk solution to $\mathcal{I}$, and let $U \subseteq V$. Then the U-sequences of W constitute a walk solution $W[U]$ to the induced Multi-Path TSP instance $\mathcal{I}[U, W]$. Moreover, for every $U' \subseteq U$, we have $(\mathcal{I}[U, W])[U', W[U]] = \mathcal{I}[U', W]$.*

Proof. This follows immediately from the definition. □

We will solve various instances on subsets of V independently, including those induced by an optimum solution. When we stitch solutions on disjoint subsets together, we need more information than just the terminal pairs. To this end, we define:

Definition 16.16 (border protocol). Let $G = (V, E)$ be an undirected graph, $s, t \in V$, and $U \subseteq V$ a subset of vertices.

A *border protocol* B for U and the terminal pair $\{s, t\}$ in G is a sequence $v_0, w_0, v_1, w_1, \ldots, v_l, w_l$ of (not necessarily distinct) vertices with $l \geq 0$, $v_0 = s$, $w_l = t$, such that

- v_j and w_j are both in U or both in $V \setminus U$ for each $j \in \{0, \ldots, l\}$, and
- $\{w_{j-1}, v_j\}$ is an edge in $\delta_G(U)$ for all $j \in \{1, \ldots, l\}$.

We denote the multi-set of these l edges by $\delta_B(U)$. For a multi-set B of border protocols $B_1, \ldots, B_r$, we write $\delta_B(U) = \delta_{B_1}(U) \,\dot\cup \cdots \dot\cup\, \delta_{B_r}(U)$.

A border protocol for U in G is also a border protocol for $V \setminus U$ in G. Figure 16.6 shows an example.

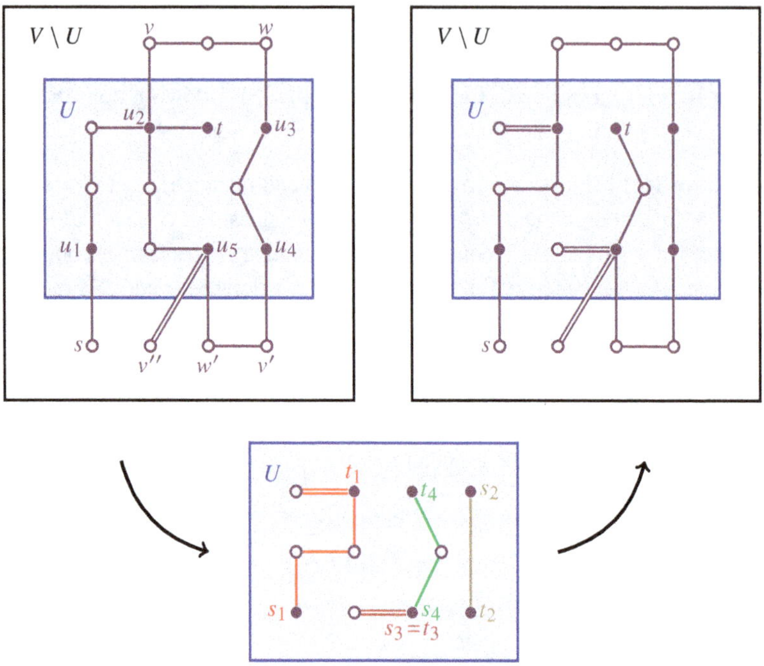

Figure 16.6 On the top left, we see a Multi-Path TSP instance $\mathcal{I}$ with vertex set V, a terminal pair $\{s, t\}$, an $\{s, t\}$-tour, and a subset U of V. Other edges of this instance are not shown. If W is the sequence of vertices that we get when traversing this $\{s, t\}$-tour so that u_1, u_2, u_3, u_4, u_5 are visited in this order, we obtain the induced terminal pairs $\{s_1, t_1\} = \{u_1, u_2\}$, $\{s_2, t_2\} = \{u_3, u_4\}$, $\{s_3, t_3\} = \{u_5, u_5\}$, and $\{s_4, t_4\} = \{u_5, t\}$. The bottom picture sketches the induced Multi-Path TSP instance $\mathcal{I}[U, \{W\}]$ and a solution to that instance. Replacing the edges inside U by this solution yields the $\{s, t\}$-tour shown on the top right. The sequence $s, s, u_1, u_2, v, w, u_3, u_4, v', w', u_5, u_5, v'', v'', u_5, t$ is the border protocol induced by W for U and the terminal pair $\{s, t\}$.

According to Definition 16.14, the multi-set of terminal pairs $T[U, B]$ induced by a multi-set B of border protocols for U contains the pair $\{v_j, w_j\}$ for all border protocols $v_0, w_0, \ldots, v_l, w_l$ in B and all j for which $v_j, w_j \in U$. The induced instance $\mathcal{I}[U, B]$ contains these pairs as terminal pairs.

Lemma 16.17. *Let $\mathcal{I} = (G, c, k, \{s_1, t_1\}, \ldots, \{s_k, t_k\})$ be a Multi-Path TSP instance with the graph $G = (V, E)$, and let $U \subseteq V$ be a subset of vertices. Let B_i be a border protocol for U and the terminal pair $\{s_i, t_i\}$ in G ($i = 1, \ldots, k$), and let B be the multi-set of these k border protocols. Let W_U be a walk solution* *to $\mathcal{I}[U, B]$ and $W_{V\setminus U}$ a walk solution to $\mathcal{I}[V \setminus U, B]$. Then W_U, $W_{V\setminus U}$, and $\delta_B(U)$ can be composed to a walk solution to $\mathcal{I}$.*

Proof. Let $i \in \{1, \ldots, k\}$, and let B_i be the border protocol $v_0, w_0, v_1, w_1, \ldots, v_l, w_l$ with $v_0 = s_i$, $w_l = t_i$. For all j such that $v_j, w_j \in U$, the induced instance $\mathcal{I}[U, \{B_i\}]$ (and hence $\mathcal{I}[U, B]$) contains a terminal pair $\{v_j, w_j\}$; thus the walk solution W_U contains a walk from v_j to w_j. For all j such that $v_j, w_j \in V \setminus U$, the induced instance $\mathcal{I}[V \setminus U, \{B_i\}]$ (and hence $\mathcal{I}[V \setminus U, B]$) contains a terminal pair $\{v_j, w_j\}$; thus the walk solution $W_{V\setminus U}$ contains a walk from v_j to w_j. By inserting these walks, we transform B_i into a walk from s_i to t_i in G. Doing this for $i = 1, \ldots, k$ yields a walk solution to $\mathcal{I}$ because each vertex is visited by at least one of the walks. □

Given a walk solution W and a subset $U \subseteq V$, the multi-set of *border protocols induced by* W for U is the multi-set B that we get by listing, for each sequence in W, the first and the last vertex of each U-sequence and each $(V \setminus U)$-sequence. Then $T[U, B] = T[U, W]$ and $T[V \setminus U, B] = T[V \setminus U, W]$.

The idea will be to guess the border protocols for U induced by an optimum (walk) solution for sets U for which this solution contains only few edges of $\delta(U)$. We will next explain which cuts exactly we will consider.

16.7 The Case of Long Total Distance

Complementing Lemma 16.13, we now deal with the case of long total distance. Consider an instance $(G, c, k, \{s_1, t_1\}, \ldots, \{s_k, t_k\})$ with total distance more than $\frac{\varepsilon}{3\alpha} \cdot \mathrm{OPT}$ (for some constants $\varepsilon > 0$, $\alpha > 1$, and $k \in \mathbb{N}$). Without loss of generality, $s_i \neq t_i$ for $i = 1, \ldots, \bar{k}$ and $s_i = t_i$ for $i = \bar{k} + 1, \ldots, k$ for some $\bar{k} \in \{1, \ldots, k\}$. (The case $\bar{k} = 0$ is covered by Section 16.5.)

As in Section 16.1 we define a chain of cuts, but now one for each terminal pair. Let $G = (V, E)$ and $n = |V|$. For $i = 1, \ldots, \bar{k}$, let $f^i(v) := \min\{\mathrm{dist}_{(G,c)}(s_i, v), \mathrm{dist}_{(G,c)}(s_i, t_i)\}$ for $v \in V$, and order the vertices so that $v_1^i = s_i$, $v_n^i = t_i$, and $f^i(v_1^i) \leq f^i(v_2^i) \leq \cdots \leq f^i(v_n^i)$. Consider the chain of cuts $\delta(U_j^i)$ for $j = 1, \ldots, n-1$, where $U_j^i = \{v_1^i, \ldots, v_j^i\}$, and set $y^i(U_j^i) = f^i(v_{j+1}^i) - f^i(v_j^i)$.

Let $(V_1^*, F_1^*), \ldots, (V_k^*, F_k^*)$ be an optimum solution, and let F^* be the disjoint union of $F_1^*, \ldots, F_k^*$. Call a cut *small* if it contains at most K edges of F^*, where we set $K = \frac{6\alpha k}{\varepsilon}$. Since ε, α, and k are constants, K is a constant, too. For $i = 1, \ldots, \bar{k}$, let

$$\Lambda^i = \left\{ j \in \{1, \ldots, n-1\} : |F^* \cap \delta(U_j^i)| \leq K \right\}$$

be the set of indices of small cuts in the chain for terminal pair $\{s_i, t_i\}$. Our goal is again to "guess" the edges in these small cuts – that is, the set

$$L^* = \bigcup_{i=1}^{\bar{k}} \bigcup_{j \in \Lambda^i} (F^* \cap \delta(U_j^i)). \tag{16.6}$$

See Figure 16.7 for an example. Very similarly to (16.2), we first show:

Lemma 16.18. $c(L^*) \geq \frac{\varepsilon}{6\alpha} \cdot \text{OPT}$.

Proof. First note (exactly as in inequality (16.3)) that for any edge set F and any $i = 1, \ldots, \bar{k}$,

$$c(F) \geq \sum_{j=1}^{n-1} |F \cap \delta(U_j^i)| \cdot y^i(U_j^i).$$

Applying this inequality first to F^* and then to $L^* \cap F_i^*$ (using $|L^* \cap F_i^* \cap \delta(U_j^i)| = |F_i^* \cap \delta(U_j^i)| \geq 1$ for $j \in \Lambda^i$), we get

$$\begin{aligned} c(F^*) &\geq \sum_{j=1}^{n-1} |F^* \cap \delta(U_j^i)| \cdot y^j(U_j^i) \\ &\geq K \sum_{j=1}^{n-1} y^i(U_j^i) - K \cdot \sum_{j \in \Lambda^i} y^i(U_j^i) \\ &\geq K \sum_{j=1}^{n-1} y^i(U_j^i) - K \cdot c(L^* \cap F_i^*) \\ &= K \cdot \text{dist}_{(G,c)}(s_i, t_i) - K \cdot c(L^* \cap F_i^*). \end{aligned}$$

For $i = \bar{k} + 1, \ldots, k$ we have $\text{dist}_{(G,c)}(s_i, t_i) = 0$, and thus this inequality holds trivially. Summing over $i = 1, \ldots, k$ yields

$$K \cdot c(L^*) \geq K \cdot \sum_{i=1}^{k} \text{dist}_{(G,c)}(s_i, t_i) - k \cdot c(F^*).$$

Using $\sum_{i=1}^{k} \text{dist}_{(G,c)}(s_i, t_i) > \frac{\varepsilon}{3\alpha} \cdot \text{OPT} = \frac{2k}{K} \cdot \text{OPT}$ and $c(F^*) = \text{OPT}$, we obtain $K \cdot c(L^*) > k \cdot \text{OPT}$, which completes the proof (using again $K = \frac{6\alpha k}{\varepsilon}$). □

So the edges in L^* amount to a constant fraction of the total cost of F^*. We will now again set up a dynamic program to guess these edges, but here this is more complicated than in Section 16.1. Consider the grid formed by all the cuts $\delta(U_i^j)$ and the subgrid formed by the small cuts (cf. Figure 16.7).

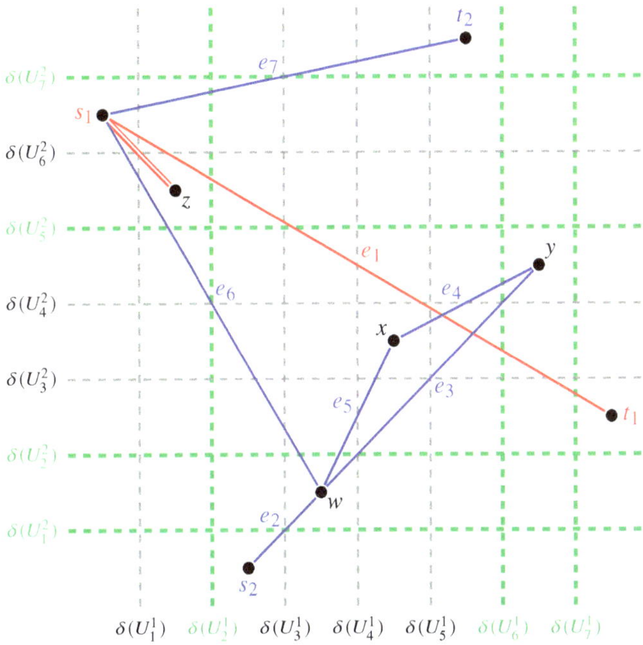

Figure 16.7 A Multi-Path TSP instance with eight vertices and two terminal pairs. The horizontal order corresponds to the terminal pair $\{s_1, t_1\}$ and the vertical order to the terminal pair $\{s_2, t_2\}$. Suppose the red edges (F_1^*) and the blue edges (F_2^*) form an optimum solution, and let $K = 3$. Then the small cuts $\delta(U^i_j)$ (those with at most K edges in $F_1^* \dot{\cup} F_2^*$) are indicated by thick green dashed lines. In this example, all edges except the two parallel red edges in the upper left are contained in a small cut.
A boxed set with box definition $((U^1_2, U^1_6), U^2_2)$ is indicated by the light green area: It contains the vertices in the area left of $\delta(U^1_2)$, as well as those in the area left of $\delta(U^1_6)$ and below $\delta(U^2_2)$.
The solution (F_1^*, F_2^*) corresponds to a $d_\emptyset$-d_V-path in the digraph (D, A), which has vertices with the boxed sets $\emptyset$, U^1_2, $U^1_2 \cup (U^1_6 \cap U^2_1)$, $U^1_2 \cup (U^1_6 \cap U^2_2)$, $U^1_2 \cup (U^1_6 \cap U^2_5)$, U^1_6, U^1_7, V.

We would like to proceed column by column in this subgrid (from left to right) and, within a column, row by row (from bottom to top). Of course, we know the complete grid, but not the subgrid.

Recall that $U^i_j = \{v^i_1, \ldots, v^i_j\}$ for all $i = 1, \ldots, \bar{k}$ and all $j = 0, \ldots, n$. Let $\mathcal{U}^i := \{U^i_j : j = 0, \ldots, n\}$ be the chain corresponding to the terminal pair $\{s_i, t_i\}$.

Definition 16.19 (boxed vertex set)**.** We say that a vertex set $U \subseteq V$ is *boxed* if $U = \emptyset$ or U has the form

$$U := \bigcup_{i=1}^{\bar{k}} \left(\bigcap_{h=1}^{i-1} Y^h \cap X^i \right),$$

where $X^i, Y^i \in \mathcal{U}^i$ with $X^i \subsetneq Y^i$ for $i = 1, \ldots, \bar{k} - 1$ and $X^{\bar{k}} \in \mathcal{U}^{\bar{k}} \setminus \emptyset$. We also call $\left((X^1, Y^1), (X^2, Y^2), \ldots, (X^{\bar{k}-1}, Y^{\bar{k}-1}), X^{\bar{k}}\right)$ a *box definition* of U.

See Figure 16.7 for an example.

A boxed set can have several box definitions. Note that V is a boxed set, one of its box definitions is $\left((\emptyset, V), \ldots, (\emptyset, V), V\right)$. If

$$\left((X^1, Y^1), (X^2, Y^2), \ldots, (X^{\bar{k}-1}, Y^{\bar{k}-1}), X^{\bar{k}}\right)$$

is a box definition of a set U, then U is the union of the disjoint sets

$$\begin{aligned} &X^1, \\ &(Y^1 \setminus X^1) \cap X^2, \\ &(Y^1 \setminus X^1) \cap (Y^2 \setminus X^2) \cap X^3, \\ &\ldots \\ &(Y^1 \setminus X^1) \cap \cdots \cap (Y^{\bar{k}-1} \setminus X^{\bar{k}-1}) \cap X^{\bar{k}}. \end{aligned}$$

The number of box definitions (and hence the number of boxed sets) is at most $1 + n^{2\bar{k}-1}$. Our goal is to guess the boxed vertex sets that are defined by small cuts, and for each such vertex set U, we also need to guess the induced terminal pairs. Due to the structure of our boxed sets, we will be able to restrict the number of induced terminal pairs in any boxed set.

Definition 16.20 (candidate)**.** A *candidate* d consists of a boxed set $U(d)$ and a multi-set $T(d)$ of at most $k(K + 1)$ terminal pairs in $U(d)$.

We denote by D the set of all candidates. There are at most $(1 + n^{2k-1}) \cdot n^{2k(K+1)} \leq n^{2k(K+2)}$ candidates. Since k and K are constants, we can enumerate all candidates in polynomial time. Note that there is exactly one candidate $d_\emptyset$ with $U(d_\emptyset) = \emptyset$ (and $T(d_\emptyset) = \emptyset$), corresponding to the "empty instance." Let d_V be the candidate with $U(d_V) = V$ and $T(d_V) = \{\{s_1, t_1\}, \ldots, \{s_k, t_k\}\}$, corresponding to the original instance that we want to solve.

We construct an acyclic digraph (D, A) on the vertex set D. An arc from d to d' will exist only if $U(d) \subsetneq U(d')$, so such an arc corresponds to expanding the boxed set. To this end, we will need to solve MULTI-PATH TSP instances on the vertex set $U(d') \setminus U(d)$ with at most $2k(K + 1)$ terminal pairs.

To compute solutions to such instances, we will call a β-approximation algorithm $\mathcal{A}$ for MULTI-PATH TSP. Let $\mathcal{A}(\mathcal{I})$ denote the walk solution obtained by running $\mathcal{A}$ on the MULTI-PATH TSP instance $\mathcal{I}$. If this instance $\mathcal{I}$ is infeasible, then we write $c(\mathcal{A}(\mathcal{I})) = \infty$.

Consider a pair of candidates d and d' with $U(d) \subsetneq U(d')$. For each multi-set B of border protocols for $U(d)$ in $G[U(d')]$, one for each terminal pair $\{s', t'\} \in T(d')$, such that $T(d) = T[U(d), B]$, the digraph contains an arc from d to d', and the weight of the arc is

$$w((d, d')) := c\left(\mathcal{A}\left(\mathcal{I}\left[U(d') \setminus U(d), B\right]\right)\right) + c\left(\delta_B(U(d))\right).$$

With the arc and its weight, we can also store B and the solution that $\mathcal{A}$ produces for the instance $\mathcal{I}[U(d') \setminus U(d), B]$. Figure 16.8 shows an example.

Lemma 16.21. *The number of multi-sets B of border protocols that we consider for a pair (d, d') of candidates (and thus the number of parallel arcs (d, d')) is at most $(n+1)^{6k(K+1)}$. Moreover, $T[U(d') \setminus U(d), B]$ contains at most $2k(K+1)$ terminal pairs for any such B.*

Proof. Since each of $T(d)$ and $T(d')$ contains at most $k(K+1)$ terminal pairs, $T[U(d') \setminus U(d), B]$ can contain at most $2k(K+1)$ terminal pairs. So the total length of B is at most $6k(K+1)$. □

So we can construct (D, A) in polynomial time, with $O(n^{10k(K+2)})$ calls to $\mathcal{A}$. (This bound can be improved – see Exercise 16.8 – but this is not important.) We will find a shortest $d_\emptyset$-d_V-path in this digraph with respect to the arc weights w. From this, we get a solution as follows:

Lemma 16.22. *Given a $d_\emptyset$-d_V-path P in (D, A), one can construct a solution to $\mathcal{I}$ of cost at most $w(P)$ in polynomial time.*

Proof. We show that for all $d' \in D \setminus \{d_\emptyset\}$ and for any $d_\emptyset$-d'-path P', we can construct a walk solution to the instance $\mathcal{I} = (G[U(d')], c, T(d'))$ of cost at most $w(P')$ in polynomial time. For $d' = d_V$, this implies the lemma.

We use induction on the number of arcs in P'. If P' consists of a single arc $(d_\emptyset, d')$, we obtain the solution $\mathcal{A}(\mathcal{I})$ of cost $w(P') = w((d_\emptyset, d'))$.

Otherwise, let (d, d') be the last arc of a $d_\emptyset$-d'-path P', let B be the multi-set of border protocols corresponding to that arc, and let P be the $d_\emptyset$-d-path that results from P' by removing that last arc. By the induction hypothesis, we can construct a walk solution to the instance $(G[U(d)], c, T(d)) = \mathcal{I}[U(d), B]$ of cost at most $w(P)$ in polynomial time. Moreover, we have computed $\mathcal{A}\left(\mathcal{I}[U(d') \setminus U(d), B]\right)$. Together with $\delta_B(U(d))$, these walk solutions can be

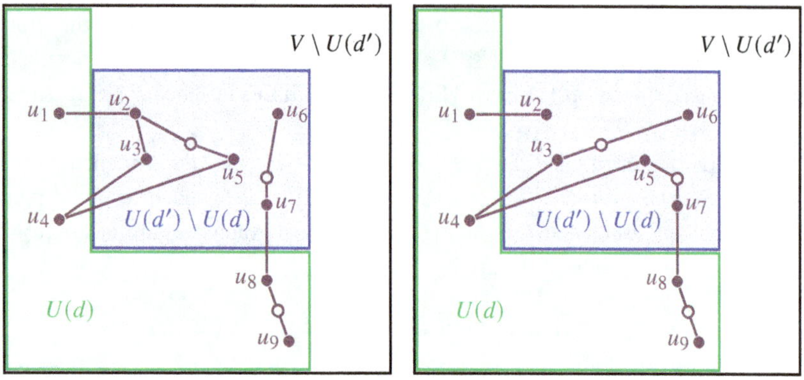

Figure 16.8 Illustration of two parallel arcs (d, d') in the digraph (D, A). In this example, the boxed set $U(d')$ is shown in green and blue; its terminal pairs in $T(d')$ are $\{u_1, u_2\}$ and $\{u_6, u_9\}$. The boxed set $U(d)$ is shown in green, and its terminal pairs in $T(d)$ are $\{u_1, u_1\}$, $\{u_4, u_4\}$, and $\{u_8, u_9\}$. The left and the right part of the figure show two different ways of extending the same solution to the Multi-Path TSP instance $(G[U(d)], c, T(d))$ to a solution to the Multi-Path TSP instance $(G[U(d')], c, T(d'))$. These two different extensions can be obtained by considering two different arcs from d to d', corresponding to two different multi-sets of border protocols for $U(d)$ in $G[U(d')]$. The solution on the left is obtained by considering the multi-set of border protocols that consists of $u_1, u_1, u_2, u_3, u_4, u_4, u_5, u_2$ and u_6, u_7, u_8, u_9, then computing a solution to the induced instance with vertex set $U(d') \setminus U(d)$ and terminal pairs $\{u_2, u_3\}$, $\{u_5, u_2\}$, $\{u_6, u_7\}$, and applying Lemma 16.17. The solution on the right is obtained by considering the multi-set of border protocols that consists of u_1, u_1, u_2, u_2 and $u_6, u_3, u_4, u_4, u_5, u_7, u_8, u_9$, then computing a solution to the induced instance with vertex set $U(d') \setminus U(d)$ and terminal pairs $\{u_2, u_2\}$, $\{u_6, u_3\}$, $\{u_5, u_7\}$, and applying Lemma 16.17.

composed to a solution to the instance $\mathcal{I}$ by Lemma 16.17. Its total cost is

$$w(P) + c\left(\mathcal{A}\left(\mathcal{I}[U(d') \setminus U(d), B]\right)\right) + c(\delta_B(U(d))) = w(P'),$$

as required. □

Moreover, we need to show that a short $d_\emptyset$-d_V-path exists. We construct one from an optimum solution:

Lemma 16.23. *Let F^* be an optimum solution, and let L^* be defined as in* (16.6)*. Then there exists a $d_\emptyset$-d_V-path in (D, A) of total weight at most $c(L^*) + \beta \cdot c(F^* \setminus L^*)$.*

Proof. Given F^*, we first define a chain of boxed sets such that their box definitions correspond to small cuts only, and such that for each edge e in a small cut, at least one of these boxed sets contains exactly one endpoint of e. As before, for $i = 1, \dots, \bar{k}$, let

$$\Lambda^i = \{j \in \{1, \dots, n-1\} : |F^* \cap \delta(U_j^i)| \le K\}$$

be the set of indices of small cuts in the chain for terminal pair $\{s_i, t_i\}$, and let $\mathcal{Y}^i := \{U_j^i : j \in \Lambda^i \cup \{n\}\}$. The set $\mathcal{Y} := \mathcal{Y}^1 \times \dots \times \mathcal{Y}^{\bar{k}}$ is totally ordered lexicographically by the subset relation. For each $\lambda = (Y^1, \dots, Y^{\bar{k}}) \in \mathcal{Y}$, consider the boxed set $U(\lambda)$ with box definition

$$((Y_-^1, Y^1), (Y_-^2, Y^2), \dots, (Y_-^{\bar{k}-1}, Y^{\bar{k}-1}), Y^{\bar{k}}),$$

where Y_-^i denotes the largest element of $\{\emptyset\} \cup \mathcal{Y}^i$ that is a proper subset of Y^i.

By definition, we have:

$$\text{If } \lambda, \lambda' \in \mathcal{Y} \text{ and } \lambda \text{ is lexicographically smaller than } \lambda', \text{ then } U(\lambda) \subseteq U(\lambda'). \tag{16.7}$$

Moreover, the endpoints of an edge in a small cut belong to different boxes of the grid induced by the small cuts (cf. Figure 16.7). If κ and λ are two lexicographically consecutive vectors in $\mathcal{Y}$ and $\lambda = (Y^1, \dots, Y^{\bar{k}})$, then

$$U(\lambda) \setminus U(\kappa) = \bigcap_{i=1}^{\bar{k}} (Y^i \setminus Y_-^i);$$

this is a box in the grid defined by the small cuts. Since we add one box at each step, we have:

$$\text{For each edge } e \text{ that belongs to a small cut, there is a } \lambda \in \mathcal{Y} \text{ with } e \in \delta(U(\lambda)). \tag{16.8}$$

By (16.7), $\mathcal{U} := \emptyset \cup \{U(\lambda) : \lambda \in \mathcal{Y}\}$ is a chain. If W^* is a walk solution with footprint F^*, then consider, for each $U \in \mathcal{U}$, the candidate d with $U(d) = U$ and $T(d) = T[U, W^*]$. Note that $T[U, W^*]$ contains at most $k(K+1)$ terminal pairs because U is a boxed set induced by small cuts, which implies $|\delta_{F^*}(U)| \le (2k-1)K$. Hence, these candidates form a path P from $d_\emptyset$ to d_V in (D, A), where we use the arc (d, d') associated with the border protocols B for $U(d)$ in $G[U(d')]$ that are induced by $W^*[U(d')]$. (So, for each $U(d')$-sequence in W^*, we list the first and the last vertex of each $U(d)$-subsequence and each $(U(d') \setminus U(d))$-subsequence.) Then indeed $T[U(d), B] = T[U(d), W^*]$ and $T[U(d') \setminus U(d), B] = T[U(d') \setminus U(d), W^*]$.

By Lemma 16.15, the instance $I[U(d') \setminus U(d), B]$ has a solution that consists of the $(U(d') \setminus U(d))$-sequences in W^*, and the solution that $\mathcal{A}$ finds is at most β times more expensive. Moreover, $\delta_B(U(d)) = \delta_{F^*}(U(d)) \setminus \delta_{F^*}(U(d'))$. Hence, the total weight of the path P is at most $\beta \cdot c(F^* \setminus L^*) + c(L^*)$, as required. □

We conclude the following booster theorem:

Theorem 16.24. *Let $\varepsilon > 0$, $\alpha > 1$, $\beta > \alpha$, and $k \in \mathbb{N}$ be constants. Suppose we have an α-approximation algorithm for* Symmetric *TSP and a β-approximation algorithm for* Multi-Path *TSP instances with up to $\frac{16\alpha}{\varepsilon}k^2$ terminal pairs. Then we also have a $\max\{\alpha + \varepsilon, (\beta - (\alpha - 1)\frac{\varepsilon}{6\alpha})\}$-approximation algorithm for* Multi-Path *TSP instances with up to k terminal pairs.*

Proof. By Proposition 16.12, we may assume $\varepsilon < 2$. Given a Multi-Path TSP instance $\mathcal{I}$ with up to k terminal pairs, we run both algorithms: the one for short total distance (Lemma 16.13) and the one for long total distance (see below) and output the better result. If the total distance is at most $\frac{\varepsilon}{3\alpha} \cdot \mathrm{OPT}$, Lemma 16.13 guarantees that we are done.

Otherwise, we need the β-approximation algorithm $\mathcal{A}$ for Multi-Path TSP instances with up to $2k(K+1)$ terminal pairs; recall that $K = \frac{6\alpha k}{\varepsilon}$ and note that $2k(K+1) = 2k(\frac{6\alpha}{\varepsilon}k + 1) < \frac{16\alpha}{\varepsilon}k^2$. Construct the digraph (D, A) with arc weights w as before, calling $\mathcal{A}$ at most $O(n^{10k(K+2)})$ times. Find a shortest $d_\emptyset$-d_V-path P in (D, A) and derive a solution to $\mathcal{I}$ of cost at most $w(P)$ by Lemma 16.22.

By Lemmas 16.23 and 16.18, we have

$$\begin{aligned} w(P) &\leq c(L^*) + \beta \cdot c(F^* \setminus L^*) \\ &= \beta \cdot \mathrm{OPT} - (\beta - 1) \cdot c(L^*) \\ &\leq \beta \cdot \mathrm{OPT} - (\alpha - 1) \cdot c(L^*) \\ &\leq \left(\beta - (\alpha - 1)\tfrac{\varepsilon}{6\alpha}\right) \cdot \mathrm{OPT}. \end{aligned}$$ □

Applying this booster theorem repeatedly yields the main result of this chapter:

Corollary 16.25. *Let $\varepsilon > 0$, $\alpha > 1$, and $k \in \mathbb{N}$ be constants. If there is an α-approximation algorithm for* Symmetric *TSP, then there is an $(\alpha + \varepsilon)$-approximation algorithm for* Multi-Path *TSP instances with up to k terminal pairs.*

Proof. By Proposition 16.12, we may assume $\alpha < 3$. Let

$$M := \left\lceil \frac{3 - \alpha - \varepsilon}{(\alpha - 1)\frac{\varepsilon}{6\alpha}} \right\rceil;$$

note that this is a constant. Starting with a 3-approximation algorithm (Proposition 16.12) for general Multi-Path TSP instances (in particular those with up to $(\frac{16\alpha}{\varepsilon})^{(2^M - 1)} k^{(2^M)}$ terminal pairs, which is still a constant number) and applying Theorem 16.24 at most M times yields an $(\alpha + \varepsilon)$-approximation algorithm for Multi-Path TSP instances with up to k terminal pairs. □

The following is the special case $k = 1$.

Theorem 16.26 (Traub, Vygen, and Zenklusen [2022]). *Let $\alpha > 1$ and $\varepsilon > 0$ be constants. If there is an α-approximation algorithm for* Symmetric *TSP, then there is an $(\alpha + \varepsilon)$-approximation algorithm for* Path *TSP.* □

This yields in particular:

Corollary 16.27. *There is an α-approximation algorithm for* Path *TSP for some $\alpha < \frac{3}{2}$.*

Proof. This follows directly from combining Theorems 16.26 and 11.20. □

We summarize again the state of the art for Path TSP in Table 16.1.

If the given Multi-Path TSP instance has unit edge weights, all generated instances of Symmetric TSP will be unit-weight instances as well (see Section 16.5). Hence, we conclude:

Theorem 16.28 (Traub, Vygen, and Zenklusen [2022]). *Let $\alpha > 1$ and $\varepsilon > 0$ be constants. If there is an α-approximation algorithm for* Graph *TSP, then there is an $(\alpha + \varepsilon)$-approximation algorithm for* Graph Path *TSP.* □

Corollary 16.29. *For any $\varepsilon > 0$, there is a $(\frac{7}{5} + \varepsilon)$-approximation algorithm for* Graph Path *TSP.*

Proof. This follows from Theorems 13.21 and 16.28. □

It is an interesting open question whether Theorem 16.26 can be extended to the Asymmetric TSP:

Open Problem 16.30. Is it true that any α-approximation algorithm for Asymmetric TSP implies an $(\alpha + \varepsilon)$-approximation algorithm for Asymmetric Path TSP?

The best-known black-box reduction loses a factor of 2 (see Theorem 9.8). Another open question addresses the T-Tour Problem:

Open Problem 16.31. Is it true that any α-approximation algorithm for Symmetric TSP implies an $(\alpha + \varepsilon)$-approximation algorithm for the T-Tour Problem?

We only know this for the special case when $|T|$ is bounded by a constant (see Exercise 16.9).

Table 16.1 *Approximation ratios and upper bounds on the integrality ratio for* Path TSP *in the order of their discovery. The integrality ratio refers to the* LP (14.1) *for instances that satisfy the triangle inequality. The ratio* $\frac{3}{2}+\varepsilon$ *holds for any* $\varepsilon > 0$. *The last result follows from the reduction from* Path TSP *to* Symmetric TSP *by Traub, Vygen, and Zenklusen [2022].*

Approximation Ratio	Integrality Ratio	Year	Reference	Chapter
$\frac{5}{3}$	–	1990	Hoogeveen [1991]	14.1
$\frac{1+\sqrt{5}}{2}$	$\frac{1+\sqrt{5}}{2}$	2011	An, Kleinberg, and Shmoys [2015]	15.2
$\frac{8}{5}$	$\frac{8}{5}$	2012	Sebő [2013]	15.2
1.599	1.599	2015	Vygen [2016]	–
1.566	1.566	2015	Gottschalk and Vygen [2018]	15.3
$\frac{26}{17}$	$\frac{26}{17}$	2016	Sebő and van Zuylen [2019]	15.4
$\frac{3}{2}+\varepsilon$	–	2017	Traub and Vygen [2019a]	–
	1.528	2018	Traub and Vygen [2019b], Zhong [2020]	–
$\frac{3}{2}$	–	2018	Zenklusen [2019]	16.2
$\frac{3}{2}-10^{-36}$	–	2022	Karlin, Klein, and Oveis Gharan [2023]	10–11, 16

Exercises

16.1 Let $G = (V, E)$ be an undirected graph with edge weights $c : E \to \mathbb{R}_{\geq 0}$. Show that y as defined in the proof of Theorem 16.1 defines the nonzero entries of an optimum solution to the dual to the shortest path LP

$$\min\{c(x) : x(\delta(U)) \geq 1 \ (s \in U \subseteq V \setminus \{t\}),\ x_e \geq 0 \ (e \in E)\}.$$

Conclude that this shortest path LP always has an integral optimum solution.

16.2 Show that the constraints $x(Q)+\chi^{\{v_1,w_1\}}(Q)+\chi^{\{v_2,w_2\}}(Q) \geq 3$ in (16.5) can be omitted for all $Q \in \mathcal{B}^a$ with $\{v_1, w_1\} \in Q$ or $\{v_2, w_2\} \in Q$.

16.3 Describe a linear-time algorithm that decides whether a given Multi-Path TSP instance has a solution.

16.4 Let $\mathcal{I} = (G, c, \{s_1, t_1\}, \ldots, \{s_k, t_k\})$ be a Multi-Path TSP instance, $G = (V, E)$, and F be a multi-subset of E. Show that there exists a solution $(V_1, F_1), \ldots, (V_k, F_k)$ to I such that F is the disjoint union of $F_1, \ldots, F_k$ if and only if

- a vertex has odd degree in (V, F) if and only if it appears an odd number of times in the list $s_1, t_1, s_2, t_2, \ldots, s_k, t_k$ of terminals;
- for every vertex, there is a terminal in the same connected component of (V, F);
- there are pairwise edge-disjoint paths P_i from s_i to t_i $(i = 1, \ldots, k)$ in (V, F).

Note: It is *NP*-hard to check the third condition, even if the other two conditions are satisfied (Middendorf and Pfeiffer [1993]). For constant k, Robertson and Seymour [1995] devised a polynomial-time algorithm.

16.5 An instance of Φ-TSP consists of an undirected graph $G = (V, E)$ with edge weights $c : E \to \mathbb{R}_{\geq 0}$ and an interface $\Phi = (I, T, \mathscr{C})$, consisting of a subset $I \subseteq V$, a subset $T \subseteq I$, and a partition $\mathscr{C}$ of I. A Φ-tour is a multi-subset F of E such that $\text{odd}(F) = T$, the graph $(V, F)/I$ is connected, and for all $C \in \mathscr{C}$, the vertices in C lie in the same connected component of (V, F). Φ-TSP asks for a minimum-weight Φ-tour in (G, c) (or to decide that no Φ-tour exists).

(a) Devise a linear-time algorithm to decide whether a given instance has a feasible solution.

(b) Assume a 2-approximation for the Steiner Forest Problem, which asks for a minimum-weight edge set F such that for all $C \in \mathscr{C}$, the vertices in C lie in the same connected component of (V, F). Show that this implies a 7-approximation algorithm for Φ-TSP.

Note: Agrawal, Klein, and Ravi [1995] indeed devised a 2-approximation for the Steiner Forest Problem. Traub, Vygen, and Zenklusen [2022] devised a 4-approximation algorithm for Φ-TSP.

16.6 Let $\varepsilon > 0$ and $\alpha > 1$ be constants such that we have an α-approximation algorithm for Symmetric TSP, and consider Φ-TSP (cf. Exercise 16.5) restricted to instances (G, c, Φ) with $\Phi = (I, T, \mathscr{C})$ where $|I|$ is bounded by a constant and there exists a T-join of cost at most $\frac{\varepsilon}{2\alpha}\text{OPT}$. Show that then there is an $(\alpha + \varepsilon)$-approximation algorithm for such Φ-TSP instances.

Hint: Mimic the proof of Lemma 16.13.

(Traub, Vygen, and Zenklusen [2022])

16.7 Now consider a Φ-TSP instance (G, c, Φ) with $\Phi = (I, T, \mathscr{C})$ (cf. Exercise 16.5) for which every T-join costs at least $\frac{\varepsilon}{2\alpha}\text{OPT}$. Show that then there is a laminar family $\mathcal{L}$ of subsets of V such that, for an optimum

Φ-tour F^* and $K = \frac{4\alpha}{\varepsilon}$, the set

$$L^* := \{e \in F^* : \text{there is an } L \in \mathcal{L} \text{ with } e \in \delta(L),\ |F^* \cap \delta(L)| \leq K\}$$

satisfies $c(L^*) \geq \frac{\varepsilon}{4\alpha}\text{OPT}$.
Hint: Use Lemma 4.15.
Note: Traub, Vygen, and Zenklusen [2022] set up a dynamic program to guess L^* and obtained a booster theorem similar to Theorem 16.24.

16.8 Show that the digraph (D, A) can be reduced to only $O(n^{3k(K+2)})$ arcs (for constant k and K) so that Lemma 16.23 still holds.

16.9 Let $k \in \mathbb{N}$ be a constant. Show that Corollary 16.25 implies that an α-approximation algorithm for Symmetric TSP implies an $(\alpha + \varepsilon)$-approximation algorithm for instances of the T-Tour Problem with $|T| \leq k$.

17

Further Results, Related Problems

In this chapter, we mention further results on the approximability of variants or special cases of the traveling salesman problem. We will also briefly mention a few important related problems for which the best-known approximation algorithms use a TSP approximation algorithm as a subroutine. In particular, we discuss inapproximability results, geometric special cases, the minimum 2-edge-connected spanning subgraph problem, the prize-collecting TSP, the a priori TSP, and capacitated vehicle routing.

17.1 Inapproximability

For the problems studied in this book, better and better approximation algorithms have been found, and this is likely to continue. It is natural to ask whether there is a limit or whether there is an *approximation scheme*: a $(1 + \varepsilon)$-approximation algorithm for every $\varepsilon > 0$, with a polynomial running time for each fixed ε.

Unless $P = NP$, the answer is no, except for some special cases. Papadimitriou and Yannakakis [1993] showed that Symmetric TSP (and even the very special case 1-2-TSP; see Section 17.3) is *MAXSNP*-hard. Today, the term *APX*-hard is more commonly used. We do not define these terms here but just say that, by the results of Arora et al. [1998], they imply that there is no approximation scheme unless $P = NP$.

In fact, concrete lower bounds have been obtained for many problems. Unless $P = NP$, no approximation algorithms exist with better ratio than

- $\frac{75}{74}$ for Asymmetric TSP (Karpinski, Lampis, and Schmied [2015], improving on Papadimitriou and Vempala [2006]);
- $\frac{123}{122}$ for Symmetric TSP (Karpinski, Lampis, and Schmied [2015], improving on Lampis [2014]);

- $\frac{685}{684}$ for GRAPH TSP (Karpinski and Schmied [2015]), even for subcubic graphs.

With the current techniques, it seems difficult to prove much larger lower bounds, but it would be very interesting.

Open Problem 17.1. Improve the lower bounds on the approximability of any of the studied problems substantially.

Some important special cases that do have an approximation scheme are discussed in Section 17.3.

17.2 Two-Edge-Connected Spanning Subgraphs

For SYMMETRIC TSP, we often worked with the linear programming relaxation (2.12). The set of feasible integral solutions to this LP includes not only the incidence vectors of tours, but the incidence vectors of all 2-edge-connected spanning multi-subgraphs. Hence, it is an interesting question how well this – seemingly easier – problem can be approximated. A related (and arguably even more interesting) problem does not allow parallel edges:

Problem 17.2 (TWO-EDGE-CONNECTED SPANNING SUBGRAPH PROBLEM).

Instance: A 2-edge-connected undirected graph $G = (V, E)$ and a cost function $c : E \to \mathbb{R}_{\geq 0}$.

Task: Compute a subset $F \subseteq E$ of minimum cost such that (V, F) is 2-edge-connected.

Note that in this formulation, it is not allowed that F contains two copies of an edge.

In the unweighted case ($c(e) = 1$ for all $e \in E$), it does not make a difference whether we allow multi-subgraphs or subgraphs (i.e., whether we allow picking an edge more than once). The reason is that we can always avoid taking a second copy of an edge (see Proposition 13.3 and Exercise 2.14). In Section 13.6, we showed a $\frac{4}{3}$-approximation algorithm for the problem of finding a smallest 2-edge-connected spanning subgraph. This approximation ratio has recently been improved to $1.3 + \varepsilon$ by Garg, Grandoni, and Jabal Ameli [2023] and Kobayashi and Noguchi [2023]. Fernandes [1998] proved that the problem is *APX*-hard.

For the general (weighted) problem, it does make a difference whether we allow picking an edge twice. If we allow this, we immediately get a $\frac{3}{2}$-approximation algorithm.

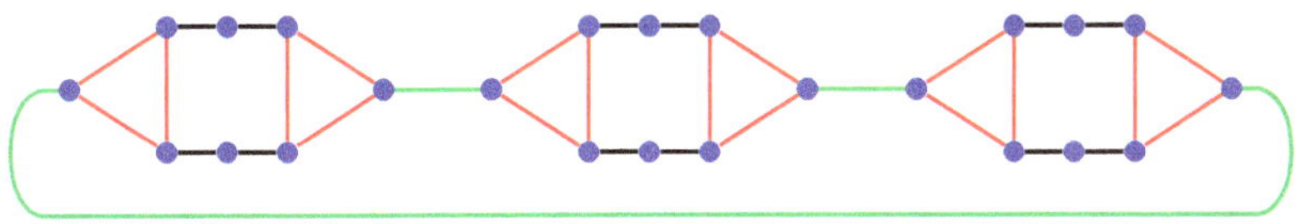

Figure 17.1 Example from Alexander, Boyd, and Elliott-Magwood [2006] showing a lower bound of $\frac{6}{5}$ on the integrality ratio of the LP (2.12). The red edges cost 2, the black edges 1, and the green edges 0. An LP solution is given by setting $x_e = \frac{1}{2}$ for the red edges and $x_e = 1$ for all other edges; its cost is $10\frac{n}{8}$. Any 2-edge-connected multi-subgraph costs at least $12\frac{n}{8}$.

Theorem 17.3 (Frederickson and Ja'ja' [1982]). *There is a $\frac{3}{2}$-approximation algorithm for the problem of finding a minimum-weight 2-edge-connected spanning multi-subgraph of a given connected undirected graph $G = (V, E)$ with weights $c : E \to \mathbb{R}_{\geq 0}$.*

Proof. Directly from Wolsey's analysis of Christofides' algorithm (cf. Theorem 2.29). □

The algorithm by Karlin, Klein, and Oveis Gharan [2022, 2023] obtains an approximation ratio of $\frac{3}{2} - 10^{-36}$.

If we do not allow picking edges more than once, the problem is apparently harder unless the edge weights satisfy the triangle inequality (in which case the problem reduces to the one in which we allow picking edges twice; cf. Exercise 17.1). In general, the best-known approximation ratio for Problem 17.2 is still 2. We need the following classical result:

Theorem 17.4 (Edmonds [1973,1975]). *Let $G = (V, E)$ be a directed graph with weights $c : E \to \mathbb{R}_{\geq 0}$ and $r \in V$. Suppose $|\delta^+(U)| \geq 2$ for all $U \subsetneq V$ with $r \in U$. Then G contains two edge-disjoint spanning arborescences rooted at r, and we can find two such arborescences with minimum total weight in polynomial time.*

In fact, this problem can be reduced to weighted matroid intersection (see, e.g., Schrijver [2003] or Korte and Vygen [2018]). Using this, the following is quite easy:

Theorem 17.5 (Khuller and Vishkin [1994]). *There is a 2-approximation algorithm for the Two-Edge-Connected Spanning Subgraph Problem.*

Proof. Let $G^{\leftrightarrow} = (V, E^{\leftrightarrow})$ be the digraph that results from orienting every edge of G both ways. Let $c((v, w)) = c(\{v, w\})$ for every edge $e = (v, w) \in E^{\leftrightarrow}$. Choose a root $r \in V$ arbitrarily, and find two edge-disjoint spanning

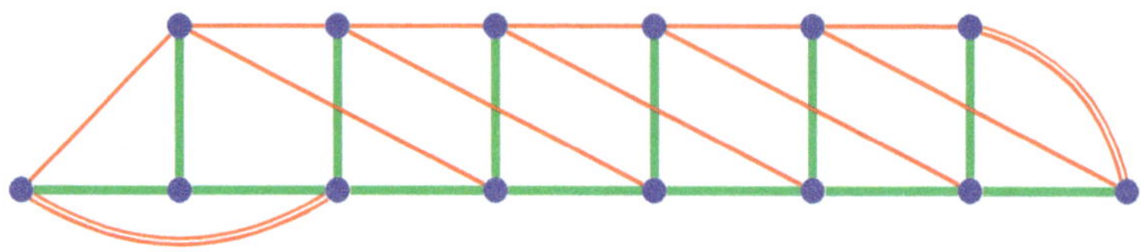

Figure 17.2 Example from Cheriyan et al. [2008] showing a lower bound of $\frac{3}{2}$ on the integrality ratio of the LP that results from (2.12) by adding the constraints $x_e \leq 1$ for all $e \in E$. We have $2k + 2$ vertices (here $k = 6$). The $2k + 1$ green edges have cost 0, and the red edges have cost 1. Setting $x_e = 1$ for the green edges, $x_e = \frac{1}{3}$ for the single red edges, and $x_e = \frac{2}{3}$ for the two bent double red edges yields a feasible LP solution of total cost $\frac{2}{3}k + 1$. There is no simple 2-edge-connected spanning subgraph with fewer than $k + 1$ red edges. Since there is a spanning tree of cost 0, this is actually an instance of the tree augmentation problem.

arborescences rooted at r in $G^{\leftrightarrow}$, say (V, A_1) and (V, A_2), so that their total weight is minimum (cf. Theorem 17.4). Let $A = A_1 \cup A_2$, and let $F = \{e = \{v, w\} : (v, w) \in A \text{ or } (w, v) \in A\}$ be the corresponding subset of E. Then for any $U \subsetneq V$ with $r \in U$, we have $|F \cap \delta(U)| \geq |A \cap \delta^+(U)| = |A_1 \cap \delta^+(U)| + |A_2 \cap \delta^+(U)| \geq 2$, so (V, F) is 2-edge-connected. Moreover, $c(F) \leq c(A_1) + c(A_2) \leq 2c(F^*)$ for any 2-edge-connected subgraph (V, F^*), because $(F^*)^{\leftrightarrow}$ contains two edge-disjoint spanning arborescences, again by Theorem 17.4. □

Still no better approximation algorithm is known. This is somewhat surprising because Jain [2001] obtained the approximation ratio 2 for much more general network design problems. A first step towards improving on Theorem 17.5 and resolving the following open problem may be the recent progress on the weighted tree augmentation problem by Traub and Zenklusen [2021,2022].

Open Problem 17.6. Find an approximation algorithm with an approximation ratio better than 2 for the Two-Edge-Connected Spanning Subgraph Problem.

By Theorem 2.29, the integrality ratio of (2.12) is at most $\frac{3}{2}$. In fact, by Theorem 2.31 and the fact that every tour is 2-edge-connected, it is at most the integrality ratio of (2.2) and hence at most $\frac{3}{2} - 10^{-36}$ (cf. Theorem 10.19). Carr and Ravi [1998] conjectured it to be $\frac{4}{3}$. An example of Alexander, Boyd, and Elliott-Magwood [2006] shows that it is at least $\frac{6}{5}$ (see Figure 17.1). The ratio restricted to unit weights is between $\frac{8}{7}$ (Boyd, Fu, and Sun [2016]) and $\frac{4}{3}$ (Theorem 13.23).

For unit weights, the integrality ratio does not change if we add the constraints $x_e \leq 1$ for all $e \in E$ to (2.12) – see Exercise 2.14 and Proposition 13.3. For general weights, the integrality ratio is different if we add $x_e \leq 1$ for all $e \in E$: It is at least $\frac{3}{2}$ (see the example by Cheriyan et al. [2008] in Figure 17.2) and at most 2 (by Jain's [2001] general result on survivable network design).

The directed analogon of Problem 17.2 is the problem of computing a minimum-weight strongly connected spanning subgraph of a given digraph (a relaxation of the ASYMMETRIC TSP). See Exercise 3.7.

17.3 Special Cases and Variants of the Symmetric TSP

Besides GRAPH TSP (cf. Chapters 12 and 13), other special cases of the SYMMETRIC TSP have been studied.

For example, in the special case of the GRAPH TSP where the instance is a k-regular graph (with k large), Vishnoi [2012] showed how to find a tour of length at most $(1 + \sqrt{64/\ln k})n$ in polynomial time. This was improved by Chiplunkar and Vishwanathan [2015] and Feige, Ravi, and Singh [2014]. For $k = 3$ (cubic graphs) and for subcubic graphs, stronger results are known (see Section 12.3).

Another well-studied special case is the 1-2-TSP, in which $c(v, w) \in \{1, 2\}$ for all $v, w \in V$. To see that this is essentially a special case of the GRAPH TSP (up to an additive constant of 1), add a vertex x and consider the graph with edge set $\{\{v, x\} : v \in V\} \cup \{\{v, w\} : v, w \in V, v \neq w, c(v, w) = 1\}$. The 1-2-TSP has an $\frac{8}{7}$-approximation algorithm (Berman and Karpinski [2006]). The integrality ratio of (2.2) for the 1-2-TSP is between $\frac{10}{9}$ (Williamson [1990]) and $\frac{5}{4}$ (Mnich and Mömke [2018], improving on Qian et al. [2015]).

Another line of research deals with geometric instances. Here, each city is associated with a point in $\mathbb{R}^d$, and the distances are ℓ_p-distances. This case is also *NP*-hard for any fixed $d \geq 2$ and any p. Arora [1998] found an approximation scheme for geometric instances based on dynamic programming. The most prominent case $d = p = 2$ is called the EUCLIDEAN TSP, for which Mitchell [1999] also devised such an approximation scheme. Rao and Smith [1998] and Bartal and Gottlieb [2013] improved the running time: For every fixed $\varepsilon > 0$, they have $(1 + \varepsilon)$-approximation algorithms that run in $O(n \log n)$ and $O(n)$ time, respectively. The dependence of the constant factor on ε was improved by Kisfaludi-Bak, Nederlof, and Węgrzycki [2021], but the running time remains large for reasonable values of ε. Bartal, Gottlieb, and Krauthgamer [2016] found a randomized approximation scheme for metric spaces with bounded doubling dimension. Disser et al. [2021] found a fully polynomial approximation scheme for instances with highway dimension 1.

Interestingly, it is not known whether the decision version of the Euclidean TSP belongs to *NP*: For a given instance of Euclidean TSP with integral coordinates, a given tour T, and an integer k, we do not know how to decide in polynomial time whether the cost of T is at most k.

Hougardy [2014] showed that even when restricted to instances of Euclidean TSP, the integrality ratio of (2.2) is at least $\frac{4}{3}$.

Similar techniques as for geometric instances apply to the Symmetric TSP in planar graphs (with nonnegative edge weights). Klein [2008] found an approximation scheme that has linear running time for every fixed $\varepsilon > 0$. An approximation scheme exists even for bounded genus graphs (Demaine, Hajiaghayi, and Mohar [2010]) and for the even-more-general class of minor-free graphs (Demaine, Hajiaghayi, and Kawarabayashi [2011], Borradaile, Le, and Wulff-Nilsen [2017]).

Many variants of Symmetric TSP have been studied, often motivated by practical applications. We will mention only a few examples here. For a more comprehensive survey, see Saller, Koehler, and Karrenbauer [2023].

Bérczi, Mnich, and Vincze [2022] studied the Many-Visits TSP, in which we are given an instance (V, c) of Symmetric TSP with Triangle Inequality plus a positive integer $r(v)$ for each $v \in V$ and ask for a closed walk that visits each city v exactly $r(v)$ times. If $r(v)$ is polynomially bounded, one can just make $r(v)$ copies of v, but in general, the problem is more difficult. Still, Bérczi, Mnich, and Vincze [2022] devised a $\frac{3}{2}$-approximation algorithm, also for the path version. Pillai and Singh [2023] devised a black-box reduction to an LP-based algorithm for Symmetric TSP.

Bender and Chekuri [2000] and Mömke [2015] devised approximation algorithms for a generalization of Symmetric TSP with Triangle Inequality where instead of the triangle inequality, it is only required that $c(u, w) \leq \beta(c(u, v) + c(v, w))$ for all $u, v, w \in V$ and some $\beta > 1$. Of course, the approximation ratios depend on β.

Chalasani, Motwani, and Rao [1997] and Frank et al. [1998] found a 2-approximation algorithm for the Bipartite TSP. Here we look for a minimum-cost Hamiltonian circuit in a complete bipartite graph $K_{n,n}$ with edge weights satisfying the quadrilateral inequality.

Other variants put constraints on the feasible tours. In the Clustered TSP, for which Guttmann-Beck, Knaan, and Stern [2018] presented a 4-approximation algorithm, a set of clusters (not necessarily disjoint subsets of V) is given, and points in the same cluster must be visited consecutively. See Exercise 17.2 for a better approximation algorithm in a special case.

Further generalizations of the TSP require a vehicle to transport goods from pickup locations to delivery locations. There are several variants, depending on whether there is only one kind of good and whether the vehicle has limited capacity; for the most general version, no constant-factor approximation algorithm is known. If there is only one type of good and the vehicle can transport one unit only, the tour must alternate between pickups and deliveries; this is equivalent to the BIPARTITE TSP mentioned earlier.

All these variants can also be formulated in the asymmetric case, but these have been studied less so far. In Sections 17.4 and 17.5, we will discuss variants of the TSP in which it is not mandatory to visit all points.

17.4 Prize-Collecting TSP and Orienteering

Several problems have been studied in which, for a given TSP instance, it is not mandatory to visit all cities. One way to model such problems is as follows. In addition to an instance (V, c) of SYMMETRIC or ASYMMETRIC TSP WITH TRIANGLE INEQUALITY, we have a penalty (or reward or prize) $p(v)$ for every element $v \in V$. Then we need to collect a certain total reward and/or we have to pay a penalty if we decide to not visit a city. This kind of problem was first suggested by Balas [1989]. Without loss of generality, we assume that the tour must start at a given depot (if this is not given, we can try all possible starting points). One of the best-studied version is (the symmetric version of) the following:

Problem 17.7 (PRIZE-COLLECTING TSP).

Instance: A finite set V (of customers), a depot $s \notin V$, and a cost function $c : (\{s\} \cup V) \times (\{s\} \cup V) \to \mathbb{R}_{\geq 0}$ that satisfies the triangle inequality. A penalty $p(v) \geq 0$ for each customer $v \in V$.

Task: Compute a circuit C containing s such that $c(E(C)) + p(V \setminus V(C))$ is minimum.

Like TSP, this has an asymmetric and a symmetric version. Following Bienstock et al. [1993] (with minor changes to adapt to the asymmetric setting), we show:

Theorem 17.8. *If there is an algorithm for* ASYMMETRIC *TSP that always computes a solution of cost at most α times the value of the Asymmetric TSP LP* (3.2), *then there is an $(\alpha + 1)$-approximation algorithm for the* ASYMMETRIC PRIZE-COLLECTING *TSP.*

Proof. Let (V, s, c, p) be an instance of the ASYMMETRIC PRIZE-COLLECTING TSP. Adapting the linear program (3.2) for the ASYMMETRIC TSP instance

$(\{s\} \cup V, c)$, we obtain an LP relaxation for the Asymmetric Prize-Collecting TSP by introducing a variable y_v $(v \in V)$ that indicates whether we visit v. Let $E = \{(v, w) : v, w \in \{s\} \cup V, v \neq w\}$.

$$\begin{array}{llcll} \min & c(x) + \sum_{v \in V} p(v)(1 - y_v) & & & \\ \text{subject to} & x(\delta(U)) & \geq & 2y_v & (v \in U \subseteq V) \\ & x(\delta^+(v)) & = & x(\delta^-(v)) & (v \in \{s\} \cup V) \\ & x_e & \geq & 0 & (e \in E) \\ & 0 \leq y_v & \leq & 1 & (v \in V). \end{array} \tag{17.1}$$

Note that this LP can be solved in polynomial time by Theorem 2.10, using Corollary 2.8 for the separation problem. Let (x, y) be an optimum solution. For a threshold $\tau \in (0, 1]$ let $S^\tau := \{v \in V : y_v \geq \tau\}$ and $x_e^\tau := \frac{x_e}{\tau}$ for $e \in E$. Then x^τ is a feasible solution to the LP

$$\begin{array}{llcll} \min & c(x) & & & \\ \text{subject to} & x(\delta(U)) & \geq & 2 & (U \subseteq V,\ S^\tau \cap U \neq \emptyset) \\ & x(\delta^+(v)) & = & x(\delta^-(v)) & (v \in \{s\} \cup V) \\ & x_e & \geq & 0 & (e \in E). \end{array} \tag{17.2}$$

Now, by Theorem 3.5, the Asymmetric TSP LP (3.2) for $(\{s\} \cup S^\tau, c)$ has value at most $c(x^\tau)$. By the assumed Asymmetric TSP algorithm, we thus get a tour for $\{s\} \cup S^\tau$ that costs at most $\alpha \cdot c(x^\tau) = \frac{\alpha}{\tau} \cdot c(x)$. Since we do not visit the customers in $V \setminus S^\tau$, we have to pay a penalty of $p(V \setminus S^\tau) \leq \sum_{v \in V} p(v) \frac{1 - y_v}{1 - \tau}$. We can thus bound the total cost by

$$\frac{\alpha}{\tau} \cdot c(x) + \frac{1}{1-\tau} \sum_{v \in V} p(v)(1 - y_v).$$

For $\tau = \frac{\alpha}{\alpha + 1}$, this yields an $(\alpha + 1)$-approximation algorithm. □

Combining Theorem 17.8 with the $(17 + \varepsilon)$-approximation algorithm for Asymmetric TSP (Theorem 8.25), we obtain an $(18 + \varepsilon)$-approximation algorithm for the Asymmetric Prize-Collecting TSP. For the symmetric case, we obtain the approximation ratio $\frac{5}{2}$ using Christofides' algorithm.

The threshold rounding algorithm in the proof of Theorem 17.8 can be improved by choosing the best τ depending on y (see Exercise 17.3). In the symmetric case, an approximation guarantee of less than 1.915 was obtained by Goemans [2009] (improving on Archer et al. [2011]) by combining this with a primal-dual algorithm that was devised by Goemans and Williamson [1995].

The threshold rounding algorithm has been improved by Blauth and Nägele [2023]: They showed how to avoid always paying the penalty for vertices below the threshold and obtained the approximation ratio of 1.774. The currently best approximation guarantee for the SYMMETRIC PRIZE-COLLECTING TSP, due to Blauth, Klein, and Nägele [2023], is 1.599.

Besides the PRIZE-COLLECTING TSP, several similar problems have been studied intensively. This includes:

- An instance of k-TSP has no penalties but an integer k and asks for a minimum-cost circuit that contains s and visits k customers.
- In the ORIENTEERING PROBLEM, we get a deadline D and ask for a circuit C with $c(E(C)) \leq D$ that contains s and maximizes $p(V(C))$.

Variants of these problems ask for a path instead of a circuit, again with a specified starting point s and possibly also with a specified endpoint. All these variants have natural applications in practice. Blum, Ravi, and Vempala [1999], Arora and Karakostas [2006], Garg [2005], Chaudhuri et al. [2003], Blum et al. [2007], Bansal et al. [2004], Nagarajan and Ravi [2011], Chekuri, Korula, and Pál [2012], and Bateni and Chuzhoy [2013] described approximation algorithms and reductions between these problems.

However, no constant-factor approximation algorithms for the asymmetric versions are known.

Open Problem 17.9. Find a constant-factor approximation algorithm for ASYMMETRIC ORIENTEERING or prove that this would imply $P = NP$.

Even more general versions of the above-mentioned problems have been studied, where every city has a deadline or a time window and must be visited by that deadline (or within its time window). Here the time when a city is visited is the cost of the initial segment of the tour, from a given starting point to that city. Another model lets the reward depend on the time when a city is visited. A special case asks that no customer shall be visited much later than its distance from the starting point (this is called bounded regret).

A related problem, although it asks for visiting all customers, is known as the MINIMUM LATENCY PROBLEM or the TRAVELING REPAIRMAN PROBLEM. It asks for a tour $s = v_0, v_1, \ldots, v_n$ that minimizes the sum of the latencies – that is, $\sum_{i=1}^{n} \sum_{j=1}^{i} c(v_{j-1}, v_j)$. Again, constant-factor approximation algorithms are known for the symmetric case (see, e.g., Chaudhuri et al. [2003]) but not for the asymmetric case (see Friggstad and Swamy [2022]).

17.5 A Priori TSP

Sometimes tours must be planned before we know which customers are to be visited. We are given an instance (V, c) of the ASYMMETRIC TSP WITH TRIANGLE INEQUALITY and a probability distribution μ on the subsets of V, and we are asked to compute an order $V = \{v_1, \ldots, v_n\}$ so that if we visit a random subset of V (the active customers, sampled according to μ) in this order, the expected total distance is as small as possible. The best-studied case is when each customer is activated independently. As it is the case in many applications, let us assume for simplicity that there is a depot s that must be visited always. Then the problem can be formulated as follows.

Problem 17.10 (A PRIORI TSP).

Instance: A finite set V (of customers), a depot $s \notin V$, and a cost function $c : (\{s\}\cup V)\times(\{s\}\cup V) \to \mathbb{R}_{\geq 0}$ that satisfies the triangle inequality. An activation probability $p(v) \in [0, 1]$ for each customer $v \in V$.

Task: Compute a circuit C with vertex set $\{s\}\cup V$ so that $\mathbb{E}_{A\sim\mu}[c(C_A)]$ is minimum, where C_A is the circuit that results from C by skipping the customers in $V\setminus A$, and $\mu(A) = \prod_{v\in A} p(v) \prod_{v\in V\setminus A}(1-p(v))$.

Only the symmetric case has been studied so far, and so we assume c to be symmetric in the following.

One possible way to design a solution is to select a subset $S \subseteq V$, compute a tour M on $\{s\} \cup S$, and insert edges $(r, v), (v, r)$ for all $v \in V \setminus S$, where $r \in \{s\} \cup S$ is chosen so that $c(r, v) + c(v, r)$ is minimum (to obtain a circuit, we can skip vertices that are visited again). Such a solution is called a *master route solution* (with master route M).

Given a master route solution and a set A of active customers, we will think that we always travel along the master route, and for all $v \in A$, we travel from the nearest vertex on the master route to v and back. See Figure 17.3. The final cost can only be cheaper than this.

We need two lower bounds. First, let OPT_A denote the minimum cost of a tour in the SYMMETRIC TSP instance induced by $\{s\} \cup A$. Then $\mathbb{E}_{A\sim\mu}[\mathrm{OPT}_A]$ is the expected cost of an a posteriori optimal tour, which is clearly a lower bound. The following lower bound is also useful:

Proposition 17.11 (Shmoys and Talwar [2008]). *For $S \subseteq V$ and $v \in V$, let*

$$D(\{s\} \cup S, v) \;:=\; \min\bigl\{2c(v,r) : r \in (\{s\} \cup S) \setminus \{v\}\bigr\}.$$

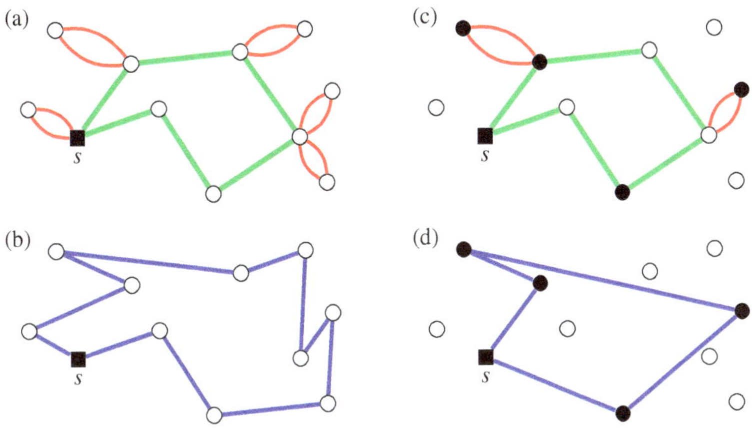

Figure 17.3 In (a), we see a master route solution with a master route (green) and connections of the other customers to that master route (red). In (b), the corresponding Hamiltonian circuit C is shown. In (d), we skip the inactive customers in C and obtain the circuit C_A on the set A of active customers (filled). In (c), the upper bound on $c(C_A)$ obtained by the master route solution is illustrated.

Then

$$\sum_{v \in V} p(v) \cdot \mathbb{E}_{A \sim \mu}\left[D(\{s\} \cup A, v)\right] \leq 2\,\mathbb{E}_{A \sim \mu}[\mathrm{OPT}_A].$$

Proof. For any $A \subseteq V$, let an optimum TSP tour for A visit $s, a_1, \ldots, a_k$ in this order, and let $a_0 := a_{k+1} := s$. Then

$$\begin{aligned}\sum_{v \in A} D(\{s\} \cup A, v) &\leq \sum_{i=1}^{k} 2 \min\{c(a_{i-1}, a_i), c(a_i, a_{i+1})\} \\ &\leq \sum_{i=1}^{k} (c(a_{i-1}, a_i) + c(a_i, a_{i+1})) \\ &\leq 2\,\mathrm{OPT}_A.\end{aligned}$$

Since the event that v belongs to A is independent of $D(\{s\} \cup A, v)$, the result follows by taking the expectation for $A \sim \mu$ on both sides of this inequality. □

The following result is essentially due to Shmoys and Talwar [2008]:

Theorem 17.12. *If there is an α-approximation algorithm for* Symmetric TSP, *then there is a randomized $(\alpha + 2)$-approximation algorithm for* Symmetric A Priori TSP. *This algorithm always computes a master route solution of cost at most $(\alpha + 2)\,\mathbb{E}_{A \sim \mu}[\mathrm{OPT}_A]$.*

Proof. Choose a random subset S of V by including each customer v independently with probability $p(v)$ – that is, a set S is chosen with probability $\mu(S) = \prod_{v\in S} p(v) \prod_{v\in V\setminus S}(1 - p(v))$. Run the α-approximation for SYMMETRIC TSP to obtain a tour M on $\{s\} \cup S$. We compute a master route solution C with master route M and output C.

The tour C_A that results from C by skipping the inactive customers (the elements of $V \setminus A$) has expected cost

$$
\begin{aligned}
\mathbb{E}_{S\sim\mu}[\mathbb{E}_{A\sim\mu}[c(C_A)]] &\le \mathbb{E}_{S\sim\mu}\left[c(M) + \sum_{v\in V\setminus S} p(v)D(\{s\}\cup S, v)\right] \\
&\le \mathbb{E}_{S\sim\mu}[\alpha\,\mathrm{OPT}_S] + \mathbb{E}_{S\sim\mu}\left[\sum_{v\in V} p(v)D(\{s\}\cup S, v)\right] \\
&= \mathbb{E}_{A\sim\mu}[\alpha\,\mathrm{OPT}_A] + \sum_{v\in V} p(v)\,\mathbb{E}_{A\sim\mu}\left[D(\{s\}\cup A, v)\right] \\
&\le (\alpha+2)\,\mathbb{E}_{A\sim\mu}[\mathrm{OPT}_A],
\end{aligned}
$$

where the last inequality follows from Proposition 17.11. □

Note that we even get expected cost $(\alpha + 2)$ times the expected cost of an a posteriori optimal tour. Theorem 17.12 yields an expected approximation ratio 4 using the double tree algorithm (Shmoys and Talwar [2008]), $\frac{7}{2}$ using Christofides' algorithm (as noted by van Ee and Sitters [2018]), and slightly better with Theorem 11.20. Blauth et al. [2023] showed that sampling less (more precisely, including a customer v into S with probability $1 - (1 - p(v))^{0.663}$ instead of $p(v)$) yields an approximation ratio better than 3.1.

Let us define the *master route ratio* to be the supremum, taken over all SYMMETRIC A PRIORI TSP instances, of the ratio of the expected cost of a best master route solution and the expected cost of an optimum A PRIORI TSP solution (where $\frac{0}{0} := 1$). Here, the expected cost of a best master route solution with master route on $\{s\} \cup S$ is

$$
\mathrm{OPT}_S + \sum_{v\in V\setminus S} p(v)D(\{s\}\cup S, v), \tag{17.3}
$$

and we take the minimum over all $S \subseteq V$.

Theorem 17.12 also shows:

Corollary 17.13. *The master route ratio is at most 3.*

Proof. Set $\alpha = 1$ in Theorem 17.12. □

Indeed, the proof reveals that the best master route solution is at most 3 times more expensive than the expected cost of an a posteriori optimal tour. Blauth et al. [2023] showed that the master route ratio is actually between 2.54 and 2.59. The lower bound is easy to see: There are examples where the best master route solution is $\frac{1}{1-e^{-1/2}}$ times more expensive than the best a priori tour (see Exercise 17.4).

Shmoys and Talwar [2008] also proposed a deterministic 8-approximation algorithm for the symmetric case, which was improved by van Zuylen [2011] to the ratio 6.5. Her algorithm is based on the method of conditional expectations. It decides for the customers one by one whether to include them into S or not, so that a pessimistic estimator of the final expected cost does not increase. To define the pessimistic estimator, we consider a fractional relaxation of finding an optimum master route solution. For $e \in E = \binom{\{s\} \cup V}{2}$, we have variables b_e to indicate whether we *buy* an edge (include it in the master route) and variables r_e^v ($v \in V$) to indicate whether we *rent* an edge if v is active (to connect v to the master route):

$$\begin{array}{llll} \min & \sum_{e \in E} c(e) \left(b_e + \sum_{v \in V} p(v) r_e^v \right) & & \\ \text{subject to} & b(\delta(U)) + r^v(\delta(U)) \geq 2 & (U \subseteq V,\ v \in U) & \quad (17.4) \\ & b_e \geq 0 & (e \in E) & \\ & r_e^v \geq 0 & (v \in V,\ e \in E). & \end{array}$$

From every feasible solution to this LP, we can construct a feasible solution to the subtour LP for any subset of customers:

Lemma 17.14. *Let $S \subseteq V$ and $x^S := b + \sum_{v \in S} r^v$. Then the value of the subtour LP (2.2) for $(\{s\} \cup S, c)$ is at most $c(x^S)$.*

Proof. Orient every edge $e = \{v, w\}$ both ways, set $x_{(v,w)} = x_{(w,v)} = \frac{1}{2} x_{\{v,w\}}$, and apply Theorem 3.5. □

Hence, if we have an algorithm that computes a tour of cost at most α times the value of (2.2), then $\mathrm{PE}^S := \alpha c(x^S) + \sum_{v \in V} p(v) D(\{s\} \cup S, v)$ is a pessimistic estimator (an upper bound on the expected cost of the resulting master route solution). Moreover, for any disjoint subsets $V_0, V_1 \subseteq V$, we can compute the conditional expectation $\mathbb{E}_{S \sim \mu}[\mathrm{PE}^S | V_1 \subseteq S \subseteq V \setminus V_0]$ efficiently. Finally, we can add a customer to V_0 or V_1 without increasing this conditional expectation (see van Zuylen [2011] for details).

We end up with a deterministic algorithm that computes a master route solution of cost at most

$$\begin{aligned}
\mathbb{E}_{S\sim\mu}[\mathrm{PE}^S] &\le \alpha\,\mathbb{E}_{S\sim\mu}[c(x^S)] + \sum_{v\in V} p(v)\,\mathbb{E}_{S\sim\mu}[D(\{s\}\cup S, v)] \\
&\le \alpha \sum_{e\in E} c(e)\left(b_e + \sum_{v\in V} p(v) r_e^v\right) + 2\,\mathbb{E}_{A\sim\mu}[\mathrm{OPT}_A] \\
&\le (3\alpha+2)\,\mathbb{E}_{A\sim\mu}[\mathrm{OPT}_A],
\end{aligned}$$

where the second inequality follows from Proposition 17.11 and the third inequality follows from the upper bound on the master route ratio (Corollary 17.13). So we get an approximation ratio 6.5 for Symmetric A Priori TSP using $\alpha = \frac{3}{2}$ (by Wolsey's analysis of Christofides' algorithm; Theorem 2.23). Since the master route ratio is actually less than 2.6, this deterministic algorithm actually has approximation ratio less than 5.9 (Blauth et al. [2023]).

The techniques presented in this section do not work in the asymmetric setting, for which essentially nothing is known.

Open Problem 17.15. Find a constant-factor approximation algorithm for Asymmetric A Priori TSP or prove that this would imply $P = NP$.

Other stochastic versions of the TSP have been studied as well – see Ganesh, Maggs, and Panigrahi [2023] for a recent example and further references.

17.6 Vehicle Routing

One of the most natural and frequent applications of the traveling salesman problem is when a vehicle must visit a certain set of locations (customers) with a tour that starts and ends at a given depot. When we have more than one vehicle, the problem generalizes in several ways.

In the first generalization, every vehicle may have its own starting position (depot), and we want to visit all customers but do not care which vehicle serves which customers. This problem is a special case of Multi-Path TSP (and Φ-TSP) and has a simple 2-approximation algorithm (contract the depots, find a minimum-cost spanning tree, double all edges). If the number of depots is bounded by a constant, Xu and Rodrigues [2015] devised a $\frac{3}{2}$-approximation algorithm, and Corollary 16.25 yields a slightly better approximation guarantee.

In another generalization, the vehicles all start at a common depot but have a capacity constraint. In the simplest case, every vehicle can serve up to Q customers; this is called the *unit-demand* special case. As for TSP, there is a symmetric and an asymmetric version. In a more general version, customers have different demands, and there is a limit on the total demand that a vehicle can serve. We scale down demands so that without loss of generality, the vehicle capacity is 1:

Problem 17.16 (CAPACITATED VEHICLE ROUTING).

Instance: A finite set V (of customers), a depot $s \notin V$, and a cost function $c : (\{s\} \cup V) \times (\{s\} \cup V) \to \mathbb{R}_{\geq 0}$ that satisfies the triangle inequality. A demand $d(v) \in [0, 1]$ for each customer $v \in V$.

Task: Compute a set of *tours*, where each tour is a circuit containing s, each customer belongs to exactly one tour, and the total demand of the customers in a tour is at most 1, such that the total cost of all tours is minimum.

Haimovich and Rinnooy Kan [1985] proposed the following algorithm for the symmetric unit-demand special case (where $d(v) = \frac{1}{Q}$ for all $v \in V$). Find a single tour through $\{s\} \cup V$ by applying an approximation algorithm for the SYMMETRIC TSP WITH TRIANGLE INEQUALITY to $(\{s\} \cup V, c)$. Note that the best tour that visits all of $\{s\} \cup V$ costs at most OPT. Then, for the best choice of $i \in \{1, \ldots, Q\}$, splitting this tour after $i, i + Q, i + 2Q, \ldots$ customers increases the cost by at most OPT as we will see in Theorem 17.18. Altinkemer and Gavish [1987] extended this idea to the general case. We need the following lower bound:

Lemma 17.17. *For every* CAPACITATED VEHICLE ROUTING *instance* (V, s, c, d), *every feasible solution costs at least* $\sum_{v \in V} d(v)(c(s, v) + c(v, s))$.

Proof. Any tour that serves a subset W of customers of total demand at most 1 (and contains s) has cost at least $c(s, v) + c(v, s)$ for every $v \in W$ (by the triangle inequality) and hence at least $\sum_{v \in W} d(v)(c(s, v) + v(v, s))$. The assertion follows by summing over all tours in a feasible solution. □

Haimovich and Rinnooy Kan [1985] and Altinkemer and Gavish [1987] formulated their results for the symmetric case, but they also work in the asymmetric case:

Theorem 17.18. *If there is an α-approximation algorithm for* SYMMETRIC/ASYMMETRIC TSP, *then there is an $(\alpha + 2)$-approximation algorithm for* SYMMETRIC/ASYMMETRIC CAPACITATED VEHICLE ROUTING. *For the unit-demand special case, we obtain an $(\alpha + 1)$-approximation.*

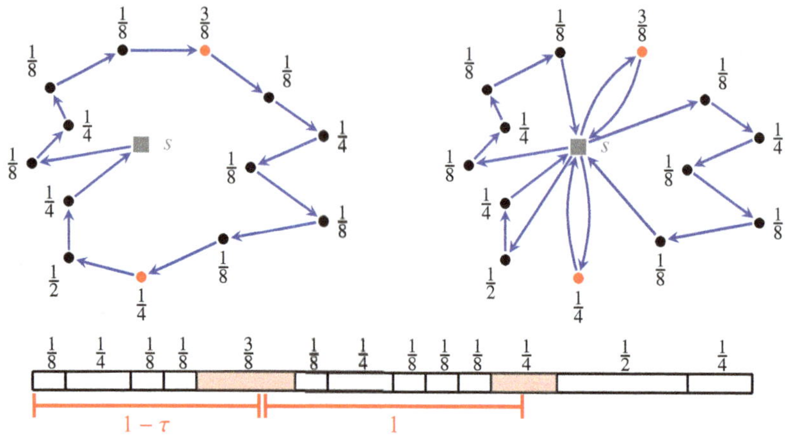

Figure 17.4 Illustration of the tour partitioning algorithm from the proof of Theorem 17.18. The numbers next to the customers show their demands. On the left, a tour serving $\{s\} \cup V$ is shown. On the right, we see the CAPACITATED VEHICLE ROUTING solution that we obtain from this tour for $\tau = \frac{1}{8}$. Below, there is an illustration of the customer demands in the order in which they appear in the tour shown on the left. The customers v_i with $i \in I$ are highlighted in red.

Proof. Let T be a tour serving $\{s\} \cup V$ of cost at most $\alpha \cdot \mathrm{OPT}$, computed by the α-approximation algorithm for SYMMETRIC/ASYMMETRIC TSP (and applying Lemma 1.8). Let $V = \{v_1, \ldots, v_n\}$ be the order in which T visits the customers. For an offset τ randomly chosen in $[0, 1)$, consider the set $I \subseteq \{1, \ldots, n\}$ of indices i for which $\lfloor \tau + \sum_{j=1}^{i-1} d(v_j) \rfloor < \lfloor \tau + \sum_{j=1}^{i} d(v_j) \rfloor$. See Figure 17.4.

If i_1 and i_2 are two consecutive indices in I, then the total demand of $v_{i_1+1}, \ldots, v_{i_2-1}$ is less than 1. Hence, by taking the piece of the tour from v_{i_1} to v_{i_2} and adding two edges from s to v_{i_1}, one edge from v_{i_1} to s, and one edge from v_{i_2} to s, we can design a tour that serves v_{i_1} only and a tour that serves $v_{i_1+1}, \ldots, v_{i_2-1}$. We proceed analogously for the piece before the first customer with index in I and the piece beginning at the last customer with index in I.

Summing up, we obtain a feasible solution of total cost at most $c(T) + \sum_{i \in I}(2c(s, v_i) + 2c(v_i, s))$. In the unit-demand case, the total demand of $v_{i_1}, \ldots, v_{i_2-1}$ is at most 1, so we do not need the separate single-customer tour and obtain a solution of total cost at most $c(T) + \sum_{i \in I}(c(s, v_i) + c(v_i, s))$.

Now a customer v_i belongs to I with probability $d(v_i)$. Hence, using Lemma 17.17, the expected total cost is at most $(\alpha + 2) \cdot \mathrm{OPT}$, and at most $(\alpha + 1) \cdot \mathrm{OPT}$ in the unit-demand case. We obtain a deterministic algorithm by trying all relevant values of τ (there are at most n) or by splitting the tour T in the cheapest possible way by dynamic programming. □

By Theorem 8.25, this yields an approximation ratio $19 + \varepsilon$ for ASYMMETRIC CAPACITATED VEHICLE ROUTING (for any $\varepsilon > 0$), as well as an approximation ratio $\frac{7}{2}$ for SYMMETRIC CAPACITATED VEHICLE ROUTING with Christofides' algorithm (Theorem 1.32), and slightly better with Theorem 11.20. Only recently, the approximation ratio for SYMMETRIC CAPACITATED VEHICLE ROUTING (as well as the unit-demand special case) has been improved by Blauth, Traub, and Vygen [2023] via a better black-box reduction to SYMMETRIC TSP. The currently best approximation ratio for CAPACITATED VEHICLE ROUTING with general demands is due to Friggstad et al. [2022] (see Exercises 17.6 and 17.7).

Unless $P = NP$, an algorithm with approximation ratio better than $\frac{3}{2}$ does not exist for SYMMETRIC CAPACITATED VEHICLE ROUTING PROBLEM: If $c(s, v) = 1$ and $c(v, w) = 0$ for all $v, w \in V$, the problem is equivalent to BIN PACKING, and to decide whether two vehicles suffice (equivalently, whether OPT ≤ 4 or OPT ≥ 6) contains the well-known *NP*-complete problem PARTITION.

Many variants have been studied. Vehicles can have capacities and different depot locations, there can be a distance constraint for every tour, and all other variants for TSP that we discussed in the previous sections can be applied to vehicle routing as well.

In practice, the situation is even more complicated. For example, travel times depend on the time of the day, one needs to schedule breaks, items can have different pickup and delivery locations as well as time windows, and vehicles can have different properties. We refer to the book edited by Toth and Vigo [2014] for a general overview on various aspects of vehicle routing and to Blauth et al. [2022] for an example of a recent heuristic that can handle all of the above and is applied successfully in practice.

Although approximation algorithms for special cases, starting with the TSP, are normally not directly applied in practice, their study has not only greatly advanced the theory of combinatorial optimization, but also our understanding of practical problems.

Exercises

17.1 Show that the problem of finding a minimum-weight 2-edge-connected spanning subgraph of a given complete undirected graph $G = (V, E)$ with weights $c : E \to \mathbb{R}_{\geq 0}$ that satisfy the triangle inequality is equivalent to the problem of finding a minimum-weight 2-edge-connected spanning multi-subgraph in a general connected graph with nonnegative weights. (Frederickson and Ja'ja' [1982]; see also Alexander, Boyd, and Elliott-Magwood [2006])

17.2 A special case of CLUSTERED TSP has been called TSP WITH BACKHAULS. Here we have a finite metric space (V, c) where V is partitioned into a set of "linehaul customers," a set of "backhaul customers," and a single depot. Starting at the depot, the linehaul customers must be visited before the backhaul customers.

(a) Obtain a $\frac{3}{2}$-approximation algorithm with running time $O(n^3)$.
Hint: Proceed similarly as Christofides' algorithm, but compose the spanning tree of two separate trees and find a matching that uses exactly one edge connecting a linehaul and a backhaul customer.
(Gendreau, Laporte, and Hertz [1997])

(b) Prove that any α-approximation algorithm for SYMMETRIC TSP implies an $(\alpha + \varepsilon)$-approximation algorithm for TSP WITH BACKHAULS.
Hint: Use the results of Chapter 16.

17.3 Show that choosing τ uniformly at random from $[e^{-1/\alpha}, 1]$ in Theorem 17.8 yields an approximation ratio $\frac{1}{1-e^{-1/\alpha}}$ for the PRIZE-COLLECTING TSP. Note that this can easily be derandomized.
(Goemans [2009])

17.4 Consider an instance (V, s, c, p) of A PRIORI TSP with k^2 customers for some $k \in \mathbb{N}$, grouped into k groups, each consisting of k customers that have distance 0 from each other. Let all other distances be 1, and let $p(v) = \frac{1}{2k}$ for all $v \in V$. Show that for this set of instances, the ratio of the best master route solution ((17.3) for the best S) and the overall best a priori solution tends to $\frac{1}{1-e^{-1/2}}$ as $k \to \infty$.
(Blauth et al. [2023])

17.5 Let ρ denote the integrality ratio of the Asymmetric TSP LP (3.2). For a CAPACITATED VEHICLE ROUTING instance (V, s, c, d), let $E = \{(v, w) : v, w \in \{s\} \cup V, v \neq w\}$. Show that the cost of an optimum solution is at most $(\rho + 2)$ times the value of the LP relaxation

$$\begin{array}{llcll} \min & c(x) & & & \\ \text{subject to} & x(\delta(U)) & \geq & 2d(U) & (\emptyset \neq U \subseteq V) \\ & x(\delta(U)) & \geq & 2 & (\emptyset \neq U \subseteq V) \\ & x(\delta^+(v)) & = & x(\delta^-(v)) & (v \in \{s\} \cup V) \\ & x_e & \geq & 0 & (e \in E) \end{array}$$

(and hence the integrality ratio of this LP is at most $\rho + 2$).

17.6 Let $0 < \delta < \frac{1}{2}$ be a constant. Consider an instance of CAPACITATED VEHICLE ROUTING. Scale up all demands by a factor $\frac{1}{1-\delta}$ and run the algorithm in the proof of Theorem 17.18. In a post-processing step, merge a single-item tour serving v_i (for $i \in I$) with the previous tour if the actual (unscaled) total demand is at most 1. Show that we obtain a solution of cost at most

$$\alpha \cdot \mathrm{OPT} + \tfrac{1}{1-\delta} \sum_{v \in V} \big(2d(v) - \min\{\delta, d(v)\}\big) \cdot \big(c(s,v) + c(v,s)\big).$$

(Friggstad et al. [2022])

17.7 Consider the following "relative greedy" algorithm for CAPACITATED VEHICLE ROUTING. Let $\delta > 0$ be a constant, and call a customer v *big* if $d(v) > \delta$. For a big customer $v \in V$, denote by

$$\sigma(v) \;:=\; \tfrac{1}{1-\delta}\big(2d(v) - \delta\big) \cdot \big(c(s,v) + c(v,s)\big).$$

Initially, let $\mathscr{C} = \emptyset$ and $V' = V$. As long as there exists a cycle C whose vertices consist of s and some big customers from V' such that $d(V(C)) \leq 1$ and $c(E(C)) \leq \sum_{v \in V(C) \setminus \{s\}} \sigma(v)$, choose such a cycle minimizing

$$\frac{c(E(C))}{\sum_{v \in V(C) \setminus \{s\}} \sigma(v)},$$

add C to $\mathscr{C}$, and remove the customers in $V(C) \setminus \{s\}$ from V'. At the end, complete $\mathscr{C}$ to a CAPACITATED VEHICLE ROUTING solution by applying the algorithm from Exercise 17.6 to V'.

(a) Prove that this algorithm can be implemented to run in polynomial time.

(b) Show that there always exists a cycle C whose vertices consist of s and some big customers from V' such that $d(V(C)) \leq 1$ and

$$\frac{c(E(C))}{\sum_{v \in V(C) \setminus \{s\}} \sigma(v)} \leq \frac{\mathrm{OPT}}{\sum_{v \in V' : d(v) > \delta} \sigma(v)},$$

where OPT denotes the cost of an optimum solution to the given instance of the CAPACITATED VEHICLE ROUTING PROBLEM.

(c) Prove that the above algorithm is an $\big(\alpha + \frac{1}{1-\delta} + \ln(2)\big)$-approximation algorithm.

Note: This is a variant of an algorithm by Friggstad et al. [2022].

18

State of the Art, Open Problems

In this short concluding chapter, we show two figures that summarize the state of the art. Moreover, we list again the open problems that we mentioned in this book.

18.1 Summary of the State of the Art

Figures 18.1 and 18.2 summarize the state of the art for ASYMMETRIC TSP and SYMMETRIC TSP, respectively, including its path versions and the special cases in unweighted graphs. These are the eight problems that played principal roles in this book. We also show what reductions are known between these problems. Of course, there is also a trivial reduction from each symmetric problem to the corresponding asymmetric problem.

We show the best-known approximation ratios and the best-known upper bounds on the integrality ratios of the natural LP relaxations. We remark that a small improvement for ASYMMETRIC (PATH) TSP can be obtained by Exercises 7.5–7.7. Moreover, a small improvement (to $\frac{3}{2} - 10^{-34}$) for SYMMETRIC TSP was announced very recently by Gurvits, Klein, and Leake [2023], and this yields essentially the same approximation ratio for PATH TSP.

The lower bounds on the approximability are very close to 1 (see Section 17.1). The lower bounds on the integrality ratios are 2 for all asymmetric problems (Theorem 3.18), $\frac{4}{3}$ for SYMMETRIC TSP and GRAPH TSP (Proposition 2.24), and $\frac{3}{2}$ for the symmetric path versions (Proposition 14.15).

The GRAPH PATH TSP is the only one of the eight problems for which we know the integrality ratio: It is exactly $\frac{3}{2}$. Nevertheless, we could obtain a better approximation ratio than $\frac{3}{2}$ (Traub and Vygen [2023]). In fact, by the results of Chapter 16, we know that (GRAPH) PATH TSP is not much harder to approximate than (GRAPH) TSP. Although we have currently the same approximation ratio for

Asymmetric TSP and Asymmetric Path TSP, and also the same upper bound on the integrality ratio, the only known black-box reductions from the latter to the former lose a factor $2+\varepsilon$ (cf. Theorem 9.8) and almost 4 (cf. Theorem 9.21), respectively.

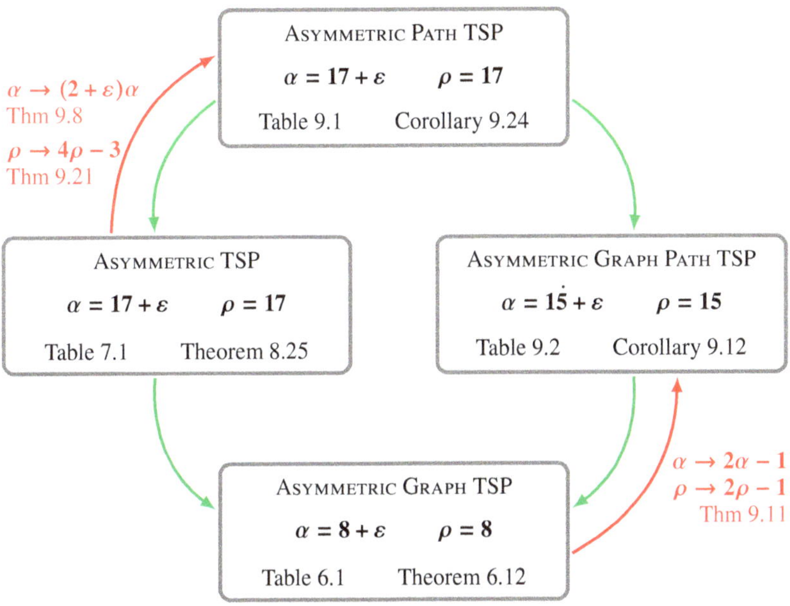

Figure 18.1 Approximation ratios (α) and upper bounds on integrality ratios (ρ) for asymmetric traveling salesman problems: the state of the art. An arrow from one problem to another means an approximation or integrality ratio bound for the former implies an approximation or integrality ratio bound for the latter (the same by a trivial reduction for the green arcs without labels). The only known lower bound on the integrality ratio of all these problems is 2 (Theorem 3.18).

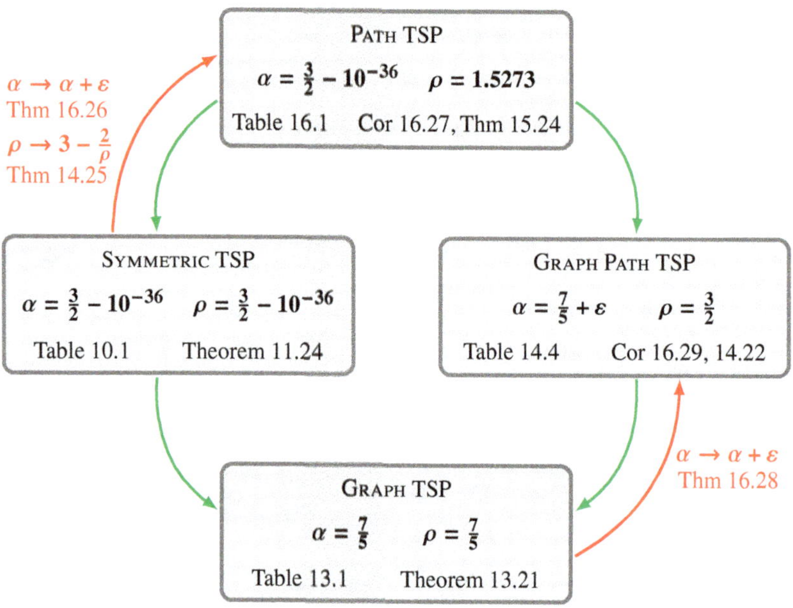

Figure 18.2 Approximation ratios (α) and upper bounds on integrality ratios (ρ) for symmetric traveling salesman problems: the state of the art. An arrow from one problem to another means an approximation or integrality ratio bound for the former implies an approximation or integrality ratio bound for the latter (the same by a trivial reduction for the green arcs without labels). As lower bounds on the integrality ratios, we have $\frac{4}{3}$ for Graph TSP and Symmetric TSP (Proposition 2.24) and $\frac{3}{2}$ for (Graph) Path TSP (Proposition 14.15).

18.2 Open Problems

We conclude this survey by listing some open research problems that we consider important. Almost all of these problems have been formulated earlier, and indeed most of them are very natural. None of them seems to be easy. However, given the remarkable progress that has been made during the last few years, one may hope that we will see some solutions soon.

Open Problem 1.33. Find an α-approximation algorithm for SYMMETRIC TSP for some $\alpha \ll \frac{3}{2}$ (say $\alpha \leq 1.49$).
(The best-known approximation ratio is $\frac{3}{2} - 10^{36}$: see Theorem 11.20.)

Open Problem 2.12. Find a (fully) combinatorial polynomial-time algorithm to solve the subtour LP (2.2) exactly.
(All known exact algorithms rely on linear programming: see Corollary 2.11.)

Open Problem 2.25. Prove or disprove that the integrality ratio of the subtour LP (2.2) is $\frac{4}{3}$.
(We know it is at least $\frac{4}{3}$ and less than $\frac{3}{2}$: see Proposition 2.24 and Theorem 10.19.)

Open Problem 2.33. Prove that there exists a polynomial-time solvable LP relaxation of the SYMMETRIC TSP WITH TRIANGLE INEQUALITY that has a smaller integrality ratio than the subtour LP.
(Adding several classes of constraints to the subtour LP is used successfully for solving TSP instances exactly, e.g. by Applegate et al. [2006], and was studied by Goemans [2006].)

Open Problem 3.19. Prove or disprove that the integrality ratio of the LP relaxation (3.2) of the ASYMMETRIC TSP is 2.
(We know it is at least 2: see Theorem 3.18; and it is at most 17: see Corollary 8.26; or actually slightly less: see Exercise 7.7.)

Open Problem 5.26. Does there exist a constant α such that for every connected undirected graph $H = (V, F)$ and any point y in the spanning tree polytope of H, there exists a spanning tree S with $|\delta_S(U)| \leq \alpha \cdot y(\delta(U))$ for all $U \subseteq V$?
(This has been called the thin tree conjecture: see Section 5.5.)

Open Problem 6.16. Devise a 2-approximation algorithm for the ASYMMETRIC GRAPH TSP.
(The best-known approximation ratio is $8 + \varepsilon$: see Theorem 6.12.)

Open Problem 8.27. Obtain an approximation ratio of $(8 + \varepsilon)$ or better for the ASYMMETRIC TSP.
(This is known only for ASYMMETRIC GRAPH TSP. The best-known approximation ratio is $17 + \varepsilon$: see Theorem 8.25; or actually slightly less: see Exercise 7.7.)

Open Problem 9.25. Does the LP relaxation (9.1) of the ASYMMETRIC PATH TSP have the same integrality ratio as the LP relaxation (3.2) of the ASYMMETRIC TSP?
(If ρ denotes the integrality ratio of (3.2), then we only know that the integrality ratio of (9.1) is at least ρ and at most $\min\{17, 4\rho - 3\}$: see Corollary 9.24 and Theorem 9.21.)

Open Problem 12.12. What is the approximation ratio of the Mömke–Svensson algorithm for GRAPH TSP?
(We know it is at least $\frac{4}{3}$ and at most $\frac{11}{9}$: see Exercise 12.3 and Theorem 12.11.)

Open Problem 13.22. Devise a $\frac{4}{3}$-approximation algorithm for GRAPH TSP.
(The best-known approximation ratio is $\frac{7}{5}$: see Theorem 13.21.)

Open Problem 15.12. What is the approximation ratio of the Best-of-Many Christofides algorithm for PATH TSP and for the T-TOUR PROBLEM?
(For $T = \emptyset$, the SYMMETRIC TSP, the answer is $\frac{3}{2}$: see Proposition 15.2. In general, we only know the upper bound $\frac{8}{5}$: see Theorem 15.11.)

Open Problem 15.25. Prove that the integrality ratio of the LP relaxation (14.1) of PATH TSP is exactly $\frac{3}{2}$.
(We know it is at least $\frac{3}{2}$ and less than 1.528: see Proposition 14.15 and Theorem 15.24. We also know that $\frac{3}{2}$ is the exact integrality ratio for GRAPH PATH TSP instances: see Corollary 14.22.)

Open Problem 15.27. Is there a $\frac{3}{2}$-approximation algorithm for the T-tour problem?
(This is known only for unweighted graphs and for constant $|T|$: see Exercise 13.8 and Exercise 16.9. For general instances, the best-known approximation ratio is $\frac{11}{7}$: see Theorem 15.26.)

Open Problem 16.30. Is it true that any α-approximation algorithm for ASYMMETRIC TSP implies an $(\alpha + \varepsilon)$-approximation algorithm for ASYMMETRIC PATH TSP?
(The best-known black-box reduction implies a $(2\alpha + \varepsilon)$-approximation algorithm only: see Theorem 9.8.)

Open Problem 16.31. Is it true that any α-approximation algorithm for SYMMETRIC TSP implies an $(\alpha+\varepsilon)$-approximation algorithm for the T-TOUR PROBLEM?
(This is known only for instances where $|T|$ is bounded by a constant: see Exercise 16.9.)

Open Problem 17.1. Improve the lower bounds on the approximability of any of the studied problems substantially.
(The current bounds are quite close to 1: see Section 17.1.)

Open Problem 17.6. Find an approximation algorithm with an approximation ratio better than 2 for the TWO-EDGE-CONNECTED SPANNING SUBGRAPH PROBLEM.
(This is only known in unweighted graphs or in complete graphs with weights satisfying the triangle inequality.)

Open Problem 17.9. Find a constant-factor approximation algorithm for ASYMMETRIC ORIENTEERING or prove that this would imply $P = NP$.
(This is known only for the symmetric special case and for the ASYMMETRIC PRIZE-COLLECTING TSP: see Section 17.4.)

Open Problem 17.15. Find a constant-factor approximation algorithm for ASYMMETRIC A PRIORI TSP or prove that this would imply $P = NP$.
(This is known only for the symmetric special case: see Section 17.5.)

Bibliography

Ageev, A.A., and Sviridenko, M.I. [2004]: Pipage rounding: A new method of constructing algorithms with proven performance guarantee. Journal on Combinatorial Optimization 8 (2004), 307–328

Agrawal, A., Klein, P.N., and Ravi, R. [1995]: When trees collide: An approximation algorithm for the generalized Steiner tree problem in networks. SIAM Journal on Computing 24 (1995), 440–456

Alexander, A., Boyd, S., and Elliott-Magwood, P. [2006]: On the integrality gap of the 2-edge connected subgraph problem. Technical Report TR-2006-04, SITE, University of Ottawa 2006

Altinkemer, K., and Gavish, B. [1987]: Heuristics for unequal weight delivery problems with a fixed error guarantee. Operations Research Letters 6 (1987), 149–158

An, H.-C., Kleinberg, R., and Shmoys, D.B. [2015]: Improving Christofides' algorithm for the s-t path TSP. Journal of the ACM 62 (2015), Article 34

An, H.-C., Kleinberg, R., and Shmoys, D.B. [2021]: Approximation algorithms for the bottleneck asymmetric traveling salesman problem. ACM Transactions on Algorithms 17 (2021), 1–12

Anari, N., Liu, K., Oveis Gharan, S., Vinzant, C., and Vuong, T.-D. [2021]: Log-concave polynomials IV: Approximate exchange, tight mixing times, and near-optimal sampling of forests. Proceedings of the 53rd Annual ACM Symposium on Theory of Computing (STOC 2021), 408–420

Anari, N., and Oveis Gharan, S. [2015]: Effective-resistance-reducing flows, spectrally thin trees, and asymmetric TSP. Proceedings of the 56th Annual IEEE Symposium on Foundations of Computer Science (FOCS 2015), 20–39

Applegate, D.L., Bixby, R., Chvátal, V., and Cook, W.J. [2006]: The Traveling Salesman Problem: A Computational Study. Princeton University Press, Princeton 2006

Archer, A., Bateni, M., Hajiaghayi, M., and Karloff, H. [2011]: Improved approximation algorithms for prize-collecting Steiner tree and TSP. SIAM Journal on Computing 30 (2011), 309–332

Arora, S. [1998]: Polynomial time approximation schemes for Euclidean traveling salesman and other geometric problems. Journal of the ACM 45 (1998), 753–782

Arora, S., and Karakostas, G. [2006]: A $2+\varepsilon$ approximation algorithm for the k-MST problem. Mathematical Programming 107 (2006), 491–504

Arora, S., Lund, C., Motwani, R., Sudan, M., and Szegedy, M. [1998]: Proof verification and hardness of approximation problems. Journal of the ACM 45 (1998), 501–555

Asadpour, A., Goemans, M.X., Mądry, A., Oveis Gharan, S., and Saberi, A. [2017]: An $O(\log n/\log\log n)$-approximation algorithm for the asymmetric traveling salesman problem. Operations Research 65 (2017), 1043–1061

Balas, E. [1989]: The prize collecting traveling salesman problem. Networks 19 (1989), 621–636

Bang-Jensen, J., Gabow, H.N., Jordán, T., and Szigeti, Z. [1999]: Edge-connectivity augmentation with partition constraints. SIAM Journal on Discrete Mathematics 12 (1999), 160–207

Bansal, N., Blum, A., Chawla, S., and Meyerson, A. [2004]: Approximation algorithms for deadline-TSP and vehicle routing with time-windows. Proceedings of the 36th Annual ACM Symposium on Theory of Computing (STOC 2004), 166–174

Barahona, F., and Conforti, M. [1987]: A construction for binary matroids. Discrete Mathematics 66 (1987), 213–218

Bartal, Y., and Gottlieb, L.-A. [2013]: A linear time approximation scheme for Euclidean TSP. Proceedings of the 54th Annual IEEE Symposium on Foundations of Computer Science (FOCS 2013), 698–706

Bartal, Y., Gottlieb, L.-A., and Krauthgamer, R. [2016]: The traveling salesman problem: Low-dimensionality implies a polynomial time approximation scheme. SIAM Journal on Computing 45 (2016), 1563–1581

Bateni, M., and Chuzhoy, J. [2013]: Approximation algorithms for the directed k-tour and k-stroll problems. Algorithmica 65 (2013), 545–561

Beideman, C., Chandrasekaran, K., and Wang, W. [2023]: Approximate minimum cuts and their enumeration. Proceedings of the 6th Symposium on Simplicity in Algorithms (SOSA 2023), 36–41

Bellman, R. [1962]: Dynamic programming treatment of the travelling salesman problem. Journal of the ACM 9 (1962), 61–63

Benczúr, A.A. [1995]: A representation of cuts within 6/5 times the edge connectivity with applications. Proceedings of the 36th Annual IEEE Symposium on Foundations of Computer Science (FOCS 1995), 92–102

Benczúr, A.A., and Goemans, M.X. [2008]: Deformable polygon representation and near-mincuts. In: Building Bridges: Between Mathematics and Computer Science (M. Grötschel and G.O.H. Katona, eds.), Bolyai Society Mathematical Studies 19, Budapest, and Springer, Berlin 2008, pp. 103–135

Bender, M.A., and Chekuri, C. [2000]: Performance guarantees for the TSP with a parameterized triangle inequality. Information Processing Letters 73 (2000), 17–21

Benoit, G., and Boyd, S. [2008]: Finding the exact integrality gap for small traveling salesman problems. Mathematics of Operations Research 33 (2008), 921–931

Bérczi, K., Mnich, M., and Vincze, R. [2022]: A 3/2-approximation for the metric many-visits path TSP. SIAM Journal on Discrete Mathematics 36 (2022), 2995–3030

Berman, P., and Karpinski, M. [2006]: 8/7-approximation algorithm for (1,2)-TSP. Proceedings of the 17th Annual ACM-SIAM Symposium on Discrete Algorithms (SODA 2006), 641–648

van Bevern, R., and Slugina, V.A. [2020]: A historical note on the 3/2-approximation algorithm for the metric traveling salesman problem. Historia Mathematica 53 (2020), 118–127

Bienstock, D., Goemans, M.X., Simchi-Levi, D., and Williamson, D. [1993]: A note on the prize collecting traveling salesman problem. Mathematical Programming 39 (1993), 413–420

Bläser, M. [2008]: A new approximation algorithm for the asymmetric TSP with triangle inequality. ACM Transactions on Algorithms 4 (2008), Article 47

Blauth, J., Held, S., Müller, D., Schlomberg, N., Traub, V., Tröbst, T., and Vygen, J. [2022]: Vehicle routing with time-dependent travel times: theory, practice, and benchmarks. arXiv:2205.00889

Blauth, J., Klein, N., and Nägele, M. [2023]: A better-than-1.6-approximation for Prize-Collecting TSP. arXiv:2308.06254. To appear in IPCO 2024

Blauth, J., and Nägele, M. [2023]: An improved approximation guarantee for Prize-Collecting TSP. Proceedings of the 55th Annual ACM Symposium on Theory of Computing (STOC 2023), 1848–1861

Blauth, J., Neuwohner, M., Puhlmann, L., and Vygen, J. [2023]: Improved guarantees for the a priori TSP. Proceedings of the 34th International Symposium on Algorithms and Computation (ISAAC 2023), Article 14

Blauth, J., Traub, V., and Vygen J. [2023]: Improving the approximation ratio for capacitated vehicle routing. Mathematical Programming 197 (2023), 451–497

Blum, A., Chawla, S., Karger, D.R., Lane, T., Meyerson, A., and Minkoff, M. [2007]: Approximation algorithms for orienteering and discounted-reward TSP. SIAM Journal on Computing 37 (2007), 653–670

Blum, A., Ravi, R., and Vempala, S. [1999]: A constant-factor approximation for the k-MST problem. Journal of Computer and System Sciences 58 (1999), 101–108

Bock, F.C. [1971]: An algorithm to construct a minimum directed spanning tree in a directed network. In: Developments in Operations Research, Volume I (B. Avi-Itzhak, ed.), Gordon and Breach, New York 1971, pp. 29–44

Borcea, J., Brändén, P., and Liggett, T. [2009]: Negative dependence and the geometry of polynomials. Journal of the American Mathematical Society 22 (2009), 521–567

Borradaile, G., Le, H., and Wulff-Nilsen, C. [2017]: Minor-free graphs have light spanners. Proceedings of the 58th Annual IEEE Symposium on Foundations of Computer Science (FOCS 2017), 767–778

Borůvka, O. [1926]: O jistém problému minimálním. Práca Moravské Přírodovědecké Společnosti 3 (1926), 37–58 [in Czech]

Bosch-Calvo, M., Grandoni, F., and Jabal Ameli, A. [2023]: A 4/3 approximation for 2-vertex-connectivity. Proceedings of the 50th International Colloquium on Automata, Languages, and Programming (ICALP 2023), Article 29

Boyd, S., and Carr, R. [2011]: Finding low cost TSP and 2-matching solutions using certain half-integer subtour vertices. Discrete Optimization 8 (2011), 525–539

Boyd, S., and Elliott-Magwood, P. [2005]: Computing the integrality gap of the asymmetric traveling salesman problem. Electronic Notes in Discrete Mathematics 19 (2005), 241–247

Boyd, S., and Elliott-Magwood, P. [2010]: Structure of the extreme points of the subtour elimination polytope of the STSP. RIMS Kôkyûroku Bessatsu B23 (2010), 33–47

Boyd, S., Fu, Y., and Sun, Y. [2016]: A $\frac{5}{4}$-approximation for subcubic 2EC using circulations and obliged edges. Discrete Applied Mathematics 209 (2016), 48–58

Boyd, S.C., and Pulleyblank, W.R. [1991]: Optimizing over the subtour polytope of the travelling salesman problem. Mathematical Programming 49 (1991), 163–187

Boyd, S., and Sebő, A. [2021]: The salesman's improved tours for fundamental classes. Mathematical Programming 186 (2021), 289–307

Boyd, S., Sitters, R., van der Ster, S., and Stougie, L. [2014]: The traveling salesman problem on cubic and subcubic graphs. Mathematical Programming 144 (2014), 227–245

van den Brand, J., Chen, L., Kyng, R., Liu, Y.P., Peng, R., Gutenberg, M.P., Sachdeva, S., and Sidford, A. [2023]: A deterministic almost-linear time algorithm for minimum-cost flow. Proceedings of the 64th Annual IEEE Symposium on Foundations of Computer Science (FOCS 2023), 503–514

Brooks, R.L., Smith, C.A.B., Stone, A.H., and Tutte, W.T. [1940]: The dissection of rectangles into squares. Duke Mathematical Journal 7 (1940), 312–340

Calinescu, G., Chekuri, C., Pál, M., and Vondrák, J. [2011]: Maximizing a monotone submodular function subject to a matroid constraint. SIAM Journal on Computing 40 (2011), 1740–1766

Carathéodory, C. [1911]: Über den Variabilitätsbereich der Fourierschen Konstanten von positiven harmonischen Funktionen. Rendiconto del Circolo Matematico di Palermo 32 (1911), 193–217 [in German]

Carr, R., and Ravi, R. [1998]: A new bound for the 2-edge connected subgraph problem. Proceedings of the 6th Conference on Integer Programming and Combinatorial Optimization (IPCO 1998), 112–125

Carr, R., and Vempala, S. [2004]: On the Held–Karp relaxation for the asymmetric and symmetric traveling salesman problems. Mathematical Programming 100 (2004), 569–587

Cen, R., Li, J., and Panigrahi, D. [2022]: Edge connectivity augmentation in near-linear time. Proceedings of the 54th Annual ACM Symposium on Theory of Computing (STOC 2022), 137–150

Chalasani, P., Motwani, R., and Rao, A. [1997]: Algorithms for robot grasp and delivery. Proceedings of the 2nd International Workshop on Algorithmic Foundations of Robotics (WAFR 1997), 347–362

Charikar, M., Goemans, M.X., and Karloff, H. [2006]: On the integrality ratio for the asymmetric traveling salesman problem. Mathematics of Operations Research 31 (2006), 245–252

Chaudhuri, K., Godfrey, B., Rao, S., and Talwar, K. [2003]: Paths, trees, and minimum latency tours. Proceedings of the 44th Annual IEEE Symposium on Foundations of Computer Science (FOCS 2003), 36–45

Chekuri, C., Korula, N., and Pál, M. [2012]: Improved algorithms for orienteering and related problems. ACM Transactions on Algorithms 8 (2012), Article 23

Chekuri, C., and Pál, M. [2007]: An $O(\log n)$ approximation ratio for the asymmetric traveling salesman path problem. Theory of Computing 3 (2007), 197–209

Chekuri, C., and Quanrud, K. [2017]: Approximating the Held–Karp bound for metric TSP in nearly-linear time. Proceedings of the 58th Annual IEEE Symposium on Foundations of Computer Science (FOCS 2017), 789–800

Chekuri, C., and Quanrud, K. [2018]: Faster approximations for Metric-TSP via linear programming. arXiv:1802.01242

Chekuri, C., Quanrud, K., and Torres, M.R. [2021]: Fast approximation algorithms for bounded degree and crossing spanning tree problems. Proceedings of the

24th International Workshop on Approximation Algorithms for Combinatorial Optimization Problems (APPROX 2021), Article 24

Chekuri, C., Quanrud, K., and Xu, C. [2020]: LP relaxation and tree packing for minimum k-cut. SIAM Journal on Discrete Mathematics 34 (2020), 1334–1353

Chekuri, C., Vondrák, J., and Zenklusen, R. [2010]: Dependent randomized rounding via exchange properties of combinatorial structures. Proceedings of the 51st Annual IEEE Symposium on Foundations of Computer Science (FOCS 2010), 575–584

Chen, L., Kyng, R., Liu, Y.P., Peng, R., Gutenberg, M.P., and Sachdeva, S. [2022]: Maximum flow and minimum-cost flow in almost-linear time. Proceedings of the 63rd Annual IEEE Symposium on Foundations of Computer Science (FOCS 2022), 612–623

Cheriyan, J., Friggstad, Z., and Gao, Z. [2015]: Approximating minimum-cost connected T-joins. Algorithmica 72 (2015), 126–147

Cheriyan, J., Karloff, H., Khandekar, R., and Könemann, J. [2008]: On the integrality ratio for tree augmentation. Operations Research Letters 36 (2008), 399–401

Cheriyan, J., Sebő, A., and Szigeti, Z. [2001]: Improving on the 1.5-approximation of a smallest 2-edge connected spanning subgraph. SIAM Journal on Discrete Mathematics 14 (2001), 170–180

Chernoff, H. [1952]: A measure of asymptotic efficiency for tests of a hypothesis based on the sum of observations. Annals of Mathematical Statistics 23 (1952), 493–507

Chiplunkar, A., and Vishwanathan, S. [2015]: Approximating the regular graphic TSP in near linear time. Proceedings of the 35th IARCS Annual Conference on Foundations of Software Technology and Theoretical Computer Science (FSTTCS 2015), 125–135

Christofides, N. [1976]: Worst-case analysis of a new heuristic for the traveling salesman problem. Technical Report 388, Graduate School of Industrial Administration, Carnegie-Mellon University, Pittsburgh 1976 [reprinted in Operations Research Forum 3 (2022), Article 20]

Cook, W.J. [2012]: In Pursuit of the Traveling Salesman: Mathematics at the Limits of Computation. Princeton University Press, Princeton 2012

Cornuéjols, G., Fonlupt, J., and Naddef, D. [1985]: The traveling salesman problem on a graph and some related integer polyhedra. Mathematical Programming 33 (1985), 1–27

Correa, J., Larré, O., and Soto, J.A. [2015]: TSP tours in cubic graphs: Beyond 4/3. SIAM Journal on Discrete Mathematics 29 (2015), 915–939

Cunningham, W.H. [1984]: Testing membership in matroid polyhedra. Journal of Combinatorial Theory B 36 (1984), 161–188

Czeller, I., and Pap, G. [2014]: A note on bounded weighted graphic metric TSP. Technical Report No. 2014-03, EGRES, Budapest 2014

Dadush, D., Végh, L.A., and Zambelli. G. [2021]: Geometric rescaling algorithms for submodular function minimization. Mathematics of Operations Research 46 (2021), 1081–1108

Dantzig, G.B., and Fulkerson, D.R. [1956]: On the max-flow min-cut theorem of networks. In: Linear Inequalities and Related Systems (H.W. Kuhn and A.W. Tucker, eds.), Princeton University Press, Princeton 1956, pp. 215–221

Dantzig, G.B., Fulkerson, D.R., and Johnson, S.M. [1954]: Solution of a large scale traveling salesman problem. Operations Research 2 (1954), 393–410

Demaine, E.D., Hajiaghayi, M., and Kawarabayashi, K. [2011]: Contraction decomposition in H-minor free graphs and algorithmic applications. Proceedings of the 43rd Annual ACM Symposium on Theory of Computing (STOC 2011), 441 – 450

Demaine, E.D., Hajiaghayi, M., and Mohar, B. [2010]: Approximation algorithms via contraction decomposition. Combinatorica 30 (2010), 533–552

Dijkstra, E.W. [1959]: A note on two problems in connexion with graphs. Numerische Mathematik 1 (1959), 269–271

Dinits, E., Karzanov, A.V., and Lomonosov, M.V. [1976]: On the structure of a family of minimal weighed cuts in a graph. In: Studies in Discrete Optimization (A.A. Fridman, ed.), Nauka, Moscow 1976, pp. 290–306 [in Russian]

Disser, Y., Feldmann, A.E., Klimm, M., and Könemann, J. [2021]: Travelling on graphs with small highway dimension. Algorithmica 83 (2021), 1352–1370

Drees, M. [2022]: Simplifying the Karlin–Klein–Oveis Gharan analysis for $(\frac{3}{2} - \varepsilon)$-approximating TSP. Master's thesis, University of Bonn, 2022

Duník, B., and Lukot'ka, R. [2018]: Cubic TSP: A 1.3-approximation. SIAM Journal on Discrete Mathematics 32 (2018), 2094–2114

Dvořák, Z., Král', D., and Mohar, B. [2017]: Graphic TSP in cubic graphs. Proceedings of the 34th Symposium on Theoretical Aspects of Computer Science (STACS 2017), Article 27

Edmonds, J. [1965a]: Paths, trees, and flowers. Canadian Journal of Mathematics 17 (1965), 449–467

Edmonds, J. [1965b]: Maximum matching and a polyhedron with (0,1) vertices. Journal of Research of the National Bureau of Standards B 69 (1965), 125–130

Edmonds, J. [1967a]: Optimum branchings. Journal of Research of the National Bureau of Standards B 71 (1967), 233–240

Edmonds, J. [1967b]: Systems of distinct representatives and linear algebra. Journal of Research of the National Bureau of Standards B 71 (1967), 241–245

Edmonds, J. [1970]: Submodular functions, matroids and certain polyhedra. In: Combinatorial Structures and Their Applications; Proceedings of the Calgary International Conference on Combinatorial Structures and Their Applications 1969 (R. Guy, H. Hanani, N. Sauer, and J. Schönheim, eds.), Gordon and Breach, New York 1970, pp. 69–87

Edmonds, J. [1973]: Edge-disjoint branchings. In: Combinatorial Algorithms (R. Rustin, ed.), Algorithmic Press, New York 1973, pp. 91–96

Edmonds, J. [1975]: Some well-solved problems in combinatorial optimization. In: Combinatorial Programming: Methods and Applications (B. Roy, ed.), Reidel, Dordrecht 1975, pp. 285–301

Edmonds, J., and Johnson, E.L. [1973]: Matching, Euler tours and the Chinese postman. Mathematical Programming 5 (1973), 88–124

Edmonds, J., and Karp, R.M. [1972]: Theoretical improvements in algorithmic efficiency for network flow problems. Journal of the ACM 19 (1972), 248–264

van Ee, M., and Sitters, R. [2018]: The a priori traveling repairman problem. Algorithmica 80 (2018), 2818–2833

Euler, L. [1736]: Solutio problematis ad geometriam situs pertinentis. Commentarii Academiae Petropolitanae 8 (1736), 128–140 [in Latin]

Feder, T., and Mihail, M. [1992]: Balanced matroids. Proceedings of the 24th Annual ACM Symposium on Theory of Computing (STOC 1992), 26–38

Feige, U., Ravi, R., and Singh, M. [2014]: Short tours through large linear forests. Proceedings of the 17th Conference on Integer Programming and Combinatorial Optimization (IPCO 2014), 273–284

Feige, U., and Singh, M. [2007]: Improved approximation algorithms for traveling salesperson tours and paths in directed graphs. Proceedings of the 10th International Workshop on Approximation Algorithms for Combinatorial Optimization Problems (APPROX 2007), 104–118

Fernandes, C.G. [1998]: A better approximation ratio for the minimum size k-edge-connected spanning subgraph problem. Journal of Algorithms 28 (1998), 105–124

Fiorini, S., Massar, S., Pokutta, S., Tiwary, H.R., and de Wolf, R. [2015]: Exponential lower bounds for polytopes in combinatorial optimization. Journal of the ACM 62 (2015), Article 17

Fleiner, T., and Frank, A. [2009]: A quick proof for the cactus representation of mincuts. EGRES Quick Proof No. 2009-03, Budapest 2009

Ford, L.R., Jr., and Fulkerson, D.R. [1956]: Maximal flow through a network. Canadian Journal of Mathematics 8 (1956) 399–404

Frank, A. [1989]: On connectivity properties of Eulerian digraphs. Annals of Discrete Mathematics 41 (1989), 179–194

Frank, A. [1993]: Conservative weightings and ear-decompositions of graphs. Combinatorica 13 (1993), 65–81

Frank, A. [1994]: On the edge-connectivity algorithm of Nagamochi and Ibaraki. Laboratoire Artemis, IMAG, Université J. Fourier, Grenoble 1994

Frank, A. [2011]: Connections in Combinatorial Optimization. Oxford University Press 2011

Frank, A., and Tardos, É. [1987]: An application of simultaneous Diophantine approximation in combinatorial optimization. Combinatorica 7 (1987), 49–65

Frank, A., Triesch, E., Korte, B., and Vygen, J. [1998]: On the bipartite travelling salesman problem. Report No. 98866, Research Institute for Discrete Mathematics, University of Bonn, 1998

Frederickson, G.N. [1979]: Approximation algorithms for some postman problems. Journal of the ACM 26 (1979), 538–554

Frederickson, G.N., and Ja'ja', J. [1982]: On the relationship between the biconnectivity augmentation and travelling salesman problems. Theoretical Computer Science 19 (1982), 189–201

Frieze, A.M., Galbiati, G., and Maffioli, F. [1982]: On the worst-case performance of some algorithms for the asymmetric traveling salesman problem. Networks 12 (1982), 23–39

Friggstad, Z., Gupta, A., and Singh. M. [2016]: An improved integrality gap for asymmetric TSP paths. Mathematics of Operations Research 41 (2016), 745–757

Friggstad, Z., Mousavi, R., Rahgoshay, M., and Salavatipour, M.R. [2022]: Improved approximations for capacitated vehicle routing with unsplittable client demands. Proceedings of the 23rd Conference on Integer Programming and Combinatorial Optimization (IPCO 2022), 251–261

Friggstad, Z., Salavatipour, M.R., and Svitkina, Z. [2013]: Asymmetric traveling salesman path and directed latency problems. SIAM Journal on Computing 42 (2013), 1596–1619

Friggstad, Z., and Swamy. C. [2022]: A constant-factor approximation for directed latency in quasi-polynomial time. Journal of Computer and System Sciences 126 (2022), 44–58

Fulkerson, D.R. [1974]: Packing rooted directed cuts in a weighted directed graph. Mathematical Programming 6 (1974), 1–13

Gabow, H.N. [1973]: Implementations of algorithms for maximum matching on nonbipartite graphs. Ph.D. thesis, Stanford University, 1973

Gale, D., Kuhn, H.W., and Tucker, A.W. [1951]: Linear programming and the theory of games. In: Activity Analysis of Production and Allocation (T.C. Koopmans, ed.), Wiley, New York 1951, pp. 317–329

Gallai, T. [1964]: Maximale Systeme unabhängiger Kanten. Magyar Tudományos Akadémia; Matematikai Kutató Intézetének Közleményei 9 (1964), 401–413

Gamarnik D., Lewenstein M., and Sviridenko M. [2005]: An improved upper bound for the TSP in cubic 3-edge-connected graphs. Operations Research Letters 33 (2005), 467–474

Ganesh, A., Maggs, B.M., and Panigrahi, D. [2023]: Robust algorithms for TSP and Steiner tree. ACM Transactions on Algorithms 19 (2023), Article 12

Gao, Z. [2013]: An LP-based $\frac{3}{2}$-approximation algorithm for the s-t path graph traveling salesman problem. Operations Research Letters 41 (2013), 615–617

Gao, Z. [2015]: On the metric s-t path traveling salesman problem. SIAM Journal on Discrete Mathematics 29 (2015), 1133–1149

Garg, M., Grandoni, F., and Jabal Ameli, A. [2023]: Improved approximation for two-edge-connectivity. Proceedings of the 34th Annual ACM-SIAM Symposium on Discrete Algorithms (SODA 2023), 2368–2410

Garg, N. [2005]: Saving an epsilon: A 2-approximation for the k-MST problem in graphs. Proceedings of the 37th Annual ACM Symposium on the Theory of Computing (STOC 2005), 396–402

Gendreau, M., Laporte, G., and Hertz, A. [1997]: An approximation algorithm for the traveling salesman problem with backhauls. Operations Research 45 (1997), 639–641

Genova, K., and Williamson, D.P. [2017]: An experimental evaluation of the Best-of-Many Christofides' algorithm for the traveling salesman problem. Algorithmica 78 (2017), 1109–1130

Goddyn, L.A. [undated]: Some open problems I like. `https://www.sfu.ca/~goddyn/Problems/problems.html` (last checked on November 6, 2023)

Goemans, M.X. [1995]: Worst-case comparison of valid inequalities for the TSP. Mathematical Programming 69 (1995), 335–349

Goemans, M.X. [2006]: Minimum bounded-degree spanning trees. Proceedings of the 47th Annual IEEE Symposium on Foundations of Computer Science (FOCS 2006), 273–282

Goemans, M.X. [2009]: Combining approximation algorithms for the prize-collecting TSP. arXiv:0910.0553

Goemans, M.X. [2012]: Thinness spurs progress. OPTIMA 90 (2012), 12–14

Goemans, M.X., and Bertsimas, D.J. [1993]: Survivable networks, linear programming relaxations and the parsimonious property. Mathematical Programming 60 (1993), 145–166

Goemans, M.X., Harvey, N.J.A., Jain, K., and Singh, M. [2009]: A randomized rounding algorithm for the asymmetric traveling salesman problem. arXiv:0909.0941

Goemans, M.X., and Ramakrishnan, V.S. [1995]: Minimizing submodular functions over families of sets. Combinatorica 15 (1995), 499–513

Goemans, M.X., and Williamson, D.P. [1995]: A general approximation technique for constrained forest problems. SIAM Journal on Computing 24 (1995), 296–317

Gottschalk, C. [2013]: Approximation algorithms for the traveling salesman problem in graphs and digraphs. Master's thesis, University of Bonn, 2013

Gottschalk, C., and Vygen, J. [2018]: Better s-t-tours by Gao trees. Mathematical Programming 172 (2018), 191–207

Grötschel, M., Lovász, L., and Schrijver, A. [1981]: The ellipsoid method and its consequences in combinatorial optimization. Combinatorica 1 (1981), 169–197

Grötschel, M., Lovász, L., and Schrijver, A. [1988]: Geometric Algorithms and Combinatorial Optimization. Springer, Berlin 1988

Guan, M. [1962]: Graphic programming using odd or even points. Chinese Mathematics 1 (1962), 273–277

Gupta, A., Lee, E., Li, J., Mucha, M., Newman, H., and Sarkar, S. [2022]: Matroid-based TSP rounding for half-integral solutions. Proceedings of the 23rd Conference on Integer Programming and Combinatorial Optimization (IPCO 2022), 305–318

Gurvits, L., Klein, N., and Leake, J. [2023]: From trees to polynomials and back again: New capacity bounds with applications to TSP. arXiv:2311.09072

Gutin, G., and Punnen, A.P., eds. [2002]: The Traveling Salesman Problem and Its Variations. Kluwer, Dordrecht 2002

Guttmann-Beck, N., Knaan, E., and Stern, M. [2018]: Approximation algorithms for not necessarily disjoint clustered TSP. Journal on Graph Algorithms and Applications 22 (2018), 555–575

Haddadan, A., and Newman, A. [2023]: Towards improving Christofides algorithm on fundamental classes by gluing convex combinations of tours. Mathematical Programming 198 (2023), 595–620

Haddadan, A., Newman, A., and Ravi, R. [2021]: Shorter tours and longer detours: Uniform covers and a bit beyond. Mathematical Programming 185 (2021), 245–273

Haimovich, M., and Rinnooy Kan, A.H.G. [1985]: Bounds and heuristics for capacitated routing problems. Mathematics of Operations Research 10 (1985), 527–542

Hall, P. [1935]: On representatives of subsets. Journal of the London Mathematical Society 10 (1935), 26–30

Heeger, K., and Vygen, J. [2017]: Two-connected spanning subgraphs with at most $\frac{10}{7}$ OPT edges. SIAM Journal on Discrete Mathematics 31 (2017), 1820–1835

Held, M., and Karp, R.M. [1962]: A dynamic programming approach to sequencing problems. Journal of the Society of Industrial and Applied Mathematics 10 (1962), 196–210

Held, M., and Karp, R.M. [1970]: The traveling-salesman problem and minimum spanning trees. Operations Research 18 (1970), 1138–1162

Henke, D. [2018]: A linear programming relaxation with cycles for the asymmetric traveling salesman problem. Master's thesis, University of Bonn, 2018

Henzinger, M., and Williamson, D.P. [1996]: On the number of small cuts in a graph. Information Processing Letters 59 (1996), 41–44

Hierholzer, C. [1873]: Über die Möglichkeit, einen Linienzug ohne Wiederholung und ohne Unterbrechung zu umfahren. Mathematische Annalen 6 (1873), 30–32 [in German]

Hirai, H. [2016]: On uncrossing games for skew-supermodular functions. Journal of the Operations Research Society of Japan 59 (2016), 218–223

Hoeffding, W. [1956]: On the distribution of the number of successes in independent trials. The Annals of Mathematical Statistics 27 (1956), 713–721

Hoffman, A.J. [1960]: Some recent applications of the theory of linear inequalities to extremal combinatorial analysis. In: Combinatorial Analysis (R.E. Bellman and M. Hall, eds.), AMS, Providence 1960, pp. 113–128

Hoffman, A.J., and Kruskal, J.B. [1956]: Integral boundary points of convex polyhedra. In: Linear Inequalities and Related Systems; Annals of Mathematical Study 38 (H.W. Kuhn and A.W. Tucker, eds.), Princeton University Press, Princeton 1956, pp. 223–246

Hoogeveen, J.A. [1991]: Analysis of Christofides' heuristic: some paths are more difficult than cycles. Operations Research Letters 10 (1991), 291–295

Hougardy, S. [2014]: On the integrality ratio of the subtour LP for Euclidean TSP. Operations Research Letters 42 (2014), 495–499

Hougardy, S., and Vygen, J. [2016]: Algorithmic Mathematics. Springer, Cham 2016

Hunkenschröder, C., Vempala, S., and Vetta, A. [2019]: A 4/3-approximation algorithm for the minimum 2-edge connected subgraph problem. ACM Transactions on Algorithms 15 (2019), Article 55

Hurkens, C.A.J., Lovász, L., Schrijver, A., and Tardos, É. [1988]: How to tidy up your set-system? In: Combinatorics (A. Hajnal, L. Lovász, and V.T. Sós, eds.), North-Holland, Amsterdam 1988, pp. 309–314

Iwata, S., Fleischer, L., and Fujishige, S. [2001]: A combinatorial strongly polynomial algorithm for minimizing submodular functions. Journal of the ACM 48 (2001), 761–777

Iwata, S., and Ravi, R. [2013]: Approximating max–min weighted T-joins. Operations Research Letters 41 (2013) 321–324

Jackson, B. [1988]: Some remarks on arc-connectivity, vertex splitting, and orientation in digraphs. Journal of Graph Theory 12 (1988), 429–436

Jain, K. [2001]: A factor 2 approximation algorithm for the generalized Steiner network problem. Combinatorica 21 (2001), 39–60

Jarník, V. [1930]: O jistém problému minimálním. Práca Moravské Přírodovědecké Společnosti 6 (1930), 57–63 [in Czech]

Jin, B., Klein, N., and Williamson, D. [2023a]: A 4/3-approximation algorithm for half-integral cycle cut instances of the TSP. Proceedings of the 24th Conference on Integer Programming and Combinatorial Optimization (IPCO 2023), 217–230

Jin, B., Klein, N., and Williamson, D. [2023b]: A lower bound for the max entropy algorithm for TSP. arXiv:2311.01950. To appear in IPCO 2024

Kaplan, H., Lewenstein, M., Shafrir, N., and Sviridenko, M. [2005]: Approximation algorithms for asymmetric TSP by decomposing directed regular multigraphs. Journal of the ACM 52 (2005), 602–626

Karger, D.R. [2000]: Minimum cuts in near-linear time. Journal of the ACM 47 (2000), 46–76

Karger, D.R., and Stein, C. [1996]: A new approach to the minimum cut problem. Journal of the ACM 43 (1996), 601–640

Karlin, A.R., Klein, N., and Oveis Gharan, S. [2020]: An improved approximation algorithm for TSP in the half integral case. Proceedings of the 52nd Annual ACM Symposium on the Theory of Computing (STOC 2020), 28–39

Karlin, A.R., Klein, N., and Oveis Gharan, S. [2021]: A (slightly) improved approximation algorithm for metric TSP. Proceedings of the 53rd Annual ACM Symposium on the Theory of Computing (STOC 2021), 32–45. Operations Research, to appear

Karlin, A.R., Klein, N., and Oveis Gharan, S. [2022]: A (slightly) improved bound on the integrality gap of the subtour LP for TSP. Proceedings of the 63rd Annual IEEE Symposium on Foundations of Computer Science (FOCS 2022), 832–843

Karlin, A.R., Klein, N., and Oveis Gharan, S. [2023]: A deterministic better-than-3/2 approximation algorithm for metric TSP. Proceedings of the 24th Conference on Integer Programming and Combinatorial Optimization (IPCO 2023), 261–274

Karp, R.M. [1972]: Reducibility among combinatorial problems. In: Complexity of Computer Computations (R.E. Miller and J.W. Thatcher, eds.), Plenum Press, New York 1972, pp. 85–103

Karp, R.M. [1978]: A characterization of the minimum cycle mean in a digraph. Discrete Mathematics 23 (1978), 309–311

Karpinski, M., and Schmied, R. [2015]: Approximation hardness of graphic TSP on cubic graphs. RAIRO Operations Research 49 (2015), 651–668

Karpinski, M., Lampis, M., and Schmied, R. [2015]: New inapproximability bounds for TSP. Journal of Computer and System Sciences 81 (2015), 1665–1677

Karzanov, A.V. [1974]: Determining a maximal flow in a network by the method of preflows. Soviet Mathematics Doklady 15 (1974), 434–437

Karzanov, A.V. [1996]: How to tidy up a symmetric set-system by use of uncrossing operations. Theoretical Computer Science 157 (1996), 215–225

Khachiyan, L.G. [1979]: A polynomial algorithm in linear programming. Doklady Akademii Nauk SSSR 244 (1979) 1093–1096 [in Russian]

Khuller, S., and Vishkin, U. [1994]: Biconnectivity approximations and graph carvings. Journal of the ACM 41 (1994), 214–235

Kirchhoff, G. [1847]: Über die Auflösung der Gleichungen, auf welche man bei der Untersuchung der linearen Verteilung galvanischer Ströme geführt wird. Annalen der Physik und Chemie 72 (1847), 497–508 [in German]

Kisfaludi-Bak, S., Nederlof, J., and Węgrzycki, K. [2021]: A Gap-ETH-tight approximation scheme for Euclidean TSP. Proceedings of the 62nd Annual IEEE Symposium on Foundations of Computer Science (FOCS 2021), 351–362

Klein, P.N. [2008]: A linear-time approximation scheme for TSP in undirected planar graphs with edge-weights. SIAM Journal on Computing 37 (2008), 1926–1952

Klein, N., and Olver, N. [2023]: Thin trees for laminar families. Proceedings of the 64th Annual IEEE Symposium on Foundations of Computer Science (FOCS 2023), 50–59

Kobayashi, Y., and Noguchi, T. [2023]: An approximation algorithm for two-edge-connected subgraph problem via triangle-free two-edge-cover. arXiv:2304.13228

Köhne, A., Traub, V., and Vygen, J. [2020]: The asymmetric traveling salesman path LP has constant integrality ratio. Mathematical Programming 183 (2020), 379–395

Korte, B., and Vygen, J. [2018]: Combinatorial Optimization: Theory and Algorithms. Springer, Berlin, Sixth Edition 2018

Kruskal, J.B. [1956]: On the shortest spanning subtree of a graph and the traveling salesman problem. Proceedings of the AMS 7 (1956), 48–50

Laekhanukit, B., Oveis Gharan, S., and Singh, M. [2012]: A rounding by sampling approach to the minimum size k-arc connected subgraph problem. Proceedings of the 39th International Colloquium on Automata, Languages, and Programming (ICALP 2012), 606–612

Lam, F., and Newman, A. [2008]: Traveling salesman path problems. Mathematical Programming 113 (2008), 39–59

Lampis, M. [2014]: Improved inapproximability for TSP. Theory of Computing 10 (2014), 217–236

Lawler, E.L., Lenstra, J.K., Rinnooy Kan, A.H.G., and Shmoys, D.B. [1985]: The Traveling Salesman Problem. Wiley, Chichester 1985

Lee, Y.T., Sidford, A., and Wong, S.C. [2015]: A faster cutting plane method and its implications for combinatorial and convex optimization. Proceedings of the 56th Annual IEEE Symposium on Foundations of Computer Science (FOCS 2015), 1049–1065

Lempel, A., Even, S., and Cederbaum, I. [1967]: An algorithm for planarity testing of graphs. In: Théorie des Graphes. Theory of Graphs (P. Rosenstiehl, ed.), Dunod, Paris 1967

Lin, S., and Kernighan, B.W. [1973]: An effective heuristic algorithm for the traveling-salesman problem. Operations Research 21 (1973), 498–516

Lovász, L. [1972]: A note on factor-critical graphs. Studia Scientiarum Mathematicarum Hungarica 7 (1972), 279–280

Lovász, L. [1976]: On some connectivity properties of Eulerian graphs. Acta Mathematica Academiae Scientiarum Hungaricae 28 (1976), 129–138

Lovász, L., and Plummer, M.D. [1986]: Matching Theory. Akadémiai Kiadó, Budapest, and North-Holland, Amsterdam 1986

Mader, W. [1982]: Konstruktion aller n-fach kantenzusammenhängenden Digraphen. European Journal of Combinatorics 3 (1982), 63–67 [in German]

Matula, D.W., and Shahrokhi, F. [1990]: Sparsest cuts and bottlenecks in graphs. Discrete Applied Mathematics 27 (1990), 113–123

Menger, K. [1927]: Zur allgemeinen Kurventheorie. Fundamenta Mathematicae 10 (1927), 96–115 [in German]

Middendorf, M., and Pfeiffer, F. [1993]: On the complexity of the disjoint paths problem. Combinatorica 13 (1993), 97–107

Minkowski, H. [1896]: Geometrie der Zahlen. Teubner, Leipzig 1896 [in German]

Mitchell, J. [1999]: Guillotine subdivisions approximate polygonal subdivisions: a simple polynomial-time approximation scheme for geometric TSP, k-MST, and related problems. SIAM Journal on Computing 28 (1999), 1298–1309

Mnich, M., and Mömke, T. [2018]: Improved integrality gap upper bounds for TSP with distances one and two. European Journal of Operational Research 266 (2018), 436–457

Mömke, T. [2015]: An improved approximation algorithm for the traveling salesman problem with relaxed triangle inequality. Information Processing Letters 115 (2015), 866–871

Mömke, T., and Svensson, O. [2016]: Removing and adding edges for the traveling salesman problem. Journal of the ACM 63 (2016), Article 2

Monma, C.L., Munson, B.S., and Pulleyblank, W.R. [1990]: Minimum-weight two-connected spanning networks. Mathematical Programming 46 (1990), 153–171

Mucha, M. [2014]: $\frac{13}{9}$-approximation for graphic TSP. Theory of Computing Systems 55 (2014), 640–657

Nagamochi, H., and Ibaraki, T. [1992]: Computing edge-connectivity in multigraphs and capacitated graphs. SIAM Journal on Discrete Mathematics 5 (1992), 54–66

Nagamochi, H., Nishimura, K., and Ibaraki, T. [1997]: Computing all small cuts in an undirected network. SIAM Journal on Discrete Mathematics 10 (1997), 469–481

Nagarajan, V., and Ravi, R. [2008]: The directed minimum latency problem. Proceedings of the 11th International Workshop on Approximation Algorithms for Combinatorial Optimization Problems (APPROX 2008), 193–206

Nagarajan, V., and Ravi, R. [2011]: The directed orienteering problem. Algorithmica 60 (2011), 1017–1030

Nash-Williams, C.S.J.A. [1961]: Edge-disjoint spanning trees of finite graphs. Journal of the London Mathematical Society 36 (1961), 445–450

von Neumann, J. [1947]: Discussion of a maximum problem. Working paper, Princeton 1947 [published in: John von Neumann, Collected Works; Vol. VI (A.H. Taub, ed.), Pergamon Press, Oxford 1963, pp. 27–28]

Newman, A. [2020]: An improved analysis of the Mömke–Svensson algorithm for graph-TSP on subquartic graphs. SIAM Journal on Discrete Mathematics 34 (2020), 865–884

Orlin, J.B. [1993]: A faster strongly polynomial minimum cost flow algorithm. Operations Research 41 (1993), 338–350

Oveis Gharan, S., and Saberi, A. [2011]: The asymmetric traveling salesman problem on graphs with bounded genus. Proceedings of the 22nd Annual ACM-SIAM Symposium on Discrete Algorithms (SODA 2011), 967–975

Oveis Gharan, S., Saberi, A., and Singh, M. [2011]: A randomized rounding approach to the traveling salesman problem. Proceedings of the 52nd Annual IEEE Symposium on Foundations of Computer Science (FOCS 2011), 550–559

Padberg, M.W., and Rao, M.R. [1982]: Odd minimum cut-sets and b-matchings. Mathematics of Operations Research 7 (1982), 67–80

Panconesi, A., and Srinivasan, A. [1997]: Randomized distributed edge coloring via an extension of the Chernoff–Hoeffding bounds. SIAM Journal on Computing 26 (1997), 350–368

Papadimitriou, C.H., and Vempala, S. [2006]: On the approximability of the traveling salesman problem. Combinatorica 26 (2006), 101–120

Papadimitriou, C.H., and Yannakakis, M. [1993]: The traveling salesman problem with distances one and two. Mathematics of Operations Research 18 (1993), 1–12

Pillai, A., and Singh, M. [2023]: Linear programming based reductions for Multiple Visit TSP and vehicle routing problems. arXiv:2308.11742

Prim, R.C. [1957]: Shortest connection networks and some generalizations. Bell System Technical Journal 36 (1957), 1389–1401

Pritchard, D. [2010]: k-edge-connectivity: approximation and LP relaxation. Proceedings of the 8th International Workshop on Approximation and Online Algorithms (WAOA 2010). 225–236

Qian, J., Schalekamp, F., Williamson, D.P., and van Zuylen, A. [2015]: On the integrality gap of the subtour LP for the 1,2-TSP. Mathematical Programming 150 (2015), 131–151

Rado, R. [1942]: A theorem on independence relations. Quarterly Journal of Mathematics 13 (1942), 83–89

Rao, S.B., and Smith, W.D. [1998]: Approximating geometric graphs via "spanners" and "banyans." Proceedings of the 30th Annual ACM Symposium on Theory of Computing (STOC 1998), 540–550

Rayleigh, J.W.S. [1871]: On the theory of resonance. Philosophical Transactions 161 (1871), 77–118

Robertson, N., and Seymour, P.D. [1995]: Graph minors. XIII. The disjoint paths problem. Journal of Combinatorial Theory B 63 (1995), 65–110

Rosenkrantz, D.J., Stearns, R.E., and Lewis, P.M. II [1977]: An analysis of several heuristics for the traveling salesman problem. SIAM Journal on Computing 6 (1977), 563–581

Sahni, S., and Gonzalez, T. [1976]: P-complete approximation problems. Journal of the ACM 23 (1976), 555–565

Saller, S., Koehler, J., and Karrenbauer, A. [2023]: A systematic review of approximability results for traveling salesman problems leveraging the TSP-T3CO definition scheme. arXiv:2311.00604

Schalekamp, F., Sebő, A., Traub, V., and van Zuylen, A. [2018]: Layers and matroids for the traveling salesman's paths. Operations Research Letters 46 (2018), 60–63

Schalekamp, F., Williamson, D.P., and van Zuylen, A. [2014]: 2-matchings, the traveling salesman problem, and the subtour LP: A proof of the Boyd–Carr conjecture. Mathematics of Operations Research 39 (2014), 403–417

Schrijver, A. [2000]: A combinatorial algorithm minimizing submodular functions in strongly polynomial time. Journal of Combinatorial Theory B 80 (2000), 346–355

Schrijver, A. [2003]: Combinatorial Optimization: Polyhedra and Efficiency. Springer, Berlin 2003

Sebő, A. [1987]: A quick proof of Seymour's theorem on T-joins. Discrete Mathematics 64 (1987), 101–103

Sebő, A. [1997]: Potentials in undirected graphs and planar multiflows. SIAM Journal on Computing 26 (1997), 582–603

Sebő, A. [2013]: Eight-fifth approximation for TSP paths. Proceedings of the 16th Conference on Integer Programming and Combinatorial Optimization (IPCO 2013), 362–374

Sebő, A., and Vygen, J. [2014]: Shorter tours by nicer ears: 7/5-approximation for graphic TSP, 3/2 for the path version, and 4/3 for two-edge-connected subgraphs. Combinatorica 34 (2014), 597–629

Sebő, A., and van Zuylen, A. [2019]: The salesman's improved paths through forests. Journal of the ACM 66 (2019), Article 28

Serdyukov, A.I. [1978]: Some extremal bypasses in graphs. Upravlyaemye Sistemy 17 (1978), 76–79 [in Russian]

Shmoys, D., and Talwar, K. [2008] A constant approximation algorithm for the *a priori* traveling salesman problem. Proceedings of the 13th Conference on Integer Programming and Combinatorial Optimization (IPCO 2008), 331–343

Shmoys, D., and Williamson, D. P. [1990]: Analyzing the Held–Karp TSP bound: A monotonicity property with application. Information Processing Letters 35 (1990), 281–285

Singh, M., and Vishnoi, N.K. [2014]: Entropy, optimization and counting. Proceedings of the 46th Annual ACM Symposium on Theory of Computing (STOC 2014), 50–59

Spielman, D.A., and Teng, S.-H. [2014]: Nearly linear time algorithms for preconditioning and solving symmetric, diagonally dominant linear systems. SIAM Journal on Matrix Analysis and Applications 35 (2014), 835–885

Steinitz, E. [1916]: Bedingt konvergente Reihen und konvexe Systeme. Journal für die reine und angewandte Mathematik 146 (1916), 1–52 [in German]

Stoer, M., and Wagner, F. [1997]: A simple min cut algorithm. Journal of the ACM 44 (1997), 585–591

Straszak, D., and Vishnoi, N.K. [2019]: Maximum entropy distributions: Bit complexity and stability. Proceedings of the 32nd Conference on Learning Theory (COLT 2019), 2861–2891

Svensson, O. [2013]: Overview of new approaches for approximating TSP. Proceedings of the 39th International Workshop on Graph-Theoretic Concepts in Computer Science (WG 2013), 5–11

Svensson, O. [2015]: Approximating ATSP by relaxing connectivity. Proceedings of the 56th Annual IEEE Symposium on Foundations of Computer Science (FOCS 2015), 1–19

Svensson, O., Tarnawski, J., and Végh, L. [2018]: Constant factor approximation for ATSP with two edge weights. Mathematical Programming 172 (2018), 371–397

Svensson, O., Tarnawski, J., and Végh, L. [2020]: A constant-factor approximation algorithm for the asymmetric traveling salesman problem. Journal of the ACM 67 (2020), Article 37

Tardos, É. [1985]: A strongly polynomial minimum cost circulation algorithm. Combinatorica 5 (1985), 247–255

Tardos, É. [1986]: A strongly polynomial algorithm to solve combinatorial linear programs. Operations Research 34 (1986), 250–256

Toth, P., and Vigo, D., eds. [2014]: Vehicle Routing: Problems, Methods, and Applications. MOS-SIAM Series on Optimization, Philadelphia, Second Edition 2014

Traub, V. [2017]: Approximating the s-t-path TSP. Master's thesis, University of Bonn, 2017

Traub, V. [2020a]: Approximation algorithms for traveling salesman problems. PhD thesis, University of Bonn, 2020

Traub, V. [2020b]: Improving on Best-of-Many-Christofides for T-tours. Operations Research Letters 48 (2020), 798–804

Traub, V., and Vygen, J. [2019a]: Approaching $\frac{3}{2}$ for the s-t-path TSP. Journal of the ACM 66 (2019), Article 14

Traub, V., and Vygen, J. [2019b]: An improved upper bound on the integrality ratio for the s-t-path TSP. Operations Research Letters 47 (2019), 225–228

Traub, V., and Vygen, J. [2022]: An improved approximation algorithm for the asymmetric traveling salesman problem. SIAM Journal on Computing 51 (2022), 139–173

Traub, V., and Vygen, J. [2023]: Beating the integrality ratio for s-t-tours in graphs. SIAM Journal on Computing 52 (2023), FOCS18-37–FOCS18-84

Traub, V., Vygen, J., and Zenklusen, R. [2022]: Reducing Path TSP to TSP. SIAM Journal on Computing 51 (2022), STOC20-24–STOC-20-53

Traub, V., and Zenklusen, R. [2021]: A better-than-2 approximation for weighted tree augmentation. Proceedings of the 62nd Annual IEEE Symposium on Foundations of Computer Science (FOCS 2021), 1–12

Traub, V., and Zenklusen, R. [2022]: Local search for weighted tree augmentation and Steiner tree. Proceedings of the 33rd Annual ACM-SIAM Symposium on Discrete Algorithms (SODA 2022), 3253–3271

Tutte, W.T. [1961]: On the problem of decomposing a graph into n connected factors. Journal of the London Mathematical Society 36 (1961), 221–230

Vazirani, V.V., and Yannakakis, M. [1992]: Suboptimal cuts: Their enumeration, weight and number. Proceedings of the 19th International Colloquium on Automata, Languages, and Programming (ICALP 1992), 366–377

Vishnoi, N.K. [2012]: A permanent approach to the traveling salesman problem. Proceedings of the 53rd Annual IEEE Symposium on Foundations of Computer Science (FOCS 2012), 76–80

Vygen, J. [2012]: New approximation algorithms for the TSP. OPTIMA 90 (2012), 1–12

Vygen, J. [2016]: Reassembling trees for the traveling salesman. SIAM Journal on Discrete Mathematics 30 (2016), 875–894

Weyl, H. [1935]: Elementare Theorie der konvexen Polyeder. Commentarii Mathematici Helvetici 7 (1935), 290–306 [in German]

Whitney, H. [1932a]: Non-separable and planar graphs. Transactions of the American Mathematical Society 34 (1932), 339–362

Whitney, H. [1932b]: Congruent graphs and the connectivity of graphs. American Journal of Mathematics 54 (1932), 150–168

Wigal, M.C., Yoo, Y., and Yu, X. [2023]: Approximating TSP walks in subcubic graphs. Journal of Combinatorial Theory B 158 (2023), 70–104

Williamson, D.P. [1990]: Analysis of the Held–Karp heuristic for the traveling salesman problem. Master's thesis, Massachusetts Institute of Technology, 1990

Williamson, D.P. [2019]: Network Flow Algorithms. Cambridge University Press, Cambridge 2019

Williamson, D.P., and Shmoys, D.B. [2011]: The Design of Approximation Algorithms. Cambridge University Press, Cambridge 2011

Wolsey, L.A. [1980]: Heuristic analysis, linear programming and branch and bound. Mathematical Programming Study 13 (1980), 121–134

Xu, Z., and Rodrigues, B. [2015]: A 3/2-approximation algorithm for the multiple TSP with a fixed number of depots. INFORMS Journal on Computing 27 (2015), 636–645

Zenklusen, R. [2019]: A 1.5-approximation for path TSP. Proceedings of the 30th Annual ACM-SIAM Symposium on Discrete Algorithms (SODA 2019), 1539–1549

Zhong, X. [2020]: Slightly improved upper bound on the integrality ratio for the s-t path TSP. Operations Research Letters 48 (2020), 627–629

van Zuylen, A. [2011]: Deterministic sampling algorithms for network design. Algorithmica 60 (2011), 110–151

van Zuylen, A. [2018]: Improved approximations for cubic bipartite and cubic TSP. Mathematical Programming 172 (2018), 399–413

Index

For EU product safety concerns, contact us at Calle de José Abascal, 56–1°, 28003 Madrid, Spain or eugpsr@cambridge.org.

www.ingramcontent.com/pod-product-compliance
Lightning Source LLC
LaVergne TN
LVHW010328220926
848484LV00004BB/858
* 9 7 8 1 0 0 9 4 4 5 4 1 2 *